Genome Editing in Drug Discovery

Genome Editing in Drug Discovery

Edited by

Marcello Maresca
AstraZeneca, BioPharmaceuticals R&D
Mölndal, Sweden

Sumit Deswal
AstraZeneca, BioPharmaceuticals R&D
Mölndal, Sweden

WILEY

This edition first published 2022.

Registered Office
John Wiley & Sons, Inc., 111 River Street, Hoboken, NJ 07030, USA

Editorial Office
111 River Street, Hoboken, NJ 07030, USA
For details of our global editorial offices, customer services, and more information about Wiley products visit us at www.wiley.com.

Wiley also publishes its books in a variety of electronic formats and by print-on-demand. Some content that appears in standard print versions of this book may not be available in other formats.

Library of Congress Cataloging-in-Publication Data

Names: Maresca, Marcello, editor. | Deswal, Sumit, editor.
Title: Genome editing in drug discovery / Marcello Maresca, AstraZeneca, BioPharmaceuticals R&D, Mölndal, Sweden, Sumit Deswal, AstraZeneca, BioPharmaceuticals R&D, Mölndal, Sweden.
Description: First edition. | Hoboken, NJ : Wiley, 2022. | Includes index.
Identifiers: LCCN 2021025952 (print) | LCCN 2021025953 (ebook) | ISBN 9781119671343 (hardback) | ISBN 9781119671381 (adobe pdf) | ISBN 9781119671398 (epub)
Subjects: LCSH: Drug development. | Genetic engineering.
Classification: LCC RM301.25 .G46 2022 (print) | LCC RM301.25 (ebook) | DDC 615.1/9–dc23
LC record available at https://lccn.loc.gov/2021025952
LC ebook record available at https://lccn.loc.gov/2021025953

Cover image: © Yurchanka Siarhei/Shutterstock
Cover design by Wiley

Set in 9.5/12.5pt STIXTwoText by Straive, Pondicherry, India

10 9 8 7 6 5 4 3 2 1

Contents

Preface

The development of CRISPR-Cas9 for genome engineering has revolutionized the field of genome editing. Many of the cell types and animal models previously very challenging to genetic engineering, can now be engineered with high efficiency and precision using CRISPR-Cas9-derived tools. This has led to the development of many novel disease models helping scientists to better understand disease biology as well as providing opportunity to test novel therapeutics. By performing large-scale functional genomics screens with CRISPR-Cas9, it is now possible to identify and validate drug targets at a much faster rate and better precision. By assessing gene function comprehensively at large scale and in relevant cell type in early stages, candidate attrition rate is reduced. Genome editing is now part of almost every step in the early part of drug discovery pipeline, from target identification and its validation to mechanistic studies in relevant disease models. In addition, genome editing is used as a promising platform for gene therapy and molecular diagnostics.

We felt a need for a book comprehensively covering these important aspects of genome editing in drug discovery and therapy. Such a book will be of very much interest for those performing genome editing in research institutes or applying genome editing to drug discovery projects at pharmaceutical industries. Such a book will also help students of molecular biology, biotechnology, biochemistry, and pharmaceutical sciences to better understand this very important technology and to exemplify how basic research on a bacterial immune system can have such a big impact in life sciences.

We would like to thank staff at Wiley, especially Jonathan Rose, who first proposed the idea of this book. We would like to thank all the authors who spent their precious time in making this book a reality. Each of the chapter is written by experts on the respective topic working in top pharmaceutical/biotechnology companies or academic institutes.

We would also like to thank colleagues and management at AstraZeneca, Mohammad Bohlooly, Steve Rees, Mike Snowden, and Mene Pangalos for supporting us and allowing us to devote our time for such an academic exercise.

Marcello Maresca and Sumit Deswal
Gothenburg, Sweden
April 2021

List of Abbreviations

AAV	Adeno-Associated Virus
ABE	Adenosine Base Editors
ADAR	Adenosine deaminases acting on RNA
ASOs	Antisense oligonucleotides
CAR	Chimeric Antigen Receptor
Cas9	CRISPR associated protein 9
CBE	Cytosine base editors
CRISPR	Clustered Regularly Interspaced Short Palindromic Repeats
CRISPRa	CRISPR activation
CRISPRi	CRISPR interference
dCas9	Dead Cas9
DSB	Double strand DNA break
FACS	Fluorescence-activated cell sorting
GEMM	Genetically Engineered Mouse Model
HDR	Homology Directed Repair
iGEM	International Genetically Engineered Machines
IPR	Intellectual Property Rights
LNP	Lipid Nano Particles
NHEJ	Non-homologous end joining
PAM	Protospacer adjacent motifs
pegRNA	Prime editing guide RNA
RNAi	RNA interference
RNP	Ribonucleoprotein
sgRNA	Single guide RNA
siRNA	Small interfering RNA
shRNA	Short hairpin RNA
TALENs	Transcription activator-like effector nucleases
TGE	Therapeutic Genome Editing
ZFNs	Zinc finger nucleases

List of Contributors

Pinar Akcakaya
Genome Engineering, Discovery Sciences, BioPharmaceuticals R&D, AstraZeneca, Gothenburg, Sweden

Nina Akrap
Discovery Biology, Discovery Sciences, BioPharmaceuticals R&D, AstraZeneca, Gothenburg, Sweden

Lukas Badertscher
Genome Engineering, Discovery Sciences, BioPharmaceuticals R&D, AstraZeneca, Gothenburg, Sweden

Sangsu Bae
Department of Chemistry and Research Institute for Convergence of Basic Sciences, Hanyang University, Seoul, South Korea

Fiona M. Behan
Functional Genomics, R&D GlaxoSmithKline, Stevenage, UK

Nupur Bhargava
Genomics and Molecular Medicine, CSIR-Institute of Genomics and Integrative Biology, New Delhi, India

Erik Oude Blenke
Advanced Drug Delivery, Pharmaceutical Sciences, AstraZeneca R&D, Molndal, Sweden

Alessandro Bonetti
Translational Genomics, Discovery Sciences, BioPharmaceuticals R&D, AstraZeneca, Molndal, Sweden

Rakesh Kantilal Chandode
Gene Therapy, Functional and Mechanistic Safety, Clinical Pharmacology and Safety Sciences, BioPharmaceuticals R&D, AstraZeneca, Gothenburg, Sweden

Aaron T. Cheng
Functional Genomics, R&D GlaxoSmithKline, Upper Providence, PA, USA

Yacine Chérifi
genOway, Lyon, France

Matthew Coelho
Wellcome Sanger Institute, Cambridge, UK

Sumit Deswal
Genome Engineering, Discovery Sciences, BioPharmaceuticals R&D, AstraZeneca, Gothenburg, Sweden

Ramy Elgendy
Translational Genomics, Discovery Sciences, BioPharmaceuticals R&D, AstraZeneca, Molndal, Sweden

Leire Escudero-Ibarz
Discovery Biology, BioPharmaceuticals R&D, AstraZeneca, Cambridge, UK

Justin Eyquem
Gladstone-UCSF Institute of Genomic Immunology, San Francisco, CA, USA
Division of Hemato-Oncology, Department of Medicine, University of California San Francisco, San Francisco, CA, USA
and
Department of Microbiology and Immunology & Parker Institute of Cancer Immunotherapy, University of California San Francisco, San Francisco, CA, USA

Alexandre Fraichard
genOway, Lyon, France

Davide Gianni
Discovery Biology, BioPharmaceuticals R&D, AstraZeneca, Cambridge, UK

Luke A. Gilbert
Departments of Urology and Cellular & Molecular Pharmacology, and UCSF Helen Diller Family Comprehensive Cancer Center, University of California San Francisco, San Francisco, CA, USA

Sangam Giri Goswami
Genomics and Molecular Medicine, CSIR-Institute of Genomics and Integrative Biology, New Delhi, India

Antje Grotz
Genome Engineering, Discovery Sciences, BioPharmaceuticals R&D, AstraZeneca, Gothenburg, Sweden

Pragya Gupta
Genomics and Molecular Medicine, CSIR-Institute of Genomics and Integrative Biology, New Delhi, India

Gue-Ho Hwang
Department of Chemistry and Research Institute for Convergence of Basic Sciences, Hanyang University, Seoul, South Korea

Shashank Jaitly
Genomics and Molecular Medicine, CSIR-Institute of Genomics and Integrative Biology, New Delhi, India

Venkata R. Krishnamurthy
Advanced Drug Delivery, Pharmaceutical Sciences, AstraZeneca R&D, Boston, MA, USA

Songyuan Li
Genome Engineering, Discovery Sciences, BioPharmaceuticals R&D, AstraZeneca, Gothenburg, Sweden

Quinn Lu
Novel Human Genetics Research Unit, R&D GlaxoSmithKline, Upper Providence, PA, USA

Alexandra Madsen
Genome Engineering, Discovery Sciences, BioPharmaceuticals R&D, AstraZeneca, Gothenburg, Sweden

Franc Mali
Faculty of Social Sciences, University of Ljubljana, Ljubljana, Slovenia

Klio Maratou
Functional Genomics, R&D GlaxoSmithKline, Stevenage, UK

Marcello Maresca
Genome Engineering, Discovery Sciences, BioPharmaceuticals R&D, AstraZeneca, Gothenburg, Sweden

Roberto Nitsch
Gene Therapy, Functional and Mechanistic Safety, Clinical Pharmacology and Safety Sciences, BioPharmaceuticals R&D, AstraZeneca, Gothenburg, Sweden

William A. Nyberg
Gladstone-UCSF Institute of Genomic Immunology, San Francisco, CA, USA
and
Division of Hemato-Oncology, Department of Medicine, University of California San Francisco, San Francisco, CA, USA

Jenna Persson
CRISPR Functional Genomics, Karolinska Institutet and SciLifeLab, Stockholm, Sweden

Martin Peterka
Genome Engineering, Discovery Sciences, BioPharmaceuticals R&D, AstraZeneca, Gothenburg, Sweden

Michelle J. Porritt
Genome Engineering, Discovery Sciences, BioPharmaceuticals R&D, AstraZeneca, Gothenburg, Sweden

Sivaprakash Ramalingam
Genomics and Molecular Medicine, CSIR-Institute of Genomics and Integrative Biology, New Delhi, India

Steve Rees
Discovery Sciences, BioPharmaceuticals R&D, AstraZeneca, Cambridge, UK

Amélie Rezza
genOway, Lyon, France

Saumyaa Saumyaa
Translational Genomics, Discovery Sciences, BioPharmaceuticals R&D, AstraZeneca, Molndal, Sweden

Bernhard Schmierer
CRISPR Functional Genomics, Karolinska Institutet and SciLifeLab, Stockholm, Sweden

Niklas Selfjord
Genome Engineering, Discovery Sciences, BioPharmaceuticals R&D, AstraZeneca, Gothenburg, Sweden

Ning Sun
Functional Genomics, R&D GlaxoSmithKline, Upper Providence, PA, USA

Saša Šviković
Genome Engineering, Discovery Sciences, BioPharmaceuticals R&D, AstraZeneca, Gothenburg, Sweden

Amir Taheri-Ghahfarokhi
Quantitative Biology, Discovery Sciences, BioPharmaceuticals R&D, AstraZeneca, Gothenburg, Sweden

Benjamin J.M. Taylor
Discovery Sciences, BioPharmaceuticals R&D, AstraZeneca, Cambridge, UK

Priya Thakur
Genomics and Molecular Medicine, CSIR-Institute of Genomics and Integrative Biology, New Delhi, India

Pierre Theurey
genOway, Lyon, France

Kader Thiam
genOway, Lyon, France

Sandra Wimberger
Discovery Biology, Discovery Sciences, BioPharmaceuticals R&D, AstraZeneca, Gothenburg, Sweden

Part 1

Introduction to Drug Discovery and Genome Editing Methods

This part provides a historical prospective on developments in the genome editing field. Since CRISPR Cas is currently the method of choice, which in true sense has revolutionized genome editing, a separate chapter on the development of this technology is provided.

1

Genome Editing in Drug Discovery

Steve Rees

Discovery Sciences, BioPharmaceuticals R&D, AstraZeneca, Cambridge, UK

1.1 Introduction

The last decade has seen the development of an unprecedented number of new technologies that are being applied to transform our understanding of disease and the subsequent success of drug discovery. The dramatic advances in DNA sequencing technology and the more recent advances in other-omics technologies including transcriptomics, proteomics, and metabolomics are enabling the understanding of disease at the genetic and cellular level to identify new drug targets, and to identify new disease biomarkers to enable disease segmentation and patient stratification in the clinic. Advances in stem cell technology together with technologies that enable the creation of tissue organoids in the laboratory are allowing the creation of complex models of disease (Lancaster et al. 2013), which together with advances in imaging technology are enabling the drug discovery scientist to better understand the efficacy and safety of potential medicines in preclinical studies. The huge increases in computational power, together with advances in Artificial Intelligence and Machine Learning are allowing drug discovery scientists to extract greater knowledge from these large-omics datasets, to improve the speed and quality of chemistry design and enable the design of improved clinical studies (Vamathevan et al. 2019). Perhaps, the most impactful of the many new technologies applied in drug discovery in the last decade has been the rapid adoption of CRISPR/Cas9 throughout the drug discovery pipeline to create engineered cellular and animal models of disease to enable the study of the role of new drug targets in disease, alongside the development of CRISPR as a medicine in it's own right or as a key tool in the creation of cell therapy medicines. Taken together, these and other new technologies have impacted every drug discovery program to enable a better understanding of the role of the drug target in disease and the design of molecules more likely to be safe and efficacious in the clinic. Alongside this, a number of new therapeutic modalities are entering the clinic including antisense onligonucleotide, mRNA, protein, and gene and cell therapies which are leading to a situation where every target becomes amenable to therapeutic manipulation. Taken together, the ability of these technologies to improve our understanding of disease, to create safer medicines and to target those medicines to the patient population most likely to benefit from them is leading to an increase in the success of drug discovery. This has been seen in an increase in the number of new molecular entities approved by the FDA with over 40 new medicines being approved each year between 2011 and 2020 compared with an average of around 20 new medicines approved each year between 2001 and 2010 (Batta et al. 2020). A number of reports also describe an increase in success of drug discovery including a recent publication from colleagues in AstraZeneca. Through the implementation of a new research strategy at AstraZeneca in 2010, success from Candidate Selection to product launch has increased from 4% to 20% while 3 projects are now started in early discovery to deliver a Candidate Drug compared with 5 projects in earlier years (Morgan et al. 2018). While this represents a huge increase in drug discovery productivity, it remains the case that the majority of projects fail with the primary cause of failure in research being due to target validation and in the clinic a lack of efficacy in Phase II clinical studies. In both cases, the root cause of failure is that the hypothesis linking the drug target to disease was incorrect and significant efforts are underway in both academia and industry to continue to increase the level of confidence the drug target at the start of, and throughout a drug discovery program to further increase drug discovery success in all therapeutic areas. Throughout this book, authors will present examples of the application of CRISPR/Cas9 to identify novel drug

Genome Editing in Drug Discovery, First Edition. Edited by Marcello Maresca and Sumit Deswal.

targets, to understand the role of these targets in disease, and to create cellular and animal model systems to allow the development of new medicines more likely to succeed in the clinic. While we remain within the first decade following the discovery of the ability of CRISPR (clustered regularly interspersed short palindromic repeats)/Cas9 systems for the precise editing of mammalian genomes (Jinek et al. 2012; Cong et al. 2013; Mali et al. 2013), this technology has become embedded throughout drug discovery research (Fellmann et al. 2017). Throughout this book, authors will discuss the current use of CRISPR/Cas9 to facilitate the development of new medicines, as a medicine in its own right, and as a highly sensitive point of care diagnostic. However, we remain in the infancy of the application of this technology, and its potential to transform our understanding and treatment of disease remains huge.

1.2 Genome Engineering

Genome engineering describes the specific introduction of new genetic elements into target genes to mediate either a disruption of the gene sequence, and consequent loss of gene expression, or a change in the protein sequence of the gene being transcribed. The ability to modify gene sequences at precise locations in the genome arose in the 1990s with the development of meganucleases and Zinc Finger Nuclease (ZFN) technology (Kim et al. 1996; Bibikova et al. 2001). ZFNs are synthetic restriction enzymes created by fusing one or more zinc finger DNA-binding domains, engineered to target a specific DNA sequence, to a DNA nuclease, typically the restriction enzyme Fok1. ZFNs are able to recognize and cut a specific site in the genome to create a double-stranded DNA break that can be repaired by non-homologous end joining (NHEJ) or Homology Directed Repair (HDR) to result in gene inactivation or gene repair following the introduction of a DNA repair template. A boost to the technology came in 2009 with the discovery of transcription activator-like effector nucleases (TALENs) (Boch et al. 2009; Moscou and Bogdanove 2009). TALENs are synthetic restriction enzymes engineered to cut specific sequences of DNA. These are created by fusing a TAL effector DNA-binding domain designed to recognize the target DNA sequence with a nuclease to mediate DNA cleavage. TALEN constructs can be introduced into cells to cut the DNA at specific sequences to create a double-stranded DNA break, that following repair by NHEJ, results in the inactivation of the target gene as a consequence of the introduction of additional DNA sequences as part of the repair process. Furthermore, TALENs can be used to change specific nucleotides within a gene, or to introduce new sequences into the genome following transfection of cells with a TALEN's construct and a DNA repair template. Following HDR at the DNA cut site, the new sequence, encoded by the DNA repair template, can be introduced into the genome, albeit at a low editing efficiency. ZFNs and TALENs technologies have been used extensively in genome engineering projects in medical research and to modify plant genomes. Furthermore, Zinc Finger technology has been used by Sangamo Therapeutics and others to create genome editing medicines. Treatments for a range of diseases are in development with the most advanced project, a treatment for hemophilia currently in Phase 3 clinical studies. However, the widespread adoption of these technologies has been limited due to the requirement for expertise in protein engineering to create a ZFN or TALEN that precisely targets a specific DNA sequence, the relatively low editing efficiencies observed, and the potential for editing at multiple sites in the genome.

1.3 CRISPR/Cas9

The seminal publications in 2013 describing the ability of CRISPR/Cas9 to edit mammalian genomes have led to an explosion in the ability to make precise genetic changes within mammalian cells and animal models using a variety of CRISPR-based editing methods (Cong et al. 2013; Mali et al. 2013). Since these publications, there has been a dramatic increase in the efficiency and variety of genome editing methods available with these methods now in widespread use in the pharmaceutical industry and academia for target and drug discovery, alongside the discovery and characterization of a number of new editing enzymes and the development of forms of Cas9 with improved editing activity (Gilbert et al. 2014; Kampmann 2018).

CRISPR/Cas is part of the bacterial immune response system where its natural role is to recognize and destroy non-host nucleic acid sequences as part of the host immune defense system (Wiedenheft et al. 2012). The commonly used laboratory CRISPR system uses the Cas9 nuclease cloned from Streptococcus pyrogenes (SpCas9) although additional Cas9 enzymes have been cloned and characterized (Acharya et al. 2019). In contrast to TALENs and Zinc Finger technology, CRISPR/Cas9 is simple to use in any laboratory. It does not require protein engineering to create a nuclease able to recognize a specific site within the genome, rather targeting of the Cas9 nuclease to specific sites within the genome is mediated through the design of a specific synthetic guide RNA (sgRNA), complementary in sequence to the region of the genome to be targeted, that positions Cas9 at the target site in the genome to result in the creation of a double-stranded DNA break. Guide RNA design is very straightforward, indeed a number of public-domain and commercial design tools have become available for the immediate design of highly specific sgRNAs that can be ordered through the Internet and delivered to the laboratory within days. When the sgRNA is introduced into cells alongside SpCas9, the sgRNA recruits the Cas9 nuclease to a specific site in the genome at which a double-stranded DNA break is introduced into the genome. This is then repaired using the cells' endogenous DNA repair machinery, commonly through a process termed NHEJ that introduces a small insertion or deletion into the target gene sequence that results in the expression of a nonfunctional protein. Through changing the sequence of the sgRNA, it is theoretically possible to target Cas9 to any site in the genome. In studies aimed at deletion of the gene of interest, the editing efficiencies seen with CRISPR/Cas9 can exceed 90% and are typically above 50% or more of transfected cells making this a highly efficient tool to study the consequences of a loss of gene function in cells and animal models of disease. Furthermore, due to the simplicity of the system, a single Cas9 enzyme can be introduced into a cell with two or more guide RNA to result in the deletion of large pieces of DNA at a single genetic loci, or the independent deletion of multiple genes in parallel, making CRISPR/Cas9 an highly valuable and flexible tool for the study of gene function in cell and animal models.

The introduction of single nucleotide changes, or the insertion of small sequences of heterologous DNA, into a gene can be mediated through the introduction into a cell of Cas9, a sgRNA and a donor DNA template that contains the new sequence. Following gene repair by HDR, the point mutation or additional gene sequence can be introduced the target gene. In contract to gene deletion which is a highly efficient process, gene editing or repair using HDR is a low-efficiency process, with typically less than 5% of cells being edited, with much work ongoing to develop methods with improved efficiency. While this is a low-efficiency process, CRISPR enables precise genetic changes to allow the study of the effect of single nucleotide changes and protein truncation on gene function, and the introduction of affinity or other epitope tags into proteins to allow the study of protein location in cells.

In further applications of the technology, variants of Cas9 have been created in which an enzymatically inactive Cas9, that is no longer able to cut the target DNA, is fused to a transcriptional activator or repressor protein (Gilbert et al. 2014; Kampmann 2018). When recruited to the promoter region of a target gene, these versions of Cas9 are able to mediate an activation or repression of gene expression. Base Editor technology has been developed to address the challenge of creating editing systems with increased efficiency for the introduction of precise genetic changes into genes (Rees and Liu 2018). Base Editors consist of a fusion protein between an enzymatically inactive (one site) Cas9nickase and adenosine or cytosine deaminase. These proteins when introduced into cells alongside a targeting sgRNA mediate the enzymatic modification of a specific nucleotide in the genome. Cytosine base editors mediate the transformation of Cytosine to Thytmidine whereas Adenosine Base Editors mediate the transition from Adenosine to Guanosine. In contrast to the introduction of random indels (insertion/deletions) into cells using Cas9, Base Editors mediate specific editing of the target nucleotide.

A further innovation in gene editing technology arose with the publication of Prime Editing (Anzalone et al. 2019). In this technique, a fusion protein is created between an SpCas9 "nickase," that rather than creating a double-stranded break in the genome acts to cut a single strand of the DNA, and a reverse transcriptase. When introduced into cells alongside a "prime editing guide" RNA (pegRNA), the pegRNA targets Cas9 to a precise position in the genome where it creates a single-stranded break in the DNA strand. In contrast to a sgRNA, the pegRNA encodes both a sequence to target the nuclease to a specific site in the genome and a template RNA sequence to be introduced into the genome. The reverse transcriptase creates a DNA copy of the pegRNA which is then introduced into the genome using the cells' DNA mismatch repair mechanism, resulting in the insertion of a short piece of DNA. This method has been used to introduce point mutations, new codons, and to insert larger DNA sequences into target genes. This method can again theoretically be used to target any sequence in the genome and in contrast to earlier editing methods can result in highly efficient genome modulation.

The huge interest and range of applications for CRISPR have led to the establishment of a series of new vendor companies able to supply CRISPR reagents, both guide RNAs and editing enzymes, for use by the laboratory scientist. This includes organizations such as Synthego and Horizon Discovery as well as the creation of capability in established reagent supply companies including Merck and Thermo Fisher. As well as supplying CRISPR reagents for use in the scientists' laboratory, these companies also offer a variety of services including the creation of CRISPR-edited cell lines and animal models and the completion of Functional Genomic screens. Through the work of these companies, CRISPR technology has become democratized for use by any laboratory competent in basic molecular and cell biology techniques. Some of these commercial reagents are discussed in Chapter 4 of this book.

1.4 Applications of CRISPR Cas9 in Drug Discovery

The ability to precisely edit genomes with CRISPR/Cas9 and other related editing systems has become integral to the identification of new drug targets and the creation of engineered cell and animal models of disease. The impact of CRISPR in drug discovery is discussed at length throughout this book and is briefly introduced here. The field of Functional Genomics has advanced with the generation of whole genome-wide CRISPR libraries and other reagents that enable the parallel deletion, upregulation, or downregulation of every gene in the genome to ask the question "what is the effect of modulating this gene on the biology of interest?" (Doench 2018). These libraries can be prepared in micro-titer plate format with each well of the plate containing a cocktail of guide RNAs designed to delete a single gene. When screened against cellular models of disease, it becomes possible to identify specific genes which when modulated affect the biology under study. In addition to use for the identification of new drug targets, Functional Genomic screens are being applied widely to address questions such as the identification of genes that when modulated enable an increase in recombinant protein expression or an increase in the productive uptake of lipid nanoparticles. A further application of Functional Genomic screens in Oncology is the screening of whole genome-wide CRISPR libraries against multiple cancer cell lines in the presence of known cancer medicines to identify genes that mediate resistance or sensitization to that medicine. Many hundred such screens have been performed at Institutes such as the Broad Institute and the Wellcome Trust Sanger Centre to create public domain databases that describe so-called sensitivity maps of cancer types to drug action (Behan et al. 2019; Cui et al. 2021). Such studies are enabling the targeting of new medicines to specific tumor types, the identification of likely resistance mechanisms to new medicines, and drug combination opportunities in the clinic.

CRISPR is widely applied to create cellular and animal models of disease, both for the identification of new drug targets and for understanding the efficacy of new drug candidates within a discovery program (Lundin et al. 2020). CRISPR is used to create specific mutations in genes to understand the effect of that mutation on gene function and to introduce molecular tags into genes to track gene expression. The latter approach has been widely adapted to characterize the efficacy of Proteolysis Targeting Chimeras (PROTACs) drugs. PROTACs are a recently discovered class of small-molecule drugs that rather than inhibiting the function of a drug target, act to degrade the target protein. To understand the efficacy of PROTAC drugs in cellular models of disease, the drug target is typically tagged with a short protein sequence that enables the creation of assays that allow PROTAC-mediated degradation of the target to be followed in real time in an immortalized cell line or animal model of disease. CRISPR has revolutionized the ability to generate transgenic animal models of disease, both reducing the timelines and number of animals required for the creation of an animal model through the ability to highly efficiently edit the genome of the single cell embryo, while again enabling the creation of complex models of disease not previously possible.

CRISPR is being widely applied in the field of CAR-T cell therapy both to enable precise insertion of the CAR, but also to identify and delete other T-cell genes to enable improved efficacy of the cell product (Liu et al. 2017). There is also huge interest in the potential of CRISPR as a medicine in its own right to correct gene mutations in rare and perhaps common diseases and a number of biotechnology companies have been established to bring CRISPR medicines to the clinic, including Editas, CRISPR Therapeutics, Beam Therapeutics, Verve Therapeutics and Intellia. The first clinical studies of medicines to treat β-thalassemia and Sickle Cell Disease started in 2019 with highly promising results in the first patients, with the first *in-vivo* gene editing clinical trials in diseases such as Transthyretin amylodosis in which CRISPR is being used to delete genes in the patient liver, due to start in 2022. Many further projects are in discovery to develop treatments for a range of diseases including α1-antitrypsin deficiency and Cystic Fibrosis.

Last and perhaps one of the most exciting applications of CRISPR in drug discovery is the potential to create highly sensitive, inexpensive, point-of-care diagnostics for the early detection of disease (Chen et al. 2018; Gootenberg et al. 2018; Myhrvold et al. 2018). It is widely accepted, particularly in Oncology, that the probability of patient survival from the disease increases with early disease detection. The creation of diagnostics that detect cancer in stage 1 rather than when symptomatic in stage 3 or 4 will transform our ability to treat and perhaps cure this disease. Two methods have been published, described as SHERLOCK and DETECTR, that offer the potential to create such sensitive DNA diagnostics. While in early development, the potential of these innovations is huge and are being applied more broadly, including for the creation of a diagnostic test for the SARS-CoV2 virus.

1.5 Concluding Comments

Since the demonstration of the ability of CRISPR/Cas systems to precisely and efficiently edit gene sequences in 2012, CRISPR has become embedded as a routine technique in molecular and cell biology laboratories across the field. New industries have been created to supply CRISPR reagents and CRISPR-edited cell and animal models to the research scientist, to develop CRISPR medicines and to create CRISPR diagnostics. The applications and impact of

CRISPR in drug discovery are discussed at length within this book. Within eight short years, CRISPR has transformed our ability to identify and characterize the role of new drug targets in disease and to create the cell and animal models integral to identify and optimize drug candidates. With the rate of innovation in this field, we can look forward to the development of novel CRISPR systems that increase the efficiency and specificity of gene editing, to the development of transformative CRISPR therapies with the potential to cure severe genetic diseases and to the invention of highly sensitive diagnostics for the early identification and subsequent cure of many common diseases. As we move through the coming decades, the opportunity for CRISPR to improve human health remains enormous.

References

Acharya, S., Mishra, A., Paul, D. et al. (2019). Francisella novicida Cas9 interrogates genomic DNA with very high specificity and can be used for mammalian genome editing. *Proc. Natl. Acad. Sci. U. S. A.* 116: 20959–20968.

Anzalone, A.V., Randolph, P.B., Davis, J.R. et al. (2019). Search-and-replace genome editing without double-strand breaks or donor DNA. *Nature* 576: 149–157.

Batta, A., Kalra, B.S., and Khirasaria, R. (2020). Trends in FDA drug approvals over last 2 decades: an observational study. *J Family Med Prim Care* 9: 105–114.

Behan, F.M., Iorio, F., Picco, G. et al. (2019). Prioritization of cancer therapeutic targets using CRISPR-Cas9 screens. *Nature* 568: 511–516.

Bibikova, M., Carroll, D., Segal, D.J. et al. (2001). Stimulation of homologous recombination through targeted cleavage by chimeric nucleases. *Mol. Cell Biol.* 21: 289–297.

Boch, J., Scholze, H., Schornack, S. et al. (2009). Breaking the code of DNA binding specificity of TAL-type III effectors. *Science* 326: 1509–1512.

Chen, J.S., Ma, E., Harrington, L.B. et al. (2018). CRISPR-Cas12a target binding unleashes indiscriminate single-stranded DNase activity. *Science* 360: 436–439.

Cong, L., Ran, F.A., Cox, D. et al. (2013). Multiplex genome engineering using CRISPR/Cas systems. *Science* 339: 819–823.

Cui, Y., Cheng, X., Chen, Q. et al. (2021). CRISP-view: a database of functional genetic screens spanning multiple phenotypes. *Nucleic. Acids. Res.* 49 (D1): D848–D854.

Doench, J.G. (2018). Am I ready for CRISPR? A user's guide to genetic screens. *Nat. Rev. Genet.* 19: 67–80.

Fellmann, C., Gowen, B.G., Lin, P.C. et al. (2017). Cornerstones of CRISPR-Cas in drug discovery and therapy. *Nat. Rev. Drug Discov.* 16: 89–100.

Gilbert, L.A., Horlbeck, M.A., Adamson, B. et al. (2014). Genome-scale CRISPR-mediated control of gene repression and activation. *Cell* 159: 647–661.

Gootenberg, J.S., Abudayyeh, O.O., Kellner, M.J. et al. (2018). Multiplexed and portable nucleic acid detection platform with Cas13, Cas12a, and Csm6. *Science* 360: 439–444.

Jinek, M., Chylinski, K., Fonfara, I. et al. (2012). A programmable dual-RNA-guided DNA endonuclease in adaptive bacterial immunity. *Science* 337: 816–821.

Kampmann, M. (2018). CRISPRi and CRISPRa Screens in Mammalian Cells for Precision Biology and Medicine. *ACS Chem. Biol.* 13: 406–416.

Kim, Y.G., Cha, J., and Chandrasegaran, S. (1996). Hybrid restriction enzymes: zinc finger fusions to Fok I cleavage domain. *Proc. Natl. Acad. Sci. U. S. A.* 93: 1156–1160.

Lancaster, M.A., Renner, M., Martin, C.A. et al. (2013). Cerebral organoids model human brain development and microcephaly. *Nature* 501: 373–379.

Liu, X., Zhang, Y., Cheng, C. et al. (2017). CRISPR-Cas9-mediated multiplex gene editing in CAR-T cells. *Cell Res.* 27: 154–157.

Lundin, A., Porritt, M.J., Jaiswal, H. et al. (2020). Development of an ObLiGaRe Doxycycline Inducible Cas9 system for pre-clinical cancer drug discovery. *Nat. Commun.* 11: 4903.

Mali, P., Yang, L., Esvelt, K.M. et al. (2013). RNA-guided human genome engineering via Cas9. *Science* 339: 823–826.

Morgan, P., Brown, D.G., Lennard, S. et al. (2018). Impact of a five-dimensional framework on R&D productivity at AstraZeneca. *Nat. Rev. Drug Discov.* 17: 167–181.

Moscou, M.J. and Bogdanove, A.J. (2009). A simple cipher governs DNA recognition by TAL effectors. *Science* 326: 1501.

Myhrvold, C., Freije, C.A., Gootenberg, J.S. et al. (2018). Field-deployable viral diagnostics using CRISPR-Cas13. *Science* 360: 444–448.

Rees, H.A. and Liu, D.R. (2018). Base editing: precision chemistry on the genome and transcriptome of living cells. *Nat. Rev. Genet.* 19: 770–788.

Vamathevan, J., Clark, D., Czodrowski, P. et al. (2019). Applications of machine learning in drug discovery and development. *Nat. Rev. Drug Discov.* 18: 463–477.

Wiedenheft, B., Sternberg, S.H., and Doudna, J.A. (2012). RNA-guided genetic silencing systems in bacteria and archaea. *Nature* 482: 331–338.

2

Historical Overview of Genome Editing from Bacteria to Higher Eukaryotes

Marcello Maresca

Genome Engineering, Discovery Sciences, BioPharmaceuticals R&D, AstraZeneca, Gothenburg, Sweden

2.1 Introduction

Molecular cloning methods have been instrumental for the establishment of the biotechnological industry. The ability to clone any DNA sequence of interest into a DNA vector has been a key technology advancement toward the generation of cellular and animal model of disease and the development of biopharmaceuticals. Traditional molecular cloning methods mostly rely on restriction enzymes-mediated digestion and ligation of the digested fragments. Classical restriction enzymes recognize a relatively short DNA sequence and as a consequence, they are too unspecific to be used directly for DNA engineering applications *in cellula*.

Novel improvements in DNA assembly methods combined with the cost reduction and with the increase in accuracy of DNA synthesis processes have led to the possibility of assembling large DNA constructs *in vitro*. Synthetic genomes will have a key role in future DNA engineering platforms but they will not be discussed in this chapter and in this book, where we will focus on *in cellula* genome engineering approaches.

In this chapter, I will give a brief description of the advancements in the precise genome editing field starting from observations of single-stranded oligonucleotides-mediated repair in yeasts to Recombineering and CRISPR-Cas9-dependent editing. These technologies have all greatly expanded the tools and methods that are used to generate disease models and to develop assays for drug discovery (Figure 2.1).

2.2 Bacterial DNA Engineering (Recombineering)

Microbes and microbial-derived systems have been extensively used for the development of novel DNA engineering tools and for the application of these tools to DNA cloning. Restriction enzymes, recombinase systems such as CRE/Lox, integrases such as ΦC31-Int, and the Cas9-CRISPR system have all microbial origin. Recombinases and integrases-based systems have been extensively used to engineer the mammalian genomes but we will not discuss them in this book that is focusing on scarless genome engineering systems. This chapter will focus on the development of Recombineering for bacterial engineering and its use in genome engineering with particular focus on applications in drug discovery.

The inspiration for the Recombineering (**recombin**ation-mediated DNA **engineering**) method came from studies in yeast where the possibility to introduce an exogenous DNA cassette in yeast genome was demonstrated in the years 1978–1979. Yeasts DNA recombination methods used homology arms to target a gene without the need of double-strand breaks (DSBs) generation or the expression of any exogenous protein (Hinnen et al. 1978; Scherer and Davis 1979). It was also demonstrated in yeasts that single-stranded oligonucleotides with short homology arms to a target sequence are able to promote genomic insertions/deletions/modifications (Bhargava et al. 1999) Unfortunately, this approach does not work efficiently in wild-type bacteria or in higher eukaryotes. This has limited the utility of single-strand oligonucleotides-mediated DNA editing of bacterial and mammalian genomes before the advent of recombineering. Murphy's and Stewart's groups showed for the first time that a portable recombination system can be introduced in *Escherichia coli* to induce recombination in bacteria in a similar fashion to the yeast recombination system but with much higher efficiency (Murphy 1998; Zhang et al. 1998). The portable cassette encodes for an exonuclease (i.e. Redα for the lambda Red system), a DNA annealing protein (i.e. Redβ), and the RecBCD inhibitor (i.e. Redγ). In particular, Stewart's group showed that this system works with very short

Genome Editing in Drug Discovery, First Edition. Edited by Marcello Maresca and Sumit Deswal.

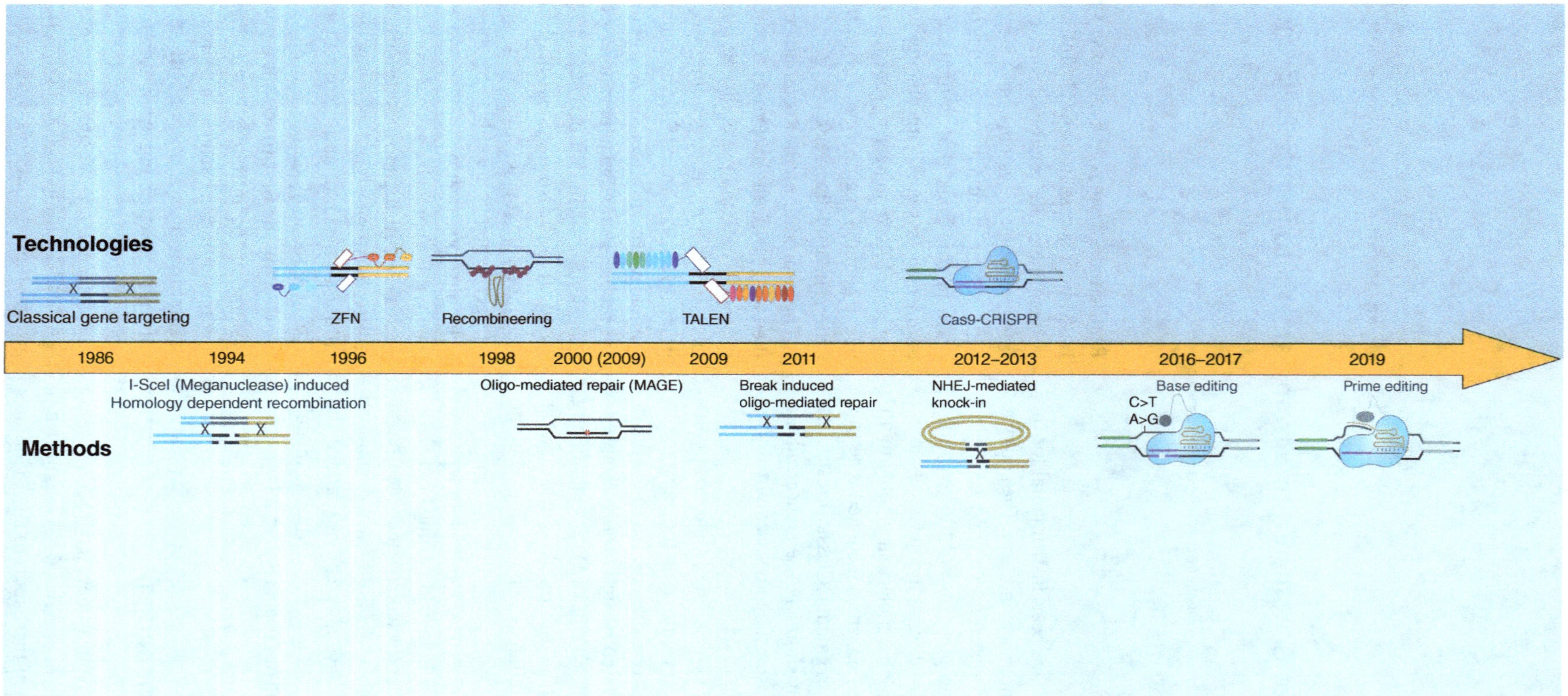

Figure 2.1 Graphical overview of genome engineering technologies (upper panel) and methods (lower panel) developed during the latest 27 years. A schematic representation of the different technologies or methods is presented, respectively, above or under the timeline arrow.

homology arms (as short as 30nt) via a peculiar mechanism of single-strand heteroduplex intermediates at the replication fork (Maresca et al. 2010). This observation paved the way to the use of Recombineering for Precise Genome Editing of Bacterial Genome and for molecular cloning strategies. Recombineering overcomes the limitation of classical restriction/ligation-based cloning because it does not require the availability of unique restriction sites in the target plasmid and it is specific enough to target the bacterial genome. Therefore, Recombineering has been extensively used for the seamless engineering of large constructs such as bacterial artificial chromosomes (BACs) and for the engineering of bacterial genome.

2.3 BAC Recombineering

The use of recombineering has been particularly important for functional genomics programmes where BAC transgenes or Gene Targeting constructs have been engineered at large scale to generate animal models of disease or to develop libraries of gene tagging. The European Conditional Mouse Mutagenesis (EUCOMM) and Knock-Out Mouse Programme (KOMP) contributed to the large-scale generation of conditional gene KO mice that have been extensively used in Drug Discovery. In particular, Skarnes and colleagues developed a large conditional knock-out mouse library in the framework of the EUCOMM programme by using a high-throughput gene-targeting pipeline based on Recombineering (Skarnes et al. 2011). This gene-targeting pipeline has been greatly facilitated by the development of a high-throughput strategy of DNA engineering where "recombineered" targeting constructs were used to engineer C57BL/6N mouse embryonic stem cell for the generation of KO mice. This mouse library has been instrumental to understand the function of genes encoded by the mammalian genomes (*In vivo*) and to validate drug targets.

Another particular relevant example of Recombineering applications in Drug Discovery/Development is the remarkable work by scientists at Regeneron Pharmaceuticals aimed to engineer a humanized mouse model producing human–mouse hybrid antibodies. Their VelociGene platform (Murphy 1998) allowed the generation of multiple knock-out/knock-in by an high-throughput recombineering & gene targeting approach where "recombineered" BACs are inserted in mESC using sequential homologous recombination steps. This led to the replacement of mouse immune genes with human orthologs (Valenzuela et al. 2003).

Finally, a very relevant application of Recombineering is the generation of tagged genes libraries. Gene tagging can potentially overcome the use of high-affinity antibodies to detect gene expression, but it is limited by the lack of faithful gene activity of tagged protein generated with the use of overexpressed cDNA vectors. BAC transgenes guarantee a quas-physiological level of gene expression maintaining transgene regulatory element and promoters, although the tagged gene is not integrated in its endogenous locus but in a so-called third allele. The generation of tagged BAC libraries at scale was greatly simplified by selecting recombineering events in liquid bacterial culture. The potential of this system is exemplified by the generation of genome-wide BAC libraries for the analysis of protein localization in *Daino Rerio* and *Caenorhabditis elegans* (Sarov et al. 2006).

2.4 Metabolic Engineering

The highly diverse chemical structures of Natural Products isolated from microbes or derived semisynthetically from natural intermediates allowed the development of a broad range of different drug activities, including antibiotics and chemotherapeutics.

Genome sequencing data facilitated by the development of Next Generation sequencing platforms indicate that microbial genomes contain an untapped resource of biosynthetic gene cluster that can be exploited to generate novel functions.

Unfortunately, most of these gene clusters are not expressed under normal laboratory growth conditions even when it is possible to grow the natural host in lab environment. In addition, the size of the Biosynthetic gene clusters (reaching up to 200kb) renders the *in vitro* manipulation of this large clusters difficult (Smanski et al. 2016).

Recombineering or Recombineering-derived strategies have therefore been an ideal method to characterize and to engineer long gene clusters. In fact, specific gene clusters can be inserted in an heterologous host to facilitate the genetic manipulation of the genes present in these clusters.

An alternative strategy would be the use of endogenous recombineering systems from different hosts to manipulate the particular genes present in the gene cluster (Yin et al. 2015).

Recombineering is rapidly becoming the method of choice to manipulate biosynthetic gene clusters but it is also increasingly used to evolve the bacterial genome as pioneered by the work of Church's group. Hang HH, Isaacs FJ et al. in their landmark paper of 2019 described the use of recombineering to accelerate bacterial genome evolution by an automated multiplex recombineering strategy that they named MAGE (Wang et al. 2009). MAGE is using a pool of ssDNA oligonucleotide coupled with the expression of a ssDNA-binding protein to install thousands of functionalized genome variants in *E. coli* genome. MAGE has been recently applied to lower Eukaryotes such as *Saccharomyces cerevisiae* (Si et al. 2017).

One of the problems associated with high-throughput Recombineering is the need to select for the functional variant because there is usually no selective advantage for the recombined bacterial cell. To overcome this limitation, several strategies have been developed to combine the precision and efficiency of Recombineering with the strong selection pressure that occurs in bacteria after the generation of DSB caused by endonucleases like Cas9/CRISPR (Jiang et al. 2013; Jiang et al. 2015; Baker et al. 2016). Despite the large success of Recombineering applied to *E. coli* genetic engineering, the engineering of large biosynthetic pathways is still inefficient in endogenous hosts where a Recombineering system is not efficient or where Homologous Recombination is not efficient. Therefore, alternative strategies need to be developed to overcome this limitation. One possibility would be to exploit alternative pathways (Su et al. 2016) of DNA recombination/repair such as non-homologous end joining (NHEJ) as previously done in mammalian cells (Maresca et al. 2013). A more promising approach relies on the recently described integration system by Transposon-encoded CRISPR-Cas (Klompe et al. 2019; Strecker et al. 2020). These strategies can facilitate genome engineering of bacteria and can possibly be implemented for the genetic manipulation of Eukaryotic genomes that are less prone to homologous recombination as described in the next section.

2.5 Genetic Engineering in Higher Eukaryotes

The genomes of Higher Eukaryotes and in particular mammalian genomes are quite resilient to genetic manipulation. It was only in the 1986 that the first gene targeting experiment to generate a functional Knock-Out of the HPRT gene was published by scientists in Capecchi's group (Thomas et al. 1986). The same targeting strategy was further optimized by Janish's group leading to the establishment of a gene targeting pipeline in mESC where potentially any gene can be modified to generate genetically engineered mice models. Despite several attempts to further optimize the method to reduce homology arms length, it became evident that this gene targeting strategy in mESC requires long stretches of homology to the target sites to obtain an optimal efficiency of recombination. High frequency of recombined cells allows the isolation of engineered clones. An increase in the homology arm length by using targeting engineered BACs could further boost this efficiency (Yang and Seed 2003). One characteristic of this canonical gene targeting strategy is that most of the targeting events result in mono-allelic insertion. Consequentially, additional gene targeting steps or animal breeding strategies are required to obtain bi-allelic targeting. As discussed above, Recombineering has recently become the method of choice to generate the complex targeting cassettes with long homology arms needed for mammalian genome manipulation.

This classical gene targeting approach works relatively well in mESC and in DT40, a chicken bursal lymphoblast cell line (Buerstedde and Takeda 1991). It is still not clear to this day the reason why this approach is difficult to implement in other cell types. However, in 2002, David Russel lab showed that it is possible to use Adeno Associated Virus (AAV) to promote gene targeting in mammalian cell lines (Hirata et al. 2002). Despite the importance of this finding, this system is still laborious and pretty inefficient in inducing bi-allelic gene targeting. A promising approach to introduce small indels/mutations by single-stranded oligonucleotide in mammalian cells has been described by Kmiec's lab using chimeric oligonucleotides, but it has not been extensively applied for the generation of cellular model due to its low efficiency and due to lack of reproducibility when targeting different loci (Cole-Strauss et al. 1996).

Experiments from Jasin's group clearly demonstrated that a targeted DSB induced by I-SceI enzyme can overcome the anti-recombinogenic feature of mammalian genome by promoting homologous recombination. This experiment was based on previous observations in budding yeast on mating-type switching (Strathern et al. 1982) and on the model of recombination dependent on DSB generation (Szostak et al. 1983). There are two important findings as result of the work from Jasin's group. The first one is that DSBs are preferentially repaired by NHEJ in mammalian cells and although it is difficult to evaluate the amount of perfect repair by precise NHEJ, imprecise NHEJ, using additional enzymes to process the broken ends of the DSBs, can result in template-independent gene disruption. The second important finding is that DSBs can promote bi-allelic DNA recombination if the break occurs in both alleles.

2.6 Targeted Endonucleases

The use of DSBs to promote homologous recombination in mammalian cells was limited by the lack of a modular site-specific endonuclease technology that could be directed to any gene of interest. The solution to this problem came from the work of Chandrasegaran's group where for the first time the engineering of Zinc Finger-based endonucleases (ZFN) was described (Kim et al. 1996). This system combines the modular design of Zinc Finger Domain to the powerful nuclease activity of the restriction enzyme FokI that requires dimerization to induce DSBs. The requirement of an obligate dimerization of FokI for an efficient cleavage guarantees high specificity to the system. The Zinc Finger DNA-binding domain was selected based on the previous work from the labs of Klug and Pabo directed to the discovery of ZF domain (Miller et al. 1985) and the understanding of its structure (Pavletich and Pabo 1991).

Following the publication of the ZFN architecture, Carroll's group showed for the first time efficient gene targeting using ZFNs in the Drosophila genome (Bibikova et al. 2002). This experiment not only resulted in efficient targeting by installation of random indels via NHEJ but it also demonstrated that ZFNs were precise enough to induce a number of DSBs in the genome compatible with embryonic development. In follow-up works, gene insertion by Homology Directed Repair after ZFN cleavage was demonstrated by Carroll and Baltimore/Porteus groups, respectively, in Drosophila (Bibikova et al. 2001) and human cells (Porteus and Baltimore 2003).

The scene was set for preclinical genome editing, that is the demonstration of gene editing at disease-relevant targets. The work from Fyodor Urnov and colleagues at Sangamo therapeutics to target the X-linked (SCID) mutation in the *IL2Rγ* led to a new era in genome engineering (Urnov et al. 2005). The other important message coming from this work was the demonstration of bi-allelic genome editing using targeted ZFNs, something that is very infrequent with classical gene targeting strategies as discussed above.

After the demonstration of preclinical genome editing, an important advancement in the field was the development of systems to engineer targeted endonucleases. These protein engineering pipelines resulted in the generation of precise endonucleases that could safely target the human genome without inducing strong toxicity as shown in the work of Joung, Kim, Cathomen & Voytas' groups (Maeder et al. 2008; Kim et al. 2009).

It resulted evident from these early experiments that NHEJ is the main pathway of repair exploited by mammalian cells to repair DSBs. Based on these observations, most of the initial preclinical genome editing approaches were directed to exploit NHEJ for the generation of loss of function alleles. Scientists at Sangamo Therapeutics in collaboration with June's group suggested that CCR5 could be an ideal target for the NHEJ-dependent loss of function approach. This strategy was based on genomics data associating CCR5 loss of function to protection from HIV infection. Their work led to the first use of genome editing in primary human cells with efficient and safe genome editing (Perez et al. 2008; Maier et al. 2013).

The application of the NHEJ-based loss of function strategy has also been important in biopharmaceuticals industry for the generation of modified CHO (Chinese Hamster Ovary cells) that are important in biopharmaceutical drug development (Santiago et al. 2008).

NHEJ-dependent knock out was also reproduced in embryos to generate Knock Out in Drosophila, Rats, Mice (Geurts et al. 2009; Carbery et al. 2010) and Zebrafish (Doyon et al. 2008). This was another important milestone in Pharmaceutical industry because it allowed to validate drug targets in different organisms beyond mice bypassing the need for stem cell manipulation and at the same time speeding up the process to generate animal model of disease. These experiments changed the concept of model organism itself. Model organism can be potentially any animal model with a sequenced genome and where it is possible to deliver gene editors. This is exemplified by the generation of transgenic models using mRNA injection or electroporation in embryos bypassing the long process required for the development of transgenic models using stem cell manipulation.

Despite the initial successes to target mammalian genomes with ZFNs, the design of these enzymes was challenging and the technology was inaccessible to most labs. The advent of TALENs, published for the first time in 2009 (Boch et al. 2009; Moscou and Bogdanove 2009), started the democratization of the genome editing field. Also thanks to improvement of TALEN design by the laboratories of Joung (Reyon et al. 2012) and Voytas (Cermak et al. 2011), it was possible to design tailored targeted nucleases for potentially any gene and for any region of interest. This technology applied to iPSC suggested novel avenues for the generation of disease models and for the validation of drug targets. As an example, the work from Cowan's group showed, for the first time, the possibility to target several genes (15) in iPSC and to study the phenotypic consequences of the gene targeting (Ding et al. 2013).

Most of the initial work with TALENs, as with ZFNs, was limited to loss of function strategy. In fact, the majority of cell lines preferentially use NHEJ to edit the genome even when an exogenous DNA donor is provided. This effect is particularly evident in non-dividing primary cells that are the main cells targeted during clinical gene editing.

A significant amount of work was devoted to increase the efficiency of Homology Directed Repair after DSB generation and the most successful approaches required the use of single-strand oligonucleotide (with similar homology arms and design to the Recombineering oligonucleotides) as DNA donor (Chen et al. 2011). Moreover, the use of DNA repair drug inhibitors (Maresca et al. 2013) (mainly targeting DNAPK, a key enzyme in the NHEJ pathway) was also proposed to boost HDR.

At the same time, a new logic of DNA engineering was presented by scientists at Novartis (Maresca et al. 2013) and at Sangamo (Cristea et al. 2013) that relied solely on NHEJ to promote gene knock-in. This technology known as ObLiGaRe (when combined to ZFNs and TALENs) or as HITI (when combined to Cas9/CRISPR) (Suzuki et al. 2016) is using NHEJ to ligate compatible sticky ends or blunt ends resulting from the simultaneous cleavage of the targeting vector and the cell genome. The technology is particularly efficient in differentiated cells, CHO cells, and Zebrafish embryos. Additional knock-in strategies were developed to use MMEJ, as an alternative pathway of DNA integration, but they are mostly efficient in transformed cells (Sakuma et al. 2016).

The development of several genome editing technologies and the limited success of functional genomics screens using RNAi set the scene for the CRISPR-Cas9 revolution. The impact of CRISPR-Cas9 in life science would not have been so meaningful without the improvements described above.

Historically, the CRISPR arrays were first described in 1987 (Ishino et al. 1987) and the association of CRISPR arrays to bacterial immunity and DNA targeting was documented in 2005/2006. In particular, Koonin's group predicted that the CRISPR system could represent an RNAi-like system and interestingly few years later it outplaced RNAi for functional genomics screening (Makarova et al. 2006). The work of Barrangou and Horvath (Barrangou et al. 2007), Sontheimer and Marraffini (Marraffini and Sontheimer 2008), and Ganeau and Moineau (Garneau et al. 2010) were all fundamental to identify CRISPR as an immune system and the CRISPR-associated proteins (Cas), particularly Cas9, as the effector of this immune system. The subsequent identification of the noncoding RNA named TracR and the validation of the *in vitro* targeted cleavage of CRISPR-Cas9 started an exciting revolution in the genome editing field. The application of CRISPR-Cas9 system to target the mammalian genome and to induce homologous recombination or NHEJ-mediated knock-in proved that this system could outcompete all the previous DNA targeting technologies. The application of CRISPR-Cas9 to engineer higher eukaryotic genomes is characterized by an extremely simplified design and a remarkable ability to induce on-target indels. Moreover, the system is compatible with being encoded in lentiviral vectors, thus facilitating applications in functional genomics screenings (Shalem et al. 2014). The biology of the CRISPR system is discussed in detail in Chapter 3 of this book and various application throughout in many of the chapters.

2.7 Novel Genome Editing Technologies

Most of the initial efforts in genome editing using CRISPR-Cas9 were mainly reproducing targeting strategies previously demonstrated with Zinc Finger Nucleases and TALENs but at the same time obtaining much better efficiency with an easier design. Several groups showed the use of CRISPR-Cas9 to promote integration by NHEJ and MMEJ. Joung and Liu's groups showed that Cas9 could be coupled to FokI to increase its specificity (Guilinger et al. 2014; Tsai et al. 2014). Moreover, transient or stable epigenome editing was demonstrated, following the first demonstration of stable epigenome editing using TALENs (Amabile et al. 2016). The efficiency of the system to induce Gene Knock-Out and Gene Knock-In has been really impressive and novel strategies to further enrich for these editing events have been developed recently (Agudelo et al. 2017; Li et al. 2021).

The real differentiator between CRISPR-Cas9 and the other genome editing technologies is the presence of an exposed single-stranded DNA after Cas9/gRNA-mediated strand invasion and target binding (Richardson et al. 2016). This presence of ssDNA is a peculiar characteristic of a D-Loop-forming enzyme and is at the base of one of the most relevant CRISPR technology, Base Editing (Komor et al. 2016; Gaudelli et al. 2017). The development of Base Editing, the precise DNA repair system that is using Cas9-recruited Deaminase to target ssDNA (discussed in Chapter 14), further accelerated the applications of Genome Editing to drug discovery and to Therapeutic Genome Editing. In few years, Base Editing and derived technologies have dramatically impacted the field of functional genomics (Hess et al. 2016; Hanna and Doench 2020) and have provided a safer and more precise alternative to HDR or NHEJ-based DNA editing *In vitro* (Webber et al. 2019) and *In vivo* (Chadwick et al. 2017; Carreras et al. 2019).

Liu's group, the same group that developed base editing, used the peculiar D-Loop induction of CRISPR-Cas9 to develop an additional technology named Prime Editing (Anzalone et al. 2020), where a Reverse Transcriptase is fused to Cas9 to prime a gRNA templated editing event. Prime Editing is still in its infancy, but it will probably expand the reach of base editing.

2.8 Conclusions

Genome Editing and particularly the CRISPR-Cas9-based technologies had a tremendous impact in the drug discovery field. Despite all the advancements, the field of Genome Editing by CRISPR-Cas is characterized by *in fieri* evolution where it is difficult to anticipate the next improvements and applications but certainly the presence of exposed ssDNAs can be used for other enzymatic functions promoting DNA integration, DNA labeling, or DNA regulation. Most of the CRISPR-Cas technologies are exploiting a common process of "proximology," i.e. proximal enzymology, that is having a tremendous impact not only in the genome editing field but more broadly in the Drug Discovery and Drug Development field, as exemplified by PROTACs and derivatives (Sakamoto et al. 2001).

References

Agudelo, D., Duringer, A., Bozoyan, L. et al. (2017). Marker-free coselection for CRISPR-driven genome editing in human cells. *Nat. Methods* 14: 615–620.

Amabile, A., Migliara, A., Capasso, P. et al. (2016). Inheritable silencing of endogenous genes by hit-and-run targeted epigenetic editing. *Cell* 167 (219–232): e14.

Anzalone, A.V., Koblan, L.W., and Liu, D.R. (2020). Genome editing with CRISPR-Cas nucleases, base editors, transposases and prime editors. *Nat. Biotechnol.* 38: 824–844.

Baker, O., Gupta, A., Obst, M. et al. (2016). RAC-tagging: recombineering and Cas9-assisted targeting for protein tagging and conditional analyses. *Sci. Rep.* 6: 25529.

Barrangou, R., Fremaux, C., Deveau, H. et al. (2007). CRISPR provides acquired resistance against viruses in prokaryotes. *Science* 315: 1709–1712.

Bhargava, J., Shashikant, C.S., Carr, J.L. et al. (1999). Direct cloning of genomic DNA by recombinogenic targeting method using a yeast-bacterial shuttle vector, pClasper. *Genomics* 62: 285–288.

Bibikova, M., Carroll, D., Segal, D.J. et al. (2001). Stimulation of homologous recombination through targeted cleavage by chimeric nucleases. *Mol. Cell Biol.* 21: 289–297.

Bibikova, M., Golic, M., Golic, K.G., and Carroll, D. (2002). Targeted chromosomal cleavage and mutagenesis in drosophila using zinc-finger nucleases. *Genetics* 161: 1169–1175.

Boch, J., Scholze, H., Schornack, S. et al. (2009). Breaking the code of DNA binding specificity of TAL-Type III effectors. *Science* 326: 1509–1512.

Buerstedde, J.M. and Takeda, S. (1991). Increased ratio of targeted to random integration after transfection of chicken B cell lines. *Cell* 67: 179–188.

Carbery, I.D., Ji, D., Harrington, A. et al. (2010). Targeted genome modification in mice using zinc-finger nucleases. *Genetics* 186: 451–459.

Carreras, A., Pane, L.S., Nitsch, R. et al. (2019). in vivo genome and base editing of a human PCSK9 knock-in hypercholesterolemic mouse model. *BMC Biol.* 17: 4.

Cermak, T., Doyle, E.L., Christian, M. et al. (2011). Efficient design and assembly of custom TALEN and other TAL effector-based constructs for DNA targeting. *Nucleic Acids Res.* 39: e82.

Chadwick, A.C., Wang, X., and Musunuru, K. (2017). in vivo base editing of PCSK9 (proprotein convertase subtilisin/kexin Type 9) as a therapeutic alternative to genome editing. *Arterioscler. Thromb. Vasc. Biol.* 37: 1741–1747.

Chen, F., Pruett-Miller, S.M., Huang, Y. et al. (2011). High-frequency genome editing using ssDNA oligonucleotides with zinc-finger nucleases. *Nat. Methods* 8: 753–755.

Cole-Strauss, A., Yoon, K., Xiang, Y. et al. (1996). Correction of the mutation responsible for sickle cell anemia by an RNA-DNA oligonucleotide. *Science* 273: 1386–1389.

Cristea, S., Freyvert, Y., Santiago, Y. et al. (2013). in vivo cleavage of transgene donors promotes nuclease-mediated targeted integration. *Biotechnol. Bioeng.* 110: 871–880.

Ding, Q., Lee, Y.K., Schaefer, E.A. et al. (2013). A TALEN genome-editing system for generating human stem cell-based disease models. *Cell Stem Cell* 12: 238–251.

Doyon, Y., Mccammon, J.M., Miller, J.C. et al. (2008). Heritable targeted gene disruption in zebrafish using designed zinc-finger nucleases. *Nat. Biotechnol.* 26: 702–708.

Garneau, J.E., Dupuis, M.E., Villion, M. et al. (2010). The CRISPR/Cas bacterial immune system cleaves bacteriophage and plasmid DNA. *Nature* 468: 67–71.

Gaudelli, N.M., Komor, A.C., Rees, H.A. et al. (2017). Programmable base editing of A*T to G*C in genomic DNA without DNA cleavage. *Nature* 551: 464–471.

Geurts, A.M., Cost, G.J., Freyvert, Y. et al. (2009). Knockout rats via embryo microinjection of zinc-finger nucleases. *Science* 325: 433.

Guilinger, J.P., Thompson, D.B., and Liu, D.R. (2014). Fusion of catalytically inactive Cas9 to FokI nuclease improves the specificity of genome modification. *Nat. Biotechnol.* 32: 577–582.

Hanna, R.E. and Doench, J.G. (2020). Design and analysis of CRISPR-Cas experiments. *Nat. Biotechnol.* 38: 813–823.

Hess, G.T., Fresard, L., Han, K. et al. (2016). Directed evolution using dCas9-targeted somatic hypermutation in mammalian cells. *Nat. Methods* 13: 1036–1042.

Hinnen, A., Hicks, J.B., and Fink, G.R. (1978). Transformation of yeast. *Proc. Natl. Acad. Sci. U. S. A.* 75: 1929–1933.

Hirata, R., Chamberlain, J., Dong, R., and Russell, D.W. (2002). Targeted transgene insertion into human chromosomes by adeno-associated virus vectors. *Nat. Biotechnol.* 20: 735–738.

Ishino, Y., Shinagawa, H., Makino, K. et al. (1987). Nucleotide sequence of the iap gene, responsible for alkaline phosphatase isozyme conversion in escherichia coli, and identification of the gene product. *J. Bacteriol.* 169: 5429–5433.

Jiang, W., Bikard, D., Cox, D. et al. (2013). RNA-guided editing of bacterial genomes using CRISPR-Cas systems. *Nat. Biotechnol.* 31: 233–239.

Jiang, W., Zhao, X., Gabrieli, T. et al. (2015). Cas9-assisted targeting of CHromosome segments CATCH enables one-step targeted cloning of large gene clusters. *Nat. Commun.* 6: 8101.

Kim, Y.G., Cha, J., and Chandrasegaran, S. (1996). Hybrid restriction enzymes: zinc finger fusions to Fok I cleavage domain. *Proc. Natl. Acad. Sci. U. S. A.* 93: 1156–1160.

Kim, H.J., Lee, H.J., Kim, H. et al. (2009). Targeted genome editing in human cells with zinc finger nucleases constructed via modular assembly. *Genome Res.* 19: 1279–1288.

Klompe, S.E., Vo, P.L.H., Halpin-Healy, T.S., and Sternberg, S.H. (2019). Transposon-encoded CRISPR-Cas systems direct RNA-guided DNA integration. *Nature* 571: 219–225.

Komor, A.C., Kim, Y.B., Packer, M.S. et al. (2016). Programmable editing of a target base in genomic DNA without double-stranded DNA cleavage. *Nature* 533: 420–424.

Li, S., Akrap, N., Cerboni, S. et al. (2021). Universal toxin-based selection for precise genome engineering in human cells. *Nat. Commun.* 12: 497.

Maeder, M.L., Thibodeau-Beganny, S., Osiak, A. et al. (2008). Rapid "open-source" engineering of customized zinc-finger nucleases for highly efficient gene modification. *Mol. Cell* 31: 294–301.

Maier, D.A., Brennan, A.L., Jiang, S. et al. (2013). Efficient clinical scale gene modification via zinc finger nuclease-targeted disruption of the HIV co-receptor CCR5. *Hum. Gene Ther.* 24: 245–258.

Makarova, K.S., Grishin, N.V., Shabalina, S.A. et al. (2006). A putative RNA-interference-based immune system in prokaryotes: computational analysis of the predicted enzymatic machinery, functional analogies with eukaryotic RNAi, and hypothetical mechanisms of action. *Biol. Direct* 1: 7.

Maresca, M., Erler, A., Fu, J. et al. (2010). Single-stranded heteroduplex intermediates in lambda red homologous recombination. *BMC Mol. Biol.* 11: 54.

Maresca, M., Lin, V.G., Guo, N., and Yang, Y. (2013). Obligate ligation-gated recombination (ObLiGaRe): custom-designed nuclease-mediated targeted integration through nonhomologous end joining. *Genome Res.* 23: 539–546.

Marraffini, L.A. and Sontheimer, E.J. (2008). CRISPR interference limits horizontal gene transfer in staphylococci by targeting DNA. *Science* 322: 1843–1845.

Miller, J., Mclachlan, A.D., and Klug, A. (1985). Repetitive zinc-binding domains in the protein transcription factor IIIA from Xenopus oocytes. *EMBO J.* 4: 1609–1614.

Moscou, M.J. and Bogdanove, A.J. (2009). A simple cipher governs DNA recognition by TAL effectors. *Science* 326: 1501.

Murphy, K.C. (1998). Use of bacteriophage lambda recombination functions to promote gene replacement in Escherichia coli. *J. Bacteriol.* 180: 2063–2071.

Pavletich, N.P. and Pabo, C.O. (1991). Zinc finger-DNA recognition: crystal structure of a Zif268-DNA complex at 2.1 A. *Science* 252: 809–817.

Perez, E.E., Wang, J., Miller, J.C. et al. (2008). Establishment of HIV-1 resistance in CD4+ T cells by genome editing using zinc-finger nucleases. *Nat. Biotechnol.* 26: 808–816.

Porteus, M.H. and Baltimore, D. (2003). Chimeric nucleases stimulate gene targeting in human cells. *Science* 300: 763.

Reyon, D., Tsai, S.Q., Khayter, C. et al. (2012). FLASH assembly of TALENs for high-throughput genome editing. *Nat. Biotechnol.* 30: 460–465.

Richardson, C.D., Ray, G.J., Dewitt, M.A. et al. (2016). Enhancing homology-directed genome editing by catalytically active and inactive CRISPR-Cas9 using asymmetric donor DNA. *Nat. Biotechnol.* 34: 339–344.

Sakamoto, K.M., Kim, K.B., Kumagai, A. et al. (2001). Protacs: chimeric molecules that target proteins to the Skp1-Cullin-F box complex for ubiquitination and degradation. *Proc. Natl. Acad. Sci. U. S. A.* 98: 8554–8559.

Sakuma, T., Nakade, S., Sakane, Y. et al. (2016). MMEJ-assisted gene knock-in using TALENs and CRISPR-Cas9 with the PITCh systems. *Nat. Protoc.* 11: 118–133.

Santiago, Y., Chan, E., Liu, P.Q. et al. (2008). Targeted gene knockout in mammalian cells by using engineered zinc-finger nucleases. *Proc. Natl. Acad. Sci. U. S. A.* 105: 5809–5814.

Sarov, M., Schneider, S., Pozniakovski, A. et al. (2006). A recombineering pipeline for functional genomics applied to Caenorhabditis elegans. *Nat. Methods* 3: 839–844.

Scherer, S. and Davis, R.W. (1979). Replacement of chromosome segments with altered DNA sequences constructed in vitro. *Proc. Natl. Acad. Sci. U. S. A.* 76: 4951–4955.

Shalem, O., Sanjana, N.E., Hartenian, E. et al. (2014). Genome-scale CRISPR-Cas9 knockout screening in human cells. *Science* 343: 84–87.

Si, T., Chao, R., Min, Y. et al. (2017). Automated multiplex genome-scale engineering in yeast. *Nat. Commun.* 8: 15187.

Skarnes, W.C., Rosen, B., West, A.P. et al. (2011). A conditional knockout resource for the genome-wide study of mouse gene function. *Nature* 474: 337–342.

Smanski, M.J., Zhou, H., Claesen, J. et al. (2016). Synthetic biology to access and expand nature's chemical diversity. *Nat. Rev. Microbiol.* 14: 135–149.

Strathern, J.N., Klar, A.J., Hicks, J.B. et al. (1982). Homothallic switching of yeast mating type cassettes is initiated by a double-stranded cut in the MAT locus. *Cell* 31: 183–192.

Strecker, J., Ladha, A., Makarova, K.S. et al. (2020). Response to Comment on "RNA-guided DNA insertion with CRISPR-associated transposases". *Science* 368: eabb2920.

Su, T., Liu, F., Gu, P. et al. (2016). A CRISPR-Cas9 assisted non-homologous end-joining strategy for one-step engineering of bacterial genome. *Sci. Rep.* 6: 37895.

Suzuki, K., Tsunekawa, Y., Hernandez-Benitez, R. et al. (2016). in vivo genome editing via CRISPR/Cas9 mediated homology-independent targeted integration. *Nature* 540: 144–149.

Szostak, J.W., Orr-Weaver, T.L., Rothstein, R.J., and Stahl, F.W. (1983). The double-strand-break repair model for recombination. *Cell* 33: 25–35.

Thomas, K.R., Folger, K.R., and Capecchi, M.R. (1986). High frequency targeting of genes to specific sites in the mammalian genome. *Cell* 44: 419–428.

Tsai, S.Q., Wyvekens, N., Khayter, C. et al. (2014). Dimeric CRISPR RNA-guided FokI nucleases for highly specific genome editing. *Nat. Biotechnol.* 32: 569–576.

Urnov, F.D., Miller, J.C., Lee, Y.L. et al. (2005). Highly efficient endogenous human gene correction using designed zinc-finger nucleases. *Nature* 435: 646–651.

Valenzuela, D.M., Murphy, A.J., Frendewey, D. et al. (2003). High-throughput engineering of the mouse genome coupled with high-resolution expression analysis. *Nat. Biotechnol.* 21: 652–659.

Wang, H.H., Isaacs, F.J., Carr, P.A. et al. (2009). Programming cells by multiplex genome engineering and accelerated evolution. *Nature* 460: 894–898.

Webber, B.R., Lonetree, C.L., Kluesner, M.G. et al. (2019). Highly efficient multiplex human T cell engineering without double-strand breaks using Cas9 base. *Nat. Commun.* 10: 5222.

Yang, Y. and Seed, B. (2003). Site-specific gene targeting in mouse embryonic stem cells with intact bacterial artificial chromosomes. *Nat. Biotechnol.* 21: 447–451.

Yin, J., Zhu, H., Xia, L. et al. (2015). A new recombineering system for Photorhabdus and Xenorhabdus. *Nucleic Acids Res.* 43: e36.

Zhang, Y., Buchholz, F., Muyrers, J.P., and Stewart, A.F. (1998). A new logic for DNA engineering using recombination in Escherichia coli. *Nat. Genet.* 20: 123–128.

3

CRISPR Cas: From Bacterial Adaptive Immunity to the Swiss Army Knife of Drug Discovery

Saša Šviković

Genome Engineering, Discovery Sciences, BioPharmaceuticals R&D, AstraZeneca, Gothenburg, Sweden

3.1 Introduction

Over the course of billions of years of evolution, prokaryotes have developed multiple ways to stay ahead of their fiercest foes, mobile genetic elements. These defense systems include a variety of nucleases, DNA-modifying enzymes, chemical inhibitors, cell death-inducing signaling pathways, and some not yet fully understood protection systems (Bernheim and Sorek 2020; Hampton et al. 2020). These systems are incredibly diverse and widespread among prokaryotes, and studies into how these systems function ultimately led to the rise of new tools that have revolutionized biomedical sciences. Similarly to how the discovery of an innate bacterial immune system mediated by restriction endonucleases has ushered a revolution in biotechnology (Roberts 2005), the relatively recent discovery of CRISPR-Cas systems, an adaptive immune component of prokaryotic defense systems, and their application in genome engineering has completely changed the way how research in life sciences is done. In this Chapter, we will address the basics of CRISPR biology, the diversity and current classification schemes, and how the mechanistic understanding of these diverse systems can give rise to novel tools for biomedical science.

3.2 CRISPR Biology in a Nutshell

CRISPR, or Clustered Regularly Interspaced Short Palindromic Repeats, represents an adaptive immune system of microbes, able to adapt in response to invasions by mobile genetic elements such as bacteriophages, plasmids, and transposons. The locus encoding for the components of the CRISPR system was discovered as an array of palindromic repeats interrupted by a 20–40 nt sequence downstream of the *iap* gene in *Escherichia coli* (Ishino et al. 1987; Nakata et al. 1989). Thanks to the increasing availability of sequences from the microbial species, the structure of CRISPR loci, with many properties relating to their function, was uncovered (Mojica et al. 2000; Jansen et al. 2002).

The identifying trait of the CRISPR locus in an array of alternating identical repeats, interleaved by unique **spacer sequence** (Figure 3.1a), with extreme interspecies diversity in sequence and size, usually in the range of 20–72 bp (Grissa et al. 2007). Bioinformatic analysis of spacer sequences within **CRISPR arrays** has shown a striking homology to various sequences found in bacteriophages and plasmids (Bolotin et al. 2005; Mojica et al. 2005; Pourcel et al. 2005), which was an early clue that CRISPR systems act as a defense mechanism against mobile genetic elements (Makarova et al. 2006). The collection of the spacers within a microbial species is extremely diverse and dynamic (Weinberger et al. 2012), as the microbes can acquire new spacers upon neutralizing the infection (Barrangou et al. 2007; Marraffini and Sontheimer 2008). The spacer acquisition represents the adaptive component of this immune system, as this confers resistance to reinfection by the same mobile genetic element.

In most species, repeat monomers vary between 23 and 47 bp in length (Godde and Bickerton 2006), and in most species consist of partially palindromic sequences, able to form stable secondary structures (Kunin et al. 2007). Related species can have similar repeat sequences, but the overall bacterial and archaeal sequence diversity of both spacers and repeats is great.

Genome Editing in Drug Discovery, First Edition. Edited by Marcello Maresca and Sumit Deswal.

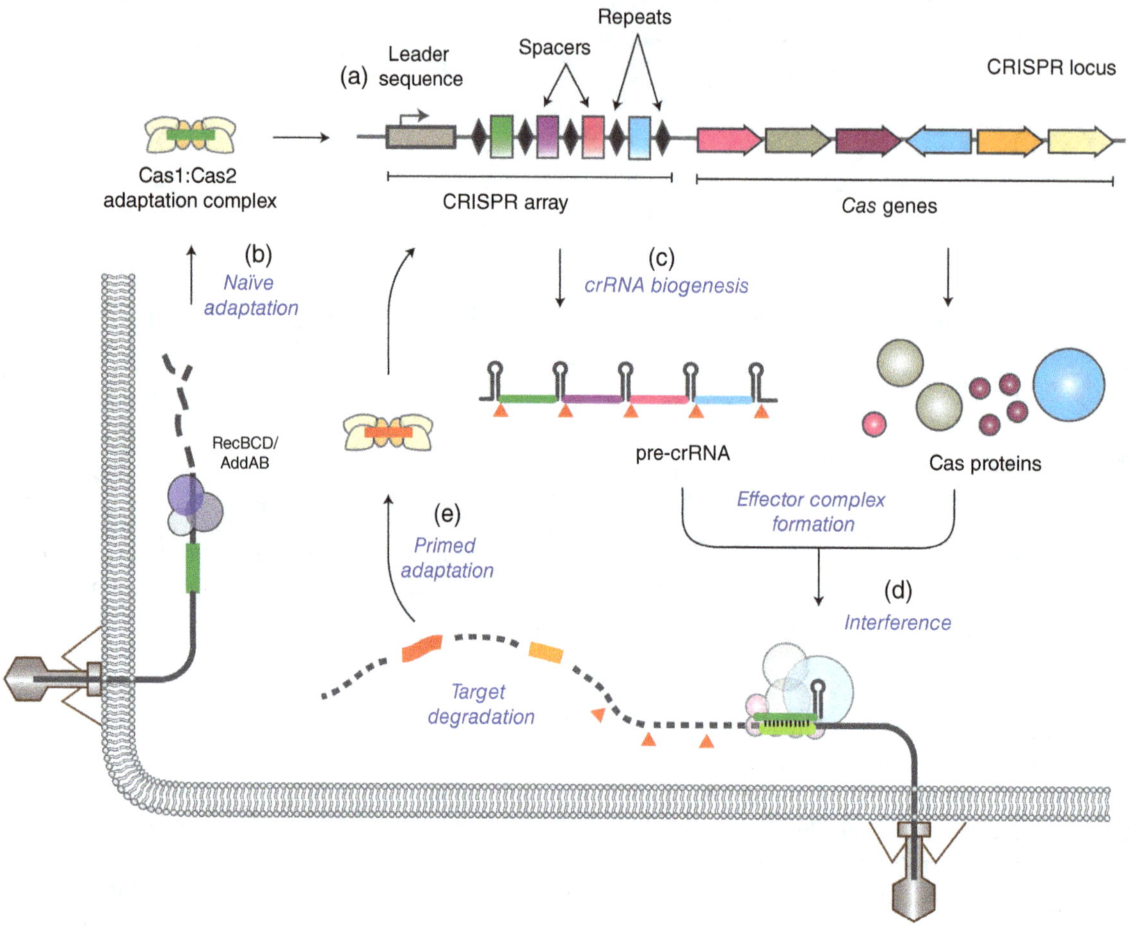

Figure 3.1 **Phases of CRISPR-mediated immune response.** (**a**) The simplified architecture of an idealized CRISPR locus, containing a CRISPR array consisting of a promoter-containing leader sequence and spacer and repeat sequences (depicted on the left) and several different *cas* genes (depicted in different colors on the right), which generate functional components of the CRISPR-Cas system. During the infection by a mobile genetic element (depicted as a bacteriophage here), invading DNA can be processed by the RecBCD/AddAB systems so that it can be integrated as a spacer by the Cas1:Cas2 adaptation complex as a part of the naïve adaptation phase (**b**). In the expression phase (**c**), the CRISPR array is transcribed into pre-crRNA, which is further processed to crRNA and paired with Cas proteins to form a functional effector complex. Upon a reinfection with phage carrying the same or similar protospacer sequence, the effector complex recognizes the sequence (**d**) and induces the degradation of the target. Suitable fragments generated during the interference phase can be used as spacers in the primed adaptation phase (**e**), updating the CRISPR array with the most recent version of suitable spacers.

Immediately adjacent to the CRISPR array is an AT-rich element, most frequently between 100 and 400 bp long (Jansen et al. 2002; Alkhnbashi et al. 2016). It was subsequently shown that this sequence, termed **leader sequence**, contains the promoter used to direct transcription of the CRISPR array (Figure 3.1a) and determines the transcriptional start site (Pougach et al. 2010; Pul et al. 2010). Transcription of the CRISPR locus and its subsequent processing ultimately gives rise to CRISPR RNA (**crRNA**), which provides specificity to the CRISPR systems and guides the degradation of the invading nucleic acids (Brouns et al. 2008). Much like the repeats, leaders are up to 80% identical within a genome, but quite dissimilar among species.

CRISPR arrays are often surrounded by a collection of conserved protein coding genes, termed CRISPR-associated or *cas* genes; these genes encode for the protein components of the CRISPR immunological systems. The groups of protein coding genes vary dramatically in composition, order, and sequence between species, underpinning the diversity of CRISPR systems, and their modes of action. Indeed, the complement of *cas* genes, together with the structure of the CRISPR locus is the basis for CRISPR classification (Makarova et al. 2019). Furthermore, many genes not evolutionary related to *cas* genes can also be found within the CRISPR loci. These genes often encode for proteins providing ancillary functions, or are evolutionary relicts and do not have a role attributed to them yet, and some loci also encode for noncoding RNA. Furthermore, there might be multiple CRISPR loci within a single genome: *E. coli* has four loci (Touchon and Rocha 2010) and *Methanocaldococcus jannaschii* has up to 18 (Lillestol et al. 2006), but rarely more than one or two are active simultaneously (Horvath et al. 2008).

All the components of CRISPR loci function together to provide an adaptive immune response. The CRISPR immune response can be divided into three phases (Figure 3.1):

1) **Adaptation (spacer acquisition, or immunization).** On a rare occasion during infection by a bacteriophage or other mobile elements, suitable pieces of invaders genome can be snatched up by the microbe's defense machinery and integrated into CRISPR array as a spacer by the adaptation machinery (Figure 3.1b). This acquisition of not previously encountered spacers is known as **naïve adaptation**. Once acquired, the spacer acts as a heritable record of immunization and will restrict mobile elements which have the exact or similar sequence through the next stages of the immune response.
2) **Expression (crRNA biogenesis).** Once acquired, the spacer can be utilized to fight future infections (Figure 3.1c). The whole CRISPR array is transcribed from the promoter located within the leader sequence, producing a pre-crRNA transcript. With the help of Cas proteins, other small RNA molecules, or the host's machinery, pre-crRNA is processed into mature crRNA, which are paired with the effector Cas proteins to form a functional **effector complex** that confers immunity.
3) **Interference (immunity)** is maintained by the assembled Cas:crRNA effector complex. Invading genomes that carry sequence complementary (or partially complementary) to one of the spacers will be recognized by base pairing of the crRNA, and subsequently degraded by the nucleolytic activity of the associated Cas proteins, thus terminating the infection (Figure 3.1d).

In some circumstances, the degraded nucleic acids can be captured by the adaptation complex and integrated as a new spacer into the CRISPR array, restarting the process. This **primed adaption** (Figure 3.1e), where an invading genome is neutralized by the previously acquired spacer and actively includes Cas systems, is several orders of magnitude more efficient than the naïve adaptation (Staals et al. 2016; Stringer et al. 2020), and acts as a magnificent example of adaptive immunity.

One caveat of an immune system relying on nucleic acid base recognition is how to discriminate between the invading genome and endogenous sequences (for example, in the CRISPR array). Nearly all CRISPR systems have a discrimination mechanism where a short sequence adjacent to the target sequence must be recognized by the effector complex to efficiently bind to and then degrade the target sequence (Garneau et al. 2010; Sashital et al. 2012; Anders et al. 2014). Similarly, these species-specific **protospacer adjacent motifs (PAM)** are recognized by the adaptation complex and are processed in such a way so they are not integrated into the CRISPR array (Datsenko et al. 2012; Wang et al. 2015; Rollie et al. 2018). The presence of PAM adjacent to the target sequence (collectively termed **protospacer**) and its absence from the CRISPR array ensure correct recognition of invading genomes as nonself and preventing cleavage of the host genome. It is important to note that the exact sequence of PAM required for the interference and adaptation stages vary dramatically between species (for example, for *Streptococcus pyogenes,* PAM is 5'-NGG-3', while for *Staphylococcus aureus,* it is 5'-NNGRRT-3', where N denotes any nucleotide and R is A or G) and different taxa of CRISPR systems (Mojica et al. 2009; Shah et al. 2013), and often can be fairly liberal (Leenay et al. 2016), allowing the immune response to be responsive even if the mutation arise within the PAM or protospacer.

3.3 The Diversity of CRISPR Systems

Thanks to the greater availability of prokaryotic genome sequences, bioinformatic analysis has revealed that CRISPR systems are extremely abundant in prokaryotes, with roughly 40% of bacterial and over 85% of archaeal species harboring these systems (Makarova et al. 2019). This diversity, conferred by the remarkable variety of the Cas protein sequences, gene composition, and architecture of the loci, underpins the differences in how each of the three phases of adaptive immunity is performed. CRISPR systems have not only evolved to use different types of nucleic acids (DNA, RNA, or both) as a substrate (Marraffini and Sontheimer 2008; Hale et al. 2009; Kazlauskiene et al. 2016), but also can target different modalities (i.e. single- or double-stranded) (Ma et al. 2015; Strutt et al. 2018) and a wide spectrum of different genomic sequences, thanks to diverse PAM requirements (Mojica et al. 2009; Gasiunas et al. 2020).

The origin and evolution of CRISPR systems are complex, but it seems that they have originated by domesticating various components of mobile genetic elements and subsequent integration with the "antediluvian" prokaryotic toxin–antitoxin defense systems (Koonin and Makarova 2019). The module of genes responsible for the adaptation phase is thought to originate from a class of transposons carrying a homologue of *cas1* gene, collectively named casposons, while the CRISPR array is thought to originate from the inverted repeats flanking the casposon. The effector module seems to have evolved from a transposon-encoded nuclease able to target DNA and RNA, which through a putative series of duplication and subsequent deletions lead to the current collection of effector enzymatic activities (Makarova et al. 2019; Koonin and Makarova 2019).

Importantly, the diversity of Cas systems across species also translates into how these systems can be used as a tool, where one can choose the most suitable CRISPR system for their target (DNA or RNA), a sequence of choice (by choosing a Cas protein with a pertinent PAM) or application (by choosing a Cas system with the desired outcome). To fully explore this untapped potential of the microbial CRISPR systems, significant efforts to establish a robust classification of CRISPR-Cas systems have been made over the past decade. As there are no universally present *cas* genes that could act as an identifying trait, CRISPR classifications have been based on multiple factors, mainly on comparison of genomic loci organization and gene repertoires involved in a particular system. The most up-to-date classification is used in this chapter (Makarova et al. 2019).

All CRISPR-Cas systems can be divided into two distinct classes based on the organization of the effector complex (Figure 3.2). In **class 1** systems, the effector complex is composed of multiple Cas proteins, where individual subunits are needed for binding crRNA and recognition of the target nucleic acids, unwinding, and nucleolytic degradation. In contrast, **class 2** systems contain a single multidomain protein, which catalyzes all of the activities necessary for the interference phase. Each of the classes can then be further divided into six **types (I–VI)** based on the organization of individual *cas* genes into functional modules, and then further classified into subtypes. *Cas* gene nomenclature also reflects their classification: currently, the set of *cas* genes involved in adaptation or crRNA biogenesis are denoted as *cas1-11* and are shared across different types. In contrast, the names of effector *cas* genes are reserved for specific types – for example, type II uses Cas9, type V uses Cas12, and type VI Cas13. Subtypes are further specified by suffices, such as Cas12a, Cas12b, Cas12c, and so on; however, as many on the genes also have older familiar names (e.g. Cpf1 is the original name for Cas12a), these will be mentioned where appropriate as a point of reference. The ancillary genes continue to be referred to by their legacy names (e.g. Csm6).

While the interference module is the prominent feature in the classification of CRISPR systems, each of the classes and types also differ mechanistically in the manners of crRNA biogenesis and acquisition of spacers. The traits of main CRISPR-Cas systems will be discussed, with some differences between different subtypes touched upon as well; however, the intricacies and finesses of further classification are beyond the scope of this Chapter and the reader is invited to consult recent excellent reviews on the topic (Hille et al. 2018; Koonin and Makarova 2019; Makarova et al. 2019; Nussenzweig and Marraffini 2020).

3.3.1 crRNA Biogenesis

Synthesis of crRNAs starts with the transcription of the CRISPR array from a promoter usually located within the leader sequence (Pul et al. 2010; Pougach et al. 2010). Processing of the pre-crRNA is specific to each of the CRISPR class, with class 1 pre-crRNAs cleaved by dedicated endoribonuclease, whereas the class 2 systems employ the same machinery that performs target destruction (Figure 3.3).

In class 1 systems, the processing is performed by either Cas6 or Cas5d ribonucleases (Nam et al. 2012; Carte et al. 2008). Both of these proteins recognize and bind to the hairpin structure formed by the palindromic sequences of the pre-cRNA, and introduce a cut immediately downstream of it (Figure 3.3a), releasing mature crRNAs (Carte et al. 2010; Haurwitz et al. 2010; Ozcan et al. 2019). Intriguingly, in CRISPR systems containing repeats which are not thermodynamically likely to form hairpin structures (namely type I-A and -B, and type III-A and -B) and, hence, lack inherent discriminatory borders between spacers, Cas6 seems to be able to identify repeat regions by restructuring them to favor the formation of a hairpin or hairpin-like structure compatible with precise cleavage that will lead to productive crRNAs (Shao et al. 2016; Sefcikova et al. 2017). Mature crRNAs of most type I systems contain part of the repeat sequence at the 5' end of the spacer and the 3' hairpin; these do not participate in recognition of the target sequence but seem to be important for the assembly of the effector complex (Jore et al. 2011). Type III crRNA, on the other hand, undergoes additional trimming that removes the hairpin structure (Hale et al. 2008). How these mature crRNAs are paired to the cognate effector complex remains unanswered.

Class 2 systems employ two different strategies to generate mature crRNAs. The first strategy employed by type V and VI is in principle similar to class 1 crRNA biogenesis (Figure 3.3b). Here, the effector nucleases, such as Cas12a (Cpf1) and Cas13, recognize the repeat hairpin structure within the pre-crRNA and cleave the RNA within or upstream of it (East-Seletsky et al. 2016; Fonfara et al. 2016).

A more elaborate strategy is used to generate mature crRNAs in type II and type V-B systems (Figure 3.3c). These two systems, exemplified by Cas9 and Cas12b (C2c1), require a second noncoding trans-activating CRISPR RNA

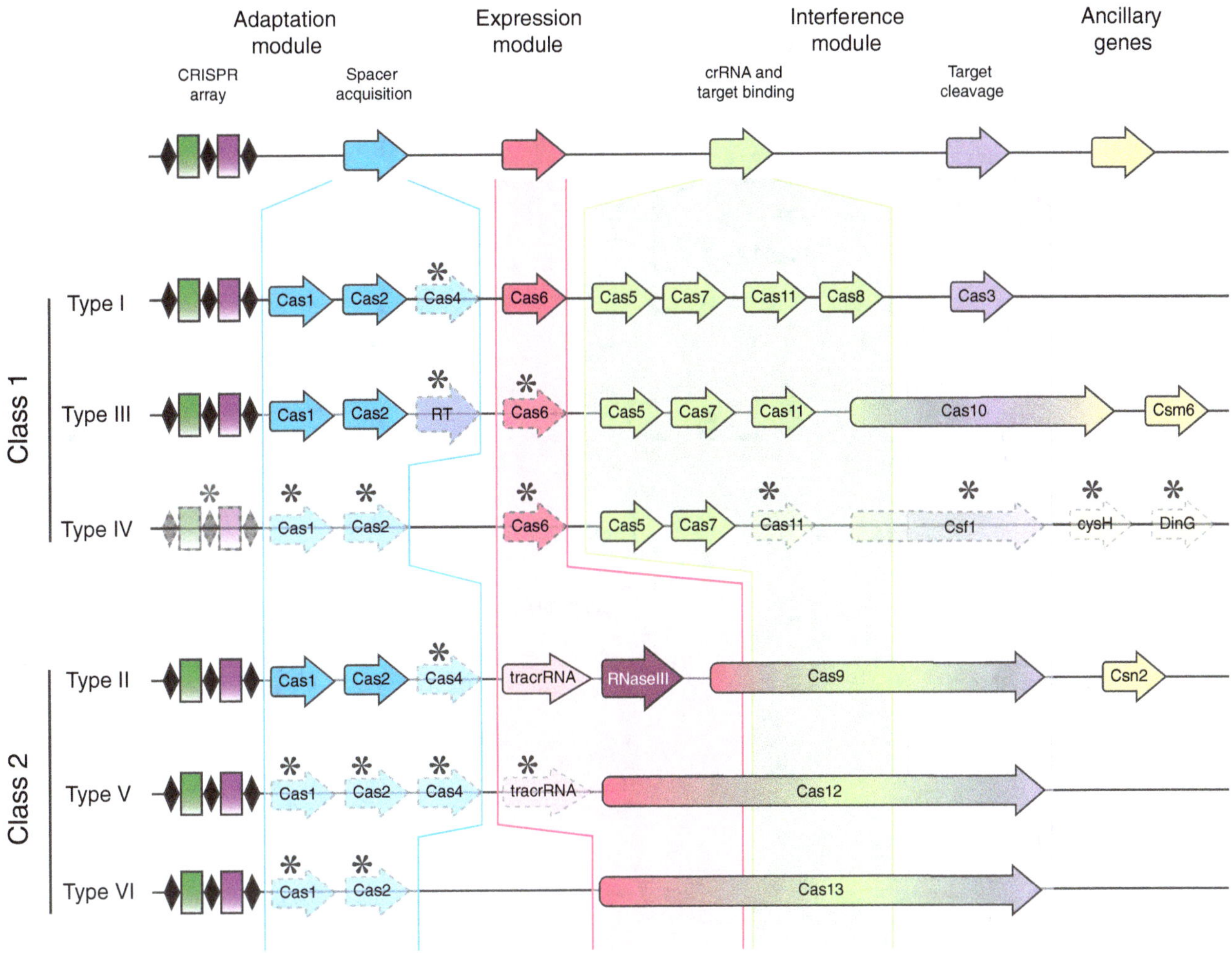

Figure 3.2 **Overview of class 1 and class 2 CRISPR systems.** General composition of various CRISPR systems across two classes and six types. The top panel shows a legend with a simplified, hypothetical CRISPR system containing key functional modules used in the immune response (adaptation, expression, interference, or ancillary). Functions of homologous genes are distinguished by color and demarcated by shaded areas. Many Cas proteins perform multiple activities in the immune response, and are as such highlighted as multicolored fusions spanning several categories. Many components of CRISPR systems are absent from specific subtypes, and are therefore represented in washout and with a dashed outline and an asterisk. Depicted CRISPR loci are schematic and do not correspond to real-life examples; hence the gene order, size, and orientation are purely didactic. Nomenclature and general module organization are from the most recent classification (Makarova et al., 2019).

(tracrRNA) to pair with the repeat regions within the pre-crRNA and form an intermediary between the crRNA and the effector protein (Deltcheva et al. 2011; Shmakov et al. 2015). The stem-loops of the tracrRNA act as a recruitment site for Cas9 and Cas12b, permitting them to form a ternary pre-crRNA:tracrRNA:Cas effector complex. The binding of the effector protein further stabilizes the interaction between pre-crRNA and tracrRNA, but also recruits cellular RNase III that cleaves the RNA:RNA duplex formed by the repeat sequences of the pre-crRNA and tracrRNA, releasing the 3' end of the crRNA (Deltcheva et al. 2011). The 5' end of the crRNA is processed further by removing the remaining repeat sequence, but the protein involved has remained elusive (Hille et al. 2018; Nussenzweig and Marraffini 2020). Once fully processed, crRNA paired with its cognate effector protein can patrol the cytosol and confer immunity to any invading DNA.

3.3.2 Interference

3.3.2.1 Class 1

3.3.2.1.1 Type I Type I systems are the most widespread CRISPR-Cas systems, and were the first to be identified (Ishino et al. 1987) and subsequently characterized. The effector complex of type I systems consists of the complex termed Cascade (CRISPR-associated complex for antiviral defense) that binds to crRNA, and the signature Cas3 protein which exhibits helicase and nuclease activity

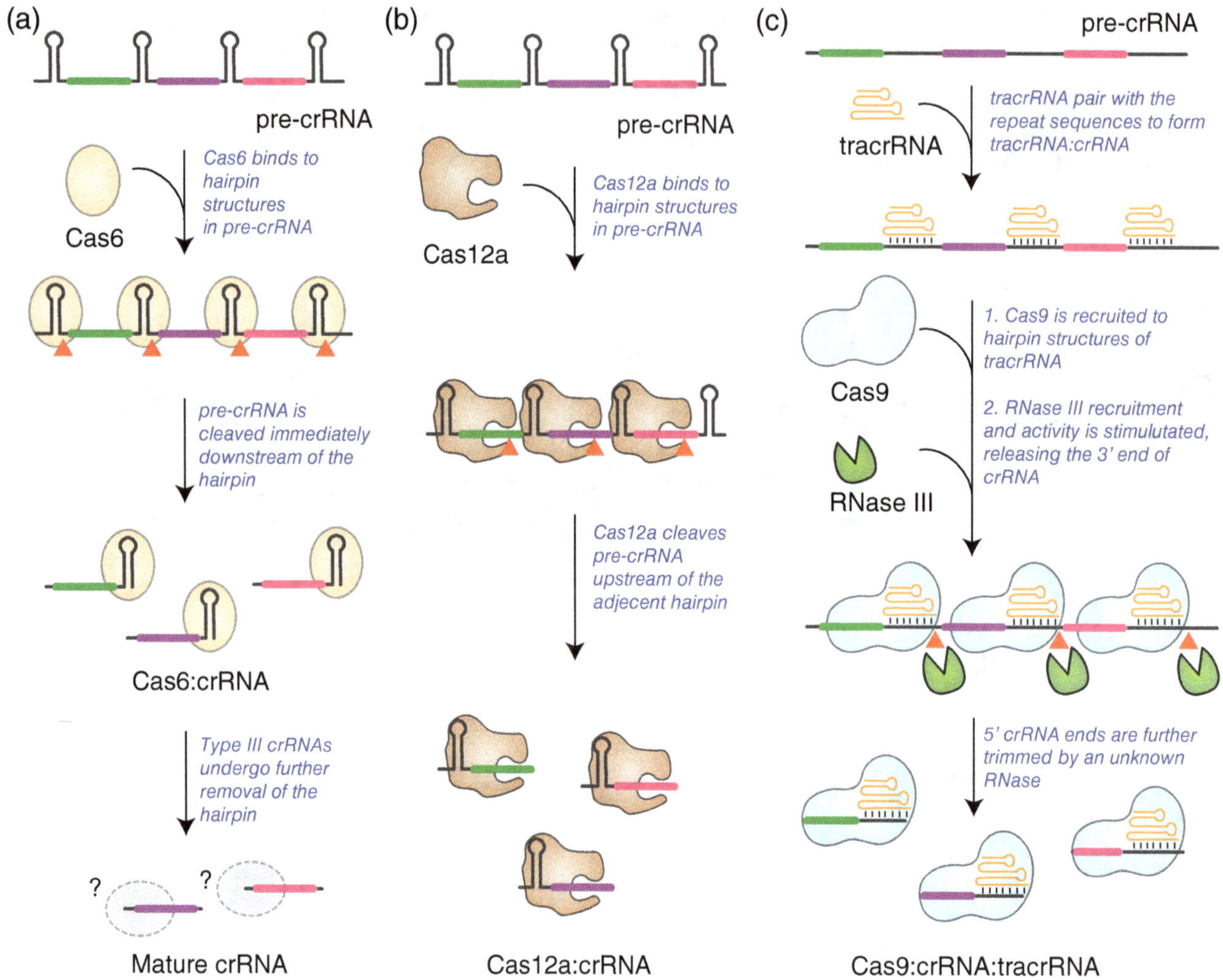

Figure 3.3 **crRNA biogenesis pathways.** (a) depicts a canonical crRNA biogenesis pathway for class 1 CRISPR systems, where Cas6 (or Cas5d in some subtypes) RNase recognizes and binds to hairpin structures in the pre-crRNA, cleaving in their vicinity to liberate mature crRNAs. Cas6 remains bound to crRNA and acts as a platform for the assembly of the rest of Cascade effector complex. Two different strategies for the maturation of crRNAs in class 2 systems are shown in panels **b** and **c**. In (**b**), some class 2 systems (represented here with Cas12a), the effector enzyme itself recognizes the repeats in pre-crRNA and cleaves the RNA 4 nucleotides upstream of the adjacent hairpin. crRNA processed in such a way remain paired with its Cas enzyme, forming a functional effector complex. In (**c**), type II systems crRNA processing requires tracrRNA which pairs with repeat regions and subsequently recruit Cas proteins and deploy cellular RNase III that liberates Cas9:crRNA:tracrRNA complex. 5' end of crRNA is further processed by an unknown cellular nuclease. Colored rectangles represent spacers, gray stem-loops repeat sequences, and red triangles cleavage sites.

(Brouns et al. 2008). The assembly of the Cascade complex initiates after the pre-crRNA processing catalyzed by the RNase activity of Cas6 protein (Figure 3.4a), which remains bound to the 3' end of the crRNA (Sashital et al. 2011). Cas6-crRNA acts as a nucleation center for the assembly of the heterododecameric Cascade complex ($(Cas5)_1$-$(Cas6)_1$-$(Cas7)_6$-$(Cas8)_1$-$(Cas11)_2$), with Cas7 proteins binding to spacer sequence (inducing a kink at every sixth nucleotide of the crRNA) and Cas5 capping the 5' end of the crRNA (Mulepati et al. 2014). Cascade complex uses its Cas8 subunit to identify PAM sequence via minor-groove interactions. Successful recognition of PAM site leads to conformational change of the Cascade complex, which in turn unwinds PAM site and permits pairing of the bound crRNA with the target DNA strand (Figure 3.4a), forming a thermodynamically stable R-loop that stabilizes the Cascade complex onto the target DNA (Hayes et al. 2016; Xiao et al. 2017). As a consequence of binding to DNA, Cascade is able to recruit Cas3, which makes a single-stranded break on the nontargeted DNA strand. The subsequent activation of the helicase domain allows Cas3 to translocate on the nontarget strand in 3'–5' direction,

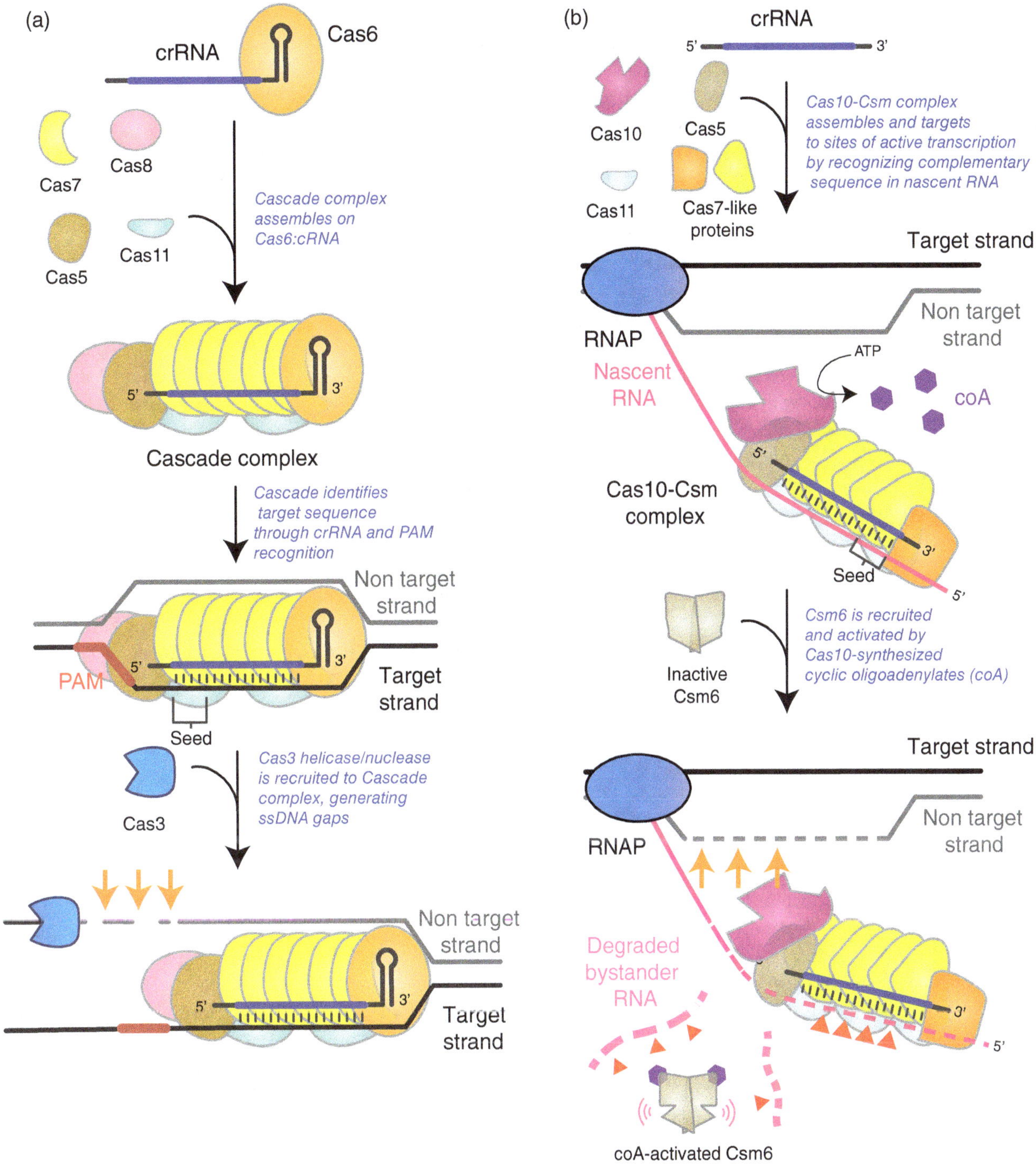

Figure 3.4 **Interference mechanism in class 1 systems.** Panel **(a)** depicts interference in type I systems. Cascade complex assembles on Cas6:crRNA (originating from crRNA biogenesis) and identifies target sequence by recognizing PAM by Cas8 subunit and protospacer by stepwise pairing with the crRNA. Successful pairing allows a Cas3 nuclease and helicase to be recruited. Cas3 unwinds and translocates on the nontarget strand while simultaneously cleaving the same strand (depicted by orange arrows), generating single-stranded gaps. The interference mechanism typical for type III systems is depicted on panel **(b)**, exemplified by Cas10-Csm complex. The effector complex assembles with hairpinless crRNA and is targeted to sites of active transcription by hybridizing the crRNA with complementary sequence in nascent RNA. Successful pairing activates RNase (red arrow heads) and nuclease activities of HD domain of Cas10, leading to on-target degradation of both the transcript and the genomic locus. Activated Cas10 is also able to convert ATP to cyclic oligoadenylates (coA), which stimulate indiscriminate RNase activity of Csm6, leading to global depletion of the cellular transcriptome.

generating 200–300 nt single-stranded DNA gap (Sinkunas et al. 2011; Sinkunas et al. 2013; Huo et al. 2014; Redding et al. 2015). While generating gaps is not likely to be sufficient to fully destroy the invading DNA, it is likely that the DNA can be degraded by one of the host's nucleolytic machinery (Lovett 2011).

3.3.2.1.2 Type III Type III CRISPR-Cas systems are based on the Cascade-like complexes (Csm or Cmr in subtypes III-A and III-B, respectively) working with their cognate hairpinless crRNAs. These complexes display overall structural similarity to Cascade complex of type I, both forming a seahorse-shaped complex (Osawa et al. 2015; Taylor et al. 2015). In the Csm/Cmr complex, the 5' end of the crRNA is bound by Cas5, with the backbone of multiple proteins belonging to Cas7 (Csm3/Csm5 and Cmr4/Cmr6/Cmr1) and Cas11 (Csm2/Cmr5) protein families. The effector complex is completed by binding of Cas10 protein (Figure 3.4b). The distinct feature of type III systems is that they can degrade both DNA and RNA, through concerted DNase activity of Cas10 and RNase activities of Csm3 and ancillary Csm6 proteins (Hale et al. 2009; Deng et al. 2013; Staals et al. 2014; Samai et al. 2015).

Type III systems confer immunity against DNA bacteriophages or plasmids (Marraffini and Sontheimer 2008; Hatoum-Aslan et al. 2014; Samai et al. 2015), but can only target the invading genome if it is actively transcribed (Deng et al. 2013; Goldberg et al. 2014). Whether Csm/Cmr complexes get recruited directly to transcribing RNA polymerase, nascent transcript, or underwound DNA generated in the wake of RNA polymerase remains a contentious topic (Elmore et al. 2016; Han et al. 2017; Liu et al. 2019b). It is clear that interference begins by the pairing of loaded crRNA to complementary nascent transcript (Figure 3.4b), stimulating the nucleolytic degradation of the nontarget DNA strand by Cas10 (Estrella et al. 2016; Liu et al. 2017d). In parallel, Cas7 subunits cleave the paired RNA at every sixth nucleotide (Tamulaitis et al. 2014; Liu et al. 2017d), and in some systems the ancillary RNase Csm6 degrades proximal RNA in an unspecific manner (Jiang et al. 2016b); this dual action of two main nucleases efficiently silences phage RNA and simultaneously disrupts the invasive genome (Figure 3.4b). Type III-A system have evolved even stronger adaptive response, where targeted binding to RNA stimulates unspecific cleavage of ssDNA by the HD domain of Cas10 (Kazlauskiene et al. 2016; Liu et al. 2017d), and the conversion of ATP to cycling oligoadenylates by its Palm polymerase domain. The cyclic oligoadenylates further stimulate the activity of the HEPN domain of the Csm6 ribonuclease (Kazlauskiene et al. 2017; Niewoehner et al. 2017), leading to an indiscriminate degradation of both host and invading RNA, causing a growth arrest which restricts invader propagation (Jiang et al. 2016b; Rostol and Marraffini 2019). Together, this gives rise to a potent, highly adaptable, and robust immune system (Pyenson et al. 2017).

3.3.2.1.3 Type IV Type IV systems are the most enigmatic category of CRISPR systems. Type IV systems are most frequently found on plasmids, conjugative plasmid elements, and seldom in phage genomes (Pinilla-Redondo et al. 2020), and typically lack genes involved in adaptation (Cas1, Cas2, and Cas4) or nucleolytic degradation Figure 3.2, and frequently even without the CRISPR arrays (Makarova et al. 2019). Subtype IV-C, a recently discovered CRISPR system of *Thermoflexia bacterium*, contains a large subunit (*LS* or *csf1*) gene with putative HD nuclease domain, providing the only suggestion that some of the type IV systems might retain nuclease activity (Makarova et al. 2019). Recently, a striking bias toward plasmid sequences as spacers (in contrast to other systems, which predominantly share homology with viral genomes) has suggested that type IV might have a role in plasmid maintenance and competition (Faure et al. 2019a; Newire et al. 2020; Pinilla-Redondo et al. 2020), whether this is true remains to be experimentally established.

Bioinformatic analyses have observed that type IV systems are frequently found together with type I systems, indicating a potential crosstalk between the two systems. Indeed, PAM, repeat, and leader sequences are nearly identical between the co-occurring systems (Pinilla-Redondo et al. 2020), suggesting that type IV systems can be viewed as minimal Cas systems that can rely on the adaptation machinery from "helper" type I system. Recently, a structure of type IV effector complex has revealed a sea cucumber-like structure, with seven Cas7-like subunits forming the backbone and interacting with crRNA (inducing a kink in the RNA every sixth nucleotide much like type I Cascade), with five Cas11 forming the belly of the structure (Zhou et al. 2020). Cas6 protein in type IV systems is responsible for processing and binding to crRNA (Ozcan et al. 2019), much like in type I. However, when components of the *Mycobacterium* sp. type IV system were co-expressed in *E. coli*, it was found that crRNA originated from a wide range of sources, including from the plasmid harboring the type IV system, but also from rRNA, tRNA, and other noncoding small RNA. This suggests that there might be indeed crosstalk between the hosts and plasmid-borne CRISPR systems; however, the apparent lack of specificity in processing/assembling crRNA–effector complex does not support a role in nuclease-mediated immunity. The fact that type IV systems are frequently accompanied by genes with a potential role in other types of bacterial defense systems, such as *cysH* or *ART* gene, supports the notion that this type of system has been co-opted for plasmid or phage maintenance, rather than their depletion (Faure et al. 2019b). Further studies are needed to unravel the role (if any) of these systems.

3.3.2.2 Class 2

When compared with class 1 systems, class 2 systems are much simpler in the sense that they comprise of a smaller number of components. Class 2 systems all have in common that the effector module is contained within a single protein, in contrast to multiple subunits of class 1 systems discussed so far. This class can be subcategorized based on the key genes involved in interference (Cas9, Cas12, and Cas13), all of which restrict the invading genomes in different ways. While this might not be optimal for a robust immune response, in particular when compared with type III systems, their simplicity makes them a better tool for genome editing purposes. While some class 1 systems have been used for gene editing, most notably in prokaryotes (Kiro et al. 2014; Li et al. 2016; Pyne et al. 2016), thanks to their simplicity (need for only one Cas protein) class 2 systems have been the easiest systems to be adapted for genome engineering. The many interesting activities of class 2 systems will be addressed below.

3.3.2.2.1 Type II The most well-known example of type II systems, and arguably of all CRISPR systems, is the *S. pyogenes* systems, epitomized by Cas9 (abbreviated as SpyCas9) for its use in gene editing in eukaryotic systems (Cong et al. 2013; DiCarlo et al. 2013; Ding et al. 2013; Friedland et al. 2013; Hwang et al. 2013; Jinek et al. 2013; Mali et al. 2013). Cas9 is a dual RNA-guided DNA endonuclease required for conferring immunity in type II systems (Barrangou et al. 2007; Gasiunas et al. 2012; Jinek et al. 2012). Apart from crRNA (required by any other CRISPR system to recognize target molecule), Cas9 also requires trans-activating crRNA (tracrRNA), a noncoding RNA that coordinates the processing of crRNA and whose hairpin Cas9 binds to (Deltcheva et al. 2011). Once bound to crRNA:tracrRNA complex, Cas9 identifies target DNA through PAM recognition, and base pairing between the crRNA and the target DNA (Figure 3.5a). If sufficient complementarity is present, *S. pyogenes* Cas9 generates a blunt DNA double-stranded break (DSB) 3 bp upstream of the PAM through concerted activity of its RuvC and HNH domains (Jinek et al. 2012). Similarly to other systems, cleavage by Cas9 initiates further degradation by the host machinery, neutralizing the invading genome (Figure 3.5a).

Detailed structural and biochemical studies of SpyCas9 protein have revealed a bilobed structure with the crRNA and target DNA accommodated in the central cleft. Cas9:crRNA:tracrRNA complex recognizes the PAM site via interaction of the GG dinucleotides and the conserved amino-acids of the C-terminal domain. Recognition of the correct PAM leads to a local unwinding of the DNA duplex, allowing the crRNA to pair with a 10–12 nt long seed sequence of the target strand (Anders et al. 2014; Sternberg et al. 2014). Successful pairing accompanied by further conformational changes prompts further invasion of the crRNA, forming a stable R-loop across the full length of crRNA (Jiang et al. 2016a; Mekler et al. 2017). This in turn induces another set of complex conformational changes, where the movement of the HNH domain leads to activation of the RuvC domain, allowing coordinated cleavage of the target and nontarget strand, respectively (Sternberg et al. 2015; Raper et al. 2018). Biochemical studies have shown that once bound to the target DNA, cleavage is performed rapidly, while the enzyme remains stably bound, indicative of single-turnover enzyme kinetics (Jinek et al. 2012; Raper et al. 2018). Of note, single-turnover kinetics of Cas9 proteins is not a universal property, as an orthogonal Cas9 from *S. aureus* (SauCas9) shows a multiple-turnover kinetics (Yourik et al. 2019). Furthermore, it seems that *in vivo* the bacterial transcriptional machinery can remove bound Cas9, making the enzyme multi-turnover, thus enhancing the immune response (Clarke et al. 2018). Recently, the eukaryotic histone chaperone complex FACT, commonly associated with active transcription (Belotserkovskaya et al. 2003), has been shown to evict Cas9 bound to DNA, improving the enzymes' kinetic and changing the editing outcome (Wang et al. 2020).

3.3.2.2.2 Type V Type V CRISPR systems contain some of the most diverse interference modules, with a total of 17 subtypes characterized so far (Burstein et al. 2017; Harrington et al. 2018; Makarova et al. 2019; Yan et al. 2019). The effector proteins (named Cas12a-k) differ fundamentally from type II systems by the domain architecture; while type II systems perform cleavage of target DNA by two different domains, Cas12 proteins have a single RuvC-like domain which is able to cleave both strands (Swarts et al. 2017). The best understood (both structurally and biochemically) are Cas12a (previously known as Cpf1) and Cas12b (C2c1) proteins. Both proteins share a similar bilobed structure with Cas9, however with only one lobe containing the catalytic RuvC domain (Liu et al. 2017a).

Cas12a effector complex has been proposed to associate unspecifically to DNA and diffuse along it until a PAM sequence is found. This initiates a local unwinding and base pairing between the crRNA and target DNA (Figure 3.5b). Cas12a performs multiple checks for target recognition during the pairing of the crRNA with the

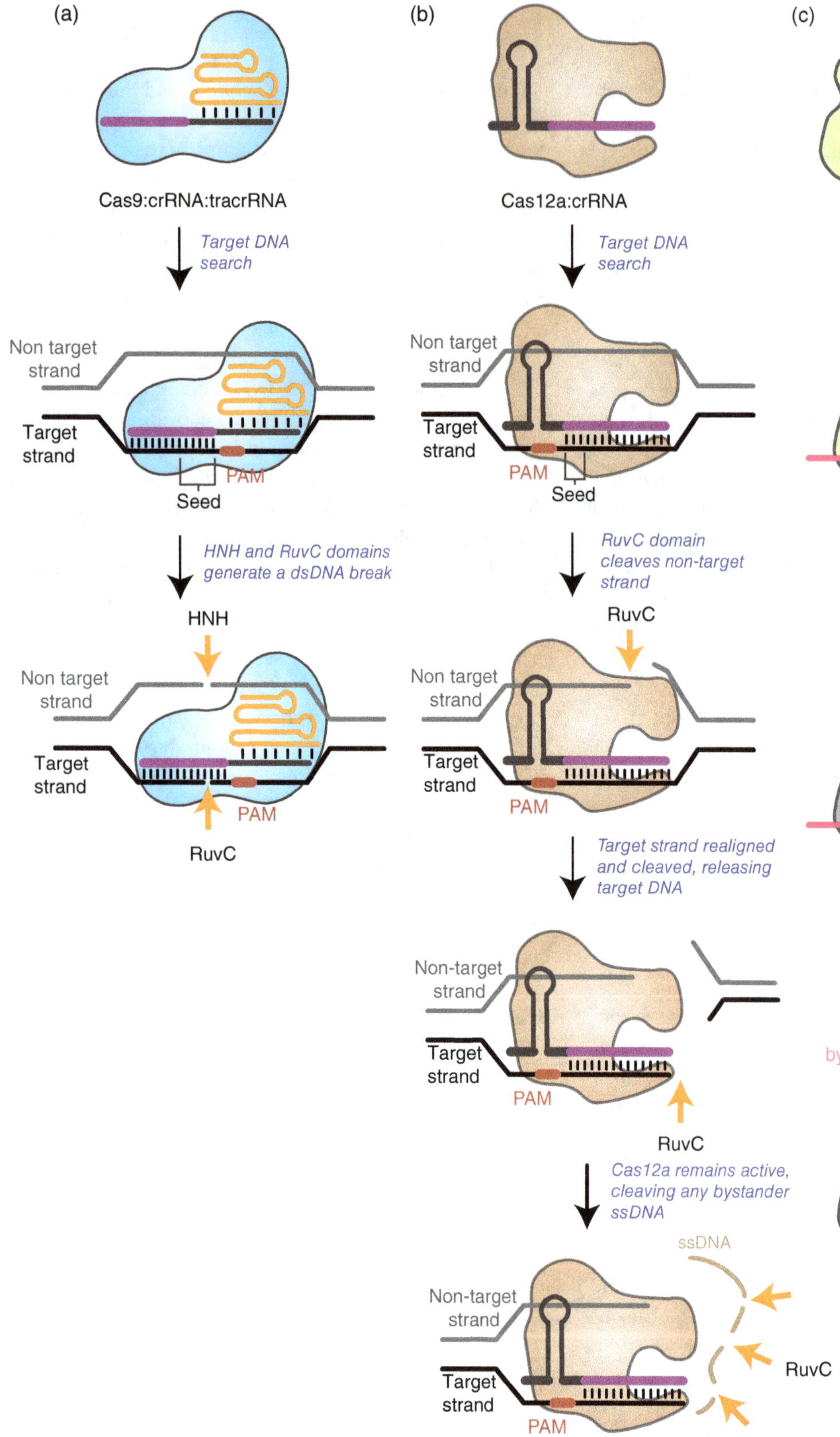

Figure 3.5 Interference mechanisms in class 2 systems. Representative mechanisms of the most commonly used class 2 systems are depicted here. In **(a)**, Cas9:crRNA:tracrRNA binds to the target sequence, activating the RuvC and HNH nuclease domain to generate a blunt double-stranded DNA break proximal to PAM site. In **(b)**, Cas12a complexed with its cognate crRNA recognized target sequence, inducing cleavage of nontarget strand by its single RuvC domain. The enzyme then rearranges to cleave the target strand, releasing the PAM-distal DNA fragment. The enzyme remains active as it still binds activating PAM-proximal sequence, exhibiting collateral activity on any ssDNA able to enter the active site. **(c)** depicts a general mechanism of Cas13 enzymes, able to target RNA by base pairing with its crRNA. Upon recognition of target RNA, Cas13 enzyme undergoes a major conformational change that assembles a composite HEPN domain, exhibiting indiscriminate ssRNA nucleolytic cleavage. Activated Cas13 is therefore able to degrade the target RNA molecule, but also any bystander RNA. Sites of DNA cleavage activity are depicted by orange arrows and RNA cleavage by red arrowheads.

3–5 nt long seed sequence proximal to the PAM, and then during the formation of the R-loop (Stella et al. 2018). Failure to form at least 17 bp long DNA:RNA hybrid will lead to rapid dissociation of the Cas12a:crRNA complex off DNA (Jeon et al. 2018, Singh et al. 2018). If a sufficiently long R-loop is formed, Cas12a is activated and proceeds to cleave the nontarget strand, and then the target strand (Jeon et al. 2018), generating a staggered DNA cut with 4–5 nt long 5' overhangs between the 18th and 23rd base from the PAM (Zetsche et al. 2015). After successful cleavage, Cas12a releases the PAM-distal DNA fragment but continues to bind to the PAM-proximal fragment (Figure 3.5b). Crucially, crRNA remains bound to the target strand, allowing the enzyme to remain in catalytically active state. This allows Cas12a proteins to exhibit an indiscriminate nuclease activity on any ssDNA it may encounter (Chen et al. 2018). This activity might be important for the complete clearance of invading genomes. Importantly, the indiscriminate ssDNase activity has been reported for many Cas12 proteins: Cas12b (Li et al. 2019a), Cas12f (Harrington et al. 2018), and Cas12c, -h, -i, and g (Yan et al. 2019). Furthermore, Cas12g was shown recently to be able to target RNA molecules, and to pose indiscriminate cleavage of ssDNA and ssRNA. Many type II proteins can target both ssDNA and dsDNA (Ma et al. 2015).

Furthermore, type V systems also contain a collection of heterogeneous and poorly characterized subtype V-U; these systems typically lack the adaptation module and contain much smaller effector proteins, making them unlikely to be functional. However, follow-up studies have identified interference functions with some, prompting their upgrade to a separate type V-F (Harrington et al. 2018). Finally, an unusual group of what are now Cas12k proteins have been experimentally shown to interact with a bacterial transposase and promote transposon integration through a crRNA-guided mechanism (Strecker et al. 2019b); this is likely to reflect the proposed evolutionary trajectory of CRISPR systems, proposed to originate as machinery important for transposition (Makarova et al. 2019).

3.3.2.2.3 Type VI The most recently fully characterized CRISPR systems belong to type VI (Meeske et al. 2019). Similar to other class 2 proteins, the type VI effector Cas13 (with representative Cas13a and Cas13b) also forms a bilobed structure with the crRNA accommodated in its central cleft (Liu et al. 2017b; Liu et al. 2017c; Slaymaker et al. 2019). Unlike Cas9 and Cas12 effectors, Cas13 lacks DNase domains such as HNH or RuvC, but contains two HEPN domains, often found in RNases (Shmakov et al. 2017; Makarova et al. 2019), allowing this enzyme to target ssRNA (Abudayyeh et al. 2016; East-Seletsky et al. 2016; Smargon et al. 2017; Zhang et al. 2018). Additionally, Cas13 proteins contain an alpha-helical domain intuitively called Helical-1 domain, which together with HEPN-2 domain participates in pre-crRNA processing. Binding of pre-crRNA in the central channel of Cas13 leads to a conformational change that activates the Helical-1 domain, stimulating it to incise the pre-crRNA generating a mature crRNA (typically with 28–30 nt spacer and 30 nt long hairpin) (East-Seletsky et al. 2016). The structure of the crRNA differs between two prominent subcategories, with the hairpin-forming repeat sequence being at the 5' or 3' end of the spacer in Cas13a or Cas13b processed RNA, respectively (East-Seletsky et al. 2016; Slaymaker et al. 2019).

Recognition of the target RNA bears similarity to other CRISPR systems (Figure 3.5c). The pairing between the crRNA and target RNA at the central seed sequence is essential for binding and then stimulating RNase activity (Abudayyeh et al. 2016; Tambe et al. 2018), whereas peripheral mismatches are tolerated to a greater extent. Further elements are also needed for activation of RNase activity of the HEPN domains. Sequence and structure of the hairpin are critical, where reducing the length under 24 nt or mutating key nucleotides significantly decreased its activity (Abudayyeh et al. 2016; Smargon et al. 2017). Analogous to PAM, Cas13 also requires protospacer flanking site (PFS) proximal to the target site; the exact sequence and position is dependent on the subtype and species. For example, Cas13a requires a 3' non-G PFS, while Cas13b need a PFS at either side of the protospacer, with the 5' being depleted of C and the 3' PFS having a consensus sequence of NAN or NNA, where N is any nucleotide (Abudayyeh et al. 2016, Smargon et al. 2017).

Productive base pairing between crRNA and target RNA triggers a conformational change bringing both of the HEPN domains in close proximity and assembling into a composite catalytic site on the outer surface of the protein (Liu et al. 2017b; Liu et al. 2017c). Activated HEPN domains can indiscriminately degrade ssRNA, both the target at single-stranded regions and any other RNA in the vicinity (Abudayyeh et al. 2016; East-Seletsky et al. 2016). This indiscriminate RNase activity was shown to be able to restrict invasion of RNA phages, such as MS2 (Abudayyeh et al. 2016). Furthermore, a recent study showed that type VI systems can confer immunity to DNA phages as well. Similarly to type III systems, indiscriminate RNase activity of Cas13 is able to induce growth arrest of the infected cells. While the growth arrest induced by type III systems is transient as Csm6 (which also contains HEPN domains) gets deactivated, Cas13 does not (likely due to persistent transcription from the phage genome, which remains intact), consequentially inducing dormancy. This way, the further expansion of the phage is prevented, achieving immunity on a population level (Meeske et al. 2019).

3.3.3 Adaptation

The last phase of the CRISPR-mediated immunity that shall be addressed here is the adaptation phase, also known as spacer acquisition. In this phase, the memory of previous infections is recorded, allowing an organism and its descendants to confer immunity to the reinvading genome. The CRISPR array acts as a genetic ledger, with spacers acting as records of infections (with the most recent usually located closest to the leader sequence). The key players in spacer acquisition are Cas1 and Cas2, present in nearly every CRISPR system. In contrast to the ubiquity of these genes, understanding of molecular mechanisms of adaptation is restricted just to type I and type II systems with many mechanistic details still missing, with more studies needed to elucidate this process in its entirety.

The acquisition of spacers upon the invasion of an infectious genome that the host has not encountered before is often termed naïve spacer acquisition. This is an incredibly rare event, estimated to occur with a frequency of 10^{-7} per infected cell (Hynes et al. 2014; Heler et al. 2015). The acquisition of spacers by Cas1:Cas2 adaptation complex in these circumstances can be viewed as accidental, as it relies on other systems to generate suitable substrate for this protein. In *E. coli*, spacer acquisition was linked to excessive replication of the viral or plasmid genome, during which the inherently unstable replication intermediates are recognized and digested by the RecBCD endonuclease complex (involved in restriction and recombination), generating small fragments that can be captured and further processed by the adaptation complex (Levy et al. 2015). A similar mechanism has been described in Gram-positive species as well, where the AddAB restriction system is able to enrich for spacers (Figure 3.6a).

The latter phase of adaptation is mediated by highly conserved Cas1:Cas2 adaptation complex. However, some CRISPR systems also require additional proteins, such as Csn2 and Cas9 of *S. pyogenes* type II system (Heler et al. 2015; Wei et al. 2015), or Cas4 of type I and type V systems (Lee et al. 2018). Furthermore, in certain species with type III and type VI systems, fusions of *cas1* and reverse transcriptase (RT) genes have been discovered (Silas et al. 2016; Gonzalez-Delgado et al. 2019; Toro et al. 2019), and then shown to be able to insert spacers originating from RNA and DNA, where the integration of RNA is followed by cDNA synthesis by the RT domain (Silas et al. 2016; Gonzalez-Delgado et al. 2019).

The adaptation complex is composed of the central Cas2 dimer flanked on either side by a Cas1 dimer (Figure 3.6d). The captured DNA fragments (termed **prespacer** DNA) are bound by the central Cas2 dimer, with the Tyr22 of the *E. coli* Cas1 acting as a protein wedge to splay open the bound dsDNA fragment (Nunez et al. 2014; Nunez et al. 2015). The splayed ends of the bound prespacer DNA can be trimmed by host's DnaQ 3'-5' exonuclease-domain containing enzymes (such as DNA polymerase III, exonuclease T, DnaQ, or in some species by Cas2-DnaQ fusion proteins) into fragments with 5 nt overhangs (Kim et al. 2020; Ramachandran et al. 2020), which are optimal for integration into CRISPR locus (Nuñez et al. 2015). In some CRISPR systems, trimming can be performed by Cas4, a RecB nuclease domain-containing protein (Lee et al. 2018). Either way, the length of the prespacer DNA is maintained at the fixed length by the distance between two Cas1 subunits flanking Cas2 dimer, ensuring a uniform length of spacers within the CRISPR array. Finally, the 3' ends are positioned into the active sites of Cas1, making them poised for catalysis at CRISPR array.

While the interactions between Cas1:Cas2 complex and protospacer DNA are through phosphate backbone rather than base-specific interactions, the selection of specific protospacers by the adaptation machinery is often nonrandom. Spacers are preferentially acquired from sequences proximal to PAM sites (Savitskaya et al. 2013), with the adaptation complex in *E. coli* type I system able to select functional prespacers by directly recognizing the PAM sequence (Datsenko et al. 2012; Swarts et al. 2012; Wang et al. 2015). Similarly, the aforementioned Cas4 also seems to have a key role in selecting prespacers (Rollie et al. 2018). Furthermore, recognizing the PAM sequence is also used to orient the prespacer into CRISPR array so that the crRNA ultimately contains the correct sequence necessary for recognition (Shiimori et al. 2018).

Crucially, while PAM is used to identify functional prespacers and position them in the correct orientation, it must be removed prior to integration into CRISPR array; otherwise, it will induce self-immunity. This is achieved by removing the PAM sequence immediately prior to integration, either by the exonucleases (Ramachandran et al. 2020) or Cas4 (Lee et al. 2019). By coupling recognition of functional PAM sites with the prespacer processing increases the chance of integrating a functional spacer that will be compatible with future interference phases.

The integration into the CRISPR array occurs more frequently at the 5' end of the locus, adjacent to the AT-rich leader sequence (Figure 3.6d). To perform an efficient and on-target integration, a major distortion of the target DNA is required. In *E. coli*, this is performed by the architectural protein integration host factor (IHF), which tightly binds to

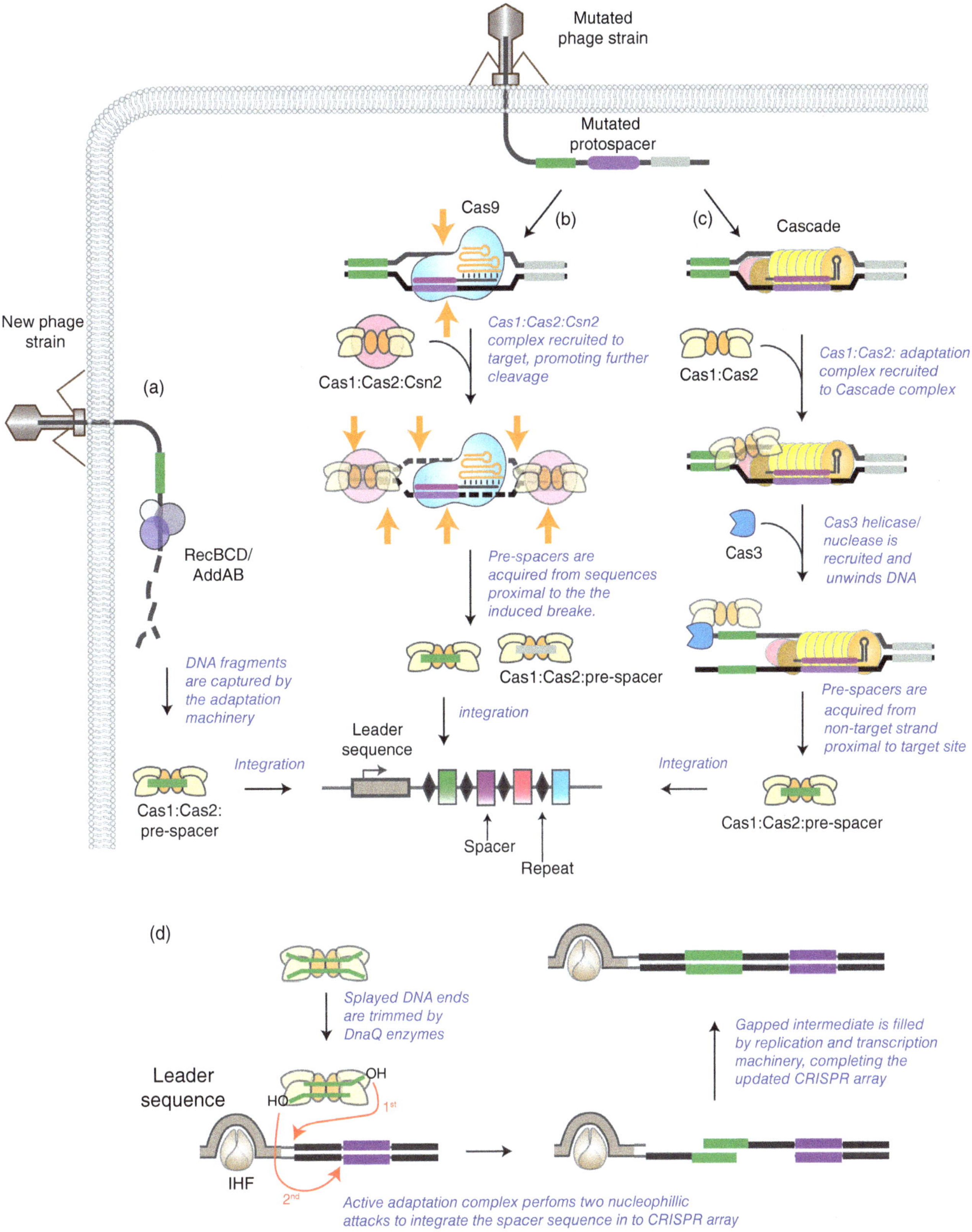

Figure 3.6 Adaptation phase(s) in CRISPR system. Depending on whether invading genomes have been encounter before, spacers can be acquired without (naïve) and with (primed) the assistance of CRISPR interference machinery. Fragments of the invading DNA generated by the RecBCD/AadAB restriction machinery can be captured by the Cas1:Cas2 adaptation complex (**a**). If a host is reinfected by previously encountered genetic element, interference machinery can recognize the pathogen and digest its genome. Successful targeting by type II effectors (**b**) leads to recruitment of Cas1:Cas2:Csn2 complex which acquires prespacer sequences proximal to the double-strand break. Type I systems (**c**) recruit Cas1:Cas2 tetramer when efficient on-target cleavage is inhibited, followed by helicase Cas3, allowing the adaptation complex to acquire prespacers from the nontarget strand. In any of the cases, the mechanism of integration into the CRISPR array is similar (**d**). Prespacer ends are processed and then integrated by two nucleophilic attacks by the 3′OH end at the first repeat sequence, with the gapped DNA filled by the host machinery. Sites of DNA cleavage activity are depicted by orange arrows.

the leader sequence of the CRISPR array (Nunez et al. 2016), inducing a major bend in the DNA (Wright et al. 2017). Active integration complex binds to the bent DNA, catalyzing the nucleophilic attack of the 3'OH group of the prespacer DNA at each end of the repeat sequence in the CRISPR array (Nuñez et al. 2015), resulting in a gapped intermediate that is ultimately resolved by the host's DNA repair and transcription machinery (Ivančić-Baće et al. 2015; Budhathoki et al. 2020). In species lacking IHF (most of Gram-positive bacteria), recognition of the first repeat instead requires a short 5 base pair-long motif termed leader-anchoring site (LAS) (McGinn and Marraffini 2016), which is directly recognized by the adaptation complex.

In order to rapidly respond to the ever-changing gallery of infectious predators, CRISPR systems have evolved a further adaptation pathway that allows rapid acquisition of new spacers through the action of already existing CRISPR-mediated immunity. This process is termed primed spacer acquisition or priming, and is at least 1000-fold more efficient than naïve adaptation (Staals et al. 2016; Stringer et al. 2020). A reinfecting strain against which the host has already been immunized will be efficiently targeted and cleaved, allowing new spacers from proximal sequences to be acquired, updating the CRISPR array with more spacers. This mechanism is particularly important in preventing infection by new phage strains that arise by mutations in the target sequence, be it PAM or the protospacer, which would make CRISPR-targeting ineffective (van Houte et al. 2016).

Priming mechanistically differs from naïve adaptation in early stages, i.e. the origin of prespacer DNA. While in naïve adaptation the prespacers originate from intermediates of nucleolytic cleavage at compromised replication forks, in primed adaptation they are generated by the effector complex. The simplest mechanism is observed in type II systems (Figure 3.6b), where primed adaptation initiates with Cas9-crRNA-tracrRNA complexes recognizing and then cleaving the invading genome's protospacer. Generated double-stranded breaks allow recruitment of the adaptation complex Cas1:Cas2:Csn2 which together with Cas9 leads to nucleolytic degradation of the bound DNA, albeit many mechanistic details are still missing (Wilkinson et al. 2019). The degradation of the invading DNA leads to the acquisition of spacers from both strands of DNA, and most frequently proximal to the DSB (Nussenzweig et al. 2019). Here, the priming is absolutely dependent on on-target cleavage by Cas9, and intuitively mutations in the seed regions or PAM abrogate spacer acquisition. As a result of this mechanism, the efficiency of spacer acquisition directly depends on the cutting efficiency mediated by a pre-acquired spacer. This means that once immunized, the host can rapidly acquire new spacers, forming a feed-forward immunization response.

In contrast, type I priming is promoted by failure to recognize a perfect target. Mutations in the protospacer will nominally decrease cleavage efficiency; however, Cascade can bind to such targets nonetheless (Blosser et al. 2015). Failure to establish a full-length R-loop, in essence failing to recognize a previously encountered target, is thought to abrogate potential off-target cleavage and instead promote Cascade complex into primed adaptation mode (Xue et al. 2016). This is mechanistically achieved by delaying the recruitment of Cas3 nuclease (Figure 3.6c), permitting recruitment of Cas1:Cas2 complex first (Redding et al. 2015). Upon subsequent binding, Cas1:Cas2:Cas3 complex is able to translocate along the nontargeted DNA strand without target degradation, with Cas1-Cas2 identifying new protospacers for integration (Hille et al. 2018). It remains unclear how prespacers are extracted when found, but it is not inconceivable that the same machinery involved in processing prespacers is involved as well.

3.4 CRISPR Systems as the Basis for New Tools in Drug Discovery

So far, how different CRISPR systems exert their function and safeguard microbes from invading genetic elements have been discussed. In the last part of this Chapter, we will turn to how the understanding of CRISPR biology has advanced biotechnology.

3.4.1 Cas Proteins for Gene Editing

The most prominent application of CRISPR systems since its discovery has been genome editing. This has stemmed from realization that Cas proteins can induce precise cuts in DNA (Marraffini and Sontheimer 2008; Garneau et al. 2010) guided by the crRNA molecule (Gasiunas et al. 2012; Jinek et al. 2012). The first protein shown to do this in a simple manner was the type II Cas9 protein of *S. pyogenes*. The simplicity of type II systems, where only one effector protein guided by crRNA:tracrRNA is required to introduce a guided DNA break, in contrast to type I systems which require a dozen of Cas proteins to form a functional Cascade complex, has allowed this system to be used in a plethora of eukaryotic systems to introduce precise genetic changes (Cong et al. 2013; DiCarlo et al. 2013; Ding et al. 2013; Friedland et al. 2013; Hwang et al. 2013; Jinek et al. 2013; Mali et al. 2013). The system has been further simplified by fusing crRNA to tracrRNA into a chimeric single-guide RNA (sgRNA) (Jinek et al. 2012), thus reducing the complexity of the toolkit needed for gene editing to just two components, where the sgRNA can very easily be replaced, allowing unprecedented flexibility in gene editing. SpyCas9-mediated gene editing has since been one of the most prolific technologies in biomedical sciences, with the application ranging from precise (epi)genome engineering, modulation of gene activity, forward genetic screens, proteomics, imaging studies, diagnostics, and therapy.

While gene editing will be addressed in detail elsewhere in this book, we would like to give the basics here. SpyCas9, guided by crRNA:tracrRNA (or more frequently sgRNA), once heterologously expressed in a cell, tissue, or organism, is guided to a specific site in the genome where upon successful recognition it will introduce a blunt-ended DSB. The repair outcome of the DSB by the host machinery can result in mutagenic events such as insertions or deletions by the nonhomologous end joining or microhomology-mediated end joining (optimal for generating genetic knockouts), or stimulate the introduction of desired DNA sequence via recombination with a donor sequence (desired outcome for model generation and therapy). It is therefore of utmost importance for Cas9 to specifically and efficiently introduce a break only at the desired site. Furthermore, the precision and efficiency of other gene editing methods, such as base editing (Komor et al. 2016), prime editing (Anzalone et al. 2019), or site-specific transposition (Chen and Wang 2019), are also in part dictated by the properties of Cas protein.

As discussed previously, Cas9 binds and identifies its target by recognizing the PAM sequence and then base pairing the crRNA initially with the seed sequence, and subsequently with the remainder of the sequence, followed by activation of the nucleolytic activities of RuvC and HNH domains. A plethora of biochemical studies have unequivocally confirmed that mismatches between seed and crRNA sequence are refractory to Cas9 editing (Jinek et al. 2012). However, mismatches outside the seed sequence (i.e. 10–20 nt away from the PAM) are well tolerated and support efficient cutting (Anderson et al. 2015), allowing Cas9 to cleave at **off-target** sites. Potential off-target mutagenesis induced by Cas proteins is a serious concern in therapy and model development, in particular, as Cas9 was shown to be able to induce large deletions and chromosomal rearrangements (Kosicki et al. 2018). In order to circumvent such problems, a number of different computational and experimental methods have been developed to identify and prevent off-target modifications (refer to Chapter 20).

3.4.1.1 Cas9 Variants

A major strategy to reduce off-target editing is to increase the specificity of Cas9 protein. To this end, SpyCas9 has been mutated at strategic residues so that mismatches between crRNA and the target strand are not compatible anymore with activating the enzyme's nucleolytic activity, generating Cas9 variants (such as eSpCas9, HypaCas9, and SpCas9-HF1) with improved specificity (Slaymaker et al. 2016; Kleinstiver et al. 2016a; Chen et al. 2017). However, improved specificity was often shown to have a negative impact on enzyme activity, decreasing the overall editing efficiency (Jones et al. 2020, Schmid-Burgk et al. 2020), likely due to the altered interaction between sgRNA and Cas9 protein.

Another major limitation of using SpyCas9 for genome editing is the PAM sequence. As discussed previously, type II effector proteins require a compatible PAM sequence downstream of the protospacer sequence. The preferred PAM for SpyCas9 is 5'-NGG-3', albeit it is able to use other GC-rich PAMs at lower efficiencies (Leenay et al. 2016). To expand the targeting repertoire, Cas9 variants have also been evolved to be able to use other PAM sequences. Initial variants relied on structural understanding of the SpyCas9 mechanism, and have therefore mutated the key residue (Arg1335) involved in recognition of the third G in the PAM, followed by compensatory mutations in the vicinity that allowed these novel proteins SpyCas9-VQR and -EQR to use the NGAN and NGNG as PAMs (Kleinstiver et al. 2015). A similar strategy was used to generate two other variants, the QQR1 which uses NAAG (Anders et al. 2016), and SpyCas9-NG requiring a dinucleotide 5'-NG-3' (Nishimasu et al. 2018). Recently, thanks to sophisticated *in vitro* evolution systems based on continuous phage-assisted evolution, multiple SpyCas9 variants that recognize non-G PAMs have been evolved; these variants can use NRN and to a lesser extent NYN PAM (Walton et al. 2020) and NRNH (where R represents A or G, Y is C or T, and H is A, C or T) (Miller et al. 2020). The development of these novel variants, albeit compromising their efficiency, brings us one step closer to a Cas system able to truly target any genomic sequence without restrictions imposed by the PAM.

3.4.1.2 Cas9 Orthologs

While *in vitro* evolution of SpyCas9 has expanded its utility, all the current variants still require a guanine in the PAM sequence or remain poorly active when using non-G PAM, making the variants of this enzyme largely unsuitable for targeting AT-rich sequences. Furthermore, while mutagenesis of Cas9 has improved its specificity, due to the structural constraints, these cannot be evolved past certain properties. Moreover, the size of the protein is of relevance for therapeutic applications, as typical vectors used for nuclease delivery (AAV) have a packaging capacity very close to the size of the SpyCas9 expression module (typically ~4.7 kb (Grieger and Samulski 2005)), making the use of this system difficult.

An alternative to *in vitro* evolution of Cas9 toward different activities, specificities, and sizes is to use Cas9 proteins originating from different species. Due to a staggering sequence variation, Cas9 from different species exhibit diverse specificities, activities, thermodynamic properties (i.e. exhibit longer stability and activity at higher temperatures), and PAM requirements. For example, in parallel with the discovery of SpyCas9 and the interference mechanisms, two orthogonal Cas9 from *Streptococcus thermophilus* (StCas9) has been described (Magadan et al. 2012; Karvelis et al. 2013), that were later shown to display editing efficiencies comparable to SpyCas9 in human cells, but with substantially lower off-target rate and longer, more diverse PAMs (NNAGAAW and NGGNG, where W is A or T) (Cong et al. 2013; Muller et al. 2016). Over the course of the last decade, a number of different, often smaller Cas9 proteins have been described and then used in genetic engineering in eukaryotic cells. These include Cas9 proteins from *S. aureus* (SauCas9) with the NNGRRT PAM (Ran et al. 2015), *Neisseria meningitidis* Nme1Cas9 with NNNNGATT PAM and lower off-target due to longer crRNAs (Esvelt et al. 2013; Hou et al. 2013; Zhang et al. 2013), and also Nme2Cas9 requiring NNNNCC (Edraki et al. 2019), *Staphylococcus auricularis* (SauriCas9) requiring NNGG in PAM (Hu et al. 2020), *Francisella novicida* (FnCas9) with NGG PAM (with rational engineering further relieving the PAM constraints to YG) (Hirano et al. 2016), *Campylobacter jejuni* (CjCas9) with NNNNRYAC (Kim et al. 2017), *Streptococcus canis* (ScCas9) with NNG (Chatterjee et al. 2018), *Geobacillus stearothermophillus* (GeoCas9) using the NNNNCRAA as a PAM (Harrington et al. 2017), and *Streptococcus macacae* (SmacCas9) requiring a NAAN in PAM site (Chatterjee et al. 2020).

Recently, a biochemical *tour de force* has identified and characterized the activity of 79 novel Cas9 proteins. Previously unknown G-, A-, T-, C-rich PAM repertoires, together with different patterns of cuts (blunt or with staggered ends), kinetics, and crRNA sequences have been described, forming the basis of a catalogue of orthologs that can be used for genome editing in the future (Gasiunas et al. 2020). Together with the ever-expanding and ever-improving number of variants with a wide range of PAM specificities (Collias and Beisel 2021), one can envisage that thanks to these efforts one will be able to choose a Cas9 protein for genome editing purposes based on the desired target sequence, unconstrained by the PAM restrictions, specificities, and activities.

3.4.1.3 The Use of Other Cas Proteins in Genome Editing

In parallel with the advent of new SpyCas9 variants and Cas9 orthologs, there was a push to examine whether other types of CRISPR-Cas9 proteins can be used for gene editing. While type I systems were essentially the first CRISPR systems to be examined in detail (Barrangou et al. 2007), the fact that the system requires a multitude of proteins (at various stoichiometry) gave little appeal to use this technology in eukaryotic systems. Type I systems are extensively used for precise engineering in microbes (Kiro et al. 2014; Li et al. 2016; Pyne et al. 2016; Xu et al. 2019), and only recently have been implemented for editing in human cells. Due to type I system's reliance on hyperactive Cas3 helicase/nuclease, editing by the Cascade complex leads to large deletions (up to 100 kb) from a single cut site (Dolan et al. 2019; Morisaka et al. 2019; Osakabe et al. 2020). Whereas using type I systems are therefore an excellent tool for deleting large segments of the genome (which could be useful for removing transgenes from model organisms or interrogating regulatory elements, for example), this system has been adapted for introducing small insertions by fusing the Cascade complex to FokI endonuclease domain. When Cas3 was omitted from and FokI-Cascade fusion expressed to target proximal sequences, DNA sequences were successfully deleted (Cameron et al. 2019). Whereas the efficiency of Cascade-FokI mediated editing is low compared with standard SpyCas9, it is on par with other genome editing approaches using catalytically inactive Cas proteins (see Section 3.4.2).

The most prominent use of other types of CRISPR systems is the Cas12a (originally named Cpf1), a member of type V systems (Zetsche et al. 2015), which have some beneficial features over Cas9 proteins. The first three described Cas12a proteins, the *Franscisella novicida* (FnCas12a), *Acidaminococcus sp.* (AsCas12a), and *Lachnospiraceae* bacterium (LbCas12a), use T-rich PAM sequences, such as TTN and TTTN, making them a complementary tool to G-rich PAM utilizing Cas9 proteins. Secondly, Cas12a proteins are able to autonomously process crRNA from pre-crRNA, unlike Cas9 proteins which require tracrRNA (Deltcheva et al. 2011). This property has been exploited to significantly improve simultaneous genome editing at multiple sites, where multiple Cas12a-compatible crRNAs are expressed as a single pre-crRNA transcript, contrasting to Cas9 multiplex editing which requires multiple expression modules, one for each sgRNA (Zetsche et al. 2017; DeWeirdt et al. 2021). The most prominent benefit of Cas12a is, however, its cleavage mechanism, where Cas12a generates staggered ends (unlike blunt ends generated by most of Cas9 proteins) outside the critical seed region. Staggered ends are particularly suitable for precise integration of DNA via NHEJ or MMEJ (Maresca et al. 2013). As Cas12a cuts the target DNA away from the seed sequence, repair via NHEJ (leading to small indels) will still support cleavage. Cas12 editing is therefore likely to promote resection of the staggered DNA break, leading to deletions spreading into the seed region or more favorably promoting HDR and MMEJ (Begemann et al. 2017; Moreno-Mateos et al. 2017; Li et al. 2018a). Finally, Cas12a shows lower tolerance to mismatches, with only the mismatches at the last 4 nt of a 23 nt-long crRNA supporting cleavage (Kleinstiver et al. 2016b), in contrast to SpyCas9 which will cut DNA even with 10 mismatches (Klein et al. 2018; Jones et al. 2020). Similarly to Cas9 systems, new Cas12a orthologs have also been described to expand the targeting spectrum by choosing Cas12a variants with convenient PAM requirements, such as those of *Coprococcus eutactus* (CeCas12a) or *F. novicida* (FnCas12a) (Aliaga Goltsman et al. 2020; Chen et al. 2020; Zetsche et al. 2020). Furthermore, AsCas12a has been evolved for higher specificity and activity (Kleinstiver et al. 2019).

Other type V systems have very recently been adapted for gene editing. Type V-B systems, epitomized by Cas12b effector proteins, were difficult to use for gene editing due to their collateral ssDNAse activity and low activity at mammalian physiological temperature. Recent engineering and repurposing efforts have allowed a set of evolved Cas12b proteins to specifically and efficiently edit eukaryotic genomes (Strecker et al. 2019a; Teng et al. 2019a; Teng et al. 2019c; Ming et al. 2020). A Cas12b worth mentioning originates from *Alicyclobacillus acidiphilus* (AaCas12b), which has unprecedented specificity where the enzyme is not able to tolerate any mismatches (Teng et al. 2018).

A recent discovery of Cas12e from *Deltaproteobacteria* (formerly known as DpbCasX) (Burstein et al. 2017) has already led to a demonstration of exquisite editing activity in human cells (Liu et al. 2019a). Very recently, a novel collection of compact type V Cas effector proteins have been identified in genomes of Biggiephage clade of huge phages. Cas12j (also known as CasΦ) was shown to possess some superb features: the size of the effector protein is very small compared with commonly used Cas9 proteins (only ~70 kDa compared to 160 kDa), a T-rich PAM (TBN, where B is G, T, or C), is able to process crRNA with its RuvC domain, generating staggered-end break end with 8–12 nt-long 5' overhang, and with potent activity in human cells (Pausch et al. 2020). With emphasis placed on the discovery of miniature class 2 effectors, a group of Cas12f proteins was recently biochemically characterized and orthologs with interesting properties, suggesting that there are many more CRISPR-Cas systems to be harnessed for gene editing (Karvelis et al. 2020).

Finally, a number of different CRISPR systems have been developed to perform Transposon-assisted site-specific integration. As discussed previously, many *cas* genes are frequently found with transposons in the same operon, with concurrent loss of essential interference machinery, such as many type IV and V systems. A spectacular example is Cas12k, which was able to support transposition of a 10 kb insert at crRNA-directed site when heterologously expressed in *E. coli* with the missing components (Strecker et al. 2019b). Similarly, the Tn*6677* transposase of *Vibrio cholerae* has co-opted the type I-F machinery for transpositions; this has very recently allowed development of highly efficient CRISPR-guided transposition in bacteria (Klompe et al. 2019; Vo et al. 2021), with the potential to be of major use in vertebrate genome editing.

3.4.2 Application of Cas Proteins Beyond Genome Editing

3.4.2.1 dCas9 Fusions

While Cas9 proteins were originally used for gene editing purposes, realization that mutating two key residues in RuvC and HNH domain (D10A and H840A, respectively) that abrogate its catalytic activity but not its binding (Jinek et al. 2012; Qi et al. 2013) generates a sequence-specific sgRNA-directed DNA binding protein has opened a new range of application of the bacterial immune system. By fusing a specific protein domain to catalytically deficient Cas9 (dCas9) allows one to direct a protein activity to specific genomic loci. This approach has been brought to life by fusing dCas9 to repressive transcriptional domains such as KRAB domain of Kox1 or the chromoshadow domain of HP1α (Figure 3.7a), producing a robust 25–100-fold reduction of targeted gene expression without any direct off-targets (Gilbert et al. 2013; Gilbert et al. 2014), generating the strategy called CRISPR interference (CRISPRi). Similarly, a complementary approach named CRISPR activation (CRISPRa) was developed by fusing transactivator domains of VP64 or NFkB (Perez-Pinera et al. 2013; Gilbert et al. 2014) (Figure 3.7a). Both of these systems have been improved further, namely by identifying protein domains with optimal activities (Chavez et al. 2016) or improving the output by recruiting multiple synergistic effectors (Tanenbaum et al. 2014; Zalatan et al. 2015). In parallel, dCas9 system for introducing specific epigenetic modifications at programmed genomic loci was also developed by fusing dCas9 to catalytic domains of DNA methyltransferases and various histone-modifying enzymes (Figure 3.7c, d), allowing one to enforce a heritable modulation of gene expression (Amabile et al. 2016; Liu et al. 2016; Vojta et al. 2016). While these systems are not as

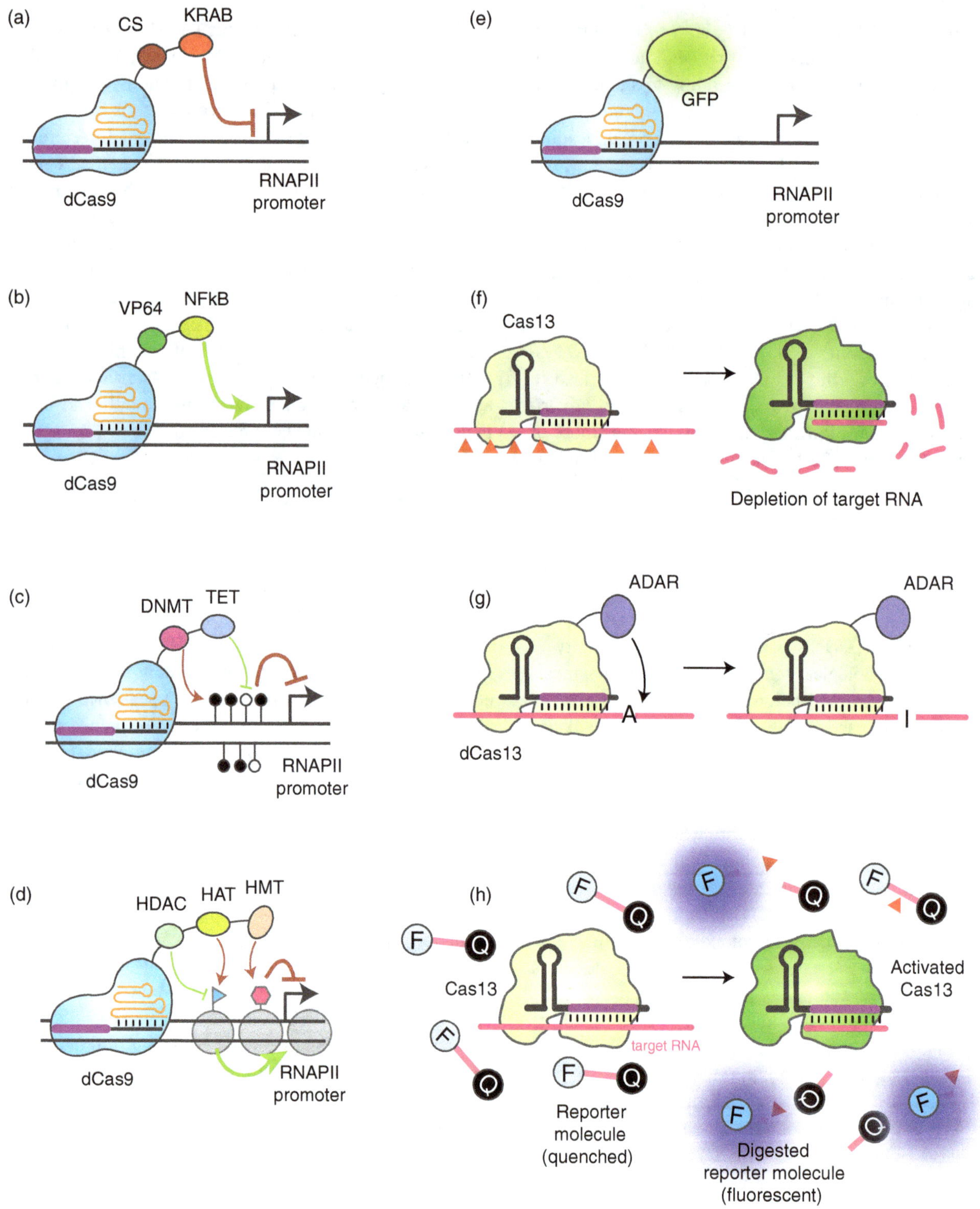

Figure 3.7 Applications of CRISPR systems beyond genome editing. Fusing catalytically inactive Cas9 (dCas9) to a transcriptional activator (**a**) or repression (**b**) domains can directly regulate gene expression. By fusing dCas9 to DNA-modifying enzymes, such as DNA methyltransferases (DNMT) or TET enzymes (**c**), or histone deacetylases (HDACs), histone acetyltransferases (HATs), or histone methyltransferases (HMTs) (**d**), one can modify the epigenetic marks leading to indirect albeit potentially permanent modulation of gene expression. If fused to a fluorescent protein, dCas9 can be used to visualize targeted locus in imaging studies (**e**). RNA targeting enzymes such as Cas13a (**f**) are able to specifically degrade target RNA molecules, or by fusing a catalytically inactive version of this enzyme to RNA-modifying enzymes (such as ADAR deaminases), one can perform RNA editing; here, deamination of adenosine to inosine (decoded as a guanosine) is depicted (**g**). Cas13-based detection methods rely on the activation of Cas13a by recognizing the target RNA molecule, which is then unspecifically able to digest the reporter RNA molecule. Cleavage of the reporter molecule separates fluorophore (F) from the quencher dye (Q), generating a diagnostic fluorescent signal (**h**). A similar approach can be used for DNA-targeting enzymes exhibiting collateral activity, such as some Cas12 enzymes.

nearly as sophisticated as CRISPRi and CRISPRa, they present a promising basis to therapeutic epigenome editing. dCas9 systems have also been used for a number of other purposes used to elucidate biology, such as imaging and proteomics Figure 3.7e; we refer the reader to excellent reviews on these recently developed systems (Xu and Qi 2019). It should be also noted that dCas9 can be used to tether DNA-modifying enzymes (such as cytidine or adenosine deaminases) which can alter the underlying sequence; this is the basis for DNA base editing, which is addressed in detail in Chapter 14.

3.4.2.2 RNA Targeting

So far, we have discussed how various CRISPR systems can be used for genome editing. The discovery of type VI systems and their ability to target RNA by Cas13 effectors has led to the rise of novel approaches to manipulate the transcriptome of a given cell without altering the underlying genetic component. As discussed in previous sections, Cas13 proteins can degrade target RNA molecules by nucleolytic activity of their HEPN domain. Heterologous expression of Cas13 orthologs, such as those of *Leptotrichia wadei* (LwaCas13a), *Leptotrichia shahii* (LshCas13a) (Abudayyeh et al. 2017), *Prevotella* sp. P5-125 (PspCas13b) (Cox et al. 2017) or *Ruminococcus flavefaciens* (RfxCas13d, also known as CasRx) (Konermann et al. 2018), and cognate crRNA in human cells leads to knockdown of specific RNA transcripts (Figure 3.7f) without substantial off-target effects usually associated with short-hairpin RNA (shRNA). The knockdown efficiency is ortholog and transcript dependent, but comparable to reduction observed with genome editing approaches (50–90%). As discussed previously, nearly all Cas13 proteins exhibit an indiscriminate RNase activity, meaning that they can degrade any bystander RNA. While this is an important mechanism of conferring population-level immunity in prokaryotes (Meeske et al. 2019), the collateral activity of tested proteins has not been shown when expressed in human cells, encouraging further use of this system in mammalian models. It should be noted that some Cas9 proteins are also able to target RNA (Sampson et al. 2013; Dugar et al. 2018; Strutt et al. 2018), but their activities have not been tested yet in human cells.

Mutating the key catalytic residues of Cas13 converted this protein to a binding-proficient but nuclease-deficient protein (dCas13). Subsequent targeting to key regulatory pre-mRNA elements (such as splicing acceptor or donor sites) permits one to alter the splicing pattern of target transcript (Konermann et al. 2018). Importantly, dCas13 can be used as a programmable RNA-binding protein (analogous to dCas9), and fusing it to a suitable protein domain allows one to modulate or analyze the properties of target RNA. For example, fusing a domain of Adenosine deaminases acting on RNA (ADAR), key enzymes involved in RNA editing, allows one to post-transcriptionally change the sequence of RNA from adenine to inosine (decoded during the translation as guanine), altering the protein primary sequence without affecting the genome (Cox et al. 2017) (Figure 3.7g).

Overall, the adaptation of RNA-targeting type VI CRISPR systems now allows researchers to manipulate the transcriptome, a powerful and complementary tool to manipulating the genome by other systems.

3.4.2.3 Biochemical Detection

Finally, we would like to point that CRISPR systems can be used for molecular diagnostics. The collateral activity of Cas13 proteins, while conceptually undesirable, has been exploited to develop highly sensitive detectors for the presence of specific RNA. This approach relies on *in vitro* activation of Cas13's indiscriminate RNase activity by an on-target recognition of specific RNA. Once activated, Cas13 is able to degrade other RNA molecules in the reaction mix; if a fluorescent RNA reporter molecule is included, then Cas13 is able to detect a specific RNA molecule with attomolar sensitivity (Gootenberg et al. 2017), and improvements claiming to bring sensitivity down to zeptomolar (10^{-21} mol/l) range (Gootenberg et al. 2018) (Figure 3.7h). This system, termed SHERLOCK, has been used since to develop fast and sensitive methods for diagnostics of a number of highly pathogenic RNA viruses (Myhrvold et al. 2018; Patchsung et al. 2020). A similar approach exploits a collateral activity of Cas12a proteins on ssDNA, allowing one to detect specific DNA (or cDNA) sequences; indeed a flurry of different variants of this concept have been developed to detect pathogen DNA or cDNA or to perform SNP profiling (Chen et al. 2018; Li et al. 2018b; Teng et al. 2019b). These developments and innovation are momentous and provide a major progress for molecular diagnostics (Li et al. 2019b).

3.5 Concluding Remarks

The discoveries made in the field of CRISPR biology in the last two decades have paved a way for efficient, cost-effective, and precise genome editing. Efforts in biochemical and structural characterization of a great number of Cas proteins expanded the toolset of CRISPR proteins one can use for genome manipulation and other purposes. Initially, SpyCas9 was the only Cas protein used for genome editing, but the development of a multiple Cas9 variants with expanded PAM targeting landscape, the discovery of Cas9 orthologs and finally characterization of new class 2 proteins, has in a manner removed constraints imposed by the biochemistry of Cas9. Thanks to an ever-expanding collections of available Cas nucleases, in the years to come we can expect to perform gene editing with the best tool for a given sequence and application, and without compromising precision and efficiency. Indeed, one can draw a parallel between the democratization of restriction enzymes for molecular cloning and CRISPR enzymes for genome editing, where we are now able to use a bespoke enzyme (rather than relying on a modest set) for diverse and complex outcomes.

Many innovative methods have been developed to mitigate pitfalls of these novel technologies: low efficiency can be enhanced by fusing the Cas proteins to more efficient nucleases (Dolan et al. 2019); specificity can be improved by careful crRNA design (Doench et al. 2016; Akcakaya et al. 2018) and enhanced by blocking unspecific sites (Coelho et al. 2020), the activity can be controlled by chemical agents (Maji et al. 2017) and optogenetically (Nihongaki et al. 2015). Whereas these tools are a powerful addition to the CRISPR arsenal, recently discovered inhibitors of CRISPR systems, the anti-CRISPR proteins (Acr) provide the basis for even tighter control of undesired activities of Cas proteins (Davidson et al. 2020; Marino et al. 2020). Acrs are, in essence, an evolutionary response of mobile genetic elements able to inhibit CRISPR systems at various steps of the immune response. Just like CRISPR systems, their inhibitors are incredibly diverse, and thus represent yet another untapped resource of wonderful tools that can be used to further improve CRISPR editing, in particular in a therapeutic setting.

Studying the biology of diverse CRISPR systems did not lead just to the development of gene editing tools, but also ways to manipulate the transcriptome and the epigenome. Furthermore, by exploiting the seemingly undesirable collateral activity of type V and type VI systems, new molecular diagnostics tools with unprecedented detection sensitivity have been developed. Together, this diverse collection of novel ways to repurpose bacterial immune systems for a bespoke application is a witness of how much more we can get by understanding and studying microbial CRISPR systems. While class 2 systems have been studied (and hence appropriated for various applications) to a great extent, the application of class 1 systems is lagging. Furthermore, other phases of CRISPR immune response are still comparably poorly characterized, for example, adaptation phases. Studying these systems and these phases might well generate new powerful tools for genome editing or something completely different. One can only eagerly wait for what the next decades are going to bring.

References

Abudayyeh, O.O., Gootenberg, J.S., Konermann, S. et al. (2016). C2c2 is a single-component programmable RNA-guided RNA-targeting CRISPR effector. *Science* 353: aaf5573.

Abudayyeh, O.O., Gootenberg, J.S., Essletzbichler, P. et al. (2017). RNA targeting with CRISPR-Cas13. *Nature* 550: 280–284.

Akcakaya, P., Bobbin, M.L., Guo, J.A. et al. (2018). in vivo CRISPR editing with no detectable genome-wide off-target mutations. *Nature* 561: 416–419.

Aliaga Goltsman, D.S., Alexander, L.M., Devoto, A.E. et al. (2020). Novel Type V-A CRISPR effectors are active nucleases with expanded targeting capabilities. *CRISPR J* 3: 454–461.

Alkhnbashi, O.S., Shah, S.A., Garrett, R.A. et al. (2016). Characterizing leader sequences of CRISPR loci. *Bioinformatics* 32: i576–i585.

Amabile, A., Migliara, A., Capasso, P. et al. (2016). Inheritable silencing of endogenous genes by hit-and-run targeted epigenetic editing. *Cell* 167: 219–232. e14.

Anders, C., Niewoehner, O., Duerst, A., and Jinek, M. (2014). Structural basis of PAM-dependent target DNA recognition by the Cas9 endonuclease. *Nature* 513: 569–573.

Anders, C., Bargsten, K., and Jinek, M. (2016). Structural plasticity of PAM recognition by engineered variants of the RNA-guided endonuclease Cas9. *Mol. Cell* 61: 895–902.

Anderson, E.M., Haupt, A., Schiel, J.A. et al. (2015). Systematic analysis of CRISPR-Cas9 mismatch tolerance reveals low levels of off-target activity. *J. Biotechnol.* 211: 56–65.

Anzalone, A.V., Randolph, P.B., Davis, J.R. et al. (2019). Search-and-replace genome editing without double-strand breaks or donor DNA. *Nature* 576: 149–157.

Barrangou, R., Fremaux, C., Deveau, H. et al. (2007). CRISPR provides acquired resistance against viruses in prokaryotes. *Science* 315: 1709–1712.

Begemann, M.B., Gray, B.N., January, E. et al. (2017). Precise insertion and guided editing of higher plant genomes using Cpf1 CRISPR nucleases. *Sci. Rep.* 7: 11606.

Belotserkovskaya, R., Oh, S., Bondarenko, V.A. et al. (2003). FACT facilitates transcription-dependent nucleosome alteration. *Science* 301: 1090–1093.

Bernheim, A. and Sorek, R. (2020). The pan-immune system of bacteria: antiviral defence as a community resource. *Nat. Rev. Microbiol.* 18: 113–119.

Blosser, T.R., Loeff, L., Westra, E.R. et al. (2015). Two distinct DNA binding modes guide dual roles of a CRISPR-Cas protein complex. *Mol. Cell* 58: 60–70.

Bolotin, A., Quinquis, B., Sorokin, A., and Ehrlich, S.D. (2005). Clustered regularly interspaced short palindrome repeats (CRISPRs) have spacers of extrachromosomal origin. *Microbiol. (Reading)* 151: 2551–2561.

Brouns, S.J., Jore, M.M., Lundgren, M. et al. (2008). Small CRISPR RNAs guide antiviral defense in prokaryotes. *Science* 321: 960–964.

Budhathoki, J.B., Xiao, Y., Schuler, G. et al. (2020). Real-time observation of CRISPR spacer acquisition by Cas1-Cas2 integrase. *Nat. Struct. Mol. Biol.* 27: 489–499.

Burstein, D., Harrington, L.B., Strutt, S.C. et al. (2017). New CRISPR-Cas systems from uncultivated microbes. *Nature* 542: 237–241.

Cameron, P., Coons, M.M., Klompe, S.E. et al. (2019). Harnessing Type I CRISPR-Cas systems for genome engineering in human cells. *Nat. Biotechnol.* 37: 1471–1477.

Carte, J., Wang, R., Li, H. et al. (2008). Cas6 is an endoribonuclease that generates guide RNAs for invader defense in prokaryotes. *Genes Dev.* 22: 3489–3496.

Carte, J., Pfister, N.T., Compton, M.M. et al. (2010). Binding and cleavage of CRISPR RNA by Cas6. *RNA* 16: 2181–2188.

Chatterjee, P., Jakimo, N., and Jacobson, J.M. (2018). Minimal PAM specificity of a highly similar SpCas9 ortholog. *Sci. Adv.* 4: eaau0766.

Chatterjee, P., Lee, J., Nip, L. et al. (2020). A Cas9 with PAM recognition for adenine dinucleotides. *Nat. Commun.* 11: 2474.

Chavez, A., Tuttle, M., Pruitt, B.W. et al. (2016). Comparison of Cas9 activators in multiple species. *Nat. Methods* 13: 563–567.

Chen, S.P. and Wang, H.H. (2019). An engineered Cas-transposon system for programmable and site-directed DNA transpositions. *CRISPR J* 2: 376–394.

Chen, J.S., Dagdas, Y.S., Kleinstiver, B.P. et al. (2017). Enhanced proofreading governs CRISPR-Cas9 targeting accuracy. *Nature* 550: 407–410.

Chen, J.S., Ma, E., Harrington, L.B. et al. (2018). CRISPR-Cas12a target binding unleashes indiscriminate single-stranded DNase activity. *Science* 360: 436–439.

Chen, P., Zhou, J., Wan, Y. et al. (2020). A Cas12a ortholog with stringent PAM recognition followed by low off-target editing rates for genome editing. *Genome Biol.* 21: 78.

Clarke, R., Heler, R., Macdougall, M.S. et al. (2018). Enhanced bacterial Immunity and mammalian genome editing via RNA-polymerase-mediated dislodging of Cas9 from double-strand DNA breaks. *Mol. Cell* 71: 42–55. e8.

Coelho, M.A., De Braekeleer, E., Firth, M. et al. (2020). CRISPR GUARD protects off-target sites from Cas9 nuclease activity using short guide RNAs. *Nat. Commun.* 11: 4132.

Collias, D. and Beisel, C.L. (2021). CRISPR technologies and the search for the PAM-free nuclease. *Nat. Commun.* 12: 555.

Cong, L., Ran, F.A., Cox, D. et al. (2013). Multiplex genome engineering using CRISPR/Cas systems. *Science* 339: 819–823.

Cox, D.B.T., Gootenberg, J.S., Abudayyeh, O.O. et al. (2017). RNA editing with CRISPR-Cas13. *Science* 358: 1019–1027.

Datsenko, K.A., Pougach, K., Tikhonov, A. et al. (2012). Molecular memory of prior infections activates the CRISPR/Cas adaptive bacterial immunity system. *Nat. Commun.* 3: 945.

Davidson, A.R., Lu, W.T., Stanley, S.Y. et al. (2020). Anti-CRISPRs: protein inhibitors of CRISPR-Cas systems. *Annu. Rev. Biochem.* 89: 309–332.

Deltcheva, E., Chylinski, K., Sharma, C.M. et al. (2011). CRISPR RNA maturation by trans-encoded small RNA and host factor RNase III. *Nature* 471: 602–607.

Deng, L., Garrett, R.A., Shah, S.A. et al. (2013). A novel interference mechanism by a Type IIIB CRISPR-Cmr module in Sulfolobus. *Mol. Microbiol.* 87: 1088–1099.

Deweirdt, P.C., Sanson, K.R., Sangree, A.K. et al. (2021). Optimization of AsCas12a for combinatorial genetic screens in human cells. *Nat. Biotechnol.* 39: 94–104.

Dicarlo, J.E., Norville, J.E., Mali, P. et al. (2013). Genome engineering in Saccharomyces cerevisiae using CRISPR-Cas systems. *Nucleic Acids Res.* 41: 4336–4343.

Ding, Q., Regan, S.N., Xia, Y. et al. (2013). Enhanced efficiency of human pluripotent stem cell genome editing through replacing TALENs with CRISPRs. *Cell Stem Cell* 12: 393–394.

Doench, J.G., Fusi, N., Sullender, M. et al. (2016). Optimized sgRNA design to maximize activity and minimize off-target effects of CRISPR-Cas9. *Nat. Biotechnol.* 34: 184–191.

Dolan, A.E., Hou, Z., Xiao, Y. et al. (2019). Introducing a spectrum of long-range genomic deletions in human embryonic stem cells using Type I CRISPR-Cas. *Mol. Cell* 74: 936–950. e5.

Dugar, G., Leenay, R.T., Eisenbart, S.K. et al. (2018). CRISPR RNA-dependent binding and cleavage of endogenous RNAs by the campylobacter jejuni Cas9. *Mol. Cell* 69: 893–905. e7.

East-Seletsky, A., O'connell, M.R., Knight, S.C. et al. (2016). Two distinct RNase activities of CRISPR-C2c2 enable guide-RNA processing and RNA detection. *Nature* 538: 270–273.

Edraki, A., Mir, A., Ibraheim, R. et al. (2019). A compact, high-accuracy Cas9 with a dinucleotide PAM for in vivo genome editing. *Mol. Cell* 73: 714–726. e4.

Elmore, J.R., Sheppard, N.F., Ramia, N. et al. (2016). Bipartite recognition of target RNAs activates DNA cleavage by the Type III-B CRISPR-Cas system. *Genes Dev.* 30: 447–459.

Estrella, M.A., Kuo, F.T., and Bailey, S. (2016). RNA-activated DNA cleavage by the Type III-B CRISPR-Cas effector complex. *Genes Dev.* 30: 460–470.

Esvelt, K.M., Mali, P., Braff, J.L. et al. (2013). Orthogonal Cas9 proteins for RNA-guided gene regulation and editing. *Nat. Methods* 10: 1116–1121.

Faure, G., Makarova, K.S., and Koonin, E.V. (2019a). CRISPR-Cas: complex functional networks and multiple roles beyond adaptive immunity. *J. Mol. Biol.* 431: 3–20.

Faure, G., Shmakov, S.A., Yan, W.X. et al. (2019b). CRISPR-Cas in mobile genetic elements: counter-defence and beyond. *Nat. Rev. Microbiol.* 17: 513–525.

Fonfara, I., Richter, H., Bratovič, M. et al. (2016). The CRISPR-associated DNA-cleaving enzyme Cpf1 also processes precursor CRISPR RNA. *Nature* 532: 517–521.

Friedland, A.E., Tzur, Y.B., Esvelt, K.M. et al. (2013). Heritable genome editing in C. elegans via a CRISPR-Cas9 system. *Nat. Methods* 10: 741–743.

Garneau, J.E., Dupuis, M.E., Villion, M. et al. (2010). The CRISPR/Cas bacterial immune system cleaves bacteriophage and plasmid DNA. *Nature* 468: 67–71.

Gasiunas, G., Barrangou, R., Horvath, P., and Siksnys, V. (2012). Cas9-crRNA ribonucleoprotein complex mediates specific DNA cleavage for adaptive immunity in bacteria. *Proc. Natl. Acad. Sci. U. S. A.* 109: E2579–E2586.

Gasiunas, G., Young, J.K., Karvelis, T. et al. (2020). A catalogue of biochemically diverse CRISPR-Cas9 orthologs. *Nat. Commun.* 11: 5512.

Gilbert, L.A., Larson, M.H., Morsut, L. et al. (2013). CRISPR-mediated modular RNA-guided regulation of transcription in eukaryotes. *Cell* 154: 442–451.

Gilbert, L.A., Horlbeck, M.A., Adamson, B. et al. (2014). Genome-scale CRISPR-mediated control of gene repression and activation. *Cell* 159: 647–661.

Godde, J.S. and Bickerton, A. (2006). The repetitive DNA elements called CRISPRs and their associated genes: evidence of horizontal transfer among prokaryotes. *J. Mol. Evol.* 62: 718–729.

Goldberg, G.W., Jiang, W., Bikard, D., and Marraffini, L.A. (2014). Conditional tolerance of temperate phages via transcription-dependent CRISPR-Cas targeting. *Nature* 514: 633–637.

Gonzalez-Delgado, A., Mestre, M.R., Martinez-Abarca, F., and Toro, N. (2019). Spacer acquisition from RNA mediated by a natural reverse transcriptase-Cas1 fusion protein associated with a Type III-D CRISPR-Cas system in Vibrio vulnificus. *Nucleic Acids Res.* 47: 10202–10211.

Gootenberg, J.S., Abudayyeh, O.O., Lee, J.W. et al. (2017). Nucleic acid detection with CRISPR-Cas13a/C2c2. *Science* 356: 438–442.

Gootenberg, J.S., Abudayyeh, O.O., Kellner, M.J. et al. (2018). Multiplexed and portable nucleic acid detection platform with Cas13, Cas12a, and Csm6. *Science* 360: 439–444.

Grieger, J.C. and Samulski, R.J. (2005). Packaging capacity of adeno-associated virus serotypes: impact of larger genomes on infectivity and postentry steps. *J. Virol.* 79: 9933–9944.

Grissa, I., Vergnaud, G., and Pourcel, C. (2007). The CRISPRdb database and tools to display CRISPRs and to generate dictionaries of spacers and repeats. *BMC Bioinform.* 8: 172.

Hale, C., Kleppe, K., Terns, R.M., and Terns, M.P. (2008). Prokaryotic silencing (psi)RNAs in Pyrococcus furiosus. *RNA* 14: 2572–2579.

Hale, C.R., Zhao, P., Olson, S. et al. (2009). RNA-guided RNA cleavage by a CRISPR RNA-Cas protein complex. *Cell* 139: 945–956.

Hampton, H.G., Watson, B.N.J., and Fineran, P.C. (2020). The arms race between bacteria and their phage foes. *Nature* 577: 327–336.

Han, W., Li, Y., Deng, L. et al. (2017). A Type III-B CRISPR-Cas effector complex mediating massive target DNA destruction. *Nucleic Acids Res.* 45: 1983–1993.

Harrington, L.B., Paez-Espino, D., Staahl, B.T. et al. (2017). A thermostable Cas9 with increased lifetime in human plasma. *Nat. Commun.* 8: 1424.

Harrington, L.B., Burstein, D., Chen, J.S. et al. (2018). Programmed DNA destruction by miniature CRISPR-Cas14 enzymes. *Science* 362: 839–842.

Hatoum-Aslan, A., Maniv, I., Samai, P., and Marraffini, L.A. (2014). Genetic characterization of antiplasmid immunity through a type III-A CRISPR-Cas system. *J. Bacteriol.* 196: 310–317.

Haurwitz, R.E., Jinek, M., Wiedenheft, B. et al. (2010). Sequence- and structure-specific RNA processing by a CRISPR endonuclease. *Science* 329: 1355–1358.

Hayes, R.P., Xiao, Y., Ding, F. et al. (2016). Structural basis for promiscuous PAM recognition in Type I-E Cascade from E. coli. *Nature* 530: 499–503.

Heler, R., Samai, P., Modell, J.W. et al. (2015). Cas9 specifies functional viral targets during CRISPR-Cas adaptation. *Nature* 519: 199–202.

Hille, F., Richter, H., Wong, S.P. et al. (2018). The Biology of CRISPR-Cas: backward and forward. *Cell* 172: 1239–1259.

Hirano, H., Gootenberg, J.S., Horii, T. et al. (2016). Structure and engineering of Francisella novicida Cas9. *Cell* 164: 950–961.

Horvath, P., Romero, D.A., Coute-Monvoisin, A.C. et al. (2008). Diversity, activity, and evolution of CRISPR loci in Streptococcus thermophilus. *J. Bacteriol.* 190: 1401–1412.

Hou, Z., Zhang, Y., Propson, N.E. et al. (2013). Efficient genome engineering in human pluripotent stem cells using Cas9 from Neisseria meningitidis. *Proc. Natl. Acad. Sci. U. S. A.* 110: 15644–15649.

Hu, Z., Wang, S., Zhang, C. et al. (2020). A compact Cas9 ortholog from Staphylococcus Auricularis (SauriCas9) expands the DNA targeting scope. *PLoS Biol.* 18: e3000686.

Huo, Y., Nam, K.H., Ding, F. et al. (2014). Structures of CRISPR Cas3 offer mechanistic insights into Cascade-activated DNA unwinding and degradation. *Nat. Struct. Mol. Biol.* 21: 771–777.

Hwang, W.Y., Fu, Y., Reyon, D. et al. (2013). Efficient genome editing in zebrafish using a CRISPR-Cas system. *Nat. Biotechnol.* 31: 227–229.

Hynes, A.P., Villion, M., and Moineau, S. (2014). Adaptation in bacterial CRISPR-Cas immunity can be driven by defective phages. *Nat. Commun.* 5: 1–6.

Ishino, Y., Shinagawa, H., Makino, K. et al. (1987). Nucleotide sequence of the iap gene, responsible for alkaline phosphatase isozyme conversion in Escherichia coli, and identification of the gene product. *J. Bacteriol.* 169: 5429–5433.

Ivančić-Baće, I., Cass, S.D., Wearne, S.J., and Bolt, E.L. (2015). Different genome stability proteins underpin primed and naïve adaptation in E. coli CRISPR-Cas immunity. *Nucleic Acids Res.* 43: 10821–10830.

Jansen, R., Van Embden, J.D.A., Gaastra, W., and Schouls, L.M. (2002). Identification of genes that are associated with DNA repeats in prokaryotes. *Mol. Microbiol.* 43: 1565–1575.

Jeon, Y., Choi, Y.H., Jang, Y. et al. (2018). Direct observation of DNA target searching and cleavage by CRISPR-Cas12a. *Nat. Commun.* 9: 2777.

Jiang, F., Taylor, D.W., Chen, J.S. et al. (2016a). Structures of a CRISPR-Cas9 R-loop complex primed for DNA cleavage. *Science* 351: 867–871.

Jiang, W., Samai, P., and Marraffini, L.A. (2016b). Degradation of phage transcripts by CRISPR-associated RNases enables Type III CRISPR-Cas Immunity. *Cell* 164: 710–721.

Jinek, M., Chylinski, K., Fonfara, I. et al. (2012). A programmable dual-RNA-guided DNA endonuclease in adaptive bacterial immunity. *Science* 337: 816–821.

Jinek, M., East, A., Cheng, A. et al. (2013). RNA-programmed genome editing in human cells. *elife* 2: e00471.

Jore, M.M., Lundgren, M., Van Duijn, E. et al. (2011). Structural basis for CRISPR RNA-guided DNA recognition by Cascade. *Nat. Struct. Mol. Biol.* 18: 529–536.

Jones, J.R.S.K., Hawkins, J.A., Johnson, N.V. et al. (2020). Massively parallel kinetic profiling of natural and engineered CRISPR nucleases. *Nat. Biotechnol.* 39: 84–93.

Karvelis, T., Gasiunas, G., Miksys, A. et al. (2013). crRNA and tracrRNA guide Cas9-mediated DNA interference in streptococcus thermophilus. *RNA Biol.* 10: 841–851.

Karvelis, T., Bigelyte, G., Young, J.K. et al. (2020). PAM recognition by miniature CRISPR-Cas12f nucleases triggers programmable double-stranded DNA target cleavage. *Nucleic Acids Res.* 48: 5016–5023.

Kazlauskiene, M., Tamulaitis, G., Kostiuk, G. et al. (2016). Spatiotemporal control of Type III-A CRISPR-Cas Immunity: coupling DNA degradation with the target RNA recognition. *Mol. Cell* 62: 295–306.

Kazlauskiene, M., Kostiuk, G., Venclovas, C. et al. (2017). A cyclic oligonucleotide signaling pathway in Type III CRISPR-Cas systems. *Science* 357: 605–609.

Kim, E., Koo, T., Park, S.W. et al. (2017). in vivo genome editing with a small Cas9 orthologue derived from Campylobacter jejuni. *Nat. Commun.* 8: 14500.

Kim, S., Loeff, L., Colombo, S. et al. (2020). Selective loading and processing of prespacers for precise CRISPR adaptation. *Nature* 579: 141–145.

Kiro, R., Shitrit, D., and Qimron, U. (2014). Efficient engineering of a bacteriophage genome using the type I-E CRISPR-Cas system. *RNA Biol.* 11: 42–44.

Klein, M., Eslami-Mossallam, B., Arroyo, D.G., and Depken, M. (2018). Hybridization kinetics explains CRISPR-Cas off-targeting rules. *Cell Rep.* 22: 1413–1423.

Kleinstiver, B.P., Prew, M.S., Tsai, S.Q. et al. (2015). Engineered CRISPR-Cas9 nucleases with altered PAM specificities. *Nature* 523: 481–485.

Kleinstiver, B.P., Pattanayak, V., Prew, M.S. et al. (2016a). High-fidelity CRISPR-Cas9 nucleases with no detectable genome-wide off-target effects. *Nature* 529: 490–495.

Kleinstiver, B.P., Tsai, S.Q., Prew, M.S. et al. (2016b). Genome-wide specificities of CRISPR-Cas Cpf1 nucleases in human cells. *Nat. Biotechnol.* 34: 869–874.

Kleinstiver, B.P., Sousa, A.A., Walton, R.T. et al. (2019). Engineered CRISPR-Cas12a variants with increased activities and improved targeting ranges for gene, epigenetic and base editing. *Nat. Biotechnol.* 37: 276–282.

Klompe, S.E., Vo, P.L.H., Halpin-Healy, T.S., and Sternberg, S.H. (2019). Transposon-encoded CRISPR-Cas systems direct RNA-guided DNA integration. *Nature* 571: 219–225.

Komor, A.C., Kim, Y.B., Packer, M.S. et al. (2016). Programmable editing of a target base in genomic DNA without double-stranded DNA cleavage. *Nature* 533: 420–424.

Konermann, S., Lotfy, P., Brideau, N.J. et al. (2018). Transcriptome engineering with RNA-targeting Type VI-D CRISPR effectors. *Cell* 173: 665–676. e14.

Koonin, E.V. and Makarova, K.S. (2019). Origins and evolution of CRISPR-Cas systems. *Philos. Trans. R. Soc. Lond. Ser. B Biol. Sci.* 374: 20180087.

Kosicki, M., Tomberg, K., and Bradley, A. (2018). Repair of double-strand breaks induced by CRISPR-Cas9 leads to large deletions and complex rearrangements. *Nat. Biotechnol.* 36: 765–771.

Kunin, V., Sorek, R., and Hugenholtz, P. (2007). Evolutionary conservation of sequence and secondary structures in CRISPR repeats. *Genome Biol.* 8: R61.

Lee, H., Zhou, Y., Taylor, D.W., and Sashital, D.G. (2018). Cas4-Dependent Prespacer Processing Ensures High-Fidelity Programming of CRISPR Arrays. *Mol. Cell* 70: 48–59. e5.

Lee, H., Dhingra, Y., and Sashital, D.G. (2019). The Cas4-Cas1-Cas2 complex mediates precise prespacer processing during CRISPR adaptation. *elife* 8.

Leenay, R.T., Maksimchuk, K.R., Slotkowski, R.A. et al. (2016). Identifying and visualizing functional PAM diversity across CRISPR-Cas systems. *Mol. Cell* 62: 137–147.

Levy, A., Goren, M.G., Yosef, I. et al. (2015). CRISPR adaptation biases explain preference for acquisition of foreign DNA. *Nature* 520: 505–510.

Li, Y., Pan, S., Zhang, Y. et al. (2016). Harnessing Type I and Type III CRISPR-Cas systems for genome editing. *Nucleic Acids Res.* 44: e34.

Li, S., Li, J., Zhang, J. et al. (2018a). Synthesis-dependent repair of Cpf1-induced double strand DNA breaks enables targeted gene replacement in rice. *J. Exp. Bot.* 69: 4715–4721.

Li, S.Y., Cheng, Q.X., Wang, J.M. et al. (2018b). CRISPR-Cas12a-assisted nucleic acid detection. *Cell Discov.* 4: 20.

Li, L., Li, S., Wu, N. et al. (2019a). HOLMESv2: a CRISPR-Cas12b-assisted platform for nucleic acid detection and DNA methylation quantitation. *ACS Synth. Biol.* 8: 2228–2237.

Li, Y., Li, S., Wang, J., and Liu, G. (2019b). CRISPR/Cas Systems towards Next-generation biosensing. *Trends Biotechnol.* 37: 730–743.

Lillestol, R.K., Redder, P., Garrett, R.A., and Brugger, K. (2006). A putative viral defence mechanism in archaeal cells. *Archaea* 2: 59–72.

Liu, X.S., Wu, H., Ji, X. et al. (2016). Editing DNA methylation in the mammalian genome. *Cell* 167: 233–247. e17.

Liu, L., Chen, P., Wang, M. et al. (2017a). C2c1-sgRNA complex structure reveals RNA-guided DNA cleavage mechanism. *Mol. Cell* 65: 310–322.

Liu, L., Li, X., Ma, J. et al. (2017b). The molecular architecture for RNA-guided RNA cleavage by Cas13a. *Cell* 170: 714–726. e10.

Liu, L., Li, X., Wang, J. et al. (2017c). Two distant catalytic sites are responsible for C2c2 RNase activities. *Cell* 168: 121–134. e12.

Liu, T.Y., Iavarone, A.T., and Doudna, J.A. (2017d). RNA and DNA targeting by a reconstituted thermus thermophilus Type III-A CRISPR-Cas system. *PLoS One* 12: e0170552.

Liu, J.J., Orlova, N., Oakes, B.L. et al. (2019a). CasX enzymes comprise a distinct family of RNA-guided genome editors. *Nature* 566: 218–223.

Liu, T.Y., Liu, J.J., Aditham, A.J. et al. (2019b). Target preference of Type III-A CRISPR-Cas complexes at the transcription bubble. *Nat. Commun.* 10: 3001.

Lovett, S.T. (2011). The DNA exonucleases of Escherichia coli. *EcoSal Plus* 4.

Ma, E., Harrington, L.B., O'connell, M.R. et al. (2015). Single-stranded DNA cleavage by Divergent CRISPR-Cas9 enzymes. *Mol. Cell* 60: 398–407.

Magadan, A.H., Dupuis, M.E., Villion, M., and Moineau, S. (2012). Cleavage of phage DNA by the streptococcus thermophilus CRISPR3-Cas system. *PLoS One* 7: e40913.

Maji, B., Moore, C.L., Zetsche, B. et al. (2017). Multidimensional chemical control of CRISPR-Cas9. *Nat. Chem. Biol.* 13: 9–11.

Makarova, K.S., Grishin, N.V., Shabalina, S.A. et al. (2006). A putative RNA-interference-based immune system in prokaryotes: computational analysis of the predicted enzymatic machinery, functional analogies with eukaryotic RNAi, and hypothetical mechanisms of action. *Biol. Direct* 1: 7.

Makarova, K.S., Wolf, Y.I., Iranzo, J. et al. (2019). Evolutionary classification of CRISPR–Cas systems: a burst of class 2 and derived variants. *Nat. Rev. Microbiol.* 18: 67–83.

Mali, P., Yang, L., Esvelt, K.M. et al. (2013). RNA-guided human genome engineering via Cas9. *Science* 339: 823–826.

Maresca, M., Lin, V.G., Guo, N., and Yang, Y. (2013). Obligate ligation-gated recombination (ObLiGaRe): custom-designed nuclease-mediated targeted integration through nonhomologous end joining. *Genome Res.* 23: 539–546.

Marino, N.D., Pinilla-Redondo, R., Csorgo, B., and Bondy-Denomy, J. (2020). Anti-CRISPR protein applications: natural brakes for CRISPR-Cas technologies. *Nat. Methods* 17: 471–479.

Marraffini, L.A. and Sontheimer, E.J. (2008). CRISPR interference limits horizontal gene transfer in staphylococci by targeting DNA. *Science* 322: 1843–1845.

Mcginn, J. and Marraffini, L.A. (2016). CRISPR-Cas systems optimize their immune response by specifying the site of spacer integration. *Mol. Cell* 64: 616–623.

Meeske, A.J., Nakandakari-Higa, S., and Marraffini, L.A. (2019). Cas13-induced cellular dormancy prevents the rise of CRISPR-resistant bacteriophage. *Nature* 570: 241–245.

Mekler, V., Minakhin, L., and Severinov, K. (2017). Mechanism of duplex DNA destabilization by RNA-guided Cas9 nuclease during target interrogation. *Proc. Natl. Acad. Sci. U. S. A.* 114: 5443–5448.

Miller, S.M., Wang, T., Randolph, P.B. et al. (2020). Continuous evolution of SpCas9 variants compatible with non-G PAMs. *Nat. Biotechnol.* 38: 471–481.

Ming, M., Ren, Q., Pan, C. et al. (2020). CRISPR-Cas12b enables efficient plant genome engineering. *Nat. Plants* 6: 202–208.

Mojica, F.J., Diez-Villasenor, C., Soria, E., and Juez, G. (2000). Biological significance of a family of regularly spaced repeats in the genomes of Archaea, Bacteria and mitochondria. *Mol. Microbiol.* 36: 244–246.

Mojica, F.J., Diez-Villasenor, C., Garcia-Martinez, J., and Soria, E. (2005). Intervening sequences of regularly spaced prokaryotic repeats derive from foreign genetic elements. *J. Mol. Evol.* 60: 174–182.

Mojica, F.J.M., Diez-Villasenor, C., Garcia-Martinez, J., and Almendros, C. (2009). Short motif sequences determine the targets of the prokaryotic CRISPR defence system. *Microbiol. (Reading)* 155: 733–740.

Moreno-Mateos, M.A., Fernandez, J.P., Rouet, R. et al. (2017). CRISPR-Cpf1 mediates efficient homology-directed repair and temperature-controlled genome editing. *Nat. Commun.* 8: 2024.

Morisaka, H., Yoshimi, K., Okuzaki, Y. et al. (2019). CRISPR-Cas3 induces broad and unidirectional genome editing in human cells. *Nat. Commun.* 10: 5302.

Mulepati, S., Heroux, A., and Bailey, S. (2014). Structural biology. Crystal structure of a CRISPR RNA-guided surveillance complex bound to a ssDNA target. *Science* 345: 1479–1484.

Muller, M., Lee, C.M., Gasiunas, G. et al. (2016). Streptococcus thermophilus CRISPR-Cas9 systems enable specific editing of the human genome. *Mol. Ther.* 24: 636–644.

Myhrvold, C., Freije, C.A., Gootenberg, J.S. et al. (2018). Field-deployable viral diagnostics using CRISPR-Cas13. *Science* 360: 444–448.

Nakata, A., Amemura, M., and Makino, K. (1989). Unusual nucleotide arrangement with repeated sequences in the Escherichia coli K-12 chromosome. *J. Bacteriol.* 171: 3553–3556.

Nam, K.I.H., Haitjema, C., Liu, X. et al. (2012). Cas5d protein processes pre-crRNA and assembles into a cascade-like interference complex in subtype I-C/Dvulg CRISPR-Cas system. *Structure* 20: 1574–1584.

Newire, E., Aydin, A., Juma, S. et al. (2020). Identification of a Type IV-A CRISPR-Cas system located exclusively on IncHI1B/IncFIB plasmids in enterobacteriaceae. *Front. Microbiol.* 11: 1937.

Niewoehner, O., Garcia-Doval, C., Rostol, J.T. et al. (2017). Type III CRISPR-Cas systems produce cyclic oligoadenylate second messengers. *Nature* 548: 543–548.

Nihongaki, Y., Kawano, F., Nakajima, T., and Sato, M. (2015). Photoactivatable CRISPR-Cas9 for optogenetic genome editing. *Nat. Biotechnol.* 33: 755–760.

Nishimasu, H., Shi, X., Ishiguro, S. et al. (2018). Engineered CRISPR-Cas9 nuclease with expanded targeting space. *Science* 361: 1259–1262.

Nunez, J.K., Kranzusch, P.J., Noeske, J. et al. (2014). Cas1-Cas2 complex formation mediates spacer acquisition during CRISPR-Cas adaptive immunity. *Nat. Struct. Mol. Biol.* 21: 528–534.

Nunez, J.K., Harrington, L.B., Kranzusch, P.J. et al. (2015). Foreign DNA capture during CRISPR-Cas adaptive immunity. *Nature* 527: 535–538.

Nuñez, J.K., Lee, A.S.Y., Engelman, A., and Doudna, J.A. (2015). Integrase-mediated spacer acquisition during CRISPR–Cas adaptive immunity. *Nature* 519: 193–198.

Nunez, J.K., Bai, L., Harrington, L.B. et al. (2016). CRISPR Immunological Memory Requires a Host Factor for Specificity. *Mol. Cell* 62: 824–833.

Nussenzweig, P.M. and Marraffini, L.A. (2020). Molecular mechanisms of CRISPR-Cas immunity in bacteria. *Annu. Rev. Genet.* 54: 93–120.

Nussenzweig, P.M., Mcginn, J., and Marraffini, L.A. (2019). Cas9 cleavage of viral genomes primes the acquisition of new immunological memories. *Cell Host Microbe* 26: 515–526. e6.

Osakabe, K., Wada, N., Miyaji, T. et al. (2020). Genome editing in plants using CRISPR type I-D nuclease. *Commun. Biol.* 3: 648.

Osawa, T., Inanaga, H., Sato, C., and Numata, T. (2015). Crystal structure of the CRISPR-Cas RNA silencing Cmr complex bound to a target analog. *Mol. Cell* 58: 418–430.

Ozcan, A., Pausch, P., Linden, A. et al. (2019). Type IV CRISPR RNA processing and effector complex formation in Aromatoleum aromaticum. *Nat. Microbiol.* 4: 89–96.

Patchsung, M., Jantarug, K., Pattama, A. et al. (2020). Clinical validation of a Cas13-based assay for the detection of SARS-CoV-2 RNA. *Nat. Biomed. Eng.* 4: 1140–1149.

Pausch, P., Al-Shayeb, B., Bisom-Rapp, E. et al. (2020). CRISPR-CasPhi from huge phages is a hypercompact genome editor. *Science* 369: 333–337.

Perez-Pinera, P., Kocak, D.D., Vockley, C.M. et al. (2013). RNA-guided gene activation by CRISPR-Cas9-based transcription factors. *Nat. Methods* 10: 973–976.

Pinilla-Redondo, R., Mayo-Munoz, D., Russel, J. et al. (2020). Type IV CRISPR-Cas systems are highly diverse and involved in competition between plasmids. *Nucleic Acids Res.* 48: 2000–2012.

Pougach, K., Semenova, E., Bogdanova, E. et al. (2010). Transcription, processing and function of CRISPR cassettes in Escherichia coli. *Mol. Microbiol.* 77: 1367–1379.

Pourcel, C., Salvignol, G., and Vergnaud, G. (2005). CRISPR elements in Yersinia pestis acquire new repeats by preferential uptake of bacteriophage DNA, and provide additional tools for evolutionary studies. *Microbiol. (Reading)* 151: 653–663.

Pul, U., Wurm, R., Arslan, Z. et al. (2010). Identification and characterization of E. coli CRISPR-cas promoters and their silencing by H-NS. *Mol. Microbiol.* 75: 1495–1512.

Pyenson, N.C., Gayvert, K., Varble, A. et al. (2017). Broad targeting specificity during bacterial Type III CRISPR-Cas immunity constrains viral escape. *Cell Host Microbe* 22: 343–353. e3.

Pyne, M.E., Bruder, M.R., MOO-Young, M. et al. (2016). Harnessing heterologous and endogenous CRISPR-Cas machineries for efficient markerless genome editing in Clostridium. *Sci. Rep.* 6: 25666.

Qi, L.S., Larson, M.H., Gilbert, L.A. et al. (2013). Repurposing CRISPR as an RNA-guided platform for sequence-specific control of gene expression. *Cell* 152: 1173–1183.

Ramachandran, A., Summerville, L., Learn, B.A. et al. (2020). Processing and integration of functionally oriented prespacers in the Escherichia coli CRISPR system depends on bacterial host exonucleases. *J. Biol. Chem.* 295: 3403–3414.

Ran, F.A., Cong, L., Yan, W.X. et al. (2015). in vivo genome editing using Staphylococcus aureus Cas9. *Nature* 520: 186–191.

Raper, A.T., Stephenson, A.A., and Suo, Z. (2018). Functional insights revealed by the kinetic mechanism of CRISPR/Cas9. *J. Am. Chem. Soc.* 140: 2971–2984.

Redding, S., Sternberg, S.H., Marshall, M. et al. (2015). Surveillance and Processing of Foreign DNA by the Escherichia coli CRISPR-Cas System. *Cell* 163: 854–865.

Roberts, R.J. (2005). How restriction enzymes became the workhorses of molecular biology. *Proc. Natl. Acad. Sci. U. S. A.* 102: 5905–5908.

Rollie, C., Graham, S., Rouillon, C., and White, M.F. (2018). Prespacer processing and specific integration in a Type I-A CRISPR system. *Nucleic Acids Res.* 46: 1007–1020.

Rostol, J.T. and Marraffini, L.A. (2019). Non-specific degradation of transcripts promotes plasmid clearance during type III-A CRISPR-Cas immunity. *Nat. Microbiol.* 4: 656–662.

Samai, P., Pyenson, N., Jiang, W. et al. (2015). Co-transcriptional DNA and RNA cleavage during Type III CRISPR-Cas immunity. *Cell* 161: 1164–1174.

Sampson, T.R., Saroj, S.D., Llewellyn, A.C. et al. (2013). A CRISPR/Cas system mediates bacterial innate immune evasion and virulence. *Nature* 497: 254–257.

Sashital, D.G., Jinek, M., and Doudna, J.A. (2011). An RNA-induced conformational change required for CRISPR RNA cleavage by the endoribonuclease Cse3. *Nat. Struct. Mol. Biol.* 18: 680–687.

Sashital, D.G., Wiedenheft, B., and Doudna, J.A. (2012). Mechanism of foreign DNA selection in a bacterial adaptive immune system. *Mol. Cell* 46: 606–615.

Savitskaya, E., Semenova, E., Dedkov, V. et al. (2013). High-throughput analysis of type I-E CRISPR/Cas spacer acquisition in E. coli. *RNA Biol.* 10: 716–725.

Schmid-Burgk, J.L., Gao, L., Li, D. et al. (2020). Highly parallel profiling of Cas9 variant specificity. *Mol. Cell* 78: 794–800. e8.

Sefcikova, J., Roth, M., Yu, G., and Li, H. (2017). Cas6 processes tight and relaxed repeat RNA via multiple mechanisms: a hypothesis. *BioEssays* 39.

Shah, S.A., Erdmann, S., Mojica, F.J., and Garrett, R.A. (2013). Protospacer recognition motifs: mixed identities and functional diversity. *RNA Biol.* 10: 891–899.

Shao, Y., Richter, H., Sun, S. et al. (2016). A non-stem-loop CRISPR RNA is processed by dual binding Cas6. *Structure* 24: 547–554.

Shiimori, M., Garrett, S.C., Graveley, B.R., and Terns, M.P. (2018). Cas4 nucleases define the PAM, length, and orientation of DNA fragments integrated at CRISPR loci. *Mol. Cell* 70: 814–824. e6.

Shmakov, S., Abudayyeh, O.O., Makarova, K.S. et al. (2015). Discovery and functional characterization of diverse class 2 CRISPR-Cas systems. *Mol. Cell* 60: 385–397.

Shmakov, S., Smargon, A., Scott, D. et al. (2017). Diversity and evolution of class 2 CRISPR-Cas systems. *Nat. Rev. Microbiol.* 15: 169–182.

Silas, S., Mohr, G., Sidote, D.J. et al. (2016). Direct CRISPR spacer acquisition from RNA by a natural reverse transcriptase-Cas1 fusion protein. *Science* 351: aad4234–aad4234.

Singh, D., Mallon, J., Poddar, A. et al. (2018). Real-time observation of DNA target interrogation and product release by the RNA-guided endonuclease CRISPR Cpf1 (Cas12a). *Proc. Natl. Acad. Sci. U. S. A.* 115: 5444–5449.

Sinkunas, T., Gasiunas, G., Fremaux, C. et al. (2011). Cas3 is a single-stranded DNA nuclease and ATP-dependent helicase in the CRISPR/Cas immune system. *EMBO J.* 30: 1335–1342.

Sinkunas, T., Gasiunas, G., Waghmare, S.P. et al. (2013). in vitro reconstitution of Cascade-mediated CRISPR immunity in Streptococcus thermophilus. *EMBO J.* 32: 385–394.

Slaymaker, I.M., Gao, L., Zetsche, B. et al. (2016). Rationally engineered Cas9 nucleases with improved specificity. *Science* 351: 84–88.

Slaymaker, I.M., Mesa, P., Kellner, M.J. et al. (2019). High-Resolution Structure of Cas13b and Biochemical Characterization of RNA Targeting and Cleavage. *Cell Rep.* 26: 3741–3751. e5.

Smargon, A.A., Cox, D.B.T., Pyzocha, N.K. et al. (2017). Cas13b Is a Type VI-B CRISPR-associated RNA-guided RNase differentially regulated by accessory proteins Csx27 and Csx28. *Mol. Cell* 65: 618–630. e7.

Staals, R.H., Zhu, Y., Taylor, D.W. et al. (2014). RNA targeting by the type III-A CRISPR-Cas Csm complex of thermus thermophilus. *Mol. Cell* 56: 518–530.

Staals, R.H., Jackson, S.A., Biswas, A. et al. (2016). Interference-driven spacer acquisition is dominant over naive and primed adaptation in a native CRISPR-Cas system. *Nat. Commun.* 7: 12853.

Stella, S., Mesa, P., Thomsen, J. et al. (2018). Conformational activation promotes CRISPR-Cas12a catalysis and resetting of the endonuclease activity. *Cell* 175: 1856–1871. e21.

Sternberg, S.H., Redding, S., Jinek, M. et al. (2014). DNA interrogation by the CRISPR RNA-guided endonuclease Cas9. *Nature* 507: 62–67.

Sternberg, S.H., Lafrance, B., Kaplan, M., and Doudna, J.A. (2015). Conformational control of DNA target cleavage by CRISPR-Cas9. *Nature* 527: 110–113.

Strecker, J., Jones, S., Koopal, B. et al. (2019a). Engineering of CRISPR-Cas12b for human genome editing. *Nat. Commun.* 10: 212.

Strecker, J., Ladha, A., Gardner, Z. et al. (2019b). RNA-guided DNA insertion with CRISPR-associated transposases. *Science* 365: 48–53.

Stringer, A.M., Cooper, L.A., Kadaba, S. et al. (2020). Characterization of primed adaptation in the Escherichia coli type I-E CRISPR-Cas system. *bioRxiv*.

Strutt, S.C., Torrez, R.M., Kaya, E. et al. (2018). RNA-dependent RNA targeting by CRISPR-Cas9. *elife* 7.

Swarts, D.C., Mosterd, C., Van Passel, M.W., and Brouns, S.J. (2012). CRISPR interference directs strand specific spacer acquisition. *PLoS One* 7: e35888.

Swarts, D.C., Van Der Oost, J., and Jinek, M. (2017). Structural basis for guide RNA processing and seed-dependent DNA targeting by CRISPR-Cas12a. *Mol. Cell* 66: 221–233. e4.

Tambe, A., East-Seletsky, A., Knott, G.J. et al. (2018). RNA binding and HEPN-nuclease activation are decoupled in CRISPR-Cas13a. *Cell Rep.* 24: 1025–1036.

Tamulaitis, G., Kazlauskiene, M., Manakova, E. et al. (2014). Programmable RNA shredding by the Type III-A CRISPR-Cas system of Streptococcus thermophilus. *Mol. Cell* 56: 506–517.

Tanenbaum, M.E., Gilbert, L.A., Qi, L.S. et al. (2014). A protein-tagging system for signal amplification in gene expression and fluorescence imaging. *Cell* 159: 635–646.

Taylor, D.W., Zhu, Y., Staals, R.H. et al. (2015). Structures of the CRISPR-Cmr complex reveal mode of RNA target positioning. *Science* 348: 581–585.

Teng, F., Cui, T., Feng, G. et al. (2018). Repurposing CRISPR-Cas12b for mammalian genome engineering. *Cell Discov.* 4: 63.

Teng, F., Cui, T., Gao, Q. et al. (2019a). Artificial sgRNAs engineered for genome editing with new Cas12b orthologs. *Cell Discov.* 5: 23.

Teng, F., Guo, L., Cui, T. et al. (2019b). CDetection: CRISPR-Cas12b-based DNA detection with sub-attomolar sensitivity and single-base specificity. *Genome Biol.* 20: 132.

Teng, F., Li, J., Cui, T. et al. (2019c). Enhanced mammalian genome editing by new Cas12a orthologs with optimized crRNA scaffolds. *Genome Biol.* 20: 15.

Toro, N., Mestre, M.R., Martinez-Abarca, F., and Gonzalez-Delgado, A. (2019). Recruitment of reverse transcriptase-Cas1 fusion proteins by Type VI-A CRISPR-Cas systems. *Front. Microbiol.* 10: 2160.

Touchon, M. and Rocha, E.P. (2010). The small, slow and specialized CRISPR and anti-CRISPR of escherichia and salmonella. *PLoS One* 5: e11126.

Van Houte, S., Ekroth, A.K., Broniewski, J.M. et al. (2016). The diversity-generating benefits of a prokaryotic adaptive immune system. *Nature* 532: 385–388.

Vo, P.L.H., Ronda, C., Klompe, S.E. et al. (2021). CRISPR RNA-guided integrases for high-efficiency, multiplexed bacterial genome engineering. *Nat. Biotechnol.* 39: 480–489.

Vojta, A., Dobrinic, P., Tadic, V. et al. (2016). Repurposing the CRISPR-Cas9 system for targeted DNA methylation. *Nucleic Acids Res.* 44: 5615–5628.

Walton, R.T., Christie, K.A., Whittaker, M.N., and Kleinstiver, B.P. (2020). Unconstrained genome targeting with near-PAMless engineered CRISPR-Cas9 variants. *Science* 368: 290–296.

Wang, J., Li, J., Zhao, H. et al. (2015). Structural and mechanistic basis of PAM-dependent spacer acquisition in CRISPR-Cas systems. *Cell* 163: 840–853.

Wang, A.S., Chen, L.C., Wu, R.A. et al. (2020). The histone chaperone FACT induces Cas9 multi-turnover behavior and modifies genome manipulation in human cells. *Mol. Cell* 79: 221–233. e5.

Wei, Y., Chesne, M.T., Terns, R.M., and Terns, M.P. (2015). Sequences spanning the leader-repeat junction mediate CRISPR adaptation to phage in Streptococcus thermophilus. *Nucleic Acids Res.* 43: 1749–1758.

Weinberger, A.D., Sun, C.L., Plucinski, M.M. et al. (2012). Persisting viral sequences shape microbial CRISPR-based immunity. *PLoS Comput. Biol.* 8: e1002475.

Wilkinson, M., Drabavicius, G., Silanskas, A. et al. (2019). Structure of the DNA-bound spacer capture complex of a Type II CRISPR-Cas system. *Mol. Cell* 75: 90–101. e5.

Wright, A.V., Liu, J.-J., Knott, G.J. et al. (2017). Structures of the CRISPR genome integration complex. *Science* 357: 1113–1118.

Xiao, Y., Luo, M., Hayes, R.P. et al. (2017). Structure basis for directional R-loop formation and substrate handover mechanisms in Type I CRISPR-Cas system. *Cell* 170: 48–60. e11.

Xu, X. and Qi, L.S. (2019). A CRISPR-dCas toolbox for genetic engineering and synthetic biology. *J. Mol. Biol.* 431: 34–47.

Xu, Z., Li, M., Li, Y. et al. (2019). Native CRISPR-Cas-mediated genome editing enables dissecting and sensitizing clinical multidrug-resistant P. aeruginosa. *Cell Rep.* 29: 1707–1717. e3.

Xue, C., Whitis, N.R., and Sashital, D.G. (2016). Conformational control of cascade interference and priming activities in CRISPR immunity. *Mol. Cell* 64: 826–834.

Yan, W.X., Hunnewell, P., Alfonse, L.E. et al. (2019). Functionally diverse type V CRISPR-Cas systems. *Science* 363: 88–91.

Yourik, P., Fuchs, R.T., Mabuchi, M. et al. (2019). Staphylococcus aureus Cas9 is a multiple-turnover enzyme. *RNA* 25: 35–44.

Zalatan, J.G., Lee, M.E., Almeida, R. et al. (2015). Engineering complex synthetic transcriptional programs with CRISPR RNA scaffolds. *Cell* 160: 339–350.

Zetsche, B., Gootenberg, J.S., Abudayyeh, O.O. et al. (2015). Cpf1 is a single RNA-guided endonuclease of a class 2 CRISPR-Cas system. *Cell* 163: 759–771.

Zetsche, B., Heidenreich, M., Mohanraju, P. et al. (2017). Multiplex gene editing by CRISPR-Cpf1 using a single crRNA array. *Nat. Biotechnol.* 35: 31–34.

Zetsche, B., Abudayyeh, O.O., Gootenberg, J.S. et al. (2020). A Survey of genome editing activity for 16 Cas12a orthologs. *Keio J. Med.* 69: 59–65.

Zhang, Y., Heidrich, N., Ampattu, B.J. et al. (2013). Processing-independent CRISPR RNAs limit natural transformation in Neisseria meningitidis. *Mol. Cell* 50: 488–503.

Zhang, B., Ye, W., Ye, Y. et al. (2018). Structural insights into Cas13b-guided CRISPR RNA maturation and recognition. *Cell Res.* 28: 1198–1201.

Zhou, Y., Bravo, J.P.K., Taylor, H.N. et al. (2020). Structure of a type IV CRISPR-Cas effector complex. *bioRxiv*.

4

Commercially Available Reagents and Contract Research Services for CRISPR-Based Studies

Klio Maratou[1], Aaron T. Cheng[2], Fiona M. Behan[1], Ning Sun[2], and Quinn Lu[3]

[1] *Functional Genomics, R&D GlaxoSmithKline, Stevenage, UK*
[2] *Functional Genomics, R&D GlaxoSmithKline, Upper Providence, PA, USA*
[3] *Novel Human Genetics Research Unit, R&D GlaxoSmithKline, Upper Providence, PA, USA*

4.1 Introduction

CRISPR genome editing technologies provide versatile tools for the genetic manipulation and screening of genes and pathways in mammalian cells and in model animals. Their applications in drug discovery are broad, including target discovery, target validation, mechanism of action, and target engagement studies (Lu et al. 2017). Since first described, many improvements and novel applications of the technology have been reported. The technologies include gene knock out (KO) via non-homologous end joining (NHEJ) following CRISPR-mediated double-stranded break (DSB), gene knock in (KI) for SNP/mutation generation or gene insertion/gene tagging via homology-directed repair (HDR) of DSB, and transcriptional or epigenetic regulation via fusion of an inactivated CRISPR Cas protein ("endonuclease dead" dCas9) with a transcriptional activator (CRISPR activation, CRISPRa) for upregulation of gene expression, with a transcriptional repressor (CRISPR interference, CRISPRi) for downregulation of gene expression, or with an epigenetic modifying enzyme (CRISPRepi) for DNA methylation or post-translational modifications of histone proteins. While these platforms were largely developed by academic labs, their rapid implementation in biomedical and pharmaceutical research settings has been greatly facilitated by reagent providers and contract research organizations (CROs) that offer state-of-the-art reagents and contract services. In fact, it is now routine practice for scientists to use commercially available CRISPR reagents for in-house studies and/or to consider working with a CRO on a specific genome editing project. This article aims to summarize our learnings and provide guidance on the selection of appropriate reagents and/or CROs for CRISPR-based studies.

4.2 CRISPR Resources and Reagents for Bespoke Editing and Genetic Screening

4.2.1 Publicly Available Resources

CRISPR technologies have been quoted as being the "Swiss army knife" equivalent for the science world, due to the ease of their design and generation – as compared with zinc-finger nucleases (ZFNs) and transcription activator-like effector nucleases (TALENS) – and the diversity of their applications (Doench 2018; Mans et al. 2015). Furthermore, the "open-source" culture of the CRISPR field has helped accelerate its continued innovation (Zhang 2019). This has been facilitated by numerous resources like web-based tutorials, publicly available webinars, databases for data deposition, and annual CRISPR meetings. For example, a freely available resource called the Open Repository for CRISPR Screens (ORCS) (https://orcs.thebiogrid.org) was developed by the Biological General Repository for Interaction Datasets (BioGRID) (Oughtred et al. 2019), which enables researchers to search, filter, and download CRISPR screen datasets. Version 1.0.3 of the ORCS contains 895 CRISPR screens from 3 major model organism species and 629 cell lines. In addition, when novel methodologies are published, the associated constructs often become available through repositories like Addgene, an international nonprofit repository (Kamens 2015), to which over 350 labs have contributed. Reagent sharing has the advantage that it enables vigorous testing of the new technology and encourages rapid further improvements. However, Addgene is not accessible to for-profit organizations; thus, for industry/biotech institutions, the only access to reagents is via commercial providers.

Genome Editing in Drug Discovery, First Edition. Edited by Marcello Maresca and Sumit Deswal.

There are many commercial providers of CRISPR reagents and major suppliers will be discussed further below. However, it is important to highlight that for pharma and biotech companies, before any work can begin and reagents purchased, it is important to ensure that there is appropriate product use license agreement to allow freedom to operate. In the case of CRISPR, this poses challenges as the foundational CRISPR patents are still under dispute for the United States and Europe with Jennifer Doudna of the University of California, Berkeley, and Emmanuelle Charpentier (formerly of the University of Vienna) competing for patent rights against Feng Zhang of the Broad Institute of MIT and Harvard University (Cohen 2019). Moreover, according to a recent patent review, beyond the foundational patents, 2072 additional patent families associated with CRISPR were filed up to a 31 December 2017 priority date (Martin-Laffon et al. 2019).

4.2.2 Cas9 Enzymes

Commercial providers have developed tools and solutions for every step of the CRISPR genome editing workflow and these are listed in Table 4.1. The most popular form of editing involves use of the *Streptococcus pyogenes* Cas9 (SpCas9) enzyme, which can be delivered to cells in one of five formats: (i) as protein (in complex with gRNA), (ii) expressed through a plasmid expression vector, (iii) via lentiviral particles, (iv) via AAV particles, or (v) via mRNA. The format of choice depends on the cell type and type of experiment that is to be performed. For example, before conducting screening experiments in cancer cell lines, one needs to generate cells stably expressing Cas9. In this case, Cas9 lentiviral particles are the format of choice, as they facilitate Cas9 integration into the genome, which allows for stable, long-term expression of Cas9 driven by promoter sequences contained within the vector. Multiple lentiviral construct options are available. For example, Horizon Discovery offers the choice of CMV, EF1a, PGK, and CAG promoters. Also, Merck offers a range of antibiotic resistance or fluorescent reporters in their lentiviruses (blasticidin, hygromycin, neomycin, zeocin, or GFP/RFP fluorophores). Other providers like Takara Bio offer the choice between constitutive or Tet-inducible Cas9 expression. Therefore, the choice of vendor may depend on the type of construct that best suits your experimental design. In addition, if the desired construct is not available off-the-shelf, several companies offer custom vector generation (Table 4.1), thus there is a lot of flexibility. However, there is a trade-off. Off-the-shelf products arrive quickly and tend to be less expensive,

Table 4.1 Major providers of CRISPR reagents.

Company	Location	Reagents
Agilent	Santa Clara, CA	Oligo library synthesis; synthetic gRNA and libraries
Cellecta	Mountain View, CA	Cas9 and gRNA expression vectors, lentiviral gRNA libraries
GenScript	Jiangning, China	Cas9 nuclease; Cas9 and gRNA expression vectors; lentiviral gRNA libraries
Horizon Discovery	Cambridge, United Kingdom	Cas9 nucleases; Cas9 and gRNA expression vectors; synthetic, plasmid, and lentiviral gRNA and libraries
Integrated DNA Technologies	Coralville, IA	Oligo library synthesis; Cas9 nucleases; Cpf1 nuclease; Cas9 and gRNA expression vectors; synthetic gRNA libraries
Merck KGaA	Darmstadt, Germany	Cas9 nucleases; Cas9 and gRNA expression vectors; synthetic, plasmid, and lentiviral gRNA and libraries
New England Biolabs	Ipswich, MA	Cas9 nucleases
Oxgene	Oxford, United Kingdom	Plasmid and lentiviral gRNA libraries
Synthego	Redwood City, CA	Synthetic gRNA and libraries
System Biosciences	Palo Alto, CA	Cas9 and gRNA expression vectors
Takara Bio	Mountain View, CA	Cas9 nucleases; Cas9 and gRNA expression vectors; synthetic, plasmid, and lentiviral gRNA and libraries
ThermoFisher	Carlsbad, CA	Cas9 nucleases; Cas9 and gRNA expression vectors; synthetic and lentiviral gRNA and libraries
Transomic Technologies	Huntsville, AL	Cas9 and gRNA expression vectors; plasmid and lentiviral gRNA libraries
Twist Bioscience	San Francisco, CA	Oligo library synthesis

whereas custom reagents need time for production and will typically cost more.

For many editing experiments including one-off editing and gene KO for arrayed screening, use of Cas9/gRNA ribonucleoprotein complex (RNP) is preferred, as it has been reported to demonstrate higher efficacy as compared with other delivery formats, together with simpler protocols and reduced toxicity, because RNP pipelines do not require delivery of foreign DNA, or enrichment/selection. Moreover, use of RNP has been reported to help reduce off-target effects (OTEs), because unlike Cas9 delivered as plasmid or lentivirus which remains expressed beyond 72hours, Cas9 introduced as RNP is cleared rapidly by 48hours (Kim et al. 2014; Liang et al. 2015). For therapeutic genome editing applications, several improved Cas9 enzymes have been described which improve on-target editing specificity and/or reduce off-target editing. Two of these, eSpCas9 (Slaymaker et al. 2016) and high-fidelity (HiFi) Cas9 (Vakulskas et al. 2018) are commercially available through Merck and IDT.

4.2.3 Cas9 Alternatives

Although SpCas9 is the most popular nuclease, Cas enzymes derived from other bacterial species have been described for editing applications. Commercially available alternatives to SpCas9 include the *Acidaminococcus* and *Lachnospiraceae* Cas12a (also known as Cpf1), from IDT and NEB respectively (Zetsche et al. 2015), the *Staphylococcus aureus* SaCas9, from Takara Bio (Ran et al. 2015), and the *Francisella novicida* FnCas9 from Merck (Acharya et al. 2019). The primary difference between Cas9 nucleases derived from different bacteria is in the protospacer adjacent motif (PAM) sequence that they require for binding and cleavage. For example, SaCas9 recognizes a longer PAM, 5'-NNGRRT-3', compared with 5'-NGG-3' for SpCas9. In addition, SaCas9 is about 1 kb smaller in size than SpCas9, so it can be packaged into viral vectors more easily, offering possibilities for non-integrative AAV-mediated gene therapy (De Caneva et al. 2019; Ginn et al. 2020).

Cas12a has recently emerged as an interesting alternative to SpCas9, due to its ability to target T-rich motifs with the PAM, typically 5′-TTTV-3′, located upstream of the spacer. This makes Cas12a attractive as an epigenome editing platform, because it can target regions around transcription starting sites, which are inaccessible to SpCas9 (Tak et al. 2017). Although its potential in human research has yet to be fully realized, Cas12a has shown remarkable versatility in genome editing across a range of model organisms, including mice, porcine (female embryos), frogs (Xenopus), zebrafish, bacteria, and plants (Safari et al. 2019).

4.2.4 Guide RNA Formats and Reagents

Guide RNA (gRNA) is an essential part of the CRISPR system, as it serves to direct the Cas nuclease to specific genomic locations defined by its complementation to the target DNA sequence of interest. In the case of SpCas9, the gRNA is composed of two parts: a) a 20-nucleotide sequence complementary to the target DNA named CRISPR RNA (crRNA), and b) a 67-nucleotide sequence which serves as a binding scaffold for the Cas nuclease, named trans-activating CRISPR RNA (tracrRNA). These two essential gRNA components can be kept separate as two-piece reagents, or manufactured as a simpler alternative that combines both the crRNA and tracrRNA elements into a chimeric single-guide RNA molecule (sgRNA).

Guide RNA design is critical to achieving efficient gene knockout. In the first few years after the emergence of CRISPR, multiple groups studied gRNA design and found that while many gRNAs will cut on-target with a reasonably high rate, a substantial portion will produce a low or zero cutting rate, or alternatively bind promiscuously in the genome, which can lead to off-target mutagenesis (Fu et al. 2013; Kim et al. 2019; Wienert et al. 2019). To address these issues, research focused on identifying the sequence and structural features that contribute to effective (and ineffective) gRNAs has led to noticeable improvements to the system (Filippova et al. 2019; Wu and Yin 2019). The production and use of chemically modified gRNAs, which are more resistant to degradation by cellular RNases, is now the norm, with different providers offering their own proprietary modification solutions. Major providers of gRNAs, for bespoke or library screening applications, are listed in Table 4.1.

As in the case of the CRISPR enzymes, there is high flexibility with regard to gRNA formats, and reagents can be purchased as synthetic single- or two-component gRNA for use with Cas9 RNP complexes, or within a plasmid or lentiviral vector backbone usually expressed under the control of a human U6 promoter. Determining the most appropriate type of gRNA for your experiment will depend on your particular application and cell type. As a general rule of thumb, for single-target KOs, we recommend the RNP approach and in our hands single- and two-component gRNAs appear to work equally well. Choice of format is more dependent on delivery time, sample handling automation, and cost considerations. Selecting good-quality gRNAs is critical for the success of your CRISPR experiments and now all major commercial providers offer predesigned gRNAs (single or pools). They have spent years optimizing their designs with proprietary algorithms and extensive internal R&D testing, so their gRNAs generally work well for simple gene KOs.

It is still best practice though to purchase a minimum of 4 gRNAs and test their editing efficiencies before selecting the best two or three for your actual experiment, as demonstrated by Martufi and colleagues (Martufi et al. 2019). In addition, for gene KO experiments exploring target biology, to achieve a high level of editing we would recommend a strategy described by Seki and Rutz (Seki and Rutz 2018), where a combination of two or three gRNAs against the same target are added in one transfection. The rationale behind this is that it increases the chances of knocking down protein expression efficiently, as gene editing will occur in multiple regions of the gene in addition to the indel mutations induced by each gRNA. This strategy has been adopted by Synthego and Horizon Discovery as part of their arrayed CRISPR libraries portfolio and has the added benefit of alleviating the need for testing multiple gRNAs individually. Joberty and colleagues further demonstrated that if gRNA pairs target sequences in close proximity (40–300bp apart), then they work in synergy and edit close to 100% of the targeted alleles. They hypothesize that the binding of the most efficient (driver) gRNA (in RNP complex format) alters the local chromatin context, which in turn helps recruit a less efficient (helper) gRNA in its vicinity (Joberty et al. 2020).

For situations where a gRNA needs to be positioned to a particular genomic location, for example, to enable KIs, or to edit genomes other than human or mouse, custom gRNA design is still required. An abundance of freely available computational tools have been generated by academic institutions, although some, like the popular CHOPCHOP (Labun et al. 2019), are reserved for nonprofit and academic use only. All major commercial providers also offer custom gRNA design tools at their websites. To assist researchers in making better-informed decisions, several recent publications have sought to benchmark gRNA design and make recommendations on the best tools available (Bradford and Perrin 2019a; Bradford and Perrin 2019b; Liu et al. 2020). Nevertheless, these studies advise caution in choosing the right tool as experimental datasets used to build models for predicting gRNA specificity or efficiency are disparate. Moreover, some tools are specific to a particular organism or genome build.

4.2.5 CRISPR Libraries: CRISPR KO, CRISPRa, CRISPRi

There are several critical decisions to consider when choosing reagents from commercial vendors for CRISPR screening applications. The execution of screens is logistically challenging, and the scale of resources required (cost, time, expertise, staffing) is substantially greater than single-gene perturbation experiments. Key factors to consider when designing genome-wide CRISPR screens are summarized by the following formula describing the multiplier effect of Model × Assay × Perturbation (M × A × P). With the biological question under investigation in mind, the correct combination of these factors is critical for the success of the screen. The following section focuses mainly on the key considerations for sourcing reagents for the "Perturbation" component of this formula.

The most fundamental question to be answered is which flavor of CRISPR technology is most appropriate; CRISPRko, CRISPRi, or CRISPRa? As a general rule of thumb, CRISPRko generates truly null alleles and will link loss of a target to a particular phenotype. It is applicable for the majority of screens. Conversely, CRISPRi and CRISPRa may be better suited to studying genes that manifest different phenotypes at different gene doses and essential genes that cannot tolerate complete knockout or to examine noncoding regions (Borys and Younger 2020; Doench 2018; le Sage et al. 2020). The selection of an appropriate fusion domain to dead Cas9 is an important consideration when using CRISPRa as there are several options (e.g. SAM or VP64) which may need to be tested for activity in your cellular model of interest (Doench 2018; Sanson et al. 2018).

Pooled genome-wide libraries are available for all three applications. CRISPRko libraries are the most mature, as they have been most widely used by the community to date. CRISPRko libraries are still undergoing rapid evolution, with improvements in performance of the most recent KO libraries over their predecessors being equal to the level of improvement that was observed between the first CRISPR libraries when compared with RNAi screens (Sanson et al. 2018). The number of gRNAs in a pooled library targeting each gene is an important consideration. Initially, genome-wide libraries were designed with 4–6 gRNAs targeting each gene (Doench et al. 2016; Tzelepis et al. 2016) with the Toronto Knock-Out library composed of 12 gRNAs per gene (Hart et al. 2015). The total size of the library is extremely important as in order to generate high-quality data, adequate "coverage" must be maintained at all times throughout the experimental process. "Coverage" refers to the number of cells that contain each individual gRNA and is generally denoted as 100× or 500× to represent that each gRNA is present in 100 or 500 cells within the pool. Based on the Poisson distribution, aiming for 30% transduction efficiency results in ~1 gRNA vector entering each cell. As an example, if your genome-wide library is 100 000 vectors total (20 000 genes with 5 gRNA per gene) and you are aiming for 500× coverage with 30% transduction efficiency, you must transduce 165 million cells for each replicate of the experiment (multiple replicates are recommended). At each stage of the process, it is important to balance logistical considerations with generating the best-quality data. For many cell models, it would not be possible to acquire a large number of cells (e.g. 3D organoids, primary cells), so a decision is necessary regarding decreasing coverage or using a smaller library. A smaller library can be obtained by designing a bespoke, focused gRNA set, but if genome-wide scale is required, minimal libraries have been designed with only 2 gRNAs targeting each gene. In generating such libraries, two highly efficient gRNAs are selected for each gene based on experimental data, reducing the library size to <40 000 gRNAs resulting in a significant reduction of cell requirements (Goncalves et al. 2020).

Dual or multi-gRNA libraries are commonly used in genetic screens to achieve KO of two genes simultaneously. Dual CRISPRi libraries have been successfully used to create genetic interaction maps (Horlbeck et al. 2018). There are several methods to deliver multiple gRNAs simultaneously to a cell including use of "Big Papi" (Najm et al. 2018), Cas12a (Sanson et al. 2020), dual promoter (Erard et al. 2017), and single promoter with tRNA separator strategies (Zhao et al. 2019). In addition, CRISPRa and CRISPRko have been successfully delivered simultaneously to cells allowing for incredible versatility for experimentally fine-tuning genetic perturbations to determine effects on phenotypes of interest (Boettcher et al. 2018).

All technologies are also available in arrayed formats. Pooled libraries are generally applicable in relatively simple assay systems, for example, where gRNA depletion is used as a surrogate for cell death resulting from KO of the target gene (Behan et al. 2019), whereas a major advantage of arrayed screening approaches is that they are much more amenable to multiple read outs, such as multiplex bead-based cytokine assays, flow cytometry, or high content imaging (de Groot et al. 2018; Metzakopian et al. 2017; Strezoska et al. 2017). Moreover, arrayed screens are more suitable to screening of primary cells, if a large-enough number of cells can be cultured. While an arrayed approach has the potential to give more detailed insight into the phenotypic changes associated with the perturbation, execution of arrayed screens at scale requires specialized liquid handling equipment and the ability to manage, process, and analyze large amounts of data. Another disadvantage of arrayed screening is the substantial increase in cost relating to all aspects of the process, from infrastructure to reagents and consumables to data management. This increased cost limits the scale and throughput potential of arrayed screens as genome-wide screens would be too large; focused, bespoke libraries are more suited to this format. The trade-off required when choosing between arrayed and pooled screens may not be an issue for much longer as encouraging work emerging from the Blainey lab indicates that imaging-based readouts of pooled screens is possible, however not yet at scale (Feldman et al. 2019).

4.3 *In vivo* CRISPR Screening

4.3.1 Pooled *In vivo* CRISPR Screening in Rodent Models

While most CRISPR screens to date have been conducted in cell-based *in vitro* models, *in vivo* CRISPR screens have the advantage of directly interrogating gene functions and revealing molecular mechanisms in the native context of animals. In 2015, the first *in vivo* CRISPR screen to identify loss-of-function mutations that promote tumor growth and metastasis was published (Chen et al. 2015). In this study, the authors mutagenized a nonmetastatic mouse cancer cell line using a genome-wide gRNA library and the resulting pooled cell library was subcutaneously transplanted into immunocompromised mice. By sequencing the enriched gRNAs in the late-stage primary tumors and lung metastases, they discovered and subsequently validated a small set of genes whose disruptions drove tumor growth and metastasis *in vivo*. Similar *in vivo* CRISPR screening approaches using xenograft mouse models have been applied to identify tumor suppressors (Katigbak et al. 2016; Song et al. 2017; Takeda et al. 2019), oncogenes (Kodama et al. 2017), and synthetic lethal drug targets (Yau et al. 2017). CRISPR screening has also been applied in syngeneic models, which have full murine immunity and comprehensive stroma, providing a more relevant setting to assess gene functions in tumor immunity and immunotherapy response. In 2017, Manguso and colleagues discovered previously unidentified cancer immunotherapy targets by transducing a focused library into mouse melanoma tumor cells and comparing the gRNA library representation in implanted tumors from immunotherapy-treated wild-type animals to tumors growing in Tcra−/− mice (Manguso et al. 2017). *In vivo* screening approaches have also been deployed for the identification of genes affecting T cell proliferation and antitumor function (Dong et al. 2019), metabolism-associated factors in T-cell-mediated antitumor immunity (Wei et al. 2019), membrane targets in CD8+ murine models of glioblastoma (Ye et al. 2019), and for the facile perturbation of innate and adaptive immune cells *in vivo* without *ex vivo* manipulation of these mature lineages (LaFleur et al. 2019).

4.3.2 Considerations for Practice of *in vivo* CRISPR Screening

There are practical considerations to bear in mind with regard to reagents and materials for the design and execution of *in vivo* CRISPR screens. Due to the limited number of cells that can be injected into each animal, it is challenging to perform genome-wide *In vivo* screens with appropriate library coverage. Therefore, most researchers have chosen to use a focused mini-pool gRNA library to ensure high screen quality. While off-the-shelf sub-pool gRNA libraries are available from major commercial vendors, custom-designed gRNA libraries can be constructed by CROs like Merck, Horizon Discovery, and GenScript. To make your own gRNA libraries, Twist Bioscience, CustomArray (now part of GenScript), Agilent, and IDT are the major suppliers for oligonucleotide library synthesis. It is easier to conduct pooled *in vivo* CRISPR screens using cells from transgenic mice constitutively expressing Cas9, as this bypasses the need for endonuclease delivery/selection into target cells. Several different Cas9 knock-in mice are available from The Jackson Laboratory, but licenses for research use must be obtained from the original inventors for industry/biotech institutions. T cell receptor transgenic mice from The Jackson Laboratory and other suppliers are commonly used for adoptive T cell transfer experiments, which allow for T-cell-based *in vivo* CRISPR screening. Major CROs that offer *in vivo* genetic screens are included in Table 4.2.

4.3.3 Next-Generation *in vivo* CRISPR Screening

The CRISPR toolkit for genome editing has been rapidly evolving and next-generation CRISPR technologies have been applied for genetic screening *in vivo*. For

Table 4.2 Genome editing service providers.

Company	Location	Technologies	Services
Applied Stem Cell	Milpitas, CA	CRISPR/Cas9	Cell engineering: cell lines, ES/iPSCs
ATCC	Manassas, VA	CRISPR/Cas9	Cell engineering: cell lines
BioGene	Shirley, NY	CRISPR/Cas9	Cell engineering: cell lines, iPSCs, primary cells; rodent models; plants models
Canopy Biosciences	St. Louis, MO	CRISPR/Cas9	Cell engineering: cell lines, ES/iPSCs
Cellecta	Mountain View, CA	CRISPR/Cas9	genetic screening
Cellular Dynamics International	Madison Wisconsin	CRISPR/Cas9	ES/iPSC engineering
Charles River Laboratories	Wilmington, MA	CRISPR/Cas9	Cell engineering; rodent models
Crown Bioscience	Santa Clara, CA	CRISPR/Cas9	Rodent models
DefiniGEN	Cambridge, UK	CRISPR/Cas9	iPSC engineering
Evotec	Hamburg, Germany	CRISPR/Cas9	*In vitro* genetic screening
genOway	Lyon, France	CRISPR/Cas9	Cell engineering: cell lines, rodent models
GenScript	Jiangning, China	CRISPR/Cas9	Bacterial engineering; cell line engineering
Horizon Discovery	Cambridge, United Kingdom	CRISPR/Cas9	Cell engineering: cell lines, ES/iPSCs; *in vitro* and *in vivo* genetic screening
Merck KGaA	Darmstadt, Germany	CRISPR/Cas9, ZFN	Cell engineering: cell lines, iPSCs, primary cells
Oxgene	Oxford, United Kingdom	CRISPR/Cas9	Cell line engineering
Synthego	Redwood City, CA	CRISPR/Cas9	Cell engineering: cell lines, iPSCs, primary cells; *in vitro* genetic screening
System Biosciences	Palo Alto, CA	CRISPR/Cas9	Cell line engineering
ThermoFisher	Carlsbad, CA	CRISPR/Cas9, TALEN	Cell line engineering
Ubigene	Guangzhou Science City, China	CRISPR/Cas9	Bacterial engineering, fungal engineering, cell line engineering
WuXi AppTec	Shanghai, China	CRISPR/Cas9	Rodent models; *in vitro* genetic screening

example, *in vivo* screens based on CRISPRa (Braun et al. 2016; Ebright et al. 2020), CRISPRi (Li et al. 2020), exon excision (Thomas et al. 2020), and Cas12a/Cpf1-mediated double gene knockout (Chow et al. 2019) have been reported. Most recently, pooled *in vivo* CRISPR screening has been combined with single-cell RNA-seq analysis so that the relationships between genotypes to *in vivo* phenotypes can be mapped at single-cell resolution (Giladi et al. 2018; Jaitin et al. 2016). We anticipate this approach to be coupled with diverse functional readouts across a broad spectrum of animal models and continue to transform biomedical research and drug discovery.

4.4 Working with Service Providers for Outsourcing CRISPR Studies

This section covers CRISPR outsourcing solutions, but the same principles apply to other forms of gene editing technologies like TALENs and ZFNs. Although the tools and methods for application of genome editing technologies are becoming more accessible, some labs may choose to outsource partial or entire experiments on a fee-for-service basis. Genome editing CROs have invested considerable efforts on internal development and have amassed great knowledge on how the technology works, which translates to a high competence to design genome editing projects and troubleshoot problems. In addition, CROs enable researchers to conduct research without the need to purchase and maintain expensive equipment (e.g. flow sorter, or next-generation sequencing instruments). CROs are also used when there is time pressure to deliver while internal resource is in short supply, or in cases where there is limited or no internal expertise. Such a business model has several advantages and disadvantages which need to be considered before proceeding with this option (Pichler and Turner 2007).

4.4.1 Critical Steps for Outsourcing

A number of companies and academic institutions offer genome editing services. Providers can be found through websites including Science Exchange (https://www.scienceexchange.com) and Biocompare (https://www.biocompare.com) and we provide a non-exhaustive list in Table 4.2. It is important to base your choice on the reputation and track record of the company, otherwise you run the risk of contracting a CRO that will use your project for its own capability development (Pichler and Turner 2007). If possible, it is advisable to visit the CRO facility to inspect their laboratory set-up, view the equipment used, and discuss case studies that the company has completed. Also, you will need to perform due diligence checks. For example, discuss if the CRO holds foundational CRISPR licenses and understand their workflows with regard to data integrity (i.e. how they record and track their data). The next step is to set up a Confidential Disclosure Agreement (CDA) so that you are able to freely exchange information about the project you want to conduct and codevelop a detailed experimental plan. As in the case of sourcing reagents, when discussing genetic screens or cell line generation experiments with a CRO, it is important to focus on the formula Model × Assay × Perturbation (M × A × P), as clearly defining all three will ensure success. Following, discussions on the science, economics, and timelines need to be considered. Sometimes, providers may offer optimistic timelines that work if everything goes well according to plan and no repeats need to be made. However, this may not always be the case, so it is important to discuss this and ensure that timelines are realistic. If possible, reach out to other groups that have used the same provider and learn how long the execution of their project took. Alongside timeline discussions, you will need to build milestone checkpoints as part of the service agreement. The importance of milestone checkpoints is to provide you (the client) with the opportunity to review the data at appropriate progress checks and terminate the programme if it is unsatisfactory. Also, one should mutually agree on the format and frequency of updates on the project and ensure that the CRO is willing to provide not only analyzed, but also raw data, so that you can independently review and assess the quality of the results. If material transfer is needed (i.e. a host cell line), a material transfer agreement (MTA) should be part of the service contract. Figure 4.1 outlines key steps involved in working with CRO on CRISPR genome editing or screen projects.

4.5 Considerations on Experimental Design and Controls Required when Outsourcing

For most experiments, CRISPR outsourcing falls into three broad categories: (i) CRISPR genetic screening; (ii) custom cell model generation; (iii) animal model generation. Animal model generation is out of scope for this chapter. For several commonly used cell lines, KO clones may already be commercially available, e.g. the human haploid HAP1 collection comprising over 7500 knockout cell lines. However, in the majority of cases, a custom edit will need to be generated. So far, we have provided a general description of the overall process of working with CROs. The remaining discussion will focus on specific

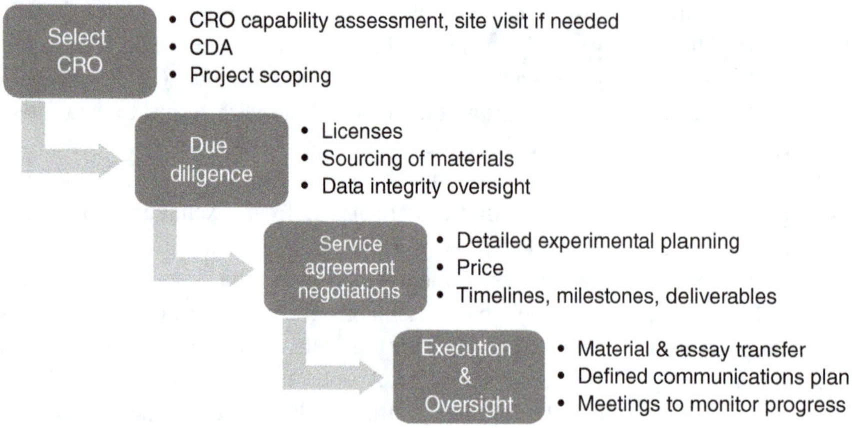

Figure 4.1 Process workflow for project externalization of genome editing to CROs.

aspects of the experimental design that are critical to discuss when outsourcing cell line generation or CRISPR genetic screens.

4.5.1 Considerations on Selecting the Appropriate Cellular Host

Selecting appropriate host cells for genetic manipulation is a key aspect of experimental design and applies both to CRISPR screens and custom cell model generation. A trade-off may need to be considered between physiological relevance (tissue origin and/or differentiation state) and technical feasibility. This is because most primary cells can only be cultured in the laboratory for a limited number of days and therefore researchers often choose immortalized cell lines as the next best alternative to conduct their research. For example, Yeung and colleagues used the THP-1 cell line as a proxy for macrophages and performed a genome-wide screen to identify loss-of-function mutations conferring resistance to Salmonella uptake (Yeung et al. 2019). Shang and colleagues used Jurkat cells as proxy for primary T cells and performed a genome-wide screen to identify genes that regulate T cell activation upon anti-T cell receptor (TCR) stimulation. Recently, strategies like gRNA lentiviral infection with Cas9 protein electroporation (SLICE) were developed to overcome the requirement for stably expressing Cas9 cells and have enabled genetic screens to occur in primary human cells (Shifrut et al. 2018). However, not many CROs have the ability to conduct such technically demanding types of genetic screens.

When work is to be carried out in non-primary cells, several aspects should be considered to help you choose the right cellular host. These have been extensively discussed in a review by Kimberland and colleagues (Kimberland et al. 2018). They include information on chromosome ploidy, genetic stability, and clonality. For most cancer cell lines, multi-omic information can already be accessed through datasets, such as the Cancer Cell Line Encyclopedia (https://portals.broadinstitute.org/ccle/about) (Ghandi et al. 2019), the Cell Model Passports portal (https://cellmodelpassports.sanger.ac.uk/) (van der Meer et al. 2019), or the Cancer Dependency Map portals (https://depmap.org/portal/ & https://depmap.sanger.ac.uk/). This information enables researchers to bioinformatically interogate their cell line of interest and make informed decisions on how similar it is to their tissue of interest. Moreover, online tools such as Cellector faciliate the selection of the most appropriate model to use based on similarity to patient samples (https://ot-cellector.shinyapps.io/CELLector_App/). We would also recommend discussing with the CRO what assays they routinely use to validate the cell lines used in their work, and if needed draft additional testing as part of the contract research. In our experience, array comparative genomic hybridization (aCGH) plus SNP genome analysis provides a cost-effective and quick survey of the host cell lines. aCGH detects structural and numeric chromosomal alterations, while SNP analysis allows detection of polyploidy, loss of heterozygosity, and uniparental disomy (Wiszniewska et al. 2014). Such information ensures that the genetic makeup of the cell line used matches the parental cell line.

4.5.2 Considerations on the Gene/Locus to be Edited

Another key aspect to consider when planning CRISPR screens or custom cell line editing experiments is to determine if the target gene-of-interest (GOI) is expressed in the host cell and whether it is essential for cellular viability. Online tools are available and can be used (Chen et al. 2017) to help determine this. However, essentiality for some genes is context dependent (Wang et al. 2015) and as such caution on data interpretation is advised (Kimberland et al. 2018). Regardless, highly amplified genes in cancer cell lines appear to be unsuitable for gene KO studies, since simultaneous on-target cleavage of a high number of alleles may affect cell viability. This has been shown to result in false positives/negatives in library screens where cell viability was used as an enrichment step (Aguirre et al. 2016; Munoz et al. 2016). One approach to address the challenge of essentiality for CRISPR screens is to use CRISPRi, which relies on dCas9 that does not alter the sequence of genomic DNA. However, CRISPRi may not be suitable for all targets, as it can downregulate multiple genes at bidirectional promoters (Rosenbluh et al. 2017).

To separate technical artifacts from true hits and to obtain a deeper understanding of how the target(s) of interest function, one can adopt a dual screening strategy, where two complementary screens (e.g. CRISPRi and CRISPRa) are performed in parallel (Jost et al. 2017; le Sage et al. 2020). Combining screening platforms substantially augments the quality and value of data derived from the screening campaigns, as well as providing novel insights not accessible when using one technology alone. For example, whereas CRISPRko will reveal targets where complete protein removal is required for a phenotype to emerge, CRISPRi could be used to investigate levels of repression needed to elicit a phenotypic response.

4.5.3 Controlling CRISPR Off-Target Effects (OTEs) and Clonal Variations

Despite the remarkable specificity of the first and commonly used CRISPR/Cas9 system derived from *Streptococcus pyogenes* (SpCas9), varying levels of OTEs have been observed (Cho et al. 2014; Fu et al. 2013; Hsu et al. 2013; Lin et al. 2014; Pattanayak et al. 2013). These effects can be exacerbated when combined with additional sources of experimental variation, such as clonal variation in the cellular system. In cases where gene KO efficiency is high (e.g. >90%), bulk cell populations can be readily assayed without further enrichment; this is analogous to experimental approaches applied to siRNA/shRNA gene knockdown. Certain CROs offer gene KO in bulk for cell lines and ship the KO cell populations as frozen stocks (e.g. Synthego). However, when the editing efficiencies are lower and bulk assays are not feasible, especially if using iPSC or cell lines, cellular cloning will be needed. Successful on-target changes in genome-edited cells are typically monitored by targeted analysis of the genomic loci by a number of assays routinely offered by CROs, including the T7 endonuclease-based Surveyor assay (Qiu et al. 2004), or the sequencing-based TIDE assay (Brinkman et al. 2014). A comprehensive assessment of OTEs would require a genome-wide approach (Zischewski et al. 2017), which are typically resource intensive. Even mapped, it is still unclear to what extent the detected OTEs will impact interpretation of the experimental results. More practically, OTEs can be controlled by analyzing multiple independent clones that are derived either from using the same gRNAs (or a mix of 2–3 gRNAs) or ideally from using different gRNAs. Given that each gRNA has its unique off-target sites to which the OTEs are most likely to occur, different gRNAs targeting the same gene or locus thus will most likely have different OTE profiles. Further, it is a good practice to also obtain "wild type" (WT) clones for use as controls. These WT clones are derived from the same CRISPR editing reaction yet are wild type at the targeting site, but may carry indels at off-target sites. Such WT clones thus offer better controls than unedited cells since they have the potential to account for OTEs. The number of independent clones to analyze will vary dependent on a balance of quality, time, and cost. Scientifically, it depends on multiple factors including homogeneity of the parental cells, since clonal heterogeneity can lead to confounding or even invalid results (Ben-David et al. 2018). Also, editing specificity and efficiency need to be considered. For practicality, it is commonplace to analyze 3–5 independent edited clones along with 3–5 WT clones. Statistically, the more clones analyzed, the higher confidence one will have in the resulting data.

4.5.4 Deciding on Specific Quality Control Experiments on Engineered Cells

A key part of the research contract is to define quality control assays for the editing project, in addition to the prerequisite sequence confirmation of the editing site (on-target editing). While these assays are not always essential, and will require additional effort and cost, certain assays can provide additional evidence supporting the quality of the resulting edited cells.

4.5.4.1 Confirmation of Gene KO at Protein Level

Ideally, gene KO clones with frameshift mutations should be further confirmed via protein analysis (e.g. Western) for depletion of the protein at the expected protein size, as long as a specific antibody is available. While this is observed for many genes in cellular KO experiments, deviations have been observed. In a recent study by Smits and colleagues, systematic characterization of frameshift KO mutations in HAP1 cell lines revealed that one-third of the quantified targets were still detected at variable levels from low to original (Smits et al. 2019). Further studies of these frameshift KO clones, as well as other studies on genetically confirmed KO clones, revealed that residual levels of expression are either derived from introducing a premature stop codon, skipping of edited exon due to alternative mRNA splicing or from using an alternative initiation codon (Mou et al. 2017; Sharpe and Cooper 2017; Smits et al. 2019; Tuladhar et al. 2019). Dependent on the particular gene in study, and influenced by the location of the edited exon and the isoform expression profile in the cellular host, such truncated proteins may have full, partial, or even opposite activities (Smits et al. 2019; Tuladhar et al. 2019). Caution should be taken when interpreting KO effects of such clones. A practical approach to mitigate such effects is to analyze multiple independent clones derived from using gRNAs targeting different exons of the gene to obtain a concordant phenotypic observation. With a specific antibody, additional quality control assays can be carried out to analyze expression of the target protein in edited cells, including FACS analysis and cell staining. However, such assays only report on the presence/absence of the domain(s) the antibody recognizes, but not report on the size of the proteins/truncates detected, whereas Western analysis can report on both. Even with Western analysis, the results are impacted by epitope specificity of the antibody used. An N-terminal-specific antibody detects the full length and truncated proteins resulting from premature translation terminations; whereas a C-terminal-specific antibody enables detection of the full length and truncates resulted from in-frame exon skipping as well as alternative translation initiations. In cases where a heterologous reporter gene (i.e. GFP, Luc) or an epitope tag is inserted in frame with an endogenous gene via knock-in reactions, well-validated reporter/epitope-specific antibodies can be readily used in QC assays at protein levels.

4.5.4.2 Confirmation of Genetic Manipulation at RNA Level

Gene KO clones can also be QC analyzed at the mRNA level via RT-PCR or RT-qPCR. This is based on the assumption that mRNAs with frameshift indels are noncoding and are subjected to nonsense mediated decay (NMD) (Popp and Maquat 2016). While this is the case for many genes with frameshift indels in defined cells, some edited RNAs have been observed to escape NMD (Smits et al. 2019). Further analyses revealed that these NMD-independent RNAs persist in cells via mechanisms involving alternative translation initiation, in-frame exon skipping, or simply location of indel-derived premature termination codon (Popp and Maquat 2016; Tuladhar et al. 2019). As such, RT-qPCR is not commonly used as a quality control assay in gene KO experiments. Nevertheless, targeted RT-PCR followed by gel electrophoresis or DNA sequencing analysis can reveal detailed information about variant RNA species transcribed from the edited gene, which can facilitate picking of appropriate cell clones for downstream functional analysis. It should be noted that RNA-based QC assays may become essential when the edited site/sequence involves expression or function of a noncoding RNA (i.e. microRNA, lncRNA).

In designing HDR-based knock-in experiments for nucleotide mutation generation, a frequently raised question is whether to incorporate a silent codon mutation corresponding to the PAM sequence in the donor sequence. Such "PAM-silent" mutations, once incorporated into the target site, will block re-cleavage by the CRISPR nuclease. This has become a common practice for SNP/mutation generation to increase KI efficiencies especially when it is low (i.e. <5%). However, this would inevitably introduce a nucleotide mutation in the edited site, albeit as a silent codon change. This silent codon change may have an impact on mRNA (stability, secondary structure, or translation efficiency) and further confound the effect of the targeted nucleotide SNP/mutation on mRNA. It is therefore preferred not to adopt the PAM-silent strategy, especially if the editing efficiencies are greater than 10%. If used, a QC assay at the mRNA level (i.e. RT-qPCR) will need to be applied to assess the effect of the mutations (the on-target nucleotide change plus PAM-silent mutation) on mRNA among isogenic lines before functional assays are applied.

4.6 Summary

In this chapter, we reviewed important design considerations for CRISPR experiments and the major providers of reagents, and discussed ways of working with CROs for genome editing projects. Many of the considerations are equally applicable to projects carried out in-house. Regardless of where the genome editing projects are to be conducted, we suggest applying the formula Model x Assay x Perturbation, as well as the triple constraint principle – quality, time, and cost – in designing and managing execution of each project. We hope that the guidelines discussed here will help researchers strike an appropriate balance of these principles to maximize scientific impact of this exciting technology within a given budget or resource.

Acknowledgments

We would like to thank Gerard Drewes and Michelle Kimberland for critical reading of the manuscript.

References

Acharya, S., Mishra, A., Paul, D. et al. (2019). Francisella novicida Cas9 interrogates genomic DNA with very high specificity and can be used for mammalian genome editing. *Proc. Natl. Acad. Sci. U. S. A.* 116: 20959–20968.

Aguirre, A.J., Meyers, R.M., Weir, B.A. et al. (2016). Genomic copy number dictates a gene-independent cell response to CRISPR/Cas9 targeting. *Cancer Discov.* 6: 914–929.

Behan, F.M., Iorio, F., Picco, G. et al. (2019). Prioritization of cancer therapeutic targets using CRISPR-Cas9 screens. *Nature* 568: 511–516.

Ben-David, U., Siranosian, B., Ha, G. et al. (2018). Genetic and transcriptional evolution alters cancer cell line drug response. *Nature* 560: 325–330.

Boettcher, M., Tian, R., Blau, J.A. et al. (2018). Dual gene activation and knockout screen reveals directional dependencies in genetic networks. *Nat. Biotechnol.* 36: 170–178.

Borys, S.M. and Younger, S.T. (2020). Identification of functional regulatory elements in the human genome using pooled CRISPR screens. *BMC Genomics* 21: 107.

Bradford, J. and Perrin, D. (2019a). A benchmark of computational CRISPR-Cas9 guide design methods. *PLoS Comput. Biol.* 15: e1007274.

Bradford, J. and Perrin, D. (2019b). Improving CRISPR guide design with consensus approaches. *BMC Genomics* 20: 931.

Braun, C.J., Bruno, P.M., Horlbeck, M.A. et al. (2016). Versatile in vivo regulation of tumor phenotypes by dCas9-mediated transcriptional perturbation. *Proc. Natl. Acad. Sci. U. S. A.* 113: E3892–E3900.

Brinkman, E.K., Chen, T., Amendola, M., and Van Steensel, B. (2014). Easy quantitative assessment of genome editing by sequence trace decomposition. *Nucleic Acids Res.* 42: e168.

Chen, S., Sanjana, N.E., Zheng, K. et al. (2015). Genome-wide CRISPR screen in a mouse model of tumor growth and metastasis. *Cell* 160: 1246–1260.

Chen, W.H., Lu, G., Chen, X. et al. (2017). OGEE v2: an update of the online gene essentiality database with special focus on differentially essential genes in human cancer cell lines. *Nucleic Acids Res.* 45: D940–D944.

Cho, S.W., Kim, S., Kim, Y. et al. (2014). Analysis of off-target effects of CRISPR/Cas-derived RNA-guided endonucleases and nickases. *Genome Res.* 24: 132–141.

Chow, R.D., Wang, G., Ye, L. et al. (2019). in vivo profiling of metastatic double knockouts through CRISPR-Cpf1 screens. *Nat. Methods* 16: 405–408.

Cohen, J. (2019). CRISPR patent fight revived. *Science* 365: 15–16.

De Caneva, A., Porro, F., Bortolussi, G. et al. (2019). Coupling AAV-mediated promoterless gene targeting to SaCas9 nuclease to efficiently correct liver metabolic diseases. *JCI Insight* 5.

De Groot, R., Luthi, J., Lindsay, H. et al. (2018). Large-scale image-based profiling of single-cell phenotypes in arrayed CRISPR-Cas9 gene perturbation screens. *Mol. Syst. Biol.* 14: e8064.

Doench, J.G. (2018). Am i ready for CRISPR? A user's guide to genetic screens. *Nat. Rev. Genet.* 19: 67–80.

Doench, J.G., Fusi, N., Sullender, M. et al. (2016). Optimized sgRNA design to maximize activity and minimize off-target effects of CRISPR-Cas9. *Nat. Biotechnol.* 34: 184–191.

Dong, M.B., Wang, G., Chow, R.D. et al. (2019). Systematic immunotherapy target discovery using genome-scale in vivo CRISPR screens in CD8 T cells. *Cell* 178: 1189–1204.e23.

Ebright, R.Y., Lee, S., Wittner, B.S. et al. (2020). Deregulation of ribosomal protein expression and translation promotes breast cancer metastasis. *Science* 367: 1468–1473.

Erard, N., Knott, S.R.V., and Hannon, G.J. (2017). A CRISPR resource for individual, combinatorial, or multiplexed gene knockout. *Mol. Cell* 67: 348–354. e4.

Feldman, D., Singh, A., Schmid-Burgk, J.L. et al. (2019). Optical pooled screens in human cells. *Cell* 179: 787–799. e17.

Filippova, J., Matveeva, A., Zhuravlev, E., and Stepanov, G. (2019). Guide RNA modification as a way to improve CRISPR/Cas9-based genome-editing systems. *Biochimie* 167: 49–60.

Fu, Y., Foden, J.A., Khayter, C. et al. (2013). High-frequency off-target mutagenesis induced by CRISPR-Cas nucleases in human cells. *Nat. Biotechnol.* 31: 822–826.

Ghandi, M., Huang, F.W., Jane-Valbuena, J. et al. (2019). Next-generation characterization of the cancer cell line encyclopedia. *Nature* 569: 503–508.

Giladi, A., Paul, F., Herzog, Y. et al. (2018). Single-cell characterization of haematopoietic progenitors and their trajectories in homeostasis and perturbed haematopoiesis. *Nat. Cell Biol.* 20: 836–846.

Ginn, S.L., Amaya, A.K., Liao, S.H.Y. et al. (2020). Efficient in vivo editing of OTC-deficient patient-derived primary human hepatocytes. *JHEP Rep* 2: 100065.

Goncalves, E., Thomas, M., Behan, F.M. et al. (2020). Minimal genome-wide human CRISPR-Cas9 library. *bioRxiv* https://genomebiology.biomedcentral.com/articles/10.1186/s13059-021-02268-4.

Hart, T., Chandrashekhar, M., Aregger, M. et al. (2015). High-resolution CRISPR screens reveal fitness genes and genotype-specific cancer liabilities. *Cell* 163: 1515–1526.

Horlbeck, M.A., Xu, A., Wang, M. et al. (2018). Mapping the genetic landscape of human cells. *Cell* 174: 953–967. e22.

Hsu, P.D., Scott, D.A., Weinstein, J.A. et al. (2013). DNA targeting specificity of RNA-guided Cas9 nucleases. *Nat. Biotechnol.* 31: 827–832.

Jaitin, D. A., Weiner, A., Yofe, I., Lara-Astiaso, D., Keren-Shaul, H., David, E., Salame, T. M., Tanay, A., Van Oudenaarden, A. & Amit, I. 2016. Dissecting immune circuits by linking CRISPR-pooled screens with single-cell RNA-Seq. *Cell*, 167, 1883–1896 e15.

Joberty, G., Falth-Savitski, M., Paulmann, M. et al. (2020). A tandem guide RNA-based strategy for efficient CRISPR gene editing of cell populations with low heterogeneity of edited alleles. *The CRISPR J.* https://doi.org/10.1089/crispr.2019.0064.

Jost, M., Chen, Y., Gilbert, L.A. et al. (2017). Combined CRISPRi/a-based chemical genetic screens reveal that Rigosertib is a microtubule-destabilizing agent. *Mol. Cell* 68: 210–223. e6.

Kamens, J. (2015). The addgene repository: an international nonprofit plasmid and data resource. *Nucleic Acids Res.* 43: D1152–D1157.

Katigbak, A., Cencic, R., Robert, F. et al. (2016). A CRISPR/Cas9 functional screen identifies rare tumor suppressors. *Sci. Rep.* 6: 38968.

Kim, S., Kim, D., Cho, S.W. et al. (2014). Highly efficient RNA-guided genome editing in human cells via delivery of purified Cas9 ribonucleoproteins. *Genome Res.* 24: 1012–1019.

Kim, D., Luk, K., Wolfe, S.A., and Kim, J.S. (2019). Evaluating and enhancing target specificity of gene-editing nucleases and deaminases. *Annu. Rev. Biochem.* 88: 191–220.

Kimberland, M.L., Hou, W., Alfonso-Pecchio, A. et al. (2018). Strategies for controlling CRISPR/Cas9 off-target effects and biological variations in mammalian genome editing experiments. *J. Biotechnol.* 284: 91–101.

Kodama, M., Kodama, T., Newberg, J.Y. et al. (2017). in vivo loss-of-function screens identify KPNB1 as a new druggable oncogene in epithelial ovarian cancer. *Proc. Natl. Acad. Sci. U. S. A.* 114: E7301–E7310.

Labun, K., Montague, T.G., Krause, M. et al. (2019). CHOPCHOP v3: expanding the CRISPR web toolbox beyond genome editing. *Nucleic Acids Res.* 47: W171–W174.

Lafleur, M.W., Nguyen, T.H., Coxe, M.A. et al. (2019). A CRISPR-Cas9 delivery system for in vivo screening of genes in the immune system. *Nat. Commun.* 10: 1668.

Le Sage, C., Lawo, S., and Cross, B.C.S. (2020). CRISPR: a screener's guide. *SLAS Discov* 25: 233–240.

Li, K., Liu, Y., Cao, H. et al. (2020). Interrogation of enhancer function by enhancer-targeting CRISPR epigenetic editing. *Nat. Commun.* 11: 485.

Liang, X., Potter, J., Kumar, S. et al. (2015). Rapid and highly efficient mammalian cell engineering via Cas9 protein transfection. *J. Biotechnol.* 208: 44–53.

Lin, Y., Cradick, T.J., Brown, M.T. et al. (2014). CRISPR/Cas9 systems have off-target activity with insertions or deletions between target DNA and guide RNA sequences. *Nucleic Acids Res.* 42: 7473–7485.

Liu, G., Zhang, Y., and Zhang, T. (2020). Computational approaches for effective CRISPR guide RNA design and evaluation. *Comput. Struct. Biotechnol. J.* 18: 35–44.

Lu, Q., Livi, G.P., Modha, S. et al. (2017). Applications of CRISPR genome editing technology in drug target identification and validation. *Expert Opin. Drug Discovery* 12: 541–552.

Manguso, R.T., Pope, H.W., Zimmer, M.D. et al. (2017). in vivo CRISPR screening identifies Ptpn2 as a cancer immunotherapy target. *Nature* 547: 413–418.

Mans, R., Van Rossum, H.M., Wijsman, M. et al. (2015). CRISPR/Cas9: a molecular Swiss army knife for simultaneous introduction of multiple genetic modifications in Saccharomyces cerevisiae. *FEMS Yeast Res.* 15.

Martin-Laffon, J., Kuntz, M., and Ricroch, A.E. (2019). Worldwide CRISPR patent landscape shows strong geographical biases. *Nat. Biotechnol.* 37: 613–620.

Martufi, M., Good, R.B., Rapiteanu, R. et al. (2019). Single-step, high-efficiency CRISPR-Cas9 genome editing in primary human disease-derived fibroblasts. *CRISPR J* 2: 31–40.

Metzakopian, E., Strong, A., Iyer, V. et al. (2017). Enhancing the genome editing toolbox: genome wide CRISPR arrayed libraries. *Sci. Rep.* 7: 2244.

Mou, H., Smith, J.L., Peng, L. et al. (2017). CRISPR/Cas9-mediated genome editing induces exon skipping by alternative splicing or exon deletion. *Genome Biol.* 18: 108.

Munoz, D.M., Cassiani, P.J., Li, L. et al. (2016). CRISPR screens provide a comprehensive assessment of cancer vulnerabilities but generate false-positive hits for highly amplified genomic regions. *Cancer Discov.* 6: 900–913.

Najm, F.J., Strand, C., Donovan, K.F. et al. (2018). Orthologous CRISPR-Cas9 enzymes for combinatorial genetic screens. *Nat. Biotechnol.* 36: 179–189.

Oughtred, R., Stark, C., Breitkreutz, B.J. et al. (2019). The BioGRID interaction database: 2019 update. *Nucleic Acids Res.* 47: D529–D541.

Pattanayak, V., Lin, S., Guilinger, J.P. et al. (2013). High-throughput profiling of off-target DNA cleavage reveals RNA-programmed Cas9 nuclease specificity. *Nat. Biotechnol.* 31: 839–843.

Pichler, F.B. and Turner, S.J. (2007). The power and pitfalls of outsourcing. *Nat. Biotechnol.* 25: 1093–1096.

Popp, M.W. and Maquat, L.E. (2016). Leveraging rules of nonsense-mediated mRNA decay for genome engineering and personalized medicine. *Cell* 165: 1319–1322.

Qiu, P., Shandilya, H., D'alessio, J.M. et al. (2004). Mutation detection using Surveyor nuclease. *BioTechniques* 36: 702–707.

Ran, F.A., Cong, L., Yan, W.X. et al. (2015). in vivo genome editing using Staphylococcus aureus Cas9. *Nature* 520: 186–191.

Rosenbluh, J., Xu, H., Harrington, W. et al. (2017). Complementary information derived from CRISPR Cas9 mediated gene deletion and suppression. *Nat. Commun.* 8: 15403.

Safari, F., Zare, K., Negahdaripour, M. et al. (2019). CRISPR Cpf1 proteins: structure, function and implications for genome editing. *Cell Biosci.* 9: 36.

Sanson, K.R., Hanna, R.E., Hegde, M. et al. (2018). Optimized libraries for CRISPR-Cas9 genetic screens with multiple modalities. *Nat. Commun.* 9: 5416.

Sanson, K.R., Deweirdt, P.C., Sangree, A.K. et al. (2020). Optimization of AsCas12a for combinatorial genetic screens in human cells. *bioRxiv* https://www.biorxiv.org/content/10.1101/747170v1.

Seki, A. and Rutz, S. (2018). Optimized RNP transfection for highly efficient CRISPR/Cas9-mediated gene knockout in primary T cells. *J. Exp. Med.* 215: 985–997.

Sharpe, J.J. and Cooper, T.A. (2017). Unexpected consequences: exon skipping caused by CRISPR-generated mutations. *Genome Biol.* 18: 109.

Shifrut, E., Carnevale, J., Tobin, V. et al. (2018). Genome-wide CRISPR screens in primary human T cells reveal key regulators of immune function. *Cell* 175: 1958–1971. e15.

Slaymaker, I.M., Gao, L., Zetsche, B. et al. (2016). Rationally engineered Cas9 nucleases with improved specificity. *Science* 351: 84–88.

Smits, A.H., Ziebell, F., Joberty, G. et al. (2019). Biological plasticity rescues target activity in CRISPR knock outs. *Nat. Methods* 16: 1087–1093.

Song, C.Q., Li, Y., Mou, H. et al. (2017). Genome-wide CRISPR screen identifies regulators of mitogen-activated protein kinase as suppressors of liver tumors in mice. *Gastroenterology* 152: 1161–1173. e1.

Strezoska, Z., Perkett, M.R., Chou, E.T. et al. (2017). High-content analysis screening for cell cycle regulators using arrayed synthetic crRNA libraries. *J. Biotechnol.* 251: 189–200.

Tak, Y.E., Kleinstiver, B.P., Nunez, J.K. et al. (2017). Inducible and multiplex gene regulation using CRISPR-Cpf1-based transcription factors. *Nat. Methods* 14: 1163–1166.

Takeda, H., Kataoka, S., Nakayama, M. et al. (2019). CRISPR-Cas9-mediated gene knockout in intestinal tumor organoids provides functional validation for colorectal cancer driver genes. *Proc. Natl. Acad. Sci. U. S. A.* 116: 15635–15644.

Thomas, J.D., Polaski, J.T., Feng, Q. et al. (2020). RNA isoform screens uncover the essentiality and tumor-suppressor activity of ultraconserved poison exons. *Nat. Genet.* 52: 84–94.

Tuladhar, R., Yeu, Y., Tyler Piazza, J. et al. (2019). CRISPR-Cas9-based mutagenesis frequently provokes on-target mRNA misregulation. *Nat. Commun.* 10: 4056.

Tzelepis, K., Koike-Yusa, H., De Braekeleer, E. et al. (2016). A CRISPR dropout screen identifies genetic vulnerabilities and therapeutic targets in Acute Myeloid Leukemia. *Cell Rep.* 17: 1193–1205.

Vakulskas, C.A., Dever, D.P., Rettig, G.R. et al. (2018). A high-fidelity Cas9 mutant delivered as a ribonucleoprotein complex enables efficient gene editing in human hematopoietic stem and progenitor cells. *Nat. Med.* 24: 1216–1224.

Van Der Meer, D., Barthorpe, S., Yang, W. et al. (2019). Cell Model Passports-a hub for clinical, genetic and functional datasets of preclinical cancer models. *Nucleic Acids Res.* 47: D923–D929.

Wang, T., Birsoy, K., Hughes, N.W. et al. (2015). Identification and characterization of essential genes in the human genome. *Science* 350: 1096–1101.

Wei, J., Long, L., Zheng, W. et al. (2019). Targeting REGNASE-1 programs long-lived effector T cells for cancer therapy. *Nature* 576: 471–476.

Wienert, B., Wyman, S.K., Richardson, C.D. et al. (2019). Unbiased detection of CRISPR off-targets in vivo using DISCOVER-Seq. *Science* 364: 286–289.

Wiszniewska, J., Bi, W., Shaw, C. et al. (2014). Combined array CGH plus SNP genome analyses in a single assay for optimized clinical testing. *Eur. J. Hum. Genet.* 22: 79–87.

Wu, J. and Yin, H. (2019). Engineering guide RNA to reduce the off-target effects of CRISPR. *J. Genet. Genomics* 46: 523–529.

Yau, E.H., Kummetha, I.R., Lichinchi, G. et al. (2017). Genome-wide CRISPR screen for essential cell growth mediators in mutant KRAS colorectal cancers. *Cancer Res.* 77: 6330–6339.

Ye, L., Park, J.J., Dong, M.B. et al. (2019). in vivo CRISPR screening in CD8 T cells with AAV-sleeping beauty hybrid vectors identifies membrane targets for improving immunotherapy for glioblastoma. *Nat. Biotechnol.* 37: 1302–1313.

Yeung, A.T.Y., Choi, Y.H., Lee, A.H.Y. et al. (2019). A genome-wide knockout screen in human macrophages identified host factors modulating salmonella infection. *MBio* 10.

Zetsche, B., Gootenberg, J.S., Abudayyeh, O.O. et al. (2015). Cpf1 is a single RNA-guided endonuclease of a class 2 CRISPR-Cas system. *Cell* 163: 759–771.

Zhang, F. (2019). Development of CRISPR-Cas systems for genome editing and beyond. *Q. Rev. Biophys.* 52: 1–31.

Zhao, Y., Tyrishkin, K., Sjaarda, C. et al. (2019). A one-step tRNA-CRISPR system for genome-wide genetic interaction mapping in mammalian cells. *Sci. Rep.* 9. 14499.

Zischewski, J., Fischer, R., and Bortesi, L. (2017). Detection of on-target and off-target mutations generated by CRISPR/Cas9 and other sequence-specific nucleases. *Biotechnol. Adv.* 35: 95–104.

5

Computational Tools for Target Design and Analysis

Gue-Ho Hwang and Sangsu Bae

Department of Chemistry and Research Institute for Convergence of Basic Sciences, Hanyang University, Seoul, South Korea

5.1 Introduction

The CRISPR-Cas system facilitates highly efficient genome editing; thus, it has been applied in many research fields, including biological science, medicine, and gene therapy. CRISPR tools enable targeted gene editing in a guide RNA-dependent manner. However, editing activity is different for each guide RNA and RNA-guided CRISPR tools can cleave off-target sites as well as on-target sites, resulting in unwanted mutations. To address this issue, a number of web-based and command line-based programs have been developed to aid researchers in selecting appropriate target sites in a gene of interest for various CRISPR systems, such as CRISPR-Cas9, CRISPR-Cas12, base editors, and prime editing tools. Furthermore, estimating the resulting mutation rates and determining the patterns of mutations after CRISPR treatments are delivered into cells or organisms are important, but not easy. High-throughput sequencing, in addition to Sanger sequencing, is one of the most precise methods for assessing genome editing outcomes. Several programs for analysis of next-generation sequencing (NGS) data have been developed. This chapter describes current programs for target design or analysis in genome editing.

5.2 Various Types of CRISPR Effectors

To date, various CRISPR-Cas systems have been harnessed as genome editing tools and each CRISPR-Cas has its own platform with a reliable guide RNA and a protospacer adjacent motif (PAM) (Chylinski et al. 2014; Jinek et al. 2012; Makarova et al. 2015; Makarova et al. 2020; Nishimasu et al. 2014). Type II CRISPR-Cas9 and type V CRISPR-Cas12a (also known as Cpf1) endonucleases are most frequently used for genome editing (Broughton et al. 2020; Hsu et al. 2014). Regarding type II CRISPR-Cas9, several useful orthologs from different organisms recognize different PAM sequences, such as SpCas9 from *Streptococcus pyogenes* with NGG PAM (Cho et al. 2013; Mali et al. 2013), SaCas9 from *Staphylococcus aureus* with NNGRRT PAM (Ran et al. 2015), NmCas9 from *Neisseria meningitidis* with NNNNGMTT PAM (Hou et al. 2013), CjCas9 from *Campylobacter jejuni* with NNNVRTAC PAM (Kim et al. 2017a), StCas9 from *Streptococcus thermophilus* with NNAGAAW PAM (Xu et al. 2015b), and Nme2Cas9 from *N. meningitidis* with NNNNCC PAM (Edraki et al. 2019). Among these, SpCas9 was the most popular nuclease and several engineered versions have been developed for expanding target range or increasing target specificity. The Joung group developed VRER (NGCG PAM), VQR (NGA PAM) (Kleinstiver et al. 2015), the Liu group developed XCas9 (NG or NAR PAM) (Hu et al. 2018), and the Nureki group developed SpCas9-NG (NG PAM) (Nishimasu et al. 2018). The Kleinstiver group recently developed near-PAMless versions, SpCas9 (SpRY) with NRN and NYN PAMs (Walton et al. 2020).

Regarding type V CRISPR-Cas12 (also known as Cpf1), AsCas12a from *Acidaminococcus sp.* with TTTV PAM, LbCas12a from *Lachnospiraceae bacterium* with TTTN PAM (Zetsche et al. 2015), and FnCas12a from *Francisella novicida* with TTN PAM (Swiat et al. 2017) have been harnessed for targeted gene editing. The Joung group also engineered AsCas12a, generating RVR and RR, which, respectively, have TATV and TYCV PAMs (Gao et al. 2017). In addition to type II CRISPR-Cas9 and type V CRISPR-Cas12a, type VI CRISPR-Cas13 effectors have been used for RNA targeting and editing. However, type I Cascade was used for gene editing in *Zea mays* cells (Young et al. 2019). Current versions of CRISPR effectors are summarized in **Table 5.1**.

Genome Editing in Drug Discovery, First Edition. Edited by Marcello Maresca and Sumit Deswal.

Table 5.1 Type of CRISPRs and properties.

Class	Type	Nuclease property	CRISPR	PAM sequence with a required spacer length	Ref.
Class 1 (Complex effector module)	Type I	Stand-alone HD nuclease	EcoCascade	5'-**AWG-32nt**-3'	Redding et al. (2015)
Class2 (single effector module)	Type II	Cas9 (effector protein), Blunt ends cleavage	SpCas9	5'-**20nt-NGG**-3'	Cho et al. (2013); Mali et al. (2013)
			VRER SpCas9	5'-**20nt-NGCG**-3'	Kleinstiver et al. (2015)
			VQR SpCas9	5'-**20nt-NGA**-3'	Kleinstiver et al. (2015)
			XCas9 3.7 (TLIKDIV SpCas9)	5'-**20nt-NG**-3'	Hu et al. (2018)
			SpCas9-NG	5'-**20nt-NG**-3'	Nishimasu et al. (2018)
			SpG	5'-**20nt-NGN**-3'	Walton et al. (2020)
			SpRY	5'-**20nt-NRN**-3' (lesser extent 5'-**20nt-NYN**-3')	Walton et al. (2020)
			Spy-macCas9	5'-**20nt-NAAN**-3'	Liu et al. (2019)
			StCas9	5'-**18nt-NNAGAAW**-3'	Xu et al. (2015b)
			NmCas9	5'-**24nt-NNNNGMTT**-3'	Hou et al. (2013)
			SaCas9	5'-**20nt-NNGRRT**-3'	Ran et al. (2015)
			CjCas9	5'-**21nt-NNNRTAC**-3'	Kim et al. (2017a)
			Nme2Cas9	5'-**23nt-NNNNCC**-3'	Edraki et al. (2019)
	Type V	Cas12 (effector protein), Staggered cleavage	AsCas12a	5'-**TTTN-23nt**-3'	Zetsche et al. (2015)
			RVR AsCas12a	5'-**TATV-24nt**-3'	Gao et al. (2017)
			RR AsCas12a	5'-**TYCV-24nt**-3'	Gao et al. (2017)
			LbCas12a	5'-**TTTN-23nt**-3'	Zetsche et al. (2015)
			FnCas12a	5'-**TTN-23nt**-3'	Swiat et al. (2017)
			BhCas12b	5'-**TTN-23nt**-3'	Strecker et al. (2019)
			BhCas12b v4	5'-**DTTN-23nt**-3'	Strecker et al. (2019)
	Type VI	Cas13 (effector protein), Targeting RNA sequence	LshCas13a	5'-**28nt-H**-3'	Abudayyeh et al. (2016)
			LwaCas13a	**no significant**	Abudayyeh et al. (2017)
			PspCas13b		Cox et al. (2017)
			CasRx (RxfCas12d)		Kushawah et al. (2020)

5.3 Computational Tools for Target Design

In design of the target sequence, the PAM sequence associated with the selected CRISPR types should be considered primarily, as described in Table 5.1. All DNA sequences containing the PAM can be candidates for targeted genome editing. Target sequences with repeated "TTTT" sequences are typically avoided, because the TTTT sequence can cause termination by polymerase III (Ma et al. 2014), and target sequences with too low (<30%) or too high (>70%) GC contents are not preferable due to the low editing activity. In addition to the above base consideration, researchers need information for each target, including, (i) potential off-target sites, (ii) editing activity, and (iii) predicted outcomes after editing. Numerous web-based or command-line-based tools have been developed to address these issues. This section describes representative web-based tools for CRISPR target design.

5.3.1 Web-Based Tools for Generating Potential Off-Target Sites

CRISPR nucleases recognize on-target sites that are complementary with single-guide RNAs (sgRNAs). However, it is not a perfect system; therefore, CRISPR nucleases also target and induce gene editing at off-target sites that have similar sequence within a few mismatches compared to the on-target (Cho et al. 2014; Fu et al. 2013; Lin et al. 2014). For SpCas9, it is usually accepted that SpCas9 can tolerate off-target sequences with up to three mismatches or 1-nt bulge (Cho et al. 2014; Lin et al. 2014).

5.3.1.1 Cas-OFFinder and Cas-Designer

Cas-Designer (Park et al. 2015) (http://www.rgenome.net/cas-designer/) is a rapid and user-friendly program to aid users in selecting appropriate target sites for type II CRISPR nucleases. Users can use the Cas-Designer by collecting PAM type and target genome, and by writing a DNA sequence of interest such as exons. Cas-Designer rapidly finds target sequences in a given input sequence and displays all possible targets, with useful information, including cleavage positions, GC content, out-of-frame score, and potential off-target sites, including bulge-type sites. The out-of-frame score is the calculated value for possibility of frame shift at cleavage sites and can help in collection of target sequences when researchers develop knock-out cell lines. To identify all potential off-target sites with mismatches (< 3) and bulge (< 2), Cas-Designer uses Cas-OFFinder (http://www.rgenome.net/cas-offinder/) software(Bae et al. 2014b), which directly searches for all possible off-target sequences at the selected reference genome. Cas-Designer currently supports 26 different PAMs and 586 different organisms and continuously updates the PAM and organism when requested by users. Cas-Designer also provides a stand-alone version of Cas-Designer (http://www.rgenome.net/cas-designer/portable).

5.3.1.2 CHOPCHOP

CHOPCHOP (Labun et al. 2016; Labun et al. 2019; Montague et al. 2014) (https://chopchop.cbu.uib.no/) is a web tool for identification of target sequences for CRISPR-Cas and TALEN. CHOPCHOP receives reference sequence information using several formats, such as gene identifiers, genomic regions, or pasting sequences, along with advanced options for target selection. In the advanced options, users can control the conditions for target sequences (like PAM, scoring system, and target region of gene) and primers for each target sequence. On the result page, CHOPCHOP displays an interactive visualization of genes with possible target sequences. In the visualization, each target sequence is colored according to quality score and, by clicking the target sequence box, CHOPCHOP shows possible primers. CHOPCHOP also analyzes off-target sites using bowtie and shows the number of off-target sites with up to three mismatches. CHOPCHOP was updated in 2019 to version 3, which supports 250 organisms and several DNA targeting modes for knock-out, knock-in, activation (CRISPRa), repression (CRISPRi), and nanopore enrichment.

5.3.1.3 CRISPR-P 2.0

CRISPR-P (Lei et al. 2014) is a CRISPR-Cas sgRNA designing web tool for plants. After publishing CRISPR-P, CRISPR-P 2.0 (Liu et al. 2017) (http://crispr.hzau.edu.cn/CRISPR2/) was developed with updated and improved CRISPR-Cas9 technology and feedback from users. CRISPR-P 2.0 receives information for reference sequence using several formats, such as gene ID, gene name, position, or sequence. CRISPR-P 2.0 finds all possible sgRNA and comprehensive restriction enzyme cutting sites and displays the information in an interactive figure and searchers for off-target sequences with up to four mismatches and calculates the efficiency of each off-target sequence and sgRNA sequence. CRISPR-P 2.0 supports 75 plant organisms.

For design of non-reference plant genomes, researchers publishing CRISPR-P developed CRISPR-Local in 2019 (Sun et al. 2019). CRISPR-Local supports the function for building databases from FASTA and GFF files by users and designs sgRNA from the developed database. CRISPR-Local can be used in web tool and off-line versions by command-line or graphical interface mode.

5.3.1.4 CRISPOR

CRISPOR (Concordet and Haeussler 2018; Haeussler et al. 2016) (http://crispor.tefor.net/) is a design web tool for the CRISPR-Cas9 system. CRISPOR receives genomic sequences (<2000bp), reference genome (supporting 549 genomes), and PAM type. CRISPOR finds all possible sgRNA sequences and off-target sequences with up to four mismatches and displays the sgRNA sequences in the interactive figures. The advantage of CRISPOR is calculation of several efficiency scores, such as the specificity score (MIT specificity score and CFD specificity score), predicted efficiency score (Doench '16, Mor. Mateos, etc.), and outcome score (Out-of-Frame score and Lindel) for each sgRNA sequence. CRISPROR also designs the PCR primers for each on-target and off-target site. Users can use CRISPOR Batch to design sgRNA sequences from several genes without limitation in sequence length. CRISPROR Batch receives a list of gene symbols, Entrez Gene IDs, of Refseq IDs with the condition for sgRNA sequence and displays a table with possible sgRNA sequences for each gene.

5.3.1.5 CRISPRdirect

CRISPRdirect (Naito et al. 2015) (https://crispr.dbcls.jp/) is a CRISPR-Cas9 design web tool that focuses on the off-target identical to the seed sequence of the target sequence. CRISPRdirect receives PAM type, reference genome (supporting 675 genomes), and targeted sequence information, such as accession number, genome location, nucleotide sequence (<10 kbp), or sequence file. On the result page, CRISPRdirect displays a table for all possible target sequences with information on GC contents, melting temperature, presence of TTTT, restriction sites, and off-target sites for each sequence. The off-target analysis is performed in three cases: (i) 20mer + PAM, (ii) 12mer + PAM, and (iii) 8mer + PAM. The results are shown in graphical view and the highly specific target is colored green. The results can be downloaded by tab-delimited text file or JSON format.

5.3.1.6 E-CRISP

E-CRISP (Heigwer et al. 2014) (http://www.e-crisp.org/) is a web application for designing SpCas9 target sequences with experiment-oriented design parameters. E-CRISP receives reference organism (supporting 55 organisms) and target sequence information by gene symbol or FASTA sequence. In addition, E-CRISP offers three options for selecting the degree of conditions for design. On the result page, E-CRISP displays a table and graphs to show all possible target sequences. In the table, each sequence is displayed with three scores; a specificity score for the number of off-target sites, an annotation score calculated using annotation information, such as exons and CDS, and an efficacy score calculated using the property of target sequences. The results can be downloaded as a tabular report, excel, and GFF file formats. E-CRISP also designs the paired target sequence by changing the option on the main page.

5.3.1.7 CCTop

CCTop (Stemmer et al. 2015) (http://crispr.cos.uni-heidelberg.de/), a CRISPR-Cas9 design web tool, has a user-friendly and intuitive interface. CCTop basically receives targeted sequence, PAM type (supporting 17 PAMs), and reference genome (supporting 51 organisms). In addition, CCTop offers several options for target selection and off-target prediction. On the result page, CCTop displays an interactive graph and information for possible target sequences, which includes off-target sites of each target and CRISPRater score calculated reflecting the off-target information searched by bowtie. The results can be downloaded using excel or fasta format files. CCTop further offers the batch mode for design of several target regions at once by selecting on the main page (Table 5.2).

5.3.2 Computational Tools for Predicting Editing Activity

Next, in addition to having information on potential off-target sites, researchers usually select guide RNAs with high editing activities. However, predicting the efficiency for each guide RNA is not easy. In the beginning, the Zhang group (Hsu et al. 2013) provided a scoring system using the property of target sequence and off-target sequences for considering the uniqueness of a target sequence in the genome. Using high-throughput sequencing data, several groups have constructed machine learning models for predicting the activity of each guide RNA. A few programs, such as CRISPRscan (Moreno-Mateos et al. 2015) and CINDEL (Kim et al. 2017b), were developed based on a regression model. However, recently programs such as DeepCas9 (Xue et al. 2019), DeepCpf1 (Kim et al. 2018), and DeepHF (Wang et al. 2019) used neural network models to predict the efficiency.

5.3.2.1 DeepCpf1

DeepCpf1 (http://deepcrispr.info/) is a prediction web tool that calculates activity of AsCpf1 using a convolutional neural network (CNN). DeepCpf1 receives the type of cell line and target sequences including 4 bp of upstream and 6 bp of downstream sequences in FASTA format. DeepCpf1 also receives a FASTA format file containing the reference sequence, finding all possible target sequences and predicting the efficiency of each one. On the result page, DeepCpf1 provides a table of target sequences, with information such as GC contents, position, strand, and DeepCpf1 score. The results can be downloaded as a text file. DeepCpf1 also provides detailed programs through github (https://github.com/MyungjaeSong/Paired-Library).

5.3.2.2 DeepSpCas9

DeepSpCas9 (Kim et al. 2019) (http://deepcrispr.info/DeepSpCas9) is a web-based tool for predicting activity of SpCas9 for each sgRNA using CNN. To predict efficiency for each sgRNA, DeepSpCas9 uses the target sequences including 4bp upstream and 3 bp of the downstream sequences. DeepSpCas9 receives at least a 30-nt sequence in FASTA format, FASTA file, or human gene symbol on the main page. On the result page, DeepSpCas9 displays a table with all possible information, including target sequence, position, GC contents, and DeepSpCas9 score. The results can be downloaded as a text file. DeepSpCas9 also provides detailed programs through github (https://github.com/MyungjaeSong/Paired-Library).

5.3.2.3 DeepHF

DeepHF (http://www.DeepHF.com/) is a web-based tool for predicting editing activity of SpCas9 for each sgRNA using a recurrent neural network (RNN). DeepHF receives various types of SpCas9 nucleases (SpCas9_U6, SpCas9_T7, eSpCas9(1.1), SpCas9-HF1) and DNA sequences ranging from 23bp to 1000bp. On the result page, DeepHF displays a table showing possible sequences, cut position,

Table 5.2 Web-based tools for target design with potential off-target sites.

Web tools	Type of CRISPR	Off-target	Scoring system	Bulge	# PAM types	Organisms	URL
Cas-Designer	Cas9 & Cas12a	O	O	O	26	586	http://www.rgenome.net/cas-designer
CHOPCHOP	Cas9 & Cas12a	O	O	X	5	225	https://chopchop.cbu.uib.no
CRISPR-P 2.0	Cas9 & Cas12a	O	O	X	14	67(plant)	http://crispr.hzau.edu.cn/CRISPR2
CRISPOR	Cas9 & Cas12a	O	O	X	32	557	http://crispor.tefor.net
CRISPRdirect	Cas9	O	X	X	Received	675	https://crispr.dbcls.jp
E-CRISP	SpCas9	O	O	X	1	55	http://www.e-crisp.org
CCTop	Cas9 & Cas12a	O	O	X	17	51	http://crispr.cos.uni-heidelberg.de
BE-Designer	Base editors	O	X	X	13	586	http://www.rgenome.net/be-designer
PE-Designer	Prime editors	O	X	X	1	586	http://www.rgenome.net/pe-designer
PrimeDesign	Prime editors	X	X	X	1	X	http://primedesign.pinellolab.org

strand, and predicted efficiency. The results can be downloaded as a text file. DeepHF is also available through github (https://github.com/izhangcd/DeepHF).

5.3.2.4 CINDEL

CINDEL (http://big.hanyang.ac.kr/cindel) is a web tool that predicts the indel frequencies for CRISPR-Cas12 with TTTV PAM using machine learning. CINDEL receives reference sequences (>27bp) as a FASTA format text or a FASTA file. On the result page, CINDEL displays the table with possible target sequence, position, strand, GC contents, and INDEL scores. The results can be downloaded as excel, csv, or pdf files.

5.3.2.5 SSC

Sequence Scan for CRISPR (SSC) (Xu et al. 2015a) (http://cistrome.org/SSC/) is a web tool for predicting editing activity of SpCas9. SSC receives a target sequence in a FASTA text with length of spacer (19nt or 20nt) and provides options according to the purpose of the experiments, such as knockout, CRISPRi, or CRISPRa. On the result page, SSC displays the possible target list with the efficiency score. The results can be downloaded as a text file.

5.3.2.6 Prediction Web Tool by Doench et al. 2016

The web tool (Doench et al. 2016) (https://crispr.ml) is used for predicting editing activity of SpCas9 for each sgRNA using machine learning. The web tool supports two modes: "on-target only" and "on-target and off-target." When using the "on-target only" mode, the web tool receives the target sequence and shows only the on-target efficiency score. When using the "on-target and off-target" mode, the web tool receives the transcript ID and shows the possible target sequence with useful information, including annotation, the number of off-target, off-target score, and on-target score. The results can be downloaded as csv or excel files.

5.3.2.7 CRISPRscan

CRISPRscan (https://www.crisprscan.org) is a web tool for predicting editing activity of SpCas9 for each sgRNA using machine learning. CRISPRscan receives DNA sequence, reference genomes (supporting 19 organisms), and promoter type. CRISPRscan offers the option for PAM and CRISPR type, but scores are not provided for Cas12a. On the result page, CRISPRscan displays the possible target sequence with CRISPRscan score, the number of off-targets, and CFD (the score for off-targets). The results can be downloaded as a text file (Table 5.3).

5.3.3 Computational Tools for Predicting Edited Outcomes After Gene Editing

This section describes web-based tools for predicting editing outcomes after CRISPR treatment. In the beginning, the Jin-soo group developed Microhomology-Predictor by calculating the possible deletion patterns via the microhomology-mediated end joining (MMEJ) repair pathway (Bae et al. 2014a). FORECasT (Allen et al. 2019) and inDelphi (Shen et al. 2018) were recently developed for prediction of mutation patterns using machine learning.

5.3.3.1 Microhomology-Predictor

Microhomology-Predictor (http://www.rgenome.net/mich-calculator) predicts the deletion patterns after treatment of ZFNs, TALENs, or CRISPR-Cas9, by calculating the scores of possible deletion patterns produced by MMEJ. Microhomology-Predictor receives the 60~80bp sequence flanking a cleavage site. On the results page, Microhomology-Predictor displays microhomology score and out-of-frame score. When users click on the target

Table 5.3 Editing activity prediction tools.

Web tool	Type	Neural network	URL	Used by
DeepCpf1	AsCas12a	O (CNN)	http://deepcrispr.info	
DeepSpCas9	SpCas9	O (CNN)	http://deepcrispr.info/DeepSpCas9	
DeepHF	SpCas9	O (RNN)	http://www.deephf.com	
CINDEL	AsCas12a	X	http://big.hanyang.ac.kr/cindel	
SSC	SpCas9	X	http://crispr.dfci.harvard.edu/SSC	CHOPCHOP
Doench et al.	SpCas9	X	https://crispr.ml	CRISPOR
CRISPRscan	SpCas9 & AsCas12a	X	https://www.crisprscan.org	CRISPOR
DeepBaseEditor	Base editors	O (CNN)	http://deepcrispr.info/DeepBaseEditor	
DeepPE	Prime editors	O (CNN)	http://deepcrispr.info/DeepPE	

Table 5.4 Editing outcome prediction tools.

Web tool	Type	Machine learning	URL
Microhomology-Predictor	ZFNs,TALEN,CRISPR	X	http://www.rgenome.net/mich-calculator
FORECasT	Cas9	O	https://partslab.sanger.ac.uk/FORECasT
inDelphi	Cas9	O (Neural network)	https://indelphi.giffordlab.mit.edu
BE-Hive	Base editors	O (Neural network)	https://www.crisprbehive.design

sequence, Microhomology-Predictor shows all possible deletion patterns with associated pattern score for each deletion pattern. Microhomology score is the sum of the pattern score and out-of-frame score is the ratio of out-of-frame pattern scores over the sum of all pattern scores.

5.3.3.2 FORECasT

FORECasT (https://partslab.sanger.ac.uk/FORECasT) is a web tool for predicting editing outcomes derived from CRISPR-Cas9 (NGG PAM) using machine learning. FORECasT receives the target DNA sequence and position of a cleavage site. On the results page, FORECasT displays graphs showing the possible mutation patterns, predicted frequency for each mutation pattern, and the predicted frequency of frame shift. Command line tool of FORECasT is also available through github (https://partslab.sanger.ac.uk/FORECasT).

5.3.3.3 inDelphi

inDelphi (Shen et al. 2018) was developed using machine learning for predicting the spectrum of cut-site proximal genotypic products. inDelphi receives the 60bp sequence on both sides of cleavage and cell types (mESC, HCT116, HEK293, K562, and U2OS). On the results page, inDelphi shows predicted mutation patterns with expected frequencies with possibility of codon frameshift. inDelphi has a "batch mode," which provides results of all possible sgRNA sequences from insert sequences. If users select a "gene mode," inDelphi receives gene symbol with cell type and displays all possible sgRNA sequences with the prediction results. The gene mode is available only for human (hg38) and mouse (mm10) (Table 5.4).

5.3.4 Computational Tools for Assessing Editing Outcomes

After treatment of CRISPR nucleases, not all cells would be mutated and the mutated patterns would vary. To calculate the editing efficiency and to determine editing patterns, genomic DNAs are amplified and subject to Sanger sequencing or NGS. A few web-based tools are currently available for Sanger sequencing analysis (TIDE (Brinkman et al. 2014)) and NGS analysis (Cas-Analyzer (Park et al. 2017), CRISPResso2 (Clement et al. 2019), CRISPRAnalyzeR (Winter et al. 2017b), and CRISPR-Sub (Hwang et al. 2020)). High-throughput sequencing using NGS is considered more accurate and informative (Reuter et al. 2015).

5.3.4.1 TIDE

TIDE (http://shinyapps.datacurators.nl) is a web tool for estimating the frequency of insertions and deletions (indels) induced by CRISPR-Cas9, TALEN, and ZFN, with Sanger sequencing data. TIDE receives guide RNA sequences with a result file of Sanger sequencing for control and tested sample. On the result page, TIDE displays two graphs: (i) indel spectrum graphs with total efficiency and (ii) quality control graph. TIDE also provides a "batch mode" for analysis of several data files at the same time.

5.3.4.2 Cas-Analyzer

Cas-Analyzer (http://www.rgenome.net/cas-analyzer/) (Park et al. 2017) is a JavaScript-based implementation for NGS data analysis. Cas-Analyzer runs all processes in the client-side web browser; therefore, the process of uploading a huge NGS dataset to the server is not necessary. Cas-Analyzer receives a FASTQ file with reference sequence and target sequence. By clicking on the "submit" button, Cas-Analyzer shows progress of the analysis in real time. Then, the NGS data are aligned to the reference sequence and the aligned results are categorized by mutation type and sorted in descending order by count. In addition, the position and size of insertions or deletions are shown as interactive graphs on the results web page.

5.3.4.3 CRISPResso2

CRISPResso2 (https://crispresso.pinellolab.partners.org/) (Clement et al. 2019; Pinello et al. 2016) is a robust and easy-to-use analysis web tool. CRISPResso2 can analyze the NGS result of Cas9, Cas12a, base editors, and prime editors by changing the optional parameters. In addition, CRISPResso2 can analyze four data sets at the same time. CRISPResso2 basically receives FASTQ file, reference sequence, and target sequence. On the result page,

CRISPResso2 provides several graphs showing the frequency of each mutation. When "base editor" or "prime editor" input option is selected, CRISPResso2 additionally displays the pattern of each substitution and several graphs for users' convenience.

5.3.4.4 CRISPRAnalyzeR

CRISPRAnalyzeR (Winter et al. 2017a) is an end-to-end analysis platform for pooled CRISPR-Cas9 screens. CRISPRAnalyzeR receives sgRNA library FASTA files and NGS sequencing files. CRISPRAnalyzeR assesses the quality of sequencing data and evaluates the screen performance. CRISPRAnalyzeR offers eight different analysis strategies controlled by researchers at the fingertips and annotates the results with information using 26 external data resources. CRISPRAnalyzeR is available as a web tool (http://crispr-analyzer.dkfz.de/) or as a stand-alone version (https://github.com/boutroslab/CRISPRAnalyzeR).

5.3.4.5 CRISPR-Sub

Detecting the substitution induced by CRISPR-Cas9 is difficult because the substitution is also produced during PCR or NGS processes. CRISPR-Sub (http://www.rgenome.net/crispr-sub) (Hwang et al. 2020) is a statistical assessment tool for assessment of Cas9-mediated substitutions through high-throughput sequencing data. CRISPR-Sub receives reference sequence, PAM type (supporting 8 PAMs), target DNA sequence, and NGS results file for control and CRISPR-treated samples. On the result page, CRISPR-Sub displays a table including the number of each mutation pattern for control and CRISPR-treated data, two graphs for statistical information, and a heatmap of hotspot positions having high probability of CRISPR-induced substitution. At the bottom of the result page, CRISPR-Sub shows the alignment results by table. An off-line version of CRISPR-Sub is also available through github (Table 5.5).

5.3.5 Computational Tools for Base Editing and Prime Editing

Base editors, cytosine base editor (CBE) (Komor et al. 2016), and adenine base editor (ABE) (Gaudelli et al. 2018), respectively, enable C·G-to-T·A or A·T-to-G·C conversions within editing activity windows without generating a DNA double-strand break (DSB). CBEs typically convert C·G to T·A at 4th–8th nucleotides from the 5' end of the target sequence, whereas ABE usually converts A·T to G·C at 4th–7th nucleotides. Currently, there are few dedicated tools for base editors. BE-Designer was developed for designing sgRNA sequence for base editors and the prediction tools (DeepBaseEditor; DeepABE, DeepCBE, DeepCBE-CA) (Song et al. 2020) for the base editing efficiency of each target sequence were recently developed.

Prime editors that consist of a reverse transcriptase linked with nCas9 can induce the pattern of insertion, deletion, and substitution at the target region without generating DSBs and the need for additional donor DNA (Anzalone et al. 2019; Lin et al. 2020). Compared to base editors, prime editors have advantages, in terms of ability to induce mutation types. Prime editors require a prime editing guide RNA (pegRNA), which contains target sequence, primer binding sites (PBS), and reverse transcriptase template (RTT). Each target region has different optimal lengths of PBS and RTT; therefore, researchers need to consider various lengths of PBS and RTT in design of pegRNA sequences. PE-Designer (Hwang et al. 2021) and PrimeDesign (Hsu et al. 2020) are currently available for pegRNA design and PE-Analyzer (Hwang et al. 2021) is for NGS data analysis.

5.3.5.1 BE-Designer

BE-Designer (Hwang et al. 2018) (http://www.rgenome.net/be-designer/) provides a list of all possible sgRNA sequences for base editors, with useful information. BE-Designer receives a reference sequence, base editing type, PAM sequence, and reference organism (586 organisms). The base editing window can be controlled by the researcher. The result page shows a list of possible sgRNA sequences with sequence information such as editing window sequence, expected mutation of amino acid sequence, GC contents, and the number of off-targets with up to two mismatches. Base editor is commonly used to change the codon sequence; therefore, BE-Designer also shows the received reference sequence with amino acid sequences by translation. The result can be downloaded as a text file.

5.3.5.2 PE-Designer

PE-Designer (http://www.rgenome.net/pe-designer) is a user-friendly web tool for providing a list of all possible pegRNAs and nicking guide RNAs (ngRNAs) with useful information. PE-Designer receives a reference organism (586 organisms), PAM sequence, and two sequences (wild-type sequence as a reference sequence and induced sequence containing desired mutations). On the result page, PE-Designer displays a table of pegRNA sequences and interactive figures showing alignment results with possible pegRNA sequences. When selecting the target sequence of pegRNA, a list of possible pegRNA extension sequences and additional nicking gRNA sequences will be displayed as a table and an interactive figure. The length of PBS and RTT and the nicking distance can be controlled by researchers.

Table 5.5 Web tools for analysis.

Web tools	For CRISPR	For Base editors	For Prime editors	URL	Etc.
Cas-Analyzer	O	X	X	http://www.rgenome.net/cas-analyzer	In client-side
CRISPResso2	O	O	O	https://crispresso.pinellolab.partners.org	
CRISPRAnalyzeR	O	X	X	http://crispr-analyzer.dkfz.de	For CRISPR screening
TIDE	O	X	X	http://shinyapps.datacurators.nl/tide	For Sanger sequencing
CRISPR-Sub	O	X	X	http://www.rgenome.net/crispr-sub	For Substitution
BE-Analyzer	X	O	X	http://www.rgenome.net/be-analyzer	In client-side
PE-Analyzer	X	X	O	http://www.rgenome.net/pe-analyzer	In client-side

5.3.5.3 PrimeDesign

PrimeDesign (https://drugthatgene.pinellolab.partners.org/) is a user-friendly, end-to-end web tool for the design of prime editors. PrimeDesign only receives a sequence including information on the mutation sequence. PrimeDesign recommends a pegRNA and an ngRNA at the top of the page and displays a list of all possible pegRNAs at the bottom of the page with parameters such as the length of PBS, RTT, and nicking distance. After selecting the pegRNA spacer, PrimeDesign displays the pegRNA extensions and ngRNA spacers. When pegRNA extension and ngRNA spacer are selected, PrimeDesign displays edited DNA sequences and pegRNA secondary structure.

5.3.5.4 DeepBaseEditor

DeepBaseEditor (http://deepcrispr.info/DeepBaseEditor) is a web tool for predicting the efficiency of base editors using CNN. DeepBaseEditor receives the type of base editors (i.e. CBE or ABE) and target sequences as FASTA text or FASTA file. DeepBaseEditor finds all possible target sequences for the selected base editor and shows a table with a list of possible targets and the predicted efficiency for each target sequence. The results can be downloaded as an excel file. DeepBaseEditor is also available through github (https://github.com/MyungjaeSong/Paired-Library).

5.3.5.5 DeepPE

DeepPE (http://deepcrispr.info/DeepPE/) is a web tool for prediction of the optimal pegRNA sequence of prime editors using CNN. DeepPE receives the target sequence in FASTA format or FASTA file and provides the predicted efficiency of each possible target sequence. The results can be downloaded as an excel file.

5.3.5.6 BE-Hive

BE-Hive (Arbab et al. 2020) (https://www.crisprbehive.design/) is a tool for predicting editing outcomes of base editors using machine learning. BE-Hive receives target genomic DNA sequence, CRISPR protospacer, amino acid frame, and the type of base editor and cell type (supporting 21 combinations). BE-Hive shows the base editing outcomes with the expected frequency for each mutation and predicts base editing efficiency. The results can be downloaded as a scv file. BE-Hive offers a "batch mode" for selecting the most appropriate type of base editor. In batch mode, BE-Hive shows the expected efficiencies and mutation patterns of received sequences for all types of base editors. BE-Hive can also be used by python package.

5.3.5.7 BE-Analyzer

BE-Analyzer (Hwang et al. 2018) (http://www.rgenome.net/be-analyzer/) is a web tool for evaluating the NGS results for base editing experiments. BE-Analyzer was developed using JavaScript; therefore, BE-Analyzer analyzes NGS data in a client-side web browser without uploading NGS data to the server. BE-Analyzer receives FASTQ files, reference sequence, base editing type, base editing window, and target sequence. Optionally, BE-Analyzer receives FASTQ files for control data for comparing the substitutions induced by base editor with NGS or PCR errors. On the result page, BE-Analyzer displays a table for the number of each mutation and a table showing nucleotide mutations with an expected amino acid. BE-Analyzer additionally displays graphs showing the rates of nucleotide mutations and a table of alignment results.

5.3.5.8 PE-Analyzer

PE-Analyzer (http://www.rgenome.net/pe-analyzer/) is a web tool for assessing the NGS results for prime editing experiments. PE-Analyzer was also developed using JavaScript; therefore, BE-Analyzer analyzes NGS data in the client-side web browser without uploading NGS data to the server. PE-Analyzer receives FASTQ files, reference sequence, target sequence, information on desired mutations, and optionally additional nicking target sequences for PE3 or PE3b. For convenience, PE-Analyzer immediately displays received status so that users confirm their data set in real time. PE-Analyzer calculates the number of each mutation and displays the distribution graphs aligned with both reference sequences and the desired sequence. The alignment results are displayed as a table at the bottom.

5.4 Summary

The gene-editing field has shown rapid advancement since CRISPR-Cas9 nucleases were first harnessed for site-specific DNA editing in the human genome, and many CRISPR-associated tools that allow targeted gene disruption, recovery, and regulation have been developed. With rapid advancement of CRISPR-associated technologies, many useful web-based tools have also been developed for designing guide RNAs and for predicting the efficiency or mutation patterns of each target, as well as for analyzing NGS data after CRISPR treatment. Each program has pros and cons according to the initial aim of development; it is therefore necessary to understand the aim of each tool. Using relevant CRISPR genome editing tools with associated programs, researchers can make dreams come true, such as human disease control, development of drugs, better food production, and improving animals in the near future.

Funding

This research was supported by grants from the National Research Foundation of Korea (NRF) nos. 2020M3A9I4036072 and 2021M3A9H3015389 to S.B.

References

Abudayyeh, O.O., Gootenberg, J.S., Konermann, S. et al. (2016). C2c2 is a single-component programmable RNA-guided RNA-targeting CRISPR effector. *Science 353*: aaf5573.

Abudayyeh, O.O., Gootenberg, J.S., Essletzbichler, P. et al. (2017). RNA targeting with CRISPR-Cas13. *Nature 550*: 280–284.

Allen, F., Crepaldi, L., Alsinet, C. et al. (2019). Predicting the mutations generated by repair of Cas9-induced double-strand breaks. *Nat. Biotechnol.* **37**: 64–72.

Anzalone, A.V., Randolph, P.B., Davis, J.R. et al. (2019). Search-and-replace genome editing without double-strand breaks or donor DNA. *Nature 576*: 149–157.

Arbab, M., Shen, M.W., Mok, B., (2020). Determinants of Base Editing Outcomes from Target Library Analysis and Machine Learning. *Cell 182*: 463–480.

Bae, S., Kweon, J., Kim, H.S., and Kim, J.S. (2014a). Microhomology-based choice of Cas9 nuclease target sites. *Nat. Methods 11*: 705–706.

Bae, S., Park, J., and Kim, J.S. (2014b). Cas-OFFinder: a fast and versatile algorithm that searches for potential off-target sites of Cas9 RNA-guided endonucleases. *Bioinform. 30*: 1473–1475.

Brinkman, E.K., Chen, T., Amendola, M., and van Steensel, B. (2014). Easy quantitative assessment of genome editing by sequence trace decomposition. *Nucleic Acids Res. 42*: e168.

Broughton, J.P., Deng, X., Yu, G. et al. (2020). CRISPR–Cas12-based detection of SARS-CoV-2. *Nat. Biotechnol.* **38**: 870–874.

Cho, S.W., Kim, S., Kim, J.M., and Kim, J.S. (2013). Targeted genome engineering in human cells with the Cas9 RNA-guided endonuclease. *Nat. Biotechnol. 31*: 230–232.

Cho, S.W., Kim, S., Kim, Y. et al. (2014). Analysis of off-target effects of CRISPR/Cas-derived RNA-guided endonucleases and nickases. *Genome Res. 24*: 132–141.

Chylinski, K., Makarova, K.S., Charpentier, E., and Koonin, E.V. (2014). Classification and evolution of type II CRISPR-Cas systems. *Nucleic Acids Res. 42*: 6091–6105.

Clement, K., Rees, H., Canver, M.C. et al. (2019). CRISPResso2 provides accurate and rapid genome editing sequence analysis. *Nat. Biotechnol. 37*: 224–226.

Concordet, J.P. and Haeussler, M. (2018). CRISPOR: intuitive guide selection for CRISPR/Cas9 genome editing experiments and screens. *Nucleic Acids Res. 46*: W242–W245.

Cox, D.B.T., Gootenberg, J.S., Abudayyeh, O.O. et al. (2017). RNA editing with CRISPR-Cas13. *Science 358*: 1019–1027.

Doench, J.G., Fusi, N., Sullender, M. et al. (2016). Optimized sgRNA design to maximize activity and minimize off-target effects of CRISPR-Cas9. *Nat. Biotechnol. 34*: 184–191.

Edraki, A., Mir, A., Ibraheim, R. et al. (2019). A compact, high-accuracy Cas9 with a dinucleotide PAM for in vivo genome editing. *Mol. Cell 73* (714–726): e714.

Fu, Y., Foden, J.A., Khayter, C. et al. (2013). High-frequency off-target mutagenesis induced by CRISPR-Cas nucleases in human cells. *Nat. Biotechnol. 31*: 822–826.

Gao, L., Cox, D.B.T., Yan, W.X. et al. (2017). Engineered Cpf1 variants with altered PAM specificities. *Nat. Biotechnol. 35*: 789–792.

Gaudelli, N.M., Komor, A.C., Rees, H.A. et al. (2018). Publisher correction: programmable base editing of A*T to G*C in genomic DNA without DNA cleavage. *Nature 559*: E8.

Haeussler, M., Schonig, K., Eckert, H. et al. (2016). Evaluation of off-target and on-target scoring algorithms and integration into the guide RNA selection tool CRISPOR. *Genome Biol. 17*: 148.

Heigwer, F., Kerr, G., and Boutros, M. (2014). E-CRISP: fast CRISPR target site identification. *Nat. Methods 11*: 122–123.

Hou, Z., Zhang, Y., Propson, N.E. et al. (2013). Efficient genome engineering in human pluripotent stem cells using Cas9 from Neisseria meningitidis. *Proc. Natl. Acad. Sci. U. S. A. 110*: 15644–15649.

Hsu, P.D., Scott, D.A., Weinstein, J.A. et al. (2013). DNA targeting specificity of RNA-guided Cas9 nucleases. *Nat. Biotechnol. 31*: 827–832.

Hsu, P.D., Lander, E.S., and Zhang, F. (2014). Development and applications of CRISPR-Cas9 for genome engineering. *Cell 157*: 1262–1278.

Hsu, J.Y., Grünewald, J., Szalay, R. et al. (2021). PrimeDesign software for rapid and simplified design of prime editing guide RNAs. *Nat Commun. 12*: 1034.

Hu, J.H., Miller, S.M., Geurts, M.H. et al. (2018). Evolved Cas9 variants with broad PAM compatibility and high DNA specificity. *Nature 556*: 57–63.

Hwang, G.H., Park, J., Lim, K. et al. (2018). Web-based design and analysis tools for CRISPR base editing. *BMC Bioinform. 19*: 542.

Hwang, G.H., Yu, J., Yang, S. et al. (2020). CRISPR-sub: Analysis of DNA substitution mutations caused by CRISPR-Cas9 in human cells. *Comput. Struct. Biotechnol. J. 18*: 1686–1694.

Hwang, G.H., Jeong, Y.K., Habib, O. et al. (2021). PE-Designer and PE-Analyzer: web-based design and analysis tools for CRISPR prime editing. *Nucleic Acids Res 49*: W499–W504.

Jinek, M., Chylinski, K., Fonfara, I. et al. (2012). A programmable dual-RNA-guided DNA endonuclease in adaptive bacterial immunity. *Science 337*: 816–821.

Kim, E., Koo, T., Park, S.W. et al. (2017a). in vivo genome editing with a small Cas9 orthologue derived from Campylobacter jejuni. *Nat. Commun. 8*: 14500.

Kim, H.K., Song, M., Lee, J. et al. (2017b). in vivo high-throughput profiling of CRISPR-Cpf1 activity. *Nat. Methods 14*: 153–159.

Kim, H.K., Min, S., Song, M. et al. (2018). Deep learning improves prediction of CRISPR-Cpf1 guide RNA activity. *Nat. Biotechnol. 36*: 239–241.

Kim, H.K., Kim, Y., Lee, S. et al. (2019). SpCas9 activity prediction by DeepSpCas9, a deep learning-based model with high generalization performance. *Sci. Adv. 5*: eaax9249.

Kleinstiver, B.P., Prew, M.S., Tsai, S.Q. et al. (2015). Engineered CRISPR-Cas9 nucleases with altered PAM specificities. *Nature 523*: 481–485.

Komor, A.C., Kim, Y.B., Packer, M.S. et al. (2016). Programmable editing of a target base in genomic DNA without double-stranded DNA cleavage. *Nature 533*: 420–424.

Kushawah, G., Hernandez-Huertas, L., Abugattas-Nunez Del Prado, J. et al. (2020). CRISPR-Cas13d induces efficient mRNA knockdown in animal embryos. *Dev. Cell 54* (805–817): e807.

Labun, K., Montague, T.G., Gagnon, J.A. et al. (2016). CHOPCHOP v2: a web tool for the next generation of CRISPR genome engineering. *Nucleic Acids Res. 44*: W272–W276.

Labun, K., Montague, T.G., Krause, M. et al. (2019). CHOPCHOP v3: expanding the CRISPR web toolbox beyond genome editing. *Nucleic Acids Res. 47*: W171–W174.

Lei, Y., Lu, L., Liu, H.Y. et al. (2014). CRISPR-P: a web tool for synthetic single-guide RNA design of CRISPR-system in plants. *Mol. Plant 7*: 1494–1496.

Lin, Y., Cradick, T.J., Brown, M.T. et al. (2014). CRISPR/Cas9 systems have off-target activity with insertions or deletions between target DNA and guide RNA sequences. *Nucleic Acids Res. 42*: 7473–7485.

Lin, Q., Zong, Y., Xue, C. et al. (2020). Prime genome editing in rice and wheat. *Nat. Biotechnol. 38*: 582–585.

Liu, H., Ding, Y., Zhou, Y. et al. (2017). CRISPR-P 2.0: an improved CRISPR-Cas9 tool for genome editing in plants. *Mol. Plant 10*: 530–532.

Liu, Z., Shan, H., Chen, S. et al. (2019). Efficient base editing with expanded targeting scope using an engineered Spy-mac Cas9 variant. *Cell Discov. 5*: 58.

Ma, H., Wu, Y., Dang, Y. et al. (2014). Pol III promoters to express small RNAs: delineation of transcription initiation. *Mol. Ther. Nucleic Acids 3*: e161.

Makarova, K.S., Wolf, Y.I., Alkhnbashi, O.S. et al. (2015). An updated evolutionary classification of CRISPR-Cas systems. *Nat. Rev. Microbiol. 13*: 722–736.

Makarova, K.S., Wolf, Y.I., Iranzo, J. et al. (2020). Evolutionary classification of CRISPR-Cas systems: a burst of class 2 and derived variants. *Nat. Rev. Microbiol. 18*: 67–83.

Mali, P., Yang, L., Esvelt, K.M. et al. (2013). RNA-guided human genome engineering via Cas9. *Science 339*: 823–826.

Montague, T.G., Cruz, J.M., Gagnon, J.A. et al. (2014). CHOPCHOP: a CRISPR/Cas9 and TALEN web tool for genome editing. *Nucleic Acids Res. 42*: W401–W407.

Moreno-Mateos, M.A., Vejnar, C.E., Beaudoin, J.D. et al. (2015). CRISPRscan: designing highly efficient sgRNAs for CRISPR-Cas9 targeting in vivo. *Nat. Methods 12*: 982–988.

Naito, Y., Hino, K., Bono, H., and Ui-Tei, K. (2015). CRISPRdirect: software for designing CRISPR/Cas guide RNA with reduced off-target sites. *Bioinform. 31*: 1120–1123.

Nishimasu, H., Ran, F.A., Hsu, P.D. et al. (2014). Crystal structure of Cas9 in complex with guide RNA and target DNA. *Cell 156*: 935–949.

Nishimasu, H., Shi, X., Ishiguro, S. et al. (2018). Engineered CRISPR-Cas9 nuclease with expanded targeting space. *Science 361*: 1259–1262.

Park, J., Bae, S., and Kim, J.S. (2015). Cas-Designer: a web-based tool for choice of CRISPR-Cas9 target sites. *Bioinform. 31*: 4014–4016.

Park, J., Lim, K., Kim, J.S., and Bae, S. (2017). Cas-analyzer: an online tool for assessing genome editing results using NGS data. *Bioinform. 33*: 286–288.

Pinello, L., Canver, M.C., Hoban, M.D. et al. (2016). Analyzing CRISPR genome-editing experiments with CRISPResso. *Nat. Biotechnol. 34*: 695–697.

Ran, F.A., Cong, L., Yan, W.X. et al. (2015). in vivo genome editing using Staphylococcus aureus Cas9. *Nature 520*: 186–191.

Redding, S., Sternberg, S.H., Marshall, M. et al. (2015). Surveillance and processing of foreign DNA by the Escherichia coli CRISPR-Cas system. *Cell 163*: 854–865.

Reuter, J.A., Spacek, D.V., and Snyder, M.P. (2015). High-throughput sequencing technologies. *Mol. Cell 58*: 586–597.

Shen, M.W., Arbab, M., Hsu, J.Y. et al. (2018). Predictable and precise template-free CRISPR editing of pathogenic variants. *Nature 563*: 646–651.

Song, M., Kim, H.K., Lee, S. et al. (2020). Sequence-specific prediction of the efficiencies of adenine and cytosine base editors. *Nat. Biotechnol. 38*: 1037–1043.

Stemmer, M., Thumberger, T., Del Sol Keyer, M. et al. (2015). CCTop: an intuitive, flexible and reliable CRISPR/Cas9 target prediction tool. *PLoS One 10*: e0124633.

Strecker, J., Jones, S., Koopal, B. et al. (2019). Engineering of CRISPR-Cas12b for human genome editing. *Nat. Commun. 10*: 212.

Sun, J., Liu, H., Liu, J. et al. (2019). CRISPR-Local: a local single-guide RNA (sgRNA) design tool for non-reference plant genomes. *Bioinform. 35*: 2501–2503.

Swiat, M.A., Dashko, S., den Ridder, M. et al. (2017). FnCpf1: a novel and efficient genome editing tool for Saccharomyces cerevisiae. *Nucleic Acids Res. 45*: 12585–12598.

Walton, R.T., Christie, K.A., Whittaker, M.N., and Kleinstiver, B.P. (2020). Unconstrained genome targeting with near-PAMless engineered CRISPR-Cas9 variants. *Science 368*: 290–296.

Wang, D., Zhang, C., Wang, B. et al. (2019). Optimized CRISPR guide RNA design for two high-fidelity Cas9 variants by deep learning. *Nat. Commun. 10*: 4284.

Winter, J., Schwering, M., Pelz, O. et al. (2017a). CRISPRAnalyzeR: interactive analysis, annotation and documentation of pooled CRISPR screens. *biorxiv*.

Winter, J., Schwering, M., Pelz, O. et al. (2017b). CRISPRAnalyzeR: Interactive analysis, annotation and documentation of pooled CRISPR screens. *bioRxiv*.

Xu, H., Xiao, T., Chen, C.H. et al. (2015a). Sequence determinants of improved CRISPR sgRNA design. *Genome Res. 25*: 1147–1157.

Xu, K., Ren, C., Liu, Z. et al. (2015b). Efficient genome engineering in eukaryotes using Cas9 from Streptococcus thermophilus. *Cell. Mol. Life Sci. 72*: 383–399.

Xue, L., Tang, B., Chen, W., and Luo, J. (2019). Prediction of CRISPR sgRNA activity using a deep convolutional neural network. *J. Chem. Inf. Model. 59*: 615–624.

Young, J.K., Gasior, S.L., Jones, S. et al. (2019). The repurposing of type I-E CRISPR-Cascade for gene activation in plants. *Commun Biol 2*: 383.

Zetsche, B., Gootenberg, J.S., Abudayyeh, O.O. et al. (2015). Cpf1 is a single RNA-guided endonuclease of a class 2 CRISPR-Cas system. *Cell 163*: 759–771.

Part 2

Genome Editing in Disease Modeling

CRISPR Cas is making it easier to create cellular and whole-animal model systems that precisely mimic human diseases. This is enabling scientists to more accurately verify the safety and efficacy of drugs, which ensures that such models are better predictors of what will happen in patients. In this part, we discuss the general approaches used to make cellular and animal models of disease. Critical factors to consider for designing such experiments as well as practical examples from literature are discussed.

6

Genome Editing in Cellular Disease Models

Pierre Theurey, Kader Thiam, Yacine Chérifi, Alexandre Fraichard, and Amélie Rezza

genOway, Lyon, France

Translational research, i.e. the transfer of fundamental research knowledge into therapeutics, strongly if not entirely relies on the quality and relevance of preclinical models, in particular disease models. This "valley of death," as some researchers call it (Butler 2008), is a major obstacle in drug development and is the most important factor explaining the attrition phenomenon that all biomedical researchers and clinical practitioners seek to reduce. Disease models are thus essential tools that need constant optimization in order to improve and fasten drug development.

Preclinical models, *in vitro* in particular, have been used since drug discovery's inception and their relevance continues to increase with successive technological advances. One of the most "game-changing" advances has arguably been the development of efficient gene editing tools.

In this chapter, we discuss the variety of cellular disease models available today for drug development, their advantages, and limitations. We focus specifically on the improvements that gene editing allowed, the limitations that its use entails, and the critical factors to consider when used in cellular models.

6.1 Gene Editing and Disease Models in Drug Discovery

The quality and relevance of preclinical models are of paramount importance in drug discovery, as they must predict the effects, protective or deleterious, of a specific compound in patients. Disease models in particular are used to establish the mechanism of action of drugs, but also to assess their efficacy and safety. Depending on the disease, the compound, and the question to be answered, animal or cellular models can be used. However, before gene editing appeared, few relevant *in vitro* models were available. With the emergence of CRISPR (Clustered Regularly Interspaced Short Palindromic Repeats), gene editing has become accessible and easier, allowing for the development of numerous novel and relevant disease models.

6.1.1 Gene Editing Tools and the CRISPR Revolution

Targeted genetic modifications in mammalian cells have been used for a long time in biomedical research. It first started in the 1980s, with the targeted integration of a small exogenous DNA molecule, thanks to a newly discovered Homologous Recombination (HR) repair system (Smithies et al. 1985). Although this repair system allowed for specific integration, it was found to be very inefficient. It has however been optimized and routinely used in mouse embryonic stem (ES) cells to generate genetically modified mouse models (Capecchi 2005).

Since the 1990s, many efforts have been focused on increasing the efficiency of HR. Endonucleases, enzymes able to cut both strands of DNA, were shown to be particularly interesting. Indeed, by producing a double-strand break (DSB) at a desired sequence in the genome, specific DNA repair systems are triggered, leading to an increased occurrence of mutations at the targeted site. DSBs are repaired by one of the two following cellular DNA repair mechanisms: non-homologous end joining (NHEJ) which gives rise to small insertions or deletions (indels), or homology-directed repair (HDR) which can integrate an exogenous donor DNA at the cut site. Both pathways lead to the mutation of the targeted site, with HDR allowing the precise integration or replacement of a specific sequence (Komor et al. 2017).

Different endonucleases have been developed to efficiently and specifically induce DSBs in mammalian cells, such as meganucleases (Rouet et al. 1994), Zinc Finger

Genome Editing in Drug Discovery, First Edition. Edited by Marcello Maresca and Sumit Deswal.

Nucleases – ZFNs (Urnov et al. 2005) or Transcription-Activator-Like Effector Nucleases – TALENs (Miller et al. 2011). Although they significantly increase HR efficiency in mammalian cells, ZFNs and TALENs' routine use has been hampered by their difficult, lengthy, and expensive design. Indeed, for every new locus targeted with these tools, two fusion proteins must be designed, produced, and tested before the actual genetic modification can be efficiently performed. This technical limitation explains why, in 2013, when the CRISPR system was first used efficiently in human cells (Cho et al. 2013; Cong et al. 2013; Jinek et al. 2013; Mali et al. 2013b), it presented the potential to become the new go-to gold standard, as the efficient, easy to design and use gene editing tool. Indeed, this system requires just one protein, a CRISPR enzyme called Cas that can cut DNA, and a single RNA molecule – guide RNA, targeting a specific sequence. With this system, the targeting of a new locus requires the design and production of a single RNA molecule. CRISPR is originally a bacterial adaptive immune system that allows them to remember invading pathogens (Barrangou et al. 2007). By integrating in their genome a short sequence of invading DNA, bacteria can later recognize the intruders if/when they infect them again, cut their genomic DNA, to ultimately eliminate the threat. Since its discovery, the CRISPR system has been modified and adapted for use in a plethora of eukaryotic cells, reigniting or opening the possibility to perform gene editing in previously hardly modifiable models, such as non-ES cellular systems or non-mouse animal models (Doudna and Charpentier 2014).

The CRISPR system still shows limitations though. Numerous studies have shown that DSBs can also occur at untargeted sites in the genome leading to off-target mutations, and focused on lowering their occurrence (Mali et al. 2013a; Ran et al. 2013; Fu et al. 2014). Different groups have shown that the use of Cas9 enzymes can lead to unwanted genomic rearrangements at the targeted loci, suggesting that thorough characterization of CRISPR-modified material is of utmost importance (Kosicki et al. 2018; Cullot et al. 2019; Rezza et al. 2019). In addition, insertion of large exogenous DNA fragments using CRISPR is less efficient than integration of shorter molecules, and can be challenging in some models (Raveux et al. 2017; Rezza et al. 2019). Despite these drawbacks, CRISPR is the best, most efficient gene editing tool available today, and it is continuously adapted and optimized to overcome these shortcomings.

Considering its significant advantages over other endonucleases and its remarkable simplicity, CRISPR is now widely used, in one form or another, for many different fields of application. Among those, drug discovery has already greatly benefited from the emergence of gene editing, and CRISPR in particular. Virtually every stage of the preclinical drug development process, from target discovery to drug safety assessment, has been optimized and improved with the use of CRISPR (Scott 2018). Disease modeling, the generation of relevant practical models to test the efficacy, safety, and toxicity of compounds for a specific pathology, is today closely connected to the CRISPR system as it is massively used to generate *in vitro* and *in vivo* disease models.

6.1.2 Disease Modeling in Drug Discovery

The identification, study, and development of an efficient drug for the treatment of a specific pathology strongly rely on disease models. As previously mentioned, the translatability of the preclinical characteristics of a compound to the clinic is the most challenging step in drug discovery, and it mostly, if not entirely, depends on the quality of the disease models used in preclinical studies.

Historically, disease models consisted of unmodified wild-type systems, *in vitro* or *in vivo*, or overexpressing systems in which the disease-causing protein or a target of interest was exogenously expressed (Horvath et al. 2016). These models offered very limited possibilities and were of poor relevance. The emergence of gene editing tools, first with HR in ES cells to generate mouse models then with endonucleases (including CRISPR) in virtually any biological material, allowed the development of optimized, more relevant models. For example, it permitted to specifically target and insert in the host genome the disease-causing modifications identified in patients, or generate isogenic controls for patient-derived models. Interestingly, CRISPR is now globally and routinely used to generate optimized *in vivo* and *in vitro* disease models.

Animal models are very useful tools in preclinical research as they constitute comprehensive models containing all biological systems forming an organism. This includes hormonal signaling, circulating macromolecules, immune system, as well as intricate tissue-crosstalks in different organs. This level of complexity is important, if not mandatory, to study multisystemic diseases such as heart diseases and diabetes, or the interaction among tumor cells, the immune system, and the stroma in immuno-oncology. Although their development has become easier and cheaper with the availability of CRISPR, animal models still have their shortcomings as they show several experimental and ethical limitations. These include the restriction of drugs that can be tested *in vivo* due to toxicity and pharmacodynamic issues, the technical difficulties to perform some experimental measurements, such as imaging or oxigraphy for certain organs, or gene expression modulations through overexpression or si-RNA knockdown (Arias et al. 1998; Hatakeyama et al. 2016). Importantly, it can be challenging to overcome species-specific physiological and genetic differences for some disease modeling, thus leading to a major problem of translatability (Wendler and Wehling 2010). Indeed, human diseases that represent the major current challenges in public health have been repeatedly virtually "cured" in animal models, but resulted in failure in clinical trials. Last but not least, the ethical question of the use of animal models must always be asked, even more so as it has been described that disease modeling *in vivo* can sometimes induce a strong phenotype leading to overwhelming suffering (Brown and Murray 2006).

Cellular models can represent more practical relevant alternatives. One major advantage of cellular models is that human cells can be used. Also, there are a variety of *in vitro* culture methods available today that can increase the physiological relevance of a cellular model. As mentioned, before the emergence of gene editing tools, cellular models were limited to the use of non-modified cells and overexpression models with poor physiological relevance. Some diseases could simply not be recapitulated *in vitro* considering these limitations. Interestingly, a variety of optimized culture methods have been developed in the past 20 years, in parallel to gene editing tools, leading to a continued improvement of disease models (Horvath et al. 2016).

The recent discoveries and progress made in the field of gene editing have profoundly changed biomedical research. They have been used in a wide array of applications, including the improvement of disease modeling.

6.2 Variety of Cellular Disease Models and Their Improvement with Gene Editing

Cellular disease modeling is often limited by the availability of a simple and practical culture method of a cell type of interest. In the past 20 years, different innovative technological advances in cellular biology have allowed the development of new culture methods, and thus new cellular systems. During the last two decades as well, the development and optimization of gene editing tools such as CRISPR had a significant impact on disease modeling.

6.2.1 Cell Lines and Primary Cultures

Traditionally, cellular biomedical models are established cell lines or freshly isolated primary cells. Both categories have advantages but also very constraining limitations (Table 6.1).

Established cell lines have been largely used in drug discovery, particularly in early phases of drug development such as target validation, compound screening, and mechanism of action. However, the very characteristics of established cell lines represent limitations for their relevance as disease models. To obtain a cell line, cells from a donor sample must be isolated and immortalized to allow virtually infinite proliferative capacity *in vitro*, and thus the establishment of a novel cell line. This process of immortalization, as well as long-term culture, often induces the accumulation of mutations, chromosomal aberrations, or epigenetic reprogramming that alters the original physiology of the cells (Nestor et al. 2015). Cancer can seem to be the easier pathology to model *in vitro* as many cancer cell

Table 6.1 Variety of *in vitro* systems available for cellular disease modeling: characteristics, advantages, and limitations.

	Established cell lines	Primary cultures	Human iPS-derived cells	Organoids
Relevance to human/patients	+	++	++	+++
Availability	+++	+/−	++	+
Scalability	+++	+/−	++	++
Lifespan	++	+/−	++	++
Gene editing efficiency	++	+/−	++	+/−
Limitations	- altered physiology - molecular and karyotypic aberrations	- limited availability, lifespan, and scalability	- partial differentiation, immature differentiated cells	- lack of cellular components - variability: no standardized protocol
Possible improvements	- 3D culture: relevance improved for some cell lines	- 3D culture: lifespan lengthened	- organoids: enhanced scalability and improved relevance	- culture optimization and vascularization: eliminate necrosis within organoids, remove size limitation

lines are available. Indeed, cancer transformation (i.e. accumulation of mutations) can facilitate the maintenance of cells *in vitro*, and thus the establishment of cell lines. However, they often show very poor relevance to the tissue of origin as they are grown in artificial environments and lose key physiological characteristics (Horvath et al. 2016). Thus, established cell lines can represent poor cellular models, particularly for physiological studies such as cellular specific mechanisms of action, safety, and toxicology.

In order to improve the biological relevance of established cell lines as disease models, overexpression of a gene of interest has been largely used. By overexpressing the disease-causing protein in a corresponding tissue-specific cell line, numerous researchers have created improved cellular disease models (Marotta et al. 1989; Hait et al. 1993; McLeskey et al. 1996). This approach obviously has limitations as it can only apply to a subset of diseases, and does not entirely recapitulate the disease-causing mutation, as it is not present in the genome but only in an exogenous chromatin environment. Gene editing, particularly CRISPR, has allowed to overcome these shortcomings, and researchers can now specifically target sequences of interest to introduce disease-causing mutations in one or multiple genes (Burgener et al. 2019), delete integrated viral DNA as a therapeutic strategy (Ramanan et al. 2015), or invalidate a specific gene for target validation (Naylor et al. 2016). In addition, gene editing allows to model diseases caused by mRNA splicing defects such as exon skipping (Gravesteijn et al. 2020) and cryptic exon inclusion (Sanz et al. 2017). Finally, CRISPR has also been used to target noncoding regulatory sequences in cellular models (Hnisz et al. 2016). Thus, gene editing, when used even in imperfect cellular systems such as established cell lines, can allow the development of novel disease models.

Primary cultures and patient-derived cellular models have been used to provide clinically more relevant models than transformed cell lines (Pieters et al. 1990). As they are directly isolated from the organ of interest, they display more physiological characteristics than established cell lines. Their advantages can be well illustrated by primary human hepatocytes. Human hepatocytes cell lines, such as the hepatocarcinoma cell line HepG2, might be easy to culture, but these cells lack a major characteristic of human hepatocytes, Cyp inducibility, which is recognized as an indicator of toxicity in drug development and liver physiology. Indeed, none of the immortalized cell lines available today completely compare with freshly isolated hepatocytes cultured in optimal conditions, in terms of drug metabolism and Cyp inducibility (Berger et al. 2016).

Gene editing has been used in primary cultures as well, to target regulatory sequences (Flavahan et al. 2016), to modify hematopoietic stem cells and model myeloid malignancies (Heckl et al. 2014; Abrahimi et al., 2015), to ablate specific genes in human endothelial cells (Abrahimi et al. 2015), or to test antiviral therapeutic strategies (Ebina et al. 2013; Hu et al. 2014). Unfortunately, primary cultures, meaning isolated cells retaining major features of their tissue of origin, can be hard to obtain (for example, human neurons). Additionally, they have a short lifespan in culture and can thus hardly be scaled for routine use in the drug development pipeline. This often means that a new batch of primary cells must be used, if not re-isolated, almost for every experiment, rendering them very tedious to manipulate.

Although established cell lines and primary cultures represent interesting cellular models, their limitations (biological relevance, availability, scalability, or *in vitro* lifespan – see Table 6.1) make them suboptimal cellular models for drug discovery. Here again, thanks to technological advances, new types of *in vitro* systems have been optimized in the past 20 years, most remarkably 3D models such as organoids, and induced pluripotent stem (iPS) cells.

6.2.2 Organoids: a Promising 3D Culture System

In vitro culture of cells historically started with the maintenance of cells as a monolayer or in suspension, so-called "2D systems." However, more complex culture systems have been developed to lengthen cells' lifespan in culture while maintaining their original physiological features. Such systems include co-culture of different cell types, bioreactors, bioprinted tissues, but also organoids (Horvath et al. 2016).

It has been known for more than half a century that freshly isolated cells, when cultured in the right "environment," can form a smaller, simpler version of their organ of origin. These structures, called mini-organs or organoids, presenting self-organizing properties were first observed with embryonic chick tissues (Weiss and Taylor 1960), but have now been observed in many developing vertebrate organs. These 3D structures have been shown to form only under very specific spatial constraints, from stem cells giving rise to several differentiated cell types. For example, Hans Clevers and collaborators showed that they could recreate mini-guts *in vitro*, with crypts and flat villi, from intestinal epithelial stem cells (Sato et al. 2009). These multipotent cells can give rise to all epithelial cells of the intestine, absorbent, and secretory cell types that self-organize in crypts and villi to form organoids that can be kept in culture for months. Organoids can be formed from pluripotent stem cells (PSCs), embryonic or induced, or organ-specific adult stem cells, and as of today, they can be used as a model for more than 10 organs from the gut to the brain (Artegiani and Clevers 2018; Kim et al. 2020).

At their inception, organoids were mostly used by developmental biologists as a tool to study organogenesis. However, in the past 15 years, they have regained popularity in many other fields of research, and are being recognized as powerful models for various applications. They have been used as a model for infectious diseases, genetic diseases such as Cystic Fibrosis (CF), and also cancer (Artegiani and Clevers 2018). For example, researchers identified new biomarkers for rare types of cancer by using these 3D structures to expand cells from three primary liver cancers and perform gene expression analysis (Broutier et al. 2017). Interestingly, organoids have already had an impact in drug discovery. They are used as a culture method to extend *in vitro* lifespan of patient-derived samples, thus creating biobanks available for drug testing (Dekkers et al. 2016). The FDA already expanded its indication of Kalydeco, a CF drug, from 10 to 33 mutations in the CFTR gene (Ratner 2017), solely based on *in vitro* tests on patient-derived intestinal organoids, clearly demonstrating their promising preclinical potential. Organoids have also been used as cellular models for toxicity studies in different organs such as the kidney (Morizane et al. 2015) and liver (Sgodda et al. 2017). They are also useful models for drug screening, as they have already been used for pharmacodynamic profiling of tumors (Vaira et al. 2010; Rookmaaker et al. 2015). As organoids can be formed from patient-derived primary samples or patient-derived iPS cells (see next section), they represent a powerful tool for personalized medicine. They have been used successfully to model disease heterogeneity in drug screening for different pathologies such as pancreatic ductal adenocarcinoma (Huang et al. 2015), or breast cancer (Sachs et al. 2018), and they even allowed the identification of optimized treatments for patients suffering from CF (Dekkers et al. 2016; Bartfeld and Clevers 2017). Interestingly, organoids have recently been used to study how SARS-CoV2, Covid-19-causing agent, affects different tissues such as the intestine (Lamers et al. 2020), blood vessels (Monteil et al. 2020), or the brain (Song et al. 2021).

Considering their potential as relevant disease models, researchers have tried to optimize and/or modify organoids using targeted genetic modifications. As organoids are self-organizing 3D structures that must be maintained in a closely defined environment, gene editing is more challenging in these models than in 2D systems. However, several groups have successfully genetically modified organoids, thus improving further their cellular disease models, and in some cases correcting the disease-causing mutation. Indeed, CRISPR was used to correct the F508del CF-causing mutation in the *CFTR* gene in intestinal organoids (Schwank et al. 2013), proving that CRISPR can be efficiently used in these 3D structures, but also demonstrating its potential in therapeutics. Other studies have successfully used CRISPR in organoids to model colorectal cancer (Drost et al. 2015; Matano et al. 2015), polycystic kidney disease (Freedman et al. 2015), fibrotic lung disease (Strikoudis et al. 2019), or brain tumorigenesis (Bian et al. 2018).

Although organoids have an outstanding potential as cellular disease models, they still present some flaws (Kim et al. 2020): cellular components can be lacking; their structure is not always relevant to the *in vivo* organ; and standardized protocols are not systematically established leading to a variability and heterogeneity of the models (Table 6.1). Interestingly, as organoids have been shown to form from PSCs, combining organoid and iPS technologies present a lot of advantages, including the fact that CRISPR is easier and more efficient in PSCs than in most other cellular systems.

6.2.3 Human iPS Cells

As previously mentioned, the availability of patient-derived tissues can be a limiting factor for cellular systems such as primary cultures or adult-tissue-derived organoids. Scaling-up these cellular models for high-throughput applications is also often challenging. An alternative approach to obtain relevant cellular models is to differentiate PSCs into functional cell types to recreate at least part of a tissue of interest. Although they can be useful and show great potential, the routine use of human ES cells can be challenging and is closely regulated. Thus, when Yamanaka and colleagues first showed that adult differentiated cells could be reprogrammed into iPS cells, they opened the door to a plethora of applications in various fields of interest, including disease modeling and personalized medicine. Indeed, by overexpressing four transcription factors in human skin cells, they showed that adult differentiated cells could acquire pluripotency (Takahashi et al. 2007). This discovery was swiftly identified as a major medical breakthrough that could transform cell therapy, but as of today it has rather profoundly transformed preclinical research and drug discovery (Sharma et al. 2020).

Numerous groups have defined and optimized protocols to differentiate iPS cells into differentiated cell types. Human iPS cells can be used to model development and/or disease for more than 10 tissues, from the brain (Brennand et al. 2011; Israel et al. 2012), to the liver (Hannan et al. 2013; Huang et al. 2014), or the heart (Lian et al. 2013; Burridge et al. 2014). The potential unlimited availability of human neurons, hepatocytes, and cardiomyocytes from healthy donors and patients is particularly attractive for drug screening and toxicity. Human iPS-derived cell types have been extensively used to test compounds for neurodegenerative diseases (Chung et al. 2013; Xu et al. 2013), heart defects (Matsa et al. 2011; Liang et al. 2013), or liver diseases (Choi et al. 2013; Cayo et al. 2017). Interestingly, human iPS cells' potential for drug safety assessment has already been recognized: the Comprehensive *In vitro* Proarrhythmia Assay initiative's new paradigm suggests to systematically include *in vitro* assays in hiPS-derived cardiomyocytes when testing compounds' cardiac toxicity (Gintant et al. 2017). This organization, focusing on setting optimized rules to test new molecules' potential cardiotoxicity, regroups researchers, medical agencies, and nonprofit institutions from all over the world. As these cells are accessible and scalable, iPS cells represent a promising cellular model for drug discovery.

Here again, in order to optimize or create novel cellular disease models, researchers have used gene editing in human iPS cells. Importantly, gene editing allows the creation of isogenic control cell lines for patient-derived iPS cells, or mutated cells from healthy donor-derived iPS cells. Indeed, iPS cells derived from patients suffering from disease caused by identified genetic mutations are very useful disease models, but researchers need appropriate "control" iPS cells for comparison. Human iPS derived from a "healthy" tissue of the same donor or from a healthy donor's same tissue are not appropriate controls. Thus, correcting the genetic mutations causing the disease in the patient-derived iPS cells is by far the optimal way to obtain isogenic control cells. This approach has been used to create isogenic hepatocytes (Maetzel et al. 2014) or neurons (Soldner et al. 2011; Zhang et al. 2017) among many other cell types. CRISPR has also been used in iPS cells to model a range of diseases such as cardiomyopathy (Ang et al. 2016), Alzheimer's disease (Paquet et al. 2016), neonatal diabetes (Saarimäki-Vire et al. 2017), or retinitis pigmentosa (Foltz et al. 2018). As previously mentioned, iPS cells can be used to form organoids, allowing the production of a vastly greater number of cells than in 2D culture (Czerniecki et al. 2018), and in some cases enhancing their capacity to differentiate (Lancaster et al. 2013; Takebe et al. 2013; Low et al. 2019).

Considering their accessibility, their scalability, and the availability of protocols to differentiate them in many cell types, human iPS cells have a tremendous potential in personalized medicine as they have been successfully used for patient-specific disease modeling, for the study of patient-specific drug response, or patient-specific drug screening. However, these cellular models also show limitations (Table 6.1). Human iPS-derived cells often show immature characteristics, resembling more fetal than adult differentiated cells, and the efficiency of differentiation can vary between donors and individual experiments (Sharma et al. 2020).

There are, indeed, a variety of cellular disease models available today, and they all show advantages and limitations (Table 6.1). As of today, there is no single preferred system or culture technique for disease modeling, as it depends on the studied disease or organ, the biological question, and most of all, the application.

6.3 Choosing and Designing a Relevant Genetically Engineered Cellular Disease Model

As discussed in Sections 6.1 and 6.2, gene editing has greatly improved cellular models, and today researchers have to choose among a plethora of existing models and many possibilities of new designs to best fit their scientific needs. In this profusion of options, the core criterion to best choose and design a model must be its application and physiological relevance. A model being a tool to best recapitulate a physiological environment, it must be designed to answer a specific biological question, rather than the question being adapted to the model. This section will address the scientific reasoning of model creation, in particular the preliminary design assessment. The purely technical considerations of model design and generation will be discussed in Section 6.4.

6.3.1 Criteria for an Educated Choice of Cellular System

To generate the best, most relevant disease model for a specific need and application, its design must not only focus on the genetic alteration to perform, but also most importantly on selecting the appropriate *in vitro* system, or cell line, to reproduce a physiologically relevant setting (Figure 6.1). Indeed, cell lines offer the possibility to access a wide variety of cell types and species, and picking the right one is a key criterion to produce robust data for a successful project. There are different features that must be examined while selecting a cell line to generate a genetically modified cellular disease model:

- Species and translatability

  Cellular disease models are tools to recapitulate specific characteristics of human pathologies, thus human cells can appear as the best choice for any applications. Importantly, translatability can be compromised by phenotypic differences between species only revealed after genetic editing.

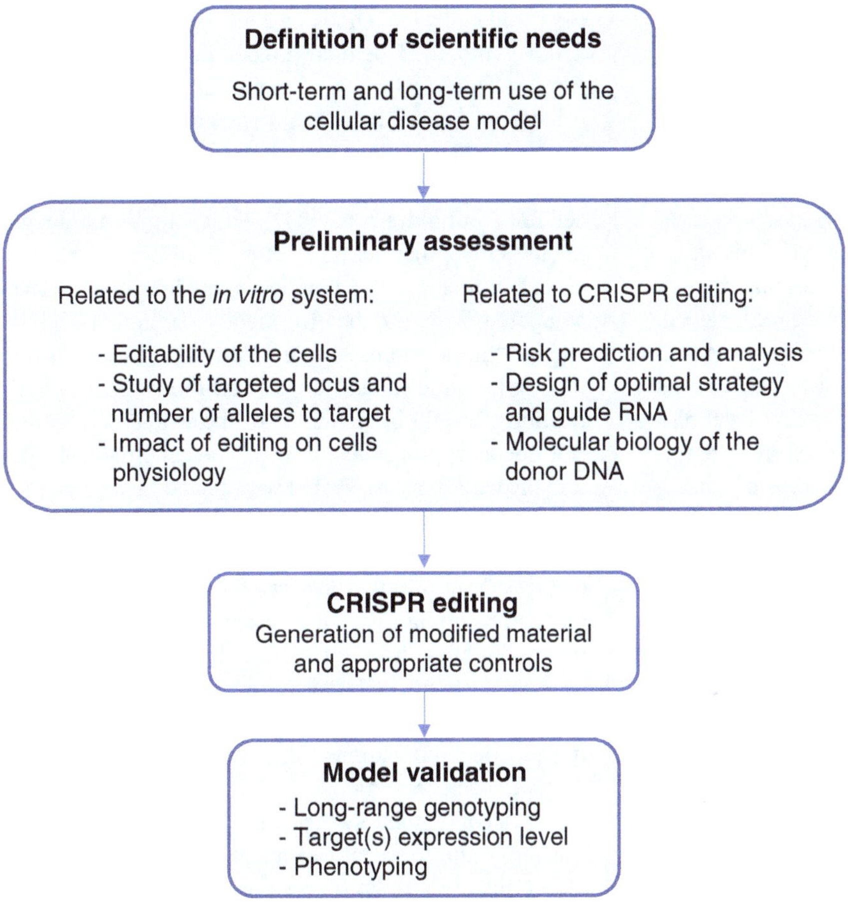

Figure 6.1 Critical steps for the generation of a relevant cellular disease model.

Indeed, when creating a murine cell line to model a human disease (e.g. cancer or neurodegenerative diseases), the inserted mutations may have a different impact on human and mouse cellular physiology, and can ultimately result in a different phenotype than expected (Dhillon et al. 2010). For molecular studies of human pathophysiology or toxicity, human cells will very often be preferred. However, experiments to set up a proof of concept, to be later confirmed *in vivo* in mouse or rat models, would rather be performed in a corresponding mouse or rat cell line.

- Cell type

Whenever possible, a relevant cellular disease model should be generated using the corresponding cell type affected in the modeled human pathology. Thus, a cell line consistent with the tissue of interest is most often preferred; for example, Huh7/HepG2 cells are preferred to HeLa cells to study liver disease (Müller and Sturla 2019). In addition to the cell type of interest, the differentiation status, or maturation state, must also be considered as it can greatly impact cellular physiology (Hart et al. 2010). Most often when designing a cellular disease model, a compromise has to be made between the optimal cellular system available (primary cells or organoids for example), and the technical feasibility and efficiency of gene editing in these particular cells.

- Targeted allele's copy number

Another important criterion to evaluate when choosing a cellular system to be genetically edited is the number of alleles of the targeted gene. Indeed, established cell lines, in particular cancer cells, can undergo extensive genomic rearrangements leading to an aneuploidy (Zasadil et al. 2013). Beyond the technical challenges that multiple alleles' targeting represents (see Section 6.4), a polyploidy of the gene to target can also impact the phenotype of mutant models. If the gene of interest is present in three copies instead of two, the ratio between mutated and wild-type alleles can be different in the edited cellular model (1 mutant and 2 wild type) compared with patients (1 mutant and 1 wild type). It is hard to predict the exact phenotype associated with this genotype, and it might differ from a true heterozygous situation, as observed in patients. Of note, a combination of mutant, knock-out (KO), and wild-type allele(s) can be an alternative to create heterozygous- and homozygous-like models in case of polyploidy.

- Target gene's expression level
  It is always preferred to confirm the expression level of a targeted gene in the specific cell line of interest. Even when this gene has been described as expressed or is a known marker of the cell type of interest, its expression can vary over passages *in vitro*. Genetic instability over long periods of time is an intrinsic characteristic of immortalized cell lines (Hughes et al. 2007). This variability can impact several features of a cellular model, including its capacity to be genetically edited. Indeed, expression patterns of multiple genes are evolving and changing over passages and successive generations (O'Driscoll et al. 2006); hence, the expression level of a targeted gene must be carefully assessed before attempting its modification. Importantly, the process of editing cells takes several passages, and the expression level of the targeted gene can thus change over the course of its modification.

As previously mentioned, in addition to the criteria described here, choosing a cell line to edit also depends on its editability, i.e. the potential efficacy of gene editing in this specific cell line. The technical aspects impacting cell line editability will be discussed in Section 6.4, but it is important to consider that designing a relevant cellular model is very often a compromise between optimal cellular characteristics and feasibility of gene editing in a specific cellular system.

6.3.2 Impact of Gene Editing Process on the Long-Term Viability and Physiology of a Cellular Model

The production of a good cellular model requires careful designing including selecting the ideal cell line and genetic strategy, but it must also include the evaluation of the potential impact of gene editing and culturing on the model's physiology in the long term (Figure 6.1). Two characteristics must then be considered:

- Cell lines' stability and drift
  As previously mentioned, genomic instability over generations is an intrinsic feature of cell lines. Indeed, it has been shown that overexpressing cellular models can exhibit variations of the transgene's expression level over time (Bailey et al. 2012). Interestingly, similar variations can be observed when editing endogenous genes (O'Driscoll et al. 2006). Genomic instability can also impact the reproducibility of results, as they can differ for experiments performed in early and late passages of the same cellular model. Thus, experiments should only be considered repetitions when performed in cells with the same number of passages. This variability must also be considered when choosing controls in experiments. If a novel modified cellular model has been generated through clonal selection, it might not be appropriate to use the parental cellular culture as a control. Indeed, the selected subpopulation, now homogeneous, can exhibit different characteristics than the heterogeneous parental cell line (Mohindra et al. 2002). For example, a homogeneous subpopulation of a tumor cell line can show altered tumor growth kinetics upon injection *in vivo*, compared with the heterogeneous parental cell line. Hence, "mock" clones, obtained through the exact same process than the edited clones but without genetic modifications, should be preferred to the wild-type parental cell line as appropriate controls. The genomic and phenotypic variations associated with the culturing of cellular models resemble a drift much like the genetic drift described in animal models. As a consequence, the same cell line from different sources (laboratories or providers) can have slightly different properties such as population doubling time, or ploidy. It is then of utmost importance to use cell lines from a validated source, and confirm all previously described criteria before performing genetic modifications.
- Impact of gene editing
  As when contemplating the generation of *in vivo* models, it is important to try and anticipate the potential impact of the inserted mutation on the cell line viability and physiology. For instance, cancer models carrying mutations in proteins involved in DNA repair mechanisms are more unstable and show an accelerated genetic drift over passages (Sousa et al. 2015). Similarly, cellular models with mutations in genes involved in cell growth and cell cycle regulation may have an altered growth rate. It is particularly important to closely monitor the impact of gene editing on cell physiology when the cellular model is to be used for a specific application dependent on a precise feature. For example, when genetically modifying cell lines used in syngeneic models, it is crucial to preserve an acceptable growth rate, but also their ability to form tumors when injected *in vivo* (Woan et al. 2015). Beyond viability and growth rate, introducing a mutation in a specific gene can have an impact on the very feasibility of the genetic modification. For instance, a mutation in a gene involved in HR may result in a decrease of HR efficiency, and thus a decrease in gene editing efficiency (Yoshino et al. 2019).

6.3.3 Disease Modeling Through Reproduction of Human Mutations *in vitro*

For genetic diseases caused by identified mutations in a given gene, the most straightforward method to produce a cellular disease model is to introduce the mutations observed in patients into an *in vitro* model. Disease-causing mutations can lead to different genetic alterations such as the introduction of a premature STOP codon, a switch of amino acid(s), or exon skipping. These modifications are associated with different outcomes at the protein level ranging from the expression of a mutated protein to no expression at all, or the expression of a truncated form of a protein. Hence, to generate a relevant cellular model, it is important to ensure that the introduced mutation will result in the same protein alteration than in patients. The mere copying of the genetic sequence can be sufficient when working with human cells, but when reproducing a disease-causing mutation in a different species, such as mouse or rat, the genetic sequence leading to the altered protein of interest can be different from the sequence observed in patients.

In addition, while faithfully reproducing the mutation observed in patients is the ultimate strategy, some critical aspects have to be considered for the scientific robustness of the end product. Here again a compromise often has to be made among the strict reproduction of the genetic sequences observed in patients, the technical optimizations that could increase gene editing efficiency, and the long-term viability of the model. For example, introducing a silent mutation to disrupt CRISPR cutting site, when the mutation of interest does not destroy it, is an unavoidable step to efficiently generate a stable mutant model (Okamoto et al. 2019). Considering this necessity, it is crucial to select a silent mutation that would have a minimal impact on the target and neighboring genes. It is thus important to optimize the silent mutation introduced to avoid CRISPR recutting, but also prevent genetic dysregulation. To do so, *in silico* predicting tools can be used to anticipate the effect of mutations on neighboring genes (Jian et al. 2014; Sung et al. 2014; Riffo-Campos et al. 2016; Yevshin et al. 2017; Khamis et al. 2018). Note that while predicting tools can be very helpful to identify potential risks of dysregulation, an empirical approach through preliminary *in vitro* testing can maximize the chances of success of the model creation. However, an unexpected impact of the editing can still exist, and a thorough validation of the genetically modified model is absolutely necessary.

To thoroughly characterize CRISPR-modified material, the target locus must be genotyped using long-range sequencing, rather than short-range characterization, as it has been described that CRISPR cuts can induce unexpected genomic rearrangements (Kosicki et al. 2018; Cullot et al. 2019; Rezza et al. 2019). Additionally, as CRISPR-induced DSBs can be repaired by introduction of donor DNA (HDR) or random insertion/deletion of nucleotides (NHEJ), the targeted alleles can exhibit different sequences while resulting in the same phenotype. For example, a triploid targeted gene in a cancer cell line can exhibit one allele that has inserted a donor DNA with a STOP codon and 2 alleles carrying indels resulting in a frameshift both leading to the appearance of a STOP codon. The modified cells are indeed invalidated for the targeted gene, although they show three different sequences for the locus of interest. When introducing a mutation of interest, such combinations can lead to a model expressing exclusively mutated protein instead of the heterozygous/wt ratio expected in a heterozygous model. In addition to carefully genotyping the targeted locus, the expression level of the gene of interest, as well as surrounding genes when necessary, should be validated, ideally at the protein level (Figure 6.1).

As previously mentioned, the robustness of a cellular disease model does not only lie in its proper validation but also in the use of appropriate controls. The very process used to perform CRISPR-editing in cells can impact their physiology, thus introducing a bias in the experimental outcome. As discussed, unmodified control "mock" clones obtained through the same procedure than the edited cells, ideally with the same number of passages, should be preferred to the unmodified parental cells as a control.

6.3.4 Alternative Cellular Models Relevant to Diseases' Pathophysiology

Producing a relevant cellular disease model can also be achieved without reproducing the exact mutations found in patients.

In some cases, the crucial role of a specific gene in the disease's pathophysiology might have been described in patients and validated *in vitro* through overexpression or silencing experiments, but the exact genetic mutation(s) involved may not have been clearly identified (Marotta et al. 1989; Hardy 2017). Alternatively, multiple mutations in a given gene may have been observed in patients which would best be recapitulated *in vitro* by a loss or gain of function model, without reproducing a specific mutation. For instance, KO cellular models can be a relevant tool to study a disease in which multiple loss of function mutations have been identified (Cacheiro et al. 2019). Similarly, an overexpressing model can be a valuable tool for gain of function mutations as well as gene duplication or expansion (Kirik and Björklund 2003). Of note, when considering a model of loss-of-function or a nonsense mutation from patients, a constitutive KO invalidating all expression of the protein of interest may not represent the best approach. Indeed, even if the canonical function of the protein is abolished in patients, the presence of a remaining inactive protein or peptide may be necessary for the pathophysiology of the disease. Although not strictly reproducing patients' mutations at the genomic level, the creation of such models must be carefully designed as it offers significant room for optimization.

The generation of a genetically modified model requires the careful study of all the isoforms of the targeted gene, to ensure their invalidation and the anticipation of the potential expression of truncated proteins. To this regard, when generating a KO model, the precise integration of a premature STOP codon through HDR of a donor oligonucleotide is often preferred to random indels. Similar biological and technical aspects must be considered when designing overexpressing cellular disease models. Indeed, some diseases are characterized by an overexpression of a specific isoform, unmodified or mutated, which should be recapitulated in a relevant cellular model. Overexpression can be achieved by random insertion of an expression cassette, as it was performed before the emergence of targeted gene editing, or by targeted insertion in a specific locus. Choosing between both genetic strategies strongly depends on the application and experimental conditions in which the cellular model will be used. Random integration of the transgene can disturb endogenous elements leading to an affected transcriptomic profile, proliferation, or physiology. The expression level of the inserted cassette also strongly depends on the site of insertion, due to a "positional effect," and the transgene can be ultimately silenced over passages. Targeted insertion in a neutral locus is more time consuming, but the integration is stable over passages and it mostly protects from positional effect (Bronson et al. 1996; Chen et al. 2011). It is important to note that, even when using a specific promoter in a neutral locus, the expression level of the gene of interest can vary from one tissue to another (Tchorz et al. 2012). Finally, and as discussed in the following paragraph, clones can prove more reliable and relevant than heterogenous, potentially unstable pools of cells.

The successful generation of a relevant cellular disease model strongly relies on the correct choice of cellular system. Additionally, carefully designing the model generation is of utmost importance when performing gene editing.

6.4 Technical Considerations of Gene Editing in Cells

As mentioned in the previous section, the first step to generate a relevant cellular disease model, or any model, should be the precise identification of the biological question it should answer, and its field of application. The technical considerations or the choice of gene editing tool to be used are only secondary. There are however critical aspects to consider when performing genetic modifications in a cellular model.

6.4.1 Technical Considerations Related to the *in vitro* System's Characteristics

There are critical aspects of gene editing that depend on the genetic targeting itself and are common to the generation of any model, *in vitro* and *in vivo*. In addition, there are also features of gene editing in cells that are linked to the very characteristics of the chosen *in vitro* system. As discussed in Section 6.2, there are a wide variety of *in vitro* models, such as cell lines, primary cultures, or organoids, and many different versions of each. For example, there are thousands of established cell lines, all with their specific phenotype and genotype. Thus, there are intrinsic parameters to a specific cell line that impact its editability and the probability of success of its genetic modification (He et al. 2016).

In order to efficiently perform gene editing in an *in vitro* model, cells need to be transfectable. Depending on the cellular model, this can be challenging and heavy empirical optimization or specific technologies might be needed. Gene editing of organoids, for example, requires the efficient introduction of CRISPR reagents in all cells, best performed with viral vectors (Nie and Hashino 2017; Artegiani and Clevers 2018). Furthermore, and as previously discussed, clonal selection of modified cells is often of interest to generate a more reliable model. This involves the technical possibility to grow cells clonally, which can be challenging for some cell types or systems. Cellular pools, a heterogeneous population exposed to CRISPR reagents, in which modified and unmodified cells are mixed, can be an alternative as it does not entail clonal growth. However, the ratio of edited vs wild-type cells within the pool will most likely be unstable over time, as an unexpected proliferation regulation can be caused by the modification, leading one or the other genotype to prevail and take over. Cellular pools are also rarely adapted to the generation of a model involving the insertion of a large fragment of DNA, as large knock-ins (KI) efficiency can be low. In that case, clonal selection of the modified cells is necessary (He et al. 2016). Although not always adapted for longer experiments, cellular pools can represent relevant tools for short-term assays, if the editing efficiency is proved high enough (Joberty et al. 2020).

To optimize CRISPR efficiency in a specific cellular system, different points must be addressed. The design of an optimal guide RNA is usually based on sequencing information from open access databases reporting a single sequence for a specific target and species. However, genetic variations between individuals and cells' genetic drift can lead to polymorphism of a gene of interest. Thus, sequencing of the region of interest, specifically in the *in vitro* system to be used, must always be considered to ensure optimal design and CRISPR cut efficiency. Similarly, and as mentioned in Section 6.3, genetic drift over passages in cell lines, or genomic rearrangements linked to tumorigenesis in cancer models can lead to a variety of allele's copy numbers (Fujikawa-Yamamoto et al. 2011). Determining the exact number of alleles to target before editing the cells can help adapting and optimizing the editing process. Indeed, homozygous targeting of a gene can be very challenging in case of high ploidy, especially if it entails the homozygous insertion of donor DNA. When more than two alleles are present, several rounds of "modification" might be necessary to target them all. As CRISPR most likely cut all alleles even if the donor DNA has not been inserted in all of them, the apparently "non-targeted" alleles most likely exhibit indels, potentially disrupting the original CRISPR cut site. In this case, a novel guide RNA might have to be designed before any additional round of modification can be performed. This strategy can prove very time and resource consuming.

As mentioned in Section 6.3, preliminary experiments deciphering the impact of mutations on the targeted gene's regulation, and empirical approaches to assess the transfectability, clonability, and ploidy of the gene of interest specifically in the *in vitro* system to be used can maximize the chances of success of the disease model creation (Figure 6.1).

6.4.2 Critical Technical Aspects Linked to CRISPR Gene Editing Tool

Regarding the optimization of the targeting strategy *per se*, cell line editing shares most of the common challenges of genetic engineering in general. When using CRISPR, several key elements need to be optimized to ensure successful editing:

- Increasing CRISPR specificity
  One of the first described and most documented limitation of the CRISPR system is the potential off-target cuttings it can induce. Indeed, targeting of the region of interest is accomplished by using guide RNAs complementary to the target DNA (protospacer), followed by a mandatory protospacer adjacent motif (PAM). A sequence homology of 3 to 5 nucleotides out of the 20 commonly used as protospacer can create off-target cutting. Different algorithms have been developed and optimized to best predict such off-targets and help design efficient guide RNAs (Hanna and Doench 2020). It is however not uncommon to have to use a guide RNA associated with predicted off-targets according to such software. In this case, the predicted off-target loci should be sequenced in the CRISPR-modified material to ensure the absence of any unwanted indels. As unpredicted off-targets could still exist, although at a very low rate (Iyer et al. 2018), several identical clones can be used as biological replicates for parallel phenotyping. Additionally, and as previously mentioned, unexpected genomic rearrangements at the targeted locus can occur (Kosicki et al. 2018; Cullot et al. 2019; Rezza et al. 2019), further demonstrating the need for in-depth characterization of materials obtained through CRISPR-mediated modifications.
- Optimizing KO models
  The generation of a CRISPR-mediated KO model can be seen as easy, and thus free from necessitating thorough designing. However, and as previously mentioned, a KO approach can be optimized in many different ways, and can thus greatly benefit from a carefully planned strategy.

  KO can be performed by deleting specific exons of the target gene, introducing a STOP codon early in its coding sequence using donor DNA, or by NHEJ-mediated indels leading to a frameshift and a premature STOP codon. Importantly, to ensure the production of a robust KO model and more precisely control the final genetic sequence of the targeted locus, the two former approaches are often preferred. Indeed, NHEJ-introduced indels have been shown to lead to the creation of alternative mRNA and proteins, caused by internal ribosomal entry and alternative splicing (Tuladhar et al. 2019). Regardless of the genetic strategy, the selection of the critical exon to target is of utmost importance. In some cases, even the most optimized strategy can result in a risk of truncated protein expression. Although not optimal, this scenario can still lead to the generation of a relevant KO model, provided that the truncated protein does not jeopardize the physiology of the model for its specific application. The safest strategy to ensure that remaining truncated proteins will not be an issue is, whenever possible, to target the exons coding for the portion of the protein responsible for its altered function in the disease of interest. For instance, membrane proteins can be efficiently invalidated by deleting their transmembrane domain(s) as they will not be addressed to the membrane properly, or enzymes' core catalytic domain can be targeted to render them unfunctional. Additionally, the selected strategy should target all or a maximum of isoforms of the gene of interest, unless an isoform-specific KO is required. Here again, careful identification and study of the different gene's potential isoforms can be very informative and can help optimizing the genetic strategy to be used to generate a relevant model (Boguski et al. 1993; Min et al. 2005).

- Optimizing KI models
  There are plenty of technical strategies to generate a CRISPR-mediated integration of donor DNA. For example, the nature of the donor DNA can be optimized depending on the size of the modification to be inserted, or the sequence of the targeted locus.

  Single-stranded (ss) donor DNAs are now routinely used, particularly to integrate short sequences of DNA. The integration efficiency of such donor DNA has been shown to greatly depend on the distance between CRISPR cutting site and the mutation to be inserted (Paquet et al. 2016). Indeed, if this distance exceeds 6 to 10 nucleotides, the proportion of mutant cells obtained can drop dramatically. This limitation can lead to the necessity to switch to a large KI strategy, even for the mere insertion of a single mutated codon, in order to secure the generation of the modified cellular model. As the efficiency of CRISPR-mediated integration has been shown to decrease with the size of the transgene (Raveux et al. 2017; Rezza et al. 2019), selection of the recombined clone through antibiotic resistance or fluorescence cell-sorting is often used as a first screening step, before further genotyping and validation (Konishi et al. 2012). Here again, the exogenous selection cassette must be carefully designed to limit the impact of its introduction on neighboring genes and cell physiology (Ryu et al. 2017).

In addition to carefully choosing a cellular system, the successful generation of a relevant cellular model depends on the efficiency of gene editing in this system. Thus, optimizing CRISPR specifically in the system of interest, through thorough preliminary assessment (empirically if needed), can improve the relevance of the final model.

6.5 Conclusion

Disease modeling and drug discovery in its entirety have greatly improved these past two decades, thanks to novel technologies. With optimized culture systems such as organoids, it is now possible to maintain cell-type specific characteristics longer *in vitro*, to recreate the original structure of some tissues with the different cellular populations found *in vivo*, or to preserve samples of interest for a virtually unlimited period of time. More advanced 3D culture methods, such as organ-on-chips or bioprinting, might soon allow the development of even more complex and adapted cellular systems. Generating a relevant cellular disease model thus includes choosing the most appropriate *in vitro* system, carefully considering their advantages and limitations (Table 6.1). Regardless of the particular *in vitro* system used, another technological progress has proved very useful for disease modeling: gene editing, and CRISPR in particular. Indeed, it is now routinely used to tag, invalidate, overexpress, or mutate any gene of interest to create better cellular disease models. As of today, and with the technical limitations of both *in vitro* cell culture and gene editing, developing a cellular disease model often requires compromising. It is then essential to carefully and thoroughly plan the generation of CRISPR-modified materials (Figure 6.1), especially disease models.

The technological advances from the past two decades have enabled researchers to choose from a profusion of novel culture systems and gene editing tools. Although it represents a definitive advantage, this variety of technical considerations should not eclipse the central attribute of any cellular disease model: its physiological relevance. The intended use and specific application of a cellular disease model should always be the leading criterion while generating the model. Thus, the delicate steps in the creation of a relevant disease model are not the experimental stages of the process, but rather its careful design, meaning the precise definition of the biological questions it will be used to answer, and the specific applications for which it is needed. New technological advances will be made to optimize, simplify, or shorten the technical phase of model generation, but its preceding careful conception will remain the most critical, sensitive part of creating a relevant disease model.

References

Abrahimi, P., Chang, W.G., Kluger, M.S. et al. (2015). Efficient gene disruption in cultured primary human endothelial cells by CRISPR/Cas9. *Circ. Res.*: 121–128. https://doi.org/10.1161/circresaha.117.306290.

Ang, Y.-S., Rivas, R.N., AJS, R. et al. (2016). Disease model of GATA4 mutation reveals transcription factor cooperativity in human cardiogenesis. *Cell*: 1734–1749. e22. doi: https://doi.org/10.1016/j.cell.2016.11.033.

Arias, C., Becerra-García, F., Arrieta, I. et al. (1998). The protein phosphatase inhibitor okadaic acid induces heat shock protein expression and neurodegeneration in rat hippocampus in vivo. *Exp. Neurol.* 153 (2): 242–254.

Artegiani, B. and Clevers, H. (2018). Use and application of 3D-organoid technology. *Hum. Mol. Genet.* 27 (R2): R99–R107.

Bailey, L.A., Hatton, D., Field, R. et al. (2012). Determination of chinese hamster ovary cell line stability and recombinant antibody expression during long-term culture. *Biotechnol. Bioeng.* 109 (8): 2093–2103.

Barrangou, R., Fremaux, C., Deveau, H. et al. (2007). CRISPR provides acquired resistance against viruses in prokaryotes. *Science* 315 (5819): 1709–1712.

Bartfeld, S. and Clevers, H. (2017). Stem cell-derived organoids and their application for medical research and patient treatment. *J. Mol. Med.* 95 (7): 729–738.

Berger, B., Donzelli, M., Maseneni, S. et al. (2016). Comparison of liver cell models using the basel phenotyping cocktail. *Front. Pharmacol.* 7: 443.

Bian, S., Repic, M., Guo, Z. et al. (2018). Genetically engineered cerebral organoids model brain tumor formation. *Nat. Methods* 15 (8): 631–639.

Boguski, M.S., Lowe, T.M., and Tolstoshev, C.M. (1993). dbEST--database for "expressed sequence tags". *Nat. Genet.* 4 (4): 332–333.

Brennand, K.J., Simone, A., Jou, J. et al. (2011). Modelling schizophrenia using human induced pluripotent stem cells. *Nature*: 221–225. https://doi.org/10.1038/nature09915.

Bronson, S.K., Plaehn, E.G., Kluckman, K.D. et al. (1996). Single-copy transgenic mice with chosen-site integration. *Proc. Natl. Acad. Sci. U. S. A.* 93 (17): 9067–9072.

Broutier, L., Mastrogiovanni, G., Verstegen, M.M. et al. (2017). Human primary liver cancer-derived organoid cultures for disease modeling and drug screening. *Nat. Med.* 23 (12): 1424–1435.

Brown, M.J. and Murray, K.A. (2006). Phenotyping of genetically engineered mice: humane, ethical, environmental, and husbandry issues. *ILAR Journal* 47 (2): 118–123.

Burgener, A.-V., Bantug, G.R., Meyer, B.J. et al. (2019). SDHA gain-of-function engages inflammatory mitochondrial retrograde signaling via KEAP1–Nrf2. *Nat. Immunol.*: 1311–1321. https://doi.org/10.1038/s41590-019-0482-2.

Burridge, P.W., Matsa, E., Shukla, P. et al. (2014). Chemically defined generation of human cardiomyocytes. *Nat. Methods*: 855–860. https://doi.org/10.1038/nmeth.2999.

Butler, D. (2008). Translational research: crossing the valley of death. *Nature* 453 (7197): 840–842.

Cacheiro, P., Haendel, M.A., Smedley, D. et al. (2019). New models for human disease from the International Mouse Phenotyping Consortium. *Mammalian Genome* 30 (5-6): 143–150.

Capecchi, M.R. (2005). Gene targeting in mice: functional analysis of the mammalian genome for the twenty-first century. *Nat. Rev. Genet.* 6 (6): 507–512.

Cayo, M.A., Mallanna, S.K., Di Furio, F. et al. (2017). A drug screen using human iPSC-derived hepatocyte-like cells reveals cardiac glycosides as a potential treatment for hypercholesterolemia. *Cell Stem Cell*: 478–489. e5. doi: https://doi.org/10.1016/j.stem.2017.01.011.

Chen, C.-M., Krohn, J., Bhattacharya, S. et al. (2011). A comparison of exogenous promoter activity at the ROSA26 locus using a ΦiC31 integrase mediated cassette exchange approach in mouse ES cells. *PLoS One* 6 (8): e23376.

Cho, S.W., Kim, S., Kim, J.M. et al. (2013). Targeted genome engineering in human cells with the Cas9 RNA-guided endonuclease. *Nat. Biotechnol.* 31 (3): 230–232.

Choi, S.M., Kim, Y., Shim, J.S. et al. (2013). Efficient drug screening and gene correction for treating liver disease using patient-specific stem cells. *Hepatology* 57 (6): 2458–2468.

Chung, C.Y., Khurana, V., Auluck, P.K. et al. (2013). Identification and rescue of -synuclein toxicity in Parkinson patient-derived neurons. *Science*: 983–987. https://doi.org/10.1126/science.1245296.

Cong, L., Ran, F.A., Cox, D. et al. (2013). Multiplex genome engineering using CRISPR/Cas systems. *Science* 339 (6121): 819–823.

Cullot, G., Boutin, J., Toutain, J. et al. (2019). CRISPR-Cas9 genome editing induces megabase-scale chromosomal truncations. *Nat. Commun.* 10 (1): 1136.

Czerniecki, S.M., Cruz, N.M., Harder, J.L. et al. (2018). High-throughput screening enhances kidney organoid differentiation from human pluripotent stem cells and enables automated multidimensional phenotyping. *Cell Stem Cell*: 929–940. e4. doi: https://doi.org/10.1016/j.stem.2018.04.022.

Dekkers, J.F., Berkers, G., Kruisselbrink, E. et al. (2016). Characterizing responses to CFTR-modulating drugs using rectal organoids derived from subjects with cystic fibrosis. *Sci. Transl. Med.* 8 (344): 344ra84.

Dhillon, K.K., Sidorova, J.M., Albertson, T.M. et al. (2010). Divergent cellular phenotypes of human and mouse cells lacking the Werner syndrome RecQ helicase. *DNA Repair* 9 (1): 11–22.

Doudna, J.A. and Charpentier, E. (2014). Genome editing. The new frontier of genome engineering with CRISPR-Cas9. *Science* 346 (6213): 1258096.

Drost, J., van Jaarsveld, R.H., Ponsioen, B. et al. (2015). Sequential cancer mutations in cultured human intestinal stem cells. *Nature* 521 (7550): 43–47.

Ebina, H., Misawa, N., Kanemura, Y. et al. (2013). Harnessing the CRISPR/Cas9 system to disrupt latent HIV-1 provirus. *Sci. Rep.* 3: 2510.

Flavahan, W.A., Drier, Y., Liau, B.B. et al. (2016). Insulator dysfunction and oncogenic activation in IDH mutant gliomas. *Nature* 529 (7584): 110–114.

Foltz, L.P., Howden, S.E., Thomson, J.A. et al. (2018). Functional assessment of patient-derived retinal pigment epithelial cells edited by CRISPR/Cas9. *Int. J. Mol. Sci.* 19 (12) https://doi.org/10.3390/ijms19124127.

Freedman, B.S., Brooks, C.R., Lam, A.Q. et al. (2015). Modelling kidney disease with CRISPR-mutant kidney organoids derived from human pluripotent epiblast spheroids. *Nat. Commun.* https://doi.org/10.1038/ncomms9715.

Fu, Y., Sander, J.D., Reyon, D. et al. (2014). Improving CRISPR-Cas nuclease specificity using truncated guide RNAs. *Nat. Biotechnol.* 32 (3): 279–284.

Fujikawa-Yamamoto, K., Miyagoshi, M., Luo, X. et al. (2011). DNA-unstable decaploid mouse H1 (ES) cells established from DNA-stable pentaploid H1 (ES) cells polyploidized using demecolcine. *Cell Prolif.* 44 (2): 111–119.

Gintant, G., Fermini, B., Stockbridge, N. et al. (2017). The evolving roles of human iPSC-derived cardiomyocytes in drug safety and discovery. *Cell Stem Cell*: 14–17. https://doi.org/10.1016/j.stem.2017.06.005.

Gravesteijn, G., Dauwerse, J.G., Overzier, M. et al. (2020). Naturally occurring NOTCH3 exon skipping attenuates NOTCH3 protein aggregation and disease severity in CADASIL patients. *Hum. Mol. Genet.* https://doi.org/10.1093/hmg/ddz285.

Hait, W.N., Choudhury, S., Srimatkandada, S. et al. (1993). Sensitivity of K562 human chronic myelogenous leukemia blast cells transfected with a human multidrug resistance cDNA to cytotoxic drugs and differentiating agents. *J. Clin. Invest.* 91 (5): 2207–2215.

Hanna, R.E. and Doench, J.G. (2020). Design and analysis of CRISPR-Cas experiments. *Nat. Biotechnol.* 38 (7): 813–823.

Hannan, N.R.F., Segeritz, C.-P., Touboul, T. et al. (2013). Production of hepatocyte-like cells from human pluripotent stem cells. *Nat. Protoc.*: 430–437. https://doi.org/10.1038/nprot.2012.153.

Hardy, J. (2017). The discovery of Alzheimer-causing mutations in the APP gene and the formulation of the "amyloid cascade hypothesis". *FEBS J.* 284 (7): 1040–1044.

Hart, S.N., Li, Y., Nakamoto, K. et al. (2010). A comparison of whole genome gene expression profiles of HepaRG cells and HepG2 cells to primary human hepatocytes and human liver tissues. *Drug Metab. Dispos.* 38 (6): 988–994.

Hatakeyama, H., Wu, S.Y., Mangala, L.S. et al. (2016). Assessment of in vivo siRNA delivery in cancer mouse models. *Methods Mol. Biol.* 1402: 189–197.

He, X., Tan, C., Wang, F. et al. (2016). Knock-in of large reporter genes in human cells via CRISPR/Cas9-induced homology-dependent and independent DNA repair. *Nucleic Acids Res.* 44 (9): e85.

Heckl, D., Kowalczyk, M.S., Yudovich, D. et al. (2014). Generation of mouse models of myeloid malignancy with combinatorial genetic lesions using CRISPR-Cas9 genome editing. *Nat. Biotechnol.*: 941–946. https://doi.org/10.1038/nbt.2951.

Hnisz, D., Weintraub, A.S., Day, D.S. et al. (2016). Activation of proto-oncogenes by disruption of chromosome neighborhoods. *Science* 351 (6280): 1454–1458.

Horvath, P., Aulner, N., Bickle, M. et al. (2016). Screening out irrelevant cell-based models of disease. *Nat. Rev. Drug Discov.* 15 (11): 751–769.

Hu, W., Kaminski, R., Yang, F. et al. (2014). RNA-directed gene editing specifically eradicates latent and prevents new HIV-1 infection. *Proc. Natl. Acad. Sci.*: 11461–11466. https://doi.org/10.1073/pnas.1405186111.

Huang, P., Zhang, L., Gao, Y. et al. (2014). Direct reprogramming of human fibroblasts to functional and expandable hepatocytes. *Cell Stem Cell* 14 (3): 370–384.

Huang, L., Holtzinger, A., Jagan, I. et al. (2015). Ductal pancreatic cancer modeling and drug screening using human pluripotent stem cell- and patient-derived tumor organoids. *Nat. Med.* 21 (11): 1364–1371.

Hughes, P., Marshall, D., Reid, Y. et al. (2007). The costs of using unauthenticated, over-passaged cell lines: how much more data do we need? *BioTechniques* 43 (5): 575. 577–8, 581–2 passim.

Israel, M.A., Yuan, S.H., Bardy, C. et al. (2012). Probing sporadic and familial Alzheimer's disease using induced pluripotent stem cells. *Nature*: 216–220. https://doi.org/10.1038/nature10821.

Iyer, V., Boroviak, K., Thomas, M. et al. (2018). No unexpected CRISPR-Cas9 off-target activity revealed by trio sequencing of gene-edited mice. *PLoS Genet.* 14 (7): e1007503.

Jian, X., Boerwinkle, E., and Liu, X. (2014). In silico tools for splicing defect prediction: a survey from the viewpoint of end users. *Genet. Med.* 16 (7): 497–503.

Jinek, M., East, A., Cheng, A. et al. (2013). RNA-programmed genome editing in human cells. *elife* 2: e00471.

Joberty, G., Fälth-Savitski, M., Paulmann, M. et al. (2020). A tandem guide RNA-based strategy for efficient CRISPR gene editing of cell populations with low heterogeneity of edited alleles. *The CRISPR journal* 3 (2): 123–134.

Khamis, A.M., Motwalli, O., Oliva, R. et al. (2018). A novel method for improved accuracy of transcription factor binding site prediction. *Nucleic Acids Res.* 46 (12): e72.

Kim, J., Koo, B.-K., and Knoblich, J.A. (2020). Human organoids: model systems for human biology and medicine. *Nat. Rev. Mol. Cell Biol.* https://doi.org/10.1038/s41580-020-0259-3.

Kirik, D. and Björklund, A. (2003). Modeling CNS neurodegeneration by overexpression of disease-causing proteins using viral vectors. *Trends Neurosci.* 26 (7): 386–392.

Komor, A.C., Badran, A.H., and Liu, D.R. (2017). CRISPR-based technologies for the manipulation of eukaryotic genomes. *Cell* 169 (3): 559.

Konishi, Y., Karnan, S., Takahashi, M. et al. (2012). A system for the measurement of gene targeting efficiency in human cell lines using an antibiotic resistance-GFP fusion gene. *BioTechniques* 53 (3): 141–152.

Kosicki, M., Tomberg, K., and Bradley, A. (2018). Repair of double-strand breaks induced by CRISPR-Cas9 leads to large deletions and complex rearrangements. *Nat. Biotechnol.* 36 (8): 765–771.

Lamers, M.M., Beumer, J., van der Vaart, J. et al. (2020). SARS-CoV-2 productively infects human gut enterocytes. *Science* 369 (6499): 50–54.

Lancaster, M.A., Renner, M., Martin, C.-A. et al. (2013). Cerebral organoids model human brain development and microcephaly. *Nature* 501 (7467): 373–379.

Lian, X., Zhang, J., Azarin, S.M. et al. (2013). Directed cardiomyocyte differentiation from human pluripotent stem cells by modulating Wnt/β-catenin signaling under fully defined conditions. *Nat. Protoc.*: 162–175. https://doi.org/10.1038/nprot.2012.150.

Liang, P., Lan, F., Lee, A.S. et al. (2013). Drug screening using a library of human induced pluripotent stem cell–derived cardiomyocytes reveals disease-specific patterns of cardiotoxicity. *Circulation*: 1677–1691. https://doi.org/10.1161/circulationaha.113.001883.

Low, J.H., Li, P., EGY, C. et al. (2019). Generation of human PSC-derived kidney organoids with patterned nephron segments and a De Novo Vascular Network. *Cell Stem Cell*: 373–387. e9. doi: https://doi.org/10.1016/j.stem.2019.06.009.

Maetzel, D., Sarkar, S., Wang, H. et al. (2014). Genetic and chemical correction of cholesterol accumulation and impaired autophagy in hepatic and neural cells derived from Niemann-Pick Type C patient-specific iPS cells. *Stem cell reports* 2 (6): 866–880.

Mali, P., Aach, J., Stranges, P.B. et al. (2013a). CAS9 transcriptional activators for target specificity screening and paired nickases for cooperative genome engineering. *Nat. Biotechnol.* 31 (9): 833–838.

Mali, P., Yang, L., Esvelt, K.M. et al. (2013b). RNA-guided human genome engineering via Cas9. *Science* 339 (6121): 823–826.

Marotta, C.A., Chou, W.G., Majocha, R.E. et al. (1989). Overexpression of amyloid precursor protein A4 (beta-amyloid) immunoreactivity in genetically transformed cells: implications for a cellular model of Alzheimer amyloidosis. *Proc. Natl. Acad. Sci. U. S. A.* 86 (1): 337–341.

Matano, M., Date, S., Shimokawa, M. et al. (2015). Modeling colorectal cancer using CRISPR-Cas9-mediated engineering of human intestinal organoids. *Nat. Med.* 21 (3): 256–262.

Matsa, E., Rajamohan, D., Dick, E. et al. (2011). Drug evaluation in cardiomyocytes derived from human induced pluripotent stem cells carrying a long QT syndrome type 2 mutation. *Eur. Heart J.*: 952–962. https://doi.org/10.1093/eurheartj/ehr073.

McLeskey, S.W., Zhang, L., Kharbanda, S. et al. (1996). Fibroblast growth factor overexpressing breast carcinoma cells as models of angiogenesis and metastasis. *Breast Cancer Res. Treat.* 39 (1): 103–117.

Miller, J.C., Tan, S., Qiao, G. et al. (2011). A TALE nuclease architecture for efficient genome editing. *Nat. Biotechnol.* 29 (2): 143–148.

Min, X.J., Butler, G., Storms, R. et al. (2005). TargetIdentifier: a webserver for identifying full-length cDNAs from EST sequences. *Nucleic Acids Res.* 33 (Web Server issue): W669–W672.

Mohindra, A., Hays, L.E., Phillips, E.N. et al. (2002). Defects in homologous recombination repair in mismatch-repair-deficient tumour cell lines. *Hum. Mol. Genet.* 11 (18): 2189–2200.

Monteil, V., Kwon, H., Prado, P. et al. (2020). Inhibition of SARS-CoV-2 infections in engineered human tissues using clinical-grade soluble human ACE2. *Cell* 181 (4): 905–913. e7.

Morizane, R., Lam, A.Q., Freedman, B.S. et al. (2015). Nephron organoids derived from human pluripotent stem cells model kidney development and injury. *Nat. Biotechnol.* 33 (11): 1193–1200.

Müller, F.A. and Sturla, S.J. (2019). Human in vitro models of nonalcoholic fatty liver disease. *Curr. Opin. Toxicol.* 16: 9–16.

Naylor, J., Suckow, A.T., Seth, A. et al. (2016). Use of CRISPR/Cas9-engineered INS-1 pancreatic β cells to define the pharmacology of dual GIPR/GLP-1R agonists. *Biochem. J.* 473 (18): 2881–2891.

Nestor, C.E., Ottaviano, R., Reinhardt, D. et al. (2015). Rapid reprogramming of epigenetic and transcriptional profiles in mammalian culture systems. *Genome Biol.* 16: 11.

Nie, J. and Hashino, E. (2017). Organoid technologies meet genome engineering. *EMBO Rep.* 18 (3): 367–376.

O'Driscoll, L., Gammell, P., McKiernan, E. et al. (2006). Phenotypic and global gene expression profile changes between low passage and high passage MIN-6 cells. *J. Endocrinol.* 191 (3): 665–676.

Okamoto, S., Amaishi, Y., Maki, I. et al. (2019). Highly efficient genome editing for single-base substitutions using optimized ssODNs with Cas9-RNPs. *Sci. Rep.* 9 (1): 4811.

Paquet, D., Kwart, D., Chen, A. et al. (2016). Efficient introduction of specific homozygous and heterozygous mutations using CRISPR/Cas9. *Nature* 533 (7601): 125–129.

Pieters, R., Loonen, A.H., Huismans, D.R. et al. (1990). in vitro drug sensitivity of cells from children with leukemia using the MTT assay with improved culture conditions. *Blood* 76 (11): 2327–2336.

Ramanan, V., Shlomai, A., DBT, C. et al. (2015). CRISPR/Cas9 cleavage of viral DNA efficiently suppresses hepatitis B virus. *Sci. Rep.* 5: 10833.

Ran, F.A., Hsu, P.D., Lin, C.-Y. et al. (2013). Double nicking by RNA-guided CRISPR Cas9 for enhanced genome editing specificity. *Cell* 154 (6): 1380–1389.

Ratner, M. (2017). FDA deems in vitro data on mutations sufficient to expand cystic fibrosis drug label. *Nat. Biotechnol.* 35 (7): 606.

Raveux, A., Vandormael-Pournin, S., and Cohen-Tannoudji, M. (2017). Optimization of the production of knock-in alleles by CRISPR/Cas9 microinjection into the mouse zygote. *Sci. Rep.* 7: 42661.

Rezza, A., Jacquet, C., Le Pillouer, A. et al. (2019). Unexpected genomic rearrangements at targeted loci associated with CRISPR/Cas9-mediated knock-in. *Sci. Rep.* 9 (1): 3486.

Riffo-Campos, Á.L., Riquelme, I., and Brebi-Mieville, P. (2016). Tools for sequence-based miRNA target prediction: What to choose? *Int. J. Mol. Sci.* 17 (12) https://doi.org/10.3390/ijms17121987.

Rookmaaker, M.B., Schutgens, F., Verhaar, M.C. et al. (2015). Development and application of human adult stem or progenitor cell organoids. *Nat. Rev. Nephrol.*: 546–554. https://doi.org/10.1038/nrneph.2015.118.

Rouet, P., Smih, F., and Jasin, M. (1994). Introduction of double-strand breaks into the genome of mouse cells by expression of a rare-cutting endonuclease. *Mol. Cell. Biol.* 14 (12): 8096–8106.

Ryu, A.H., Eckalbar, W.L., Kreimer, A. et al. (2017). Use antibiotics in cell culture with caution: genome-wide identification of antibiotic-induced changes in gene expression and regulation. *Sci. Rep.* 7 (1): 7533.

Saarimäki-Vire, J., Balboa, D., Russell, M.A. et al. (2017). An activating STAT3 mutation causes neonatal diabetes through premature induction of pancreatic differentiation. *Cell Rep.* 19 (2): 281–294.

Sachs, N., de Ligt, J., Kopper, O. et al. (2018). A living biobank of breast cancer organoids captures disease heterogeneity. *Cell* 172 (1–2): 373–386. e10.

Sanz, D.J., Hollywood, J.A., Scallan, M.F. et al. (2017). Cas9/gRNA targeted excision of cystic fibrosis-causing deep-intronic splicing mutations restores normal splicing of CFTR mRNA. *PLoS One* 12 (9): e0184009.

Sato, T., Vries, R.G., Snippert, H.J. et al. (2009). Single Lgr5 stem cells build crypt-villus structures in vitro without a mesenchymal niche. *Nature* 459 (7244): 262–265.

Schwank, G., Koo, B.-K., Sasselli, V. et al. (2013). Functional repair of CFTR by CRISPR/Cas9 in intestinal stem cell organoids of cystic fibrosis patients. *Cell Stem Cell* 13 (6): 653–658.

Scott, A. (2018). How CRISPR is transforming drug discovery. *Nature* 555 (7695): S10–S11.

Sgodda, M., Dai, Z., Zweigerdt, R. et al. (2017). A scalable approach for the generation of human pluripotent stem cell-derived hepatic organoids with sensitive hepatotoxicity features. *Stem Cells Dev.* 26 (20): 1490–1504.

Sharma, A., Sances, S., Workman, M.J. et al. (2020). Multi-lineage human iPSC-derived platforms for disease modeling and drug discovery. *Cell Stem Cell* 26 (3): 309–329.

Smithies, O., Gregg, R.G., Boggs, S.S. et al. (1985). Insertion of DNA sequences into the human chromosomal β-globin locus by homologous recombination. *Nature*: 230–234. https://doi.org/10.1038/317230a0.

Soldner, F., Laganière, J., Cheng, A.W. et al. (2011). Generation of isogenic pluripotent stem cells differing exclusively at two early onset Parkinson point mutations. *Cell* 146 (2): 318–331.

Song, E., Zhang, C., Israelow, B. et al. (2021). Neuroinvasion of SARS-CoV-2 in human and mouse brain. *J Exp Med.* 218(3):e20202135

Sousa, F.G., Matuo, R., Tang, S.-W. et al. (2015). Alterations of DNA repair genes in the NCI-60 cell lines and their predictive value for anticancer drug activity. *DNA Repair* 28: 107–115.

Strikoudis, A., Cieślak, A., Loffredo, L. et al. (2019). Modeling of fibrotic lung disease using 3D organoids derived from human pluripotent stem cells. *Cell Rep.* 27 (12): 3709–3723.e5.

Sung, M.-H., Guertin, M.J., Baek, S. et al. (2014). DNase footprint signatures are dictated by factor dynamics and DNA sequence. *Mol. Cell* 56 (2): 275–285.

Takahashi, K., Tanabe, K., Ohnuki, M. et al. (2007). Induction of pluripotent stem cells from adult human fibroblasts by defined factors. *Cell* 131 (5): 861–872.

Takebe, T., Sekine, K., Enomura, M. et al. (2013). Vascularized and functional human liver from an iPSC-derived organ bud transplant. *Nature*: 481–484. https://doi.org/10.1038/nature12271.

Tchorz, J.S., Suply, T., Ksiazek, I. et al. (2012). A modified RMCE-compatible Rosa26 locus for the expression of transgenes from exogenous promoters. *PLoS One* 7 (1): e30011.

Tuladhar, R., Yeu, Y., Tyler Piazza, J. et al. (2019). CRISPR-Cas9-based mutagenesis frequently provokes on-target mRNA misregulation. *Nat. Commun.* 10 (1): 4056.

Urnov, F.D., Miller, J.C., Lee, Y.-L. et al. (2005). Highly efficient endogenous human gene correction using designed zinc-finger nucleases. *Nature* 435 (7042): 646–651.

Vaira, V., Fedele, G., Pyne, S. et al. (2010). Preclinical model of organotypic culture for pharmacodynamic profiling of human tumors. *Proc. Natl. Acad. Sci. U. S. A.* 107 (18): 8352–8356.

Weiss, P. and Taylor, A.C. (1960). Reconstitution of complete organs from single-cell suspensions of chick embryos in advanced stages of differentiation. *Proc. Natl. Acad. Sci. U. S. A.* 46 (9): 1177–1185.

Wendler, A. and Wehling, M. (2010). The translatability of animal models for clinical development: biomarkers and disease models. *Curr. Opin. Pharmacol.* 10 (5): 601–606.

Woan, K.V., Lienlaf, M., Perez-Villaroel, P. et al. (2015). Targeting histone deacetylase 6 mediates a dual anti-melanoma effect: Enhanced antitumor immunity and impaired cell proliferation. *Mol. Oncol.* 9 (7): 1447–1457.

Xu, X., Lei, Y., Luo, J. et al. (2013). Prevention of β-amyloid induced toxicity in human iPS cell-derived neurons by inhibition of Cyclin-dependent kinases and associated cell cycle events. *Stem Cell Res.*: 213–227. https://doi.org/10.1016/j.scr.2012.11.005.

Yevshin, I., Sharipov, R., Valeev, T. et al. (2017). GTRD: a database of transcription factor binding sites identified by ChIP-seq experiments. *Nucleic Acids Res.* 45 (D1): D61–D67.

Yoshino, Y., Endo, S., Chen, Z. et al. (2019). Evaluation of site-specific homologous recombination activity of BRCA1 by direct quantitation of gene editing efficiency. *Sci. Rep.* 9 (1): 1644.

Zasadil, L.M., Britigan, E.M.C., and Weaver, B.A. (2013). 2n or not 2n: Aneuploidy, polyploidy and chromosomal instability in primary and tumor cells. *Semin. Cell Dev. Biol.* 24 (4): 370–379.

Zhang, Y., Schmid, B., Nikolaisen, N.K. et al. (2017). Patient iPSC-derived neurons for disease modeling of frontotemporal dementia with mutation in CHMP2B. *Stem cell reports* 8 (3): 648–658.

7

Utilizing CRISPR/Cas9 Technologies for *in vivo* Disease Modeling and Therapy

Lukas Badertscher and Michelle J. Porritt*

Genome Engineering, Discovery Sciences, BioPharmaceuticals R&D, AstraZeneca, Gothenburg, Sweden
**Present address: Aelian Biotechnology GmbH, Vienna, Austria*

7.1 Introduction to CRISPR/Cas9 and *in vivo* Modeling

Genetic manipulations in cells and organisms are a powerful tool to study gene function in a physiological or disease context. One strategy has been to exploit nucleases such as zinc-finger nucleases (ZNFs) (Bibikova et al. 2002), transcription activator-like effector nucleases (TALENs) (Christian et al. 2010), and clustered regularly interspaced short palindromic repeats (CRISPR) (Jinek et al. 2012) for targeted insertions and deletions (indels) or precise genome edits (Cong et al. 2013; Lee et al. 2018a; Li et al. 2011; Maresca et al. 2013; Wijshake et al. 2014). Among those nucleases, the CRISPR/Cas9 system has accelerated the generation of genetically engineered cells and animal models due to its versatility and simplicity of design (reviewed in Adli 2018).

The CRISPR system has attracted a lot of scientific attention after it was first identified as being able to edit the genome in a simple and precise manner (Deveau et al. 2010; Horvath and Barrangou 2010; Makarova et al. 2011; Ran et al. 2013b). There are currently 2 classes of CRISPR/Cas systems, each containing 3 different types (Makarova et al. 2020), with type II having only one effector protein, *Streptococcus pyogenes* CRISPR/Cas9 (SpCas9 or SpyCas9) the most widely used and the focus of this chapter. The two components of SpCas9 are: a RNA-directed Cas9 protein that is an endonuclease and an approximate 20-nucleotide single-guide RNA (sgRNA) that directs the Cas9 protein to the defined DNA target site, provided that it is adjacent to a protospacer adjacent motif (PAM) sequence (Doudna and Charpentier 2014). The highly efficient Cas9 nucleases introduce double-strand breaks (DSB) within DNA sequences in a site-specific manner (Doudna and Charpentier 2014; Jinek et al. 2012). Subsequently, the DSB is repaired by one of the four major DNA repair pathways in mammalian cells: (i) classical non-homologous end joining (NHEJ), (ii) alternative end joining (alt-EJ), (iii) single-strand annealing (SSA), and (iv) homology-directed recombination (HDR) (Taheri-Ghahfarokhi et al. 2018). In mammalian cells, DSBs are predominantly repaired by NHEJ, which is considered to be imprecise and often results in the generation of small indels instead of a precise repair (Taheri-Ghahfarokhi et al. 2018). The HDR pathway utilizes a template with long homology arms (>100 bp), generally provided endogenously by the sister chromatid or exogenously by a linear or circular DNA within the viral vector which then allows for high-fidelity repairs. The occurrence of HDR-mediated DNA repair from DSBs is extremely low in mammalian cells, ranging from HDR repair efficiencies of 0.5–20% as it requires cells to be in a specific cell cycle phase. In contrast, NHEJ-mediated repair can occur at efficiencies of 20–80%.

7.2 CRISPR Editing to Alter Gene Expression

7.2.1 Gene Knockout and Knockin to Introduce a Point Mutation or Therapeutic Gene Correction

The introduction of indels by the NHEJ repair pathway is the most common approach used to generate a gene knockout with CRISPR/Cas9 (review of indel patterns from Cas9-induced DSBs in Taheri-Ghahfarokhi et al. 2018). Thereby, a nonsense mutation results in a premature termination codon preventing protein translation.

Genome Editing in Drug Discovery, First Edition. Edited by Marcello Maresca and Sumit Deswal.

Introduction of a point mutation using CRISPR/Cas9 was initially achieved by the use of HDR. HDR-mediated genome editing requires delivering three components in the same cells: Cas9, sgRNA together with a donor template DNA (containing long homology arms). HDR has been utilized to introduce targeted point mutations (Lundin et al. 2020; Platt et al. 2014), barcoding (Winters et al. 2017), and to correct disease alleles (Yin et al. 2014) in adult mouse tissue.

Cas9 has been repurposed as an RNA-guided DNA binding domain and is catalytically dead when mutations are introduced in the RuvC and HNH nuclease domains (Jinek et al. 2012). dCas9 can be fused to other domains to be used in CRISPRa, CRISPRi, base editing, and prime editing (Xu and Qi 2019).

Base editing (BE) is a strategy of genome editing independent of HDR and dsDNA break formation and base editors consist of a catalytically dead Cas9 fused to a cytidine deaminase (rAPOBEC1) or an adenosine deaminase (TadA) (Komor et al. 2016). Three classes of base editors have been developed; Cytosine Base Editors that convert C-to-T and Adenine Base Editors that convert A-to-G and the recently described C-to-G base editors, CGBE1 and miniCGBE1 (Gaudelli et al. 2017; Kurt et al. 2021). BE has been used *in vivo* to silence *Pcsk9* and *PCSK9* gene function by introducing a nonsense mutation into the gene (Carreras et al. 2019; Chadwick et al. 2017).

Prime editing (PE) fills the gap in genome editing and enables corrections which have priorly not been feasible with BEs. PE can be used as a "search-and-replace" genome editing technology allowing targeted insertions, deletions, and all 12 possible base-to-base conversions, and combinations, base transversion, and small indels (Anzalone et al. 2019). The third version of PE is achieved by fusing an engineered dCas9 nickase with a reverse transcriptase (M-MLV Rtase) (Anzalone et al. 2019). PE utilizes an engineered prime editing guide RNA (pegRNA) and a dCas9 that nicks the target DNA, and then new genetic information is encoded into genome by a reverse transcriptase. PE results in fewer bystander mutations and undetectable off-target edits in mouse zygotes and mice (Anzalone et al. 2019; Liu et al. 2020). Liu found that besides the desirable transversion at the target sites, there was also a mixture of nucleotide conversions implying a relative low fidelity of the PE3 system in mouse embryos and that improving its fidelity is likely to be heavily investigated (Liu et al. 2020).

7.2.2 Activation and Repression of Gene Expression by CRISPRa and CRISPRi

Cas9-based activation of gene expression is an effective method to study the gain-of-function phenomena and has been used in cells to unravel drug resistance of tumors in targeted therapy and study mechanism of toxin-mediated cell death on a genomic scale (Gilbert et al. 2014; Konermann et al. 2015). For CRISPR activation (CRISPRa), initial studies fused dCas9 with VP64, which contains four copies of the strong transcriptional activator VP16 of Herpes simplex virus (Maeder et al. 2013; Mali et al. 2013; Perez-Pinera et al. 2013). Subsequently, various next-generation versions of CRISPRa systems were established by fusing either additional copies of VP16 to dCas9 (dCas9-VP160) (Cheng et al. 2013) or by utilizing transactivation complexes composed of transcription activation domains of both viral and mammalian origin (dCas9-VPR) (Chavez et al. 2015). In another strategy termed synergistic activation mediator (SAM), in addition to dVas9-VP64, the sgRNA scaffold was engineered to provide a binding platform for auxiliary effector complexes (Konermann et al. 2015). Finally, dCas9 has also been fused with a repeating polypeptide called SunTag, which can mediate strong transcriptional activation through recruitment of high copy numbers of antibody conjugated VP64 (Tanenbaum et al. 2014). While cell type and genetic context can impact the efficiency of CRISPRa, a comparative study revealed that SAM systems tend to have a slightly higher activation rate in comparison to VPR and SunTag (Chavez et al. 2016). However, all second-generation CRISPRa systems outperform dCas9-VP64. *In vivo* CRISPRa was reported by Liao and colleagues and they stimulated robust target gene activation in multiple mouse models of disease by rescuing gene expression, compensated for genetic defects (overexpression of protein) and altered cell fate by induction of transdifferentiation factors (Liao et al. 2017). Another example includes rescuing haploinsufficient mice from obesity by targeting promoters or enhancers of a gene (Matharu et al. 2019). Finally, CRISPRa has also been used in mice to activate transcription in the brain and liver (Wangensteen et al. 2018; Zhou et al. 2018).

CRISPR interference (CRISPRi) is an alternative to a gene knockout and can be used to study loss-of-function. The target gene expression is inhibited when a transcriptional repressor, dCas9, is guided in proximity to the transcription start-site. dCas9 can repress gene expression by stopping RNA polymerase progression by steric hindrance when targeted downstream of the transcription start site or regulatory elements (Jensen et al. 2017). The addition of repressor domains, such as the Krüppel-associated box (KRAB) domain, fused to dCas9 can increase the magnitude of the inhibition (Zheng et al. 2018). CRISPRi components can be introduced *in vivo* by viral vectors (Thakore et al. 2018; Zheng et al. 2018); however, this approach is useful for cell types easily targeted by viral vectors but not cell types such as hypertrophic chondrocytes and osteocytes (MacLeod et al. 2019). Silencing of the mouse *Pcsk9* gene by the administration of a dual-vector AAV8 system expressing dSaCas9KRAB and a *Pcsk9*-targeting sgRNA to adult mice resulted in significant reductions of serum *Pcsk9* and cholesterol levels (Thakore et al. 2018). Suppression of *Tnfsf11* gene via transgenic production of CRISPRi components is able to produce a phenotype comparable to that produced by gene deletion (MacLeod et al. 2019).

7.2.3 CRISPR Library Screens

Genetic library screens are powerful tools for identifying genes responsible for a phenotype; however, *in vivo* pooled screens were challenging due to the complexity of the library, limitations of virus delivery, and/or uniformity of viral transduction and complex interactions of cells within animals (reviewed in Weber et al. 2020). CRISPR library screens require delivery of dCas9 and a library of sgRNAs designed to target the research question such as pathways, oncogenic drivers, or can be genome wide, i.e. a gRNA library against 19,150 mouse protein coding genes. dCas9 fusion proteins are guided with hundreds or thousands of individual sgRNAs in a population of cells. The aim for such studies is to identify genes that influence a specific phenotype in an unbiased fashion. CRISPR screens are a powerful high-throughput assay to functionally screen a large number of genes at the same time. Since each sgRNA is stably integrated into the genome during viral infection, the guiding sequences of each sgRNA can be used as a unique "barcode." Due to the large size, libraries are usually packaged into lentivirus and ideally each cell expressing Cas9 should be transduced with only one sgRNA. After conducting a sgRNA library screen, genomic DNA (gDNA) is isolated and sgRNA regions are amplified and sequenced. After sequencing, results were analyzed to determine which sgRNAs have increased or decreased in frequency due to the selective pressure applied in the screen (Yau and Rana 2018).

When studying loss of function, CRISPR screens are an improvement over RNAi as it targets DNA and a gene knockout results in full loss of a target protein. While RNAi was adopted as a gene-silencing technique first, CRISPR has several advantages as there are less off-target effects, gene disruption is permanent, and there is a lower risk of immune response (Unniyampurath et al. 2016).

The majority of *in vivo* CRISPR screens have been used for target identification in oncology research and due to Cas9 delivery challenges, direct *in vivo* screening applications are difficult, instead, *in vivo* screens are based on cell transplantation (Weber et al. 2020). The first described *in vivo* genome-wide CRISPR/Cas9-mediated was a loss-of-function screen on tumor growth and metastasis (Chen et al. 2015). CRISPR screens to identify novel immunotherapy targets in lung adenocarcinoma (Li et al. 2020) and in a PD-1 checkpoint pathway (Manguso et al. 2017) or mediators of resistance to the chemotherapy (Braun et al. 2016).

In vivo CRISPR screens not related to oncology have identified genes involved in the maintenance of neuron viability in the adult CNS (Wertz et al. 2020), identified uncharacterized genes essential for development in *Drosophila* (Port et al. 2020), and genome-wide search identifying genes to protect beta cells against autoimmunity (Cai et al. 2020).

7.2.4 Generation of Specific Chromosomal Rearrangements Using CRISPR/Cas9

Fusion genes have been discovered in a wide range of solid tumors and they possess a great potential as therapeutic targets because of their tumor-specific expression (reviewed by Parker et al. 2013). However, despite the importance of fusion genes in cancer, modeling chromosomal rearrangements has proven challenging. The first fusion gene discovered in lung adenocarcinoma, a subtype of non-small cell lung cancer, was an intrachromosomal inversion which joined the echinoderm microtubule-associated protein-like 4 (EML4) gene to the anaplastic lymphoma receptor tyrosine kinase (ALK) (Soda et al. 2007). Two groups have developed a CRISPR/Cas9-based method to induce specific chromosomal rearrangement and generating an Eml4-Alk gene fusion *in vivo* (Blasco et al. 2014; Maddalo et al. 2014). Thereby Cas9 and two sgRNAs, each targeting an intronic region of Eml4 and Alk, respectively, were used and a concurrent DSB allowed an inversion of ~ 11 Mb on chromosome 17 in mice. Most importantly, those models reflect clinical relevance as the generated tumors are responsive to crizotinib treatment, an approved ALK kinase inhibitor. Simultaneously, another group explored the potential to induce different types of chromosomal rearrangements in HEK293 cells to generate three different fusion genes important in lung cancer using the same strategy (Choi and Meyerson 2014). It was shown that both paracentric and pericentric intrachromosomal inversions can be generated and strikingly, interchromosomal translocations could also be promoted. Nevertheless, the efficiency to induce translocations *in vivo* awaits further analysis. Together, the CRISPR/Cas9 technology has great potential for creating models with a plethora of cancer driver fusion genes and since apart from Cas9, only two sgRNAs have to be delivered, this reflects a relatively simple strategy.

7.3 Choice of Cas9 Species and/or Variant and Ortholog

There are numerous Cas9 proteins that exist naturally from differing species that have been used for gene editing, including *Streptococcus pyogenes* (SpCas9) (Chylinski et al. 2013), *Staphylococcus aureus* (SaCas9) (Ran et al. 2013a), *Neisseria meningitidis* (NmCas9) (Hou et al. 2013), *Streptococcus thermophiles* (StCas9) (Kleinstiver et al. 2015), and *Campylobacter jejuni*-derived (CjCas9) (Kim et al. 2017). Each Cas9 has a specific PAM sequence and different levels of editing activity. Due to limitations in the viral packaging capacity, nucleases, target sequences, and repair templates are often separated into different vectors, resulting in lower editing efficacy (Zetsche et al. 2015). To overcome this, there are now a series of engineered Cas9 orthologs whereby either the Cas9 has been reduced in size to fit into smaller vectors, PAM sites have become more restricted or more generous, editing is more efficient or there are reduced off-target events by weakening binding to the target DNA strand. Examples include nCas9, dCas9-FokI, Sniper-Cas9, Cas9-VQR, Cas9-EQR, Cas9-VRER, Cas9-D1135E, Cas9-QQR1, Cas9-high fidelity (Cas9-HF), SpCas9-HF1, eSpCas9 (1.0), and eSpCas9 (1.1) (summarized in Table 7.1) (Agudelo et al. 2020; Chatterjee et al. 2020a; Chatterjee et al. 2020b; Chen et al. 2019; Edraki et al. 2019; Hu et al. 2020; Kleinstiver et al. 2016; Lee et al. 2018b; Lee et al. 2019; Slaymaker et al. 2016; Tan et al. 2019). The choice of which specific Cas9 ortholog to use can improve gene editing efficiency for a given target sequence and should be considered as part of the gene editing system design.

7.4 Tissue-Specific CRISPR/Cas9 Gene Editing

7.4.1 Varying Delivery Vehicles and Choosing Different Routes of Administration

Recombinant adeno-associated virus (AAV) and recombinant adenoviral virus (AdV) have both been used to deliver the components of CRISPR/Cas9 *in vivo* (see specific sections for details). While different AAV serotypes have certain tissue tropisms, a challenge in the context of

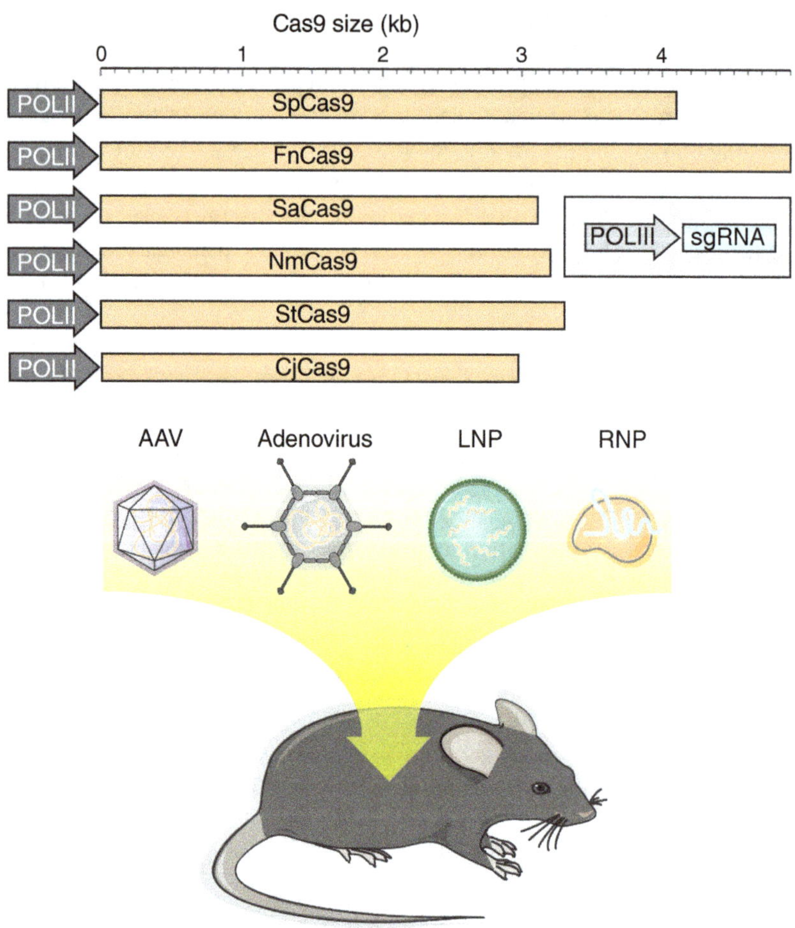

Figure 7.1 Overview of different Cas9 variants and depiction of selected delivery methods. Various Cas9 variants used for *in vivo* applications are shown based on their differences in size. Of note, the sgRNA is not drawn to scale for illustrative purposes. Commonly used delivery vehicles are indicated including adeno-associated viruses (AAV), adenoviruses, lipid nanoparticles (LNPs) containing Cas9 mRNA, and ribonucleoproteins (RNPs), a complex between Cas9 protein and the sgRNA.

Table 7.1 Comparison of Cas9 proteins.

Cas9 variant	Protein size (aa)	PAM	Proto-spacer	Properties	Refs
SpCas9	1366	NGG	20 nt	Highly efficient with simple PAM, high off-target editing	Jinek (2012)
Cas9-VQR	1372	NGAG NGAA	20 nt	Doubles SpCas9s targeting potential	Kleinstiver (2015)
Cas9-EQR	1372	NGAG	20 nt	Stringent specificity due to extension of PAM by 1 basepair	Kleinstiver (2015)
Cas9-VRER	1372	NGCG	20 nt	Gain of function mutation (T1337) near PAM	Kleinstiver (2015)
Cas9-D1135E	1372	NAG/NGA	20 nt	Point mutation increases specificity lessening use noncanonical PAMs	Kleinstiver (2015)
Cas9-QQR1	1372	NAAG	20 nt	Potential less off-target activity as the cleavage rate is slow	Anders (2016)
SpCas9-HF1	1368	NGG	20 nt	Minimized off-target editing, sgRNA sequences have 1 or 2 mismatches	Kleinstiver (2016)
eSpCas9 (1.1)	1424	NGG	20 nt	Neutralization of positive residues in the nontarget strand groove decrease off-target binding	Slaymaker (2016)
Sniper-Cas9	1372	NG	20 nt	5' mismatch tolerance allows use of truncated or extended sgRNAs	Lee (2018)
FnCas9	1629	NGG	20 nt	Can be made to recognize a 5′-YG-3′ PAM	Hirano (2016)
SaCas9	1053	NNGRRT	21 nt	Can be packaged into adeno-associated viral (AAV) vector	Ran (2015), Kleinstiver (2015)
NmCas9	1082	NNNNGATT	24 nt	Long PAM reduces the number of targetable sites at a given locus	Hou (2013), Edraki (2019)
StCas9	1121	NNAGAAW	20 nt	sgRNA protospacers of 19 nt reduce off-target editing	Muller (2016) Agudelo (2020)
CjCas9	984	NNNNACAC NNNNRYAC	22 nt	Minimal off-targeting editing, small so effectors or homology arms fit in an AAV	Kim (2017)

Cas9 is their limiting packaging capacity of ~ 4.5 Kb. Consequently, due to the large size of SpCas9, a dual vector system has to be used, whereby one virus would encode Cas9 and the other for expression of sgRNAs, which can drastically hinder editing efficiencies *in vivo*. Smaller variants of Cas9 such as SaCas9, NmeCas9, and CjCas9 have been successfully used with single vector systems (reviewed in Lau and Suh 2017) (summarized in Figure 7.1). A complementary approach would be to vary administration routes of viruses such as, for instance, using intratracheal instillation to mainly target lung or intravenous tail injection to predominantly obtain gene editing in liver. Nevertheless, undesired gene editing can often be observed in secondary tissues. An alternative for spatially controlled gene editing is to deliver CRISPR/Cas9 RNP to specific tissues by using defined lipid nanoparticle (LNP) formulations. This strategy has been employed to edit tissues including muscle, brain, liver, and lungs (Wei et al. 2020). Moreover, by complexing magnetic nanoparticles with recombinant baculoviral vectors, gene editing can be locally induced through a magnetic field *in vivo* (Zhu et al. 2019). However, finding an appropriate approach for specific needs of a project can be cumbersome and therefore additional modular regulatory elements have been implemented that can support a tight control for spatially controlled gene editing.

7.4.2 Tissue-Specific Promoters, sgRNA Processing Systems, and miRNA Response Elements

Expression of the two components of CRISPR/Cas9 is normally achieved by using RNA polymerase II (Pol-II) and RNA polymerase III (Pol-III) promoters for Cas9 and the sgRNA, respectively. In principle, tissue-specific gene editing could be achieved by restricting expression of one of the components to a desired tissue. A vast majority of tissue-specific Pol-II promoters have been described and examples include mouse albumin (Alb) promoter for liver, human podocin promoter (NPHS2) for kidney, human insulin promoter (HIP) for pancreas, human synapsin I (SYN1) promoter for brain, or human surfactant protein B (SP-B) promoter for lung. In comparison to strong viral promoters (e.g. CMV or SV40), composite promoters (CBh or CAG) or minimal promoters of human proteins such as the EFS promoter which are active in most cell types, tissue-specific promoters are often much larger in size because of their regulatory elements. For viral delivery by AAV, this can be an issue in combination with the large size of the Cas9 sequence. On the other hand, sgRNAs are primarily expressed by the Pol-III promoter of the U6 small nuclear RNA (snRNA). The U6 promoter is ubiquitously active, which would not allow a tissue-specific control, but more importantly transcripts remain in the nucleus, the required subcellular localization for complex formation with Cas9 and gene editing. Using a tissue-specific promoter for sgRNA expression would not be feasible since Pol-II products are quickly modified with a 5′ cap and 3′poly A tail, which both support through interactions with specific factors exported from the nucleus to the cytoplasm. Consequently, expression of both Cas9 and sgRNA from Pol-II promoters results in little to no gene editing activity (Shechner et al. 2015). Several approaches have been developed to release multiple sgRNAs from one transcript such as using self-cleaving ribozymes (Lee et al. 2016; Nissim et al. 2014), recruiting the endogenous t-RNA processing machinery (Xie et al. 2015), embedding the gRNA in a spliced intron (Kiani et al. 2014), or recognition sequences for an exogenously expressed ribonuclease Csy4 (Kurata et al. 2018). Although some of these strategies have relatively poor processing activities, in general they could be combined with tissue-specific Pol II-promoters. Of note, selection of a modified tRNA sequence is of central importance when using the tRNA processing system in this context as it would otherwise contain an internal Pol-III promoter that could impact the tissue-specificity (Knapp et al. 2019).

Another strategy to achieve spatial gene editing is to regulate Cas9 activity by cell type-specific endogenously expressed microRNAs (miRNAs), a large family of post-transcriptional regulators of gene expression (reviewed in Krol et al. 2010). miRNA response elements (MRE) have been placed in the Cas9 mRNA to prevent gene editing if a specific miRNA is expressed such as, for instance, miR-302 in undifferentiated hiPSC (Hirosawa et al. 2017). Nevertheless, to restrict gene editing to a specific cell-type *in vivo*, it would require finding a miRNA with high levels in all cell types other than where Cas9 should be active and therefore this strategy is not broadly applicable. This changed by the discovery of small phage-derived anti-CRIPSPR proteins which can prevent Cas9 activity in eukaryotic cells (e.g. AcrIIA4 to inhibit SpCas9) (Nakamura et al. 2019). To obtain cell-type specific editing, MREs were placed in the mRNA encoding anti-CRIPSR instead of Cas9, and therefore Cas9 would be active when a miRNA is expressed since anti-CRISPR protein levels are reduced. MiR-122 is highly expressed in liver and using a miR122-respressible anti-CRISPR protein-enabled liver-specific gene editing both in cells and mice (Hoffmann et al. 2019; Lee et al. 2019).

7.5 Advantages/Disadvantages of Cas9 Expressing Systems

Due to its large size, the delivery of Cas9 to cells represents especially for *in vivo* experiments a bottleneck for efficient gene editing. The introduction of constitutive expressing Cas9 mice, Rosa26-Cas9, reduced the number of components to be delivered to a small sgRNA and thereby increasing editing efficacy (Platt et al. 2014; Zetsche et al. 2015). Cas9 is expressed in Rosa26 loxPStoploxP-Cas9 knock-in mice and as a Cre mouse, both constitutively express Cas9 once activated (Chu et al. 2016; Platt et al. 2014). The utility of constitutive Cas9 systems can however be limited by off-target effects and inflammatory responses if Cas9 expression is not controlled. Platt states that Cas9 is not toxic in the Cre-Cas9 mouse; however, Wu highlights that constitutive Cas9 expression, as occurs post Cre induction in the loxPStoploxP-Cas9 mouse, carries unintended cellular consequences (Platt et al. 2014; Wu et al. 2014).

Cre-systems are limited by potential genotoxicity due to cryptic recombination sites in mammalian genomes (Loonstra et al. 2001), which is not observed with doxycycline- (dox) or tetracycline-inducible systems. Moreover, a Cre-based system allows only for a single activation followed by constitutive expression while an inducible system can repeatedly be turned on and off, thereby avoiding secondary effects of constitutive expressions. However, tight regulation of dox-inducible systems are challenging, so common practice is to split the two major components (Tet repressor and inducible promoter) and introduce the two different parts in different loci, thus complicating the genetic engineering process (Sun et al. 2007).

The expression of Cas9 protein in mice (independent of delivery method) evokes Cas9-specific IgM and high-affinity IgG antibody isotypes (Chew et al. 2016), indicating a humoral immune response by activation of both T-cell-independent (IgM) and T-cell-dependent (immunoglobulin type G, IgG1, IgG2a, and IgG2b subtypes) toward Cas9 detectable by ELISA (Wang et al. 2015).

Moreover, recent data in 42 different p53 wild-type cancer cells indicate that the constitutive expression of Cas9 can moderately upregulate a p53 response (Enache et al. 2020). p21 is one of the downstream targets of p53 which prevents cell cycle progression and it was shown that upon expression of Cas9, cells are enriched containing p53 loss-of-function mutations as those cells have a growth advantage. Further, in the recent review by Mirgayazova, it is stated that genome-wide CRISPR and RNAi screens reveal the mutational selection potential of CRISPR knockouts across the whole exome and CRISPR gene editing induces p53-dependent damage. This shows the importance of monitoring the oncogenic driver gene status when CRISPR/Cas9 editing *in vitro* (Mirgayazova et al. 2020) and it is also worthy of investigation *in vivo*.

7.6 Limiting and Detecting Off-Target Editing *in vivo*

Since targeting by CRISPR/Cas9 is reliant on 23 base pair matches (PAM and a guide RNA hybridizes a 20-nt DNA sequence), gene editing by CRISPR/Cas9 may produce a number of nonspecific mutations in the genome. Minimizing the off-target events is crucial for Cas9 *in vivo* application as off-target indels could affect cell viability, promote tumorigenesis, generate large deletions and genomic rearrangements, or lead to lethal genetic mutations that cause loss of gene function (reviewed in Nissim et al. 2014).

There are many ways to minimize CRISPR/Cas9 off-target effects; namely, design/choice of sgRNA, variety of Cas9, and delivery method. Optimizing sgRNA design (Dhanjal et al. 2020) and use of software such as Cas-OFFinder (an *in silico* tool to detect off-target sites) and inDelphi can be used to computationally predict the *in vivo* outcome of the repair using either nonhomologous or microhomology-mediated pathways (Shen et al. 2018). VIVO has shown that appropriately designed sgRNAs can direct efficient *in vivo* editing in mouse with no detectable off-target mutations (Akcakaya et al. 2018). Spatial and temporal control of Cas9 activity inside the cell and especially the use of conditional or inducible expression systems such as photoactivatable Cas, chemical-inducible Cas9, double Cas9 nickase, or by high-fidelity Cas9 nucleases (Kleinstiver et al. 2016; Ran et al. 2013a; Ran et al. 2013b) help minimizing off-target events.

The two methods for investigating *in vivo* off target editing are VIVO (**v**erification of ***in*** ***v**ivo* **o**ff-targets) and DISCOVER-Seq (**d**iscovery of **in** **s**itu **C**as **o**ff-targets and **ver**ification by **seq**uencing) (Akcakaya et al. 2018; Wienert et al. 2019). VIVO is a two-step system that has demonstrated CRISPR/Cas nucleases induce substantial off-target mutations *in vivo* (Akcakaya et al. 2018). DISCOVER-Seq is a universal approach for unbiased off-target identification and molecular characterization of Cas9 activity *in vivo* that relies on endogenous DNA repair factors during *in vivo* editing (Wienert et al. 2019).

Delivery of adenovirus (AdV) with Cas9 and a promiscuous sgRNA, specifically designed to target to the mouse *Pcsk9* but also to generate numerous off target sites, was intravenously administered and induced the expected genetic alterations in the mouse Pcsk9 transgene. VIVO identified 3107 off-target cleavage sites with many off-target sites containing PAM mismatches, among which NAG was the most prevalent. VIVO correlates with CIRCLE-seq, higher CIRCLE-seq read counts are in line with finding indels *in vivo* (Akcakaya et al. 2018). When this study was repeated and analyzed with DISCOVER-Seq, very-low-frequency off-targets identified by VIVO's *in vitro* protocol were not detected by *in vivo* DISCOVER-Seq, but DISCOVER-Seq's NGS-validated off-targets were not VIVO prioritized because they were lost among VIVO's nearly 3000 hits and false positives (Wienert et al. 2019). Both VIVO and DISCOVER-seq set an important standard for defining *in vivo* off-target effects of gene-editing nucleases.

7.7 Animal Species

Gene editing by CRISPR/Cas9 in numerous animal species has been reported to create disease models or to validate therapeutic targets fundamental to many areas of biology.

7.7.1 Rodents and Small Animals

7.7.1.1 Genetically Engineered Mouse Models (GEMMs)

Rather than transplanting CRISPR-modified cells, numerous groups have adapted the CRISPR/Cas9 technology for direct *in vivo* delivery, enabling specific and multiplexed genetic modification in selected tissues of adult mice.

In oncology research, there is significant investment in genetically engineered mouse models (GEMMs) where gene expression is both spatially and temporally controlled to enable more faithful modeling of equivalent human tumors. While GEMMs are powerful tools, mouse strain production is both time-consuming and costly and needs to be repeated for each new engineering. Moreover, GEMMs are expensive to maintain with each specific model requiring extensive breeding to maintain a single model. A further requirement of somatic multiplex-mutagenesis is extensive mouse intercrossing for the generation of relevant experimental cohorts of multi-allelic mutant mice (Kersten et al. 2017). In addition to long breeding programs, the tumor models have a long indolence time, which often precludes the use of these models for drug testing (Winslow et al. 2011). When these models have a latency period conducive to drug testing, due to the complex nature of breeding programs required, often only small numbers of tumor-bearing animals can be recruited onto a study at a given time, compared with more traditional xenografted or explant cancer models (Lampreht Tratar et al. 2018). To reduce cost and increase output, it is preferential to have tunable timelines to adapt tumor development to treatment time based on compound activity and mode of action (Kersten et al. 2017).

In contrast, non-germline GEMMs (nGEMMS), where somatic cells carry engineered alleles but not germline cells, provide an option to shorten production timelines and increased genomic heterogeneity (Weber et al. 2015). However, timelines to establish tumor models are still long, up to months before having tumors in which to test therapeutics (reviewed in Heyer et al. 2010). Moreover, traditionally used xenografted nGEMMs are also often immunodeficient, limiting its applicability (Kersten et al. 2017). Furthermore, cancer prone non-germline chimeric mice, produced by injecting genetically engineered embryonic stem cells into blastocysts of a mouse line, have an increased variability related to tumorigenesis between individual mice (Huijbers et al. 2011). This negatively affects standardizing treatment studies in these models and demonstrates an advantage of somatic engineering.

Instead, nGEMMs, which rely on *in vivo* delivery of editing components to generate somatic genomic changes of oncogenes and/or repressor genes, provide greater flexibility in target turnover and recapitulation of tumor microenvironment, including immune effects.

Inducible CRISPR/Cas9 systems have been developed, including Cre recombinase-mediated cassette exchange or tetracycline-inducible Cas9 mice (Chiou et al. 2015; Chu et al. 2016; Dow et al. 2015; DuPage et al. 2009; Huijbers et al. 2011; Jo et al. 2019; Katigbak et al. 2018; Platt et al. 2014; Yang et al. 2013). The most recent advance in Cas9-inducible mice is the ObLiGaRe doxycycline-inducible SpCas9 mouse or ODInCas9 mouse (Lundin et al. 2020). ODInCas9 mice demonstrate inducible Cas9 expression in all tissues and can be repeatedly induced with no Cas9 detection upon doxycycline withdrawal. Chicken β-globulin insulators downstream of Tet-On 3G, which flank the TREG3-Cas9-T2A-GFP, allow for the generation of an all-in-one system without expression in the inactive state. Delivery of sgRNAs as a single component to Cas9-expressing cells demonstrate high editing efficacy resulting in disease phenotypic readouts with no detectable edits in the absence of doxycycline.

The ODInCas9 mouse has significant potential as a universal background in which to rapidly generate stable gene modification. The ODInCas9 mouse when bred to a human *PCSK9* knockin mouse resulted in a mouse with high *PCSK9* and triglycerides that also carried an inducible Cas9. To target liver cells expressing the human *PCSK9* gene, mice were dosed with doxycycline to express Cas9 and LNP containing the validated sgRNA for *PCSK9*. 21 days post dosing, NGS was performed on genomic liver DNA and 3.5% of the NGS reads edited were covering the *PCSK9* gene. This editing resulted in a physiological decrease (20%) in circulating triglycerides (Lundin et al. 2020). Further evidence that the ODInCas9 mouse can be utilized as a universal strain is when mice are induced with 2 mg/kg doxycycline in the drinking water, and AAV9-carrying sgRNAs are delivered intravenously or by oral aspiration to the lung, the lungs developed adenomas and adenocarcinomas. AAV9 virus-bearing sgRNAs to modify the Kras oncogene and *Trp53* tumor suppressor or protein kinase *Stk11* were generated. The Kras gain of function point mutation utilized homology-directed repair (HDR) and targeted editing and indels knocked out the *Trp53* or *Stk11* (Lundin et al. 2020).

Rat – Targeted gene editing in the rat was transformed by the advent of ZFN and TALEN technologies. These allowed for the creation of mutant strains without the need for establishing embryonic stem cell cultures first. The number of germline rat mutant strains and range: knockin, knockout, fluorescent reporters as well as Cre and loxP conditional mutants, are as wide as those reported in the mouse (Filipiak et al. 2019; Ma et al. 2014; McCarron et al. 2020; Pettibone et al. 2019; Yoshimi et al. 2021). Transgenic rats that express Cre-dependent Cas9 or Cas9 nickase have been generated by Back and colleagues allowing for neuron-specific genome editing (Back et al. 2019).

Rabbit – Editing of the rabbit genome has also benefitted from the development of ZFN and TALEN technologies. Rabbits have advantages over larger animals such as sheep and pigs as they are easier to breed and are less costly. They are useful for the study of many human diseases such as atherosclerosis, cystic fibrosis, and Duchenne muscular dystrophy, for example. CRISPR/Cas9 modeling of Duchenne muscular dystrophy in rabbits is preferred to mice as they have clear clinical manifestations (Wells 2018). Since rabbits lack germline transmitting embryonic stem cells and there is a very low efficiency of somatic cell nuclear transfer, the majority of CRISPR/Cas9 models are the result of co-injection of Cas9 mRNA and sgRNA into rabbit zygotes. The generation of gene-targeted transgenic rabbits by gene editing nucleases is reviewed and described in a publication by Yang and colleagues (Yang et al. 2019).

Zebrafish – The use of zebrafish (*Danio rerio*), a small fresh water fish, in scientific research has increased considerably since the 1990s when they were first used to study embryogenesis (Driever et al. 1994). After the initial cost to install suitable housing, the small size, the large number of progeny, and the fast life cycle allow for large-scale approaches. Zebrafish have a high genetic homology with man and precise genetic editing by CRISPR/Cas9 make zebrafish a robust model for studying human disease. Injection of gRNAs and Cas9 mRNA into one-cell stage embryos yields indels at target sites with relatively high frequencies (Gagnon et al. 2014; Hwang et al. 2014). These mutations are inheritable due to mosaic targeting of the germline and establishment of mutant strains. Recent work by Meshalkina and colleagues highlighted the difference in editing when using different nucleases, LbCas12a or SpCas9, after microinjection of RNP complexes in zebrafish zygotes. The reported editing efficiencies of both nucleases were the same; however, LbCas12a editing resulted in longer deletions and more rare insertions, compared with the editing events generated by SpCas9 (Meshalkina et al. 2020).

7.7.2 Large Animal

Pig – Pigs are not a traditional model animal of choice due to their size and weight, cost, housing, and limited ability to develop targeted mutants. However, somatic gene editing by CRISPR/Cas9 is reported to ameliorate skeletal and cardiac muscle failure in pig models of Duchenne muscular dystrophy (Moretti et al. 2020). Pigs have been used to generate porcine cancer models by TP53-modification by electroporation in zygotes (Tanihara et al. 2018). Importantly, a Cre-dependent inducible Cas9-expressing pig has been reported to enable efficient *in vivo* genome editing (Wang et al. 2017).

Sheep – Sheep have been used extensively in studies on genetic editing, from "Dolly" the "cloned" sheep in 1996 to the first report of CRISPR/Cas9 knockin of a reporter protein into the Rosa26 safe harbor loci of sheep in 2016 (Wu et al. 2016). Recently, CRISPR/Cas9 has been used to create sheep models of cystic fibrosis by somatic cell nuclear transfer to induce indels and subsequent knockout of the CTRF gene and deafness by zygote microinjection of Cas9 mRNA and ssDNA guides to generate knockin of stop codons by HDR (Fan et al. 2018; Menchaca et al. 2020). Multiplex gene editing with CRISPR/Cas9 has been performed with three genes (MSTN, ASIP, and BCO2) targeted, resulting in the sheep having increased muscle development and importantly there were no detectable off-target effects (Wang et al. 2016).

Nonhuman primates – The use of CRISPR/Cas9 genome editing in the most commonly used nonhuman primates, cynomolgus, rhesus monkey, and common marmoset, is reviewed by Kang and colleagues (Kang et al. 2019). A brief summary of experimental work performed in rhesus monkey, zygotic injection of Cas9 mRNA and two sgRNAs targeting *MCPH1* into embryos resulted in knockout positive embryos, including homozygote knockout (Luo et al. 2019). Whole genome sequencing revealed no off-target editing or large deletions (Luo et al. 2019). Kumita assessed editing efficiency after delivery of Cas9 protein or mRNA to modify target genes in marmoset embryos and Cas9 protein resulted in all embryos having mosaic mutations with KI and KO of genes (Kumita et al. 2019).

7.7.3 Invertebrates

Drosophila – *Drosophila melanogaster (Drosophila)* have been used to study hereditary and genetics for over 100 years and they are the first true model organism (Rubin and Lewis 2000). There are numerous transgenic flies expressing Cas9 under the control of germline-specific or ubiquitous promoters that demonstrate NHEJ and HDR editing (Gratz et al. 2014; Kondo and Ueda 2013; Ren et al. 2013). CRISPRa resulting in physiologically target gene expression and gain-of-function phenotypes in multiple tissues with temporal control are reported (Ewen-Campen et al. 2017). Recent work by Huynh and colleagues describes *drosophila* that have both spatial control (tissue-specific Cas9 expression) and temporal control of editing by utilizing Rapamycin- or RU482-inducible Cas9 (Huynh et al. 2020). Further, Cas9 constructs were split for rapamycin-dependent gene disruption and activation (Huynh et al. 2020).

Nematodes – *Caenorhabditis elegans* (*C. elegans*) like *drosophila* has been extensively used as a model organism since the 1970s, and it was the first multicellular organism to have its whole genome sequenced as well as is the only organism to have its connectome mapped (reviewed in Brenner 1974; Consortium 1998; Schafer 2005). Initial studies using CRISPR/Cas9 to generate mutants in *C. elegans* utilized direct injection of plasmids encoding Cas9 and a sgRNA into the gonad of adult hermaphrodites (Friedland et al. 2013). NHEJ and HDR have been utilized in *C. elegans* to generate mutations creating heritable gene modifications of disease (Kim et al. 2014; Paix et al. 2014), with protocols for editing in *C. elegans* and *Caenorhabditis briggsae* provided by (Culp et al. 2020; Kim and Colaiacovo 2019). Further, conditional Cas9 mutations in *C. elegans* have been reported by Shen (Shen et al. 2014).

7.8 Delivery Systems of CRISPR/Cas9 Components *in vivo*

Currently, there are three main strategies for delivering CRISPR/Cas9 components into cells and tissues: (i) viral vectors (AdV, AAV) for DNA encoding Cas9 and sgRNA;

(ii) Cas9 with sgRNA as a preformed ribonucleoprotein (RNP) complex; (iii) lipid nanoparticles (LNPs) for either mRNA encoding Cas9 and sgRNA or loaded with a Cas9/sgRNA RNP. Among these, viral gene delivery is the closest to clinical therapeutic use, since it has been utilized in clinical trials in classical gene therapy since the 1980s (Wilson and Gilbert 2018). A limitation of viral delivery is that the Cas9 and gRNA are packaged into plasmid DNA, which is delivered via the viral vector to the target cell. This allows the CRISPR/Cas9 component coding genes to exist persistently in the target cell type possibly resulting in elevated Cas9 levels and increasing the chances of off-target activity.

7.8.1 Delivery Vehicles for CRISPR Components

7.8.1.1 Adenovirus (AdV)

AdV was the first *in vivo* gene therapy viral vector trialed in 1992 (Crystal 2014). It has a large cargo capacity, 36-kb double-stranded DNA genome. AdV can infect dividing and nondividing cells and upon transduction, the DNA remains episomal and consequently does not integrate into the genome. The most used serotype is AdV type 5, which has a high tropism for hepatocytes. AdV generate a large inflammatory response, with B cells and T helper cells activated by the viral capsid proteins, resulting in secretion of neutralizing antibody thereby decreasing the effectiveness of subsequent treatment (Shirley et al. 2020).

AdV is one of the most effective means of delivering genes *in vivo*, particularly when targeting the liver. The use of AdV for CRISPR/Cas9 delivery is reported in targeting loss-of-function *PCSK9* mutation in mouse liver (Ding et al. 2014) and in a mouse expressing human *PCSK9* that models humans carrying *PCSK9* gain-of function mutations (Akcakaya et al. 2018; Carreras et al. 2019). Carreras and colleagues compared editing of *PCSK9* by AdV delivery of either SpCas9 or base editor (BE3). They reported that SpCas9 and BE3 treatments resulted in similar editing frequencies at the human *PCSK9* locus (range 10–31% and 11–35%, respectively) (Carreras et al. 2019); however, when VIVO was used to assess genome-wide off target editing, the number of indels was significantly lower following BE compared with Cas9-mediated genome editing (Carreras et al. 2019).

7.8.1.2 Adeno-Associated Virus (AAV)

AAV is a member of the parvovirus family. There are more than 100 different AAV serotypes reported with different tissue tropisms (reviewed in Srivastava 2016; Zincarelli et al. 2008). The wild-type virus carries a ~4.5-kb single-stranded DNA genome (Chamberlain et al. 2016; Patel et al. 2019). AAVs can infect both dividing and quiescent cells as well as remain in an extrachromosomal state without integrating into the genome of the host cell. AAV vectors are a leading candidate for the delivery of CRISPR/Cas9 for therapeutic genome editing as they elicit only a mild inflammatory response in humans (Rabinowitz et al. 2019; Shirley et al. 2020). AAV vectors can integrate at a low frequency (~0.1% of total vector genomes) in man, and studies conducted in one strain of mouse found AAV insertion into the Rian locus associated with hepatocellular carcinoma formation. This locus, however, is absent in nonhuman primates and humans, and hepatocellular carcinoma induction has not been associated with AAV vectors in NHPs and humans (reviewed in Hanlon et al. 2019). The work of Hanlon and colleagues identified AAV integration in mouse brain, cochlea, and muscle (Hanlon et al. 2019). SaCas9 is 3.2 kb, approximately 1 kb smaller than SpCas9 and can be packaged within an AAV. It is noteworthy that very large doses of AAV vectors have been administered *in vivo* (10^{13} AAV vector genomes/kg) above the doses used to study immune mechanisms and immunotoxicities of AAV have been reported (Hinderer et al. 2018; Hordeaux et al. 2018).

7.8.1.3 Ribonucleoprotein (RNP) Complex

Cas9 RNP complexes are Cas9 protein bound with sgRNA. RNPs have no innate cell-penetrating activity, and direct delivery into cells requires conjugation or mixing with lipid carrier molecules which can be immunogenic, inflammatory, and toxic. A carrier-free RNP has been used to edit Kras in lung cancer. In this study, the RNP complex self-assembles and selectively targets the nucleus due to a nuclear localization sequence and a low-molecular-weight protamine (Kim et al. 2018). RNPs do not activate the cyclic GMP-AMP synthase, the cytosolic signal to trigger the immune response (Sun et al. 2013). Transgenic mice, generated by electroporation of RNPs into mouse zygotes, have resulted in 100% bi-allelic edited embryos; however, this high editing efficiency came at the cost of embryo viability (reduced to 30%) (Chen et al. 2016). An advantage of delivery of Cas9 protein and sgRNAs delivered as an RNP complex is that the editing time frame can be shortened reducing the chances of off-target editing as the Cas9 nuclease is free of exogenous DNA and will be quickly be degraded in the cell.

7.8.1.4 Lipid Nanoparticle (LNPs)

LNPs are formed by a cationic ionizable lipid (CIL), cholesterol, polyethylene glycol (PEG) lipid, and distearoylphosphatidylcholine and have been used extensively in preclinical and clinical studies to deliver mRNA (Yanez Arteta et al. 2018). The majority of LNP studies now focus on CIL as they allow high encapsulation and have low toxicity. The size of the LNP can be controlled by varying the PEG lipid component; smaller particles have a higher percentage of PEG (Yanez Arteta et al. 2018). Intravenous injection allows LNPs to accumulate in all liver cell types (Kupffer cells, Liver sinusoidal endothelial cells, and hepatocytes) in a time- and dose-dependent manner (Shi et al. 2011).

Single intravenous administration in mice of LNP with CRISPR/Cas9 components results in 60% editing and >97% knockdown of the mouse TTR protein for 12 months (Finn et al. 2018). Cas9 protein levels peak at 4 hour post-LNP administration, with Cas9 mRNA and sgRNA rapidly cleared and undetectable after 72 hour (Finn et al. 2018). Administration of the same LNP to rats resulted in 70% editing within the liver (Finn et al. 2018). LNPs are able to have RNPs as cargo, and they have been demonstrated to achieve multiplex genome editing and restorative therapeutic editing (in *Pcsk9*) in mice (Wei et al. 2020).

7.8.2 Delivery Routes for CRISPR/Cas9 Components *in vivo*

The route of administration to introduce the sgRNA/Cas9 editing components has a significant role in the ultimate outcome and accuracy of the system. The nature and number of off-target editing events can be minimized plus the possibility to reduce inflammation by delivery route. A way to overcome delivery is to use tissue-specific expression of Cas9 or sgRNA.

Intravenous, Intranasal, Oral aspiration, Intraperitoneal administration – These delivery routes are not technically challenging and allow for a large number of animals to be dosed on a single occasion. Intranasal, oral aspiration, and intratracheal delivery all target the lungs with different efficacy (Santry et al. 2017). Lundin reported as expected tumors observed in nontarget tissue (liver) by intravenous injection of AAV but tissue-specific delivery like oral aspiration restricted tumor development to the lungs (Lundin et al. 2020). The route of AAV administration is known to substantially influence liver transduction efficiency *in vivo* whereby intravenous tail vein injection of serotypes AAV1, 2, 3, 5, and 6 transduce the liver 5- to 48-fold less efficiently than portal vein injection in mice (Grimm et al. 2003). Intraperitoneal administration of AAV has been used to deliver sgRNA that targets Cas9-expressing cardiomyocytes (Carroll et al. 2016).

Hydrodynamic delivery involves rapidly injecting a large volume (approx. 8–10% body weight) in the bloodstream of an animal, usually via the tail vein in mice (Liu et al. 1999). The increase in pressure temporarily enhances permeability of endothelial and parenchymal cells allowing for proteins/cargo not normally capable of crossing a cellular membrane to enter cells. Even though this route of administration has been used to deliver plasmids carrying Cas9 and sgRNA to the heart, lungs, liver, and kidney with demonstrated genome editing (Bonamassa et al. 2011; Kamimura et al. 2015; Lee et al. 2019; Xue et al. 2014; Yin et al. 2014), it is ethically questionable. Hydrodynamic delivery can result in physiological complications such as cardiac dysfunction, increased blood pressure, and liver expansion (Bonamassa et al. 2011; Suda and Liu 2007; Suda et al. 2007).

Direct administration to target organ – Direct administration of virus, LNP, or RNP to an organ requires a large degree of technical skill. To generate pancreatic ductal adenocarcinoma GEMMs, AAVs delivered sgRNAs to the adult mouse pancreas resulting in expression of Kras G12D allele through HDR, and CRISPR-induced disruption by indels within alleles of Trp53, Lkb1, and Arid1A (Ideno et al., 2019). The systemic delivery of AAV vectors for Cas9 and sgRNA to skeletal muscle, tibialis anterior muscle of mdx mice, a model of Duchenne muscular dystrophy, resulted in removal of mutated exon 23 from the dystrophin gene and improved muscle performance (Nelson et al. 2016). Systemic delivery of RNPs to targeted organs remains challenging, thus local administration to organs such as the inner ear (Gao et al. 2018), muscle (Lee et al. 2017), and brain (Staahl et al. 2017; Wei et al. 2020) has been reported.

7.9 Concluding Remarks

The staggeringly rapid pace of technological development within the CRISPR/cas9 makes it difficult to stay abreast of all changes. Just this year, we have seen the first use of PE *in vivo*, there is a new much smaller Cas9 (PhageCas9) and delivery of RNPs has improved significantly. It is difficult to predict exactly what the future holds for CRISPR-based *in vivo* models but there are exciting times ahead. The CRISPR/Cas9 tools currently available allow the development of accurate disease models to promote better treatment options for patients.

References

Adli, M. (2018). The CRISPR tool kit for genome editing and beyond. *Nat. Commun.* 9: 1911.

Agudelo, D., Carter, S., Velimirovic, M. et al. (2020). Versatile and robust genome editing with *Streptococcus thermophilus* CRISPR1-Cas9. *Genome Res.* 30: 107–117.

Akcakaya, P., Bobbin, M.L., Guo, J.A. et al. (2018). in vivo CRISPR editing with no detectable genome-wide off-target mutations. *Nature* 561: 416–419.

Anders, C., Bargsten, K., and Jinek, M. (2016). Structural plasticity of PAM recognition by engineered variants of the RNA-guided endonuclease Cas9. *Mol. Cell.* 61 (6): 895–902.

Anzalone, A.V., Randolph, P.B., Davis, J.R. et al. (2019). Search-and-replace genome editing without double-strand breaks or donor DNA. *Nature* 576: 149–157.

Back, S., Necarsulmer, J., Whitaker, L.R. et al. (2019). Neuron-specific genome modification in the Adult Rat Brain using CRISPR-Cas9 transgenic Rats. *Neuron* 102: 105–119. e8.

Bibikova, M., Golic, M., Golic, K.G., and Carroll, D. (2002). Targeted chromosomal cleavage and mutagenesis in Drosophila using zinc-finger nucleases. *Genetics* 161: 1169–1175.

Blasco, R.B., Karaca, E., Ambrogio, C. et al. (2014). Simple and rapid in vivo generation of chromosomal rearrangements using CRISPR/Cas9 technology. *Cell Rep.* 9: 1219–1227.

Bonamassa, B., Hai, L., and Liu, D. (2011). Hydrodynamic gene delivery and its applications in pharmaceutical research. *Pharm. Res.* 28: 694–701.

Braun, C.J., Bruno, P.M., Horlbeck, M.A. et al. (2016). Versatile in vivo regulation of tumor phenotypes by dCas9-mediated transcriptional perturbation. *Proc. Natl. Acad. Sci. U. S. A.* 113: E3892–E3900.

Brenner, S. (1974). The genetics of Caenorhabditis elegans. *Genetics* 77: 71–94.

Cai, E.P., Ishikawa, Y., Zhang, W. et al. (2020). Genome-scale in vivo CRISPR screen identifies RNLS as a target for beta cell protection in type 1 diabetes. *Nat. Metab.* 9: 934–945.

Carreras, A., Pane, L.S., Nitsch, R. et al. (2019). in vivo genome and base editing of a human PCSK9 knock-in hypercholesterolemic mouse model. *BMC Biol.* 17: 4.

Carroll, K.J., Makarewich, C.A., Mcanally, J. et al. (2016). A mouse model for adult cardiac-specific gene deletion with CRISPR/Cas9. *Proc. Natl. Acad. Sci. U. S. A.* 113: 338–343.

Chadwick, A.C., Wang, X., and Musunuru, K. (2017). in vivo base editing of PCSK9 (proprotein convertase subtilisin/kexin type 9) as a therapeutic alternative to genome editing. *Arterioscler. Thromb. Vasc. Biol.* 37: 1741–1747.

Chamberlain, K., Riyad, J.M., and Weber, T. (2016). Expressing transgenes that exceed the packaging capacity of adeno-associated virus capsids. *Hum. Gene Ther. Methods* 27: 1–12.

Chatterjee, P., Jakimo, N., Lee, J. et al. (2020a). Publisher correction: an engineered ScCas9 with broad PAM range and high specificity and activity. *Nat. Biotechnol.* 38: 1154–1158.

Chatterjee, P., Lee, J., Nip, L. et al. (2020b). A Cas9 with PAM recognition for adenine dinucleotides. *Nat. Commun.* 11: 2474.

Chavez, A., Scheiman, J., Vora, S. et al. (2015). Highly efficient Cas9-mediated transcriptional programming. *Nat. Methods* 12: 326–328.

Chavez, A., Tuttle, M., Pruitt, B.W. et al. (2016). Comparison of Cas9 activators in multiple species. *Nat. Methods* 13: 563–567.

Chen, S., Lee, B., Lee, A.Y. et al. (2016). Highly efficient mouse genome editing by CRISPR ribonucleoprotein electroporation of zygotes. *J. Biol. Chem.* 291: 14457–14467.

Chen, S., Sanjana, N.E., Zheng, K. et al. (2015). Genome-wide CRISPR screen in a mouse model of tumor growth and metastasis. *Cell* 160: 1246–1260.

Chen, W., Zhang, H., Zhang, Y. et al. (2019). Molecular basis for the PAM expansion and fidelity enhancement of an evolved Cas9 nuclease. *PLoS Biol.* 17: e3000496.

Cheng, A.W., Wang, H., Yang, H. et al. (2013). Multiplexed activation of endogenous genes by CRISPR-on, an RNA-guided transcriptional activator system. *Cell Res.* 23: 1163–1171.

Chew, W.L., Tabebordbar, M., Cheng, J.K. et al. (2016). A multifunctional AAV-CRISPR-Cas9 and its host response. *Nat. Methods* 13: 868–874.

Chiou, S.H., Winters, I.P., Wang, J. et al. (2015). Pancreatic cancer modeling using retrograde viral vector delivery and in vivo CRISPR/Cas9-mediated somatic genome editing. *Genes Dev.* 29: 1576–1585.

Choi, P.S. and Meyerson, M. (2014). Targeted genomic rearrangements using CRISPR/Cas technology. *Nat. Commun.* 5: 3728.

Christian, M., Cermak, T., Doyle, E.L. et al. (2010). Targeting DNA double-strand breaks with TAL effector nucleases. *Genetics* 186: 757–761.

Chu, V.T., Weber, T., Graf, R. et al. (2016). Efficient generation of Rosa26 knock-in mice using CRISPR/Cas9 in C57BL/6 zygotes. *BMC Biotechnol.* 16: 4.

Chylinski, K., Le Rhun, A., and Charpentier, E. (2013). The tracrRNA and Cas9 families of type II CRISPR-Cas immunity systems. *RNA Biol.* 10: 726–737.

Cong, L., Ran, F.A., Cox, D. et al. (2013). Multiplex genome engineering using CRISPR/Cas systems. *Science* 339: 819–823.

Consortium, C.E.S. (1998). Genome sequence of the nematode C. elegans: a platform for investigating biology. *Science* 282: 2012–2018.

Crystal, R.G. (2014). Adenovirus: the first effective in vivo gene delivery vector. *Hum. Gene Ther.* 25: 3–11.

Culp, E., Richman, C., Sharanya, D. et al. (2020). Genome editing in the nematode Caenorhabditis briggsae using the CRISPR/Cas9 system. *Biol. Methods Protoc.* 5: bpaa003.

Deveau, H., Garneau, J.E., and Moineau, S. (2010). CRISPR/Cas system and its role in phage-bacteria interactions. *Annu. Rev. Microbiol.* 64: 475–493.

Dhanjal, J.K., Dammalapati, S., Pal, S., and Sundar, D. (2020). Evaluation of off-targets predicted by sgRNA design tools. *Genomics* 112: 3609–3614.

Ding, Q., Strong, A., Patel, K.M. et al. (2014). Permanent alteration of PCSK9 with in vivo CRISPR-Cas9 genome editing. *Circ. Res.* 115: 488–492.

Doudna, J.A. and Charpentier, E. (2014). Genome editing. The new frontier of genome engineering with CRISPR-Cas9. *Science* 346: 1258096.

Dow, L.E., Fisher, J., O'rourke, K.P. et al. (2015). Inducible in vivo genome editing with CRISPR-Cas9. *Nat. Biotechnol.* 33: 390–394.

Driever, W., Stemple, D., Schier, A., and Solnica-Krezel, L. (1994). Zebrafish: genetic tools for studying vertebrate development. *Trends Genet.* 10: 152–159.

Dupage, M., Dooley, A.L., and Jacks, T. (2009). Conditional mouse lung cancer models using adenoviral or lentiviral delivery of Cre recombinase. *Nat. Protoc.* 4: 1064–1072.

Edraki, A., Mir, A., Ibraheim, R. et al. (2019). A compact, high-accuracy Cas9 with a dinucleotide PAM for in vivo genome editing. *Mol. Cell* 73: 714–726. e4.

Enache, O.M., Rendo, V., Abdusamad, M. et al. (2020). Author correction: Cas9 activates the p53 pathway and selects for p53-inactivating mutations. *Nat. Genet.* 52: 748–749.

Ewen-Campen, B., YANG-Zhou, D., Fernandes, V.R. et al. (2017). Optimized strategy for in vivo Cas9-activation in Drosophila. *Proc. Natl. Acad. Sci. U. S. A.* 114: 9409–9414.

Fan, Z., Perisse, I.V., Cotton, C.U. et al. (2018). A sheep model of cystic fibrosis generated by CRISPR/Cas9 disruption of the CFTR gene. *JCI Insight* 3.

Filipiak, W.E., Hughes, E.D., Gavrilina, G.B. et al. (2019). Next generation transgenic rat model production. *Methods Mol. Biol.* 2018: 97–114.

Finn, J.D., Smith, A.R., Patel, M.C. et al. (2018). A single administration of CRISPR/Cas9 lipid nanoparticles achieves robust and persistent in vivo genome editing. *Cell Rep.* 22: 2227–2235.

Friedland, A.E., Tzur, Y.B., Esvelt, K.M. et al. (2013). Heritable genome editing in C. elegans via a CRISPR-Cas9 system. *Nat. Methods* 10: 741–743.

Gagnon, J.A., Valen, E., Thyme, S.B. et al. (2014). Efficient mutagenesis by Cas9 protein-mediated oligonucleotide insertion and large-scale assessment of single-guide RNAs. *PLoS One* 9: e98186.

Gao, X., Tao, Y., Lamas, V. et al. (2018). Treatment of autosomal dominant hearing loss by in vivo delivery of genome editing agents. *Nature* 553: 217–221.

Gaudelli, N.M., Komor, A.C., Rees, H.A. et al. (2017). Programmable base editing of A*T to G*C in genomic DNA without DNA cleavage. *Nature* 551: 464–471.

Gilbert, L.A., Horlbeck, M.A., Adamson, B. et al. (2014). Genome-scale CRISPR-mediated control of gene repression and activation. *Cell* 159: 647–661.

Gratz, S.J., Ukken, F.P., Rubinstein, C.D. et al. (2014). Highly specific and efficient CRISPR/Cas9-catalyzed homology-directed repair in Drosophila. *Genetics* 196: 961–971.

Grimm, D., Zhou, S., Nakai, H. et al. (2003). Preclinical in vivo evaluation of pseudotyped adeno-associated virus vectors for liver gene therapy. *Blood* 102: 2412–2419.

Hanlon, K.S., Kleinstiver, B.P., Garcia, S.P. et al. (2019). High levels of AAV vector integration into CRISPR-induced DNA breaks. *Nat. Commun.* 10: 4439.

Heyer, J., Kwong, L.N., Lowe, S.W., and Chin, L. (2010). Non-germline genetically engineered mouse models for translational cancer research. *Nat. Rev. Cancer* 10: 470–480.

Hinderer, C., Katz, N., Buza, E.L. et al. (2018). Severe toxicity in nonhuman primates and piglets following high-dose intravenous administration of an adeno-associated virus vector expressing human SMN. *Hum. Gene Ther.* 29: 285–298.

Hirano, H., Gootenberg, J.S., Horii, T. et al. (2016). Structure and engineering of *Francisella novicida* Cas9. *Cell* 164 (5): 950–961.

Hirosawa, M., Fujita, Y., Parr, C.J.C. et al. (2017). Cell-type-specific genome editing with a microRNA-responsive CRISPR-Cas9 switch. *Nucleic Acids Res.* 45: e118.

Hoffmann, M.D., Aschenbrenner, S., Grosse, S. et al. (2019). Cell-specific CRISPR-Cas9 activation by microRNA-dependent expression of anti-CRISPR proteins. *Nucleic Acids Res.* 47: e75.

Hordeaux, J., Wang, Q., Katz, N. et al. (2018). The neurotropic properties of AAV-PHP.B are limited to C57BL/6J mice. *Mol. Ther.* 26: 664–668.

Horvath, P. and Barrangou, R. (2010). CRISPR/Cas, the immune system of bacteria and archaea. *Science* 327: 167–170.

Hou, Z., Zhang, Y., Propson, N.E. et al. (2013). Efficient genome engineering in human pluripotent stem cells using Cas9 from Neisseria meningitidis. *Proc. Natl. Acad. Sci. U. S. A.* 110: 15644–15649.

Hu, Z., Wang, S., Zhang, C. et al. (2020). A compact Cas9 ortholog from Staphylococcus Auricularis (SauriCas9) expands the DNA targeting scope. *PLoS Biol.* 18: e3000686.

Huijbers, I.J., Krimpenfort, P., Berns, A., and Jonkers, J. (2011). Rapid validation of cancer genes in chimeras derived from established genetically engineered mouse models. *BioEssays* 33: 701–710.

Huynh, N., Wang, S., and King-Jones, K. (2020). Spatial and temporal control of gene manipulation in Drosophila via drug-activated Cas9 nucleases. *Insect Biochem. Mol. Biol.* 120: 103336.

Hwang, W.Y., Peterson, R.T., and Yeh, J.R. (2014). Methods for targeted mutagenesis in zebrafish using TALENs. *Methods* 69: 76–84.

Ideno, N., Yamaguchi, H., Okumura, T. et al. (2019). A pipeline for rapidly generating genetically engineered mouse models of pancreatic cancer using in vivo CRISPR-Cas9-mediated somatic recombination. *Lab. Investig.* 99: 1233–1244.

Jensen, E.D., Ferreira, R., Jakociunas, T. et al. (2017). Transcriptional reprogramming in yeast using dCas9 and combinatorial gRNA strategies. *Microb. Cell Factories* 16: 46.

Jinek, M., Chylinski, K., Fonfara, I. et al. (2012). A programmable dual-RNA-guided DNA endonuclease in adaptive bacterial immunity. *Science* 337: 816–821.

Jo, N., Sogabe, Y., Yamada, Y. et al. (2019). Platforms of in vivo genome editing with inducible Cas9 for advanced cancer modeling. *Cancer Sci.* 110: 926–938.

Kamimura, K., Yokoo, T., Abe, H. et al. (2015). Image-guided hydrodynamic gene delivery: current status and future directions. *Pharmaceutics* 7: 213–223.

Kang, Y., Chu, C., Wang, F., and Niu, Y. (2019). CRISPR/Cas9-mediated genome editing in nonhuman primates. *Dis. Model. Mech.* 12: 1–7.

Katigbak, A., Robert, F., Paquet, M., and Pelletier, J. (2018). Inducible Genome Editing with Conditional CRISPR/Cas9 Mice. *G3 (Bethesda)* 8: 1627–1635.

Kersten, K., De Visser, K.E., Van Miltenburg, M.H., and Jonkers, J. (2017). Genetically engineered mouse models in oncology research and cancer medicine. *EMBO Mol. Med.* 9: 137–153.

Kiani, S., Beal, J., Ebrahimkhani, M.R. et al. (2014). CRISPR transcriptional repression devices and layered circuits in mammalian cells. *Nat. Methods* 11: 723–726.

Kim, E., Koo, T., Park, S.W. et al. (2017). in vivo genome editing with a small Cas9 orthologue derived from Campylobacter jejuni. *Nat. Commun.* 8: 14500.

Kim, H., Ishidate, T., Ghanta, K.S. et al. (2014). A co-CRISPR strategy for efficient genome editing in Caenorhabditis elegans. *Genetics* 197: 1069–1080.

Kim, H.M. and Colaiacovo, M.P. (2019). CRISPR-Cas9-guided genome engineering in caenorhabditis elegans. *Curr. Protoc. Mol. Biol.* 129: e106.

Kim, S.M., Shin, S.C., Kim, E.E. et al. (2018). Simple in vivo gene editing via direct self-assembly of Cas9 ribonucleoprotein complexes for cancer treatment. *ACS Nano* 12: 7750–7760.

Kleinstiver, B.P., Pattanayak, V., Prew, M.S. et al. (2016). High-fidelity CRISPR-Cas9 nucleases with no detectable genome-wide off-target effects. *Nature* 529: 490–495.

Kleinstiver, B.P., Prew, M.S., Tsai, S.Q. et al. (2015). Broadening the targeting range of Staphylococcus aureus CRISPR-Cas9 by modifying PAM recognition. *Nat. Biotechnol.* 33: 1293–1298.

Knapp, D., Michaels, Y.S., Jamilly, M. et al. (2019). Decoupling tRNA promoter and processing activities enables specific Pol-II Cas9 guide RNA expression. *Nat. Commun.* 10: 1490.

Komor, A.C., Kim, Y.B., Packer, M.S. et al. (2016). Programmable editing of a target base in genomic DNA without double-stranded DNA cleavage. *Nature* 533: 420–424.

Kondo, S. and Ueda, R. (2013). Highly improved gene targeting by germline-specific Cas9 expression in Drosophila. *Genetics* 195: 715–721.

Konermann, S., Brigham, M.D., Trevino, A.E. et al. (2015). Genome-scale transcriptional activation by an engineered CRISPR-Cas9 complex. *Nature* 517: 583–588.

Krol, J., Loedige, I., and Filipowicz, W. (2010). The widespread regulation of microRNA biogenesis, function and decay. *Nat. Rev. Genet.* 11: 597–610.

Kumita, W., Sato, K., Suzuki, Y. et al. (2019). Efficient generation of Knock-in/Knock-out marmoset embryo via CRISPR/Cas9 gene editing. *Sci. Rep.* 9: 12719.

Kurata, M., Wolf, N.K., Lahr, W.S. et al. (2018). Highly multiplexed genome engineering using CRISPR/Cas9 gRNA arrays. *PLoS One* 13: e0198714.

Kurt, I.C., Zhou, R., Iyer, S. et al. (2021). CRISPR C-to-G base editors for inducing targeted DNA transversions in human cells. *Nat. Biotechnol.* 39: 41–46.

Lampreht Tratar, U., Horvat, S., and Cemazar, M. (2018). Transgenic mouse models in cancer research. *Front. Oncol.* 8: 268.

Lau, C.H. and Suh, Y. (2017). in vivo genome editing in animals using AAV-CRISPR system: applications to translational research of human disease. *F1000Res* 6: 2153.

Lee, J.K., Jeong, E., Lee, J. et al. (2018). Directed evolution of CRISPR-Cas9 to increase its specificity. *Nature Commun.* 9: 3048.

Lee, J., Mou, H., Ibraheim, R. et al. (2019). Tissue-restricted genome editing in vivo specified by microRNA-repressible anti-CRISPR proteins. *RNA* 25: 1421–1431.

Lee, J.G., Sung, Y.H., and Baek, I.J. (2018a). Generation of genetically-engineered animals using engineered endonucleases. *Arch. Pharm. Res.* 41: 885–897.

Lee, J.K., Jeong, E., Lee, J. et al. (2018b). Directed evolution of CRISPR-Cas9 to increase its specificity. *Nat. Commun.* 9: 3048.

Lee, K., Conboy, M., Park, H.M. et al. (2017). Nanoparticle delivery of Cas9 ribonucleoprotein and donor DNA in vivo induces homology-directed DNA repair. *Nat. Biomed Eng.* 1: 889–901.

Lee, R.T., Ng, A.S., and Ingham, P.W. (2016). Ribozyme mediated gRNA generation for in vitro and in vivo CRISPR/Cas9 mutagenesis. *PLoS One* 11: e0166020.

Li, F., Huang, Q., Luster, T.A. et al. (2020). in vivo epigenetic CRISPR screen identifies Asf1a as an immunotherapeutic target in Kras-Mutant Lung Adenocarcinoma. *Cancer Discov.* 10: 270–287.

Li, H., Haurigot, V., Doyon, Y. et al. (2011). in vivo genome editing restores haemostasis in a mouse model of haemophilia. *Nature* 475: 217–221.

Liao, H.K., Hatanaka, F., Araoka, T. et al. (2017). in vivo target gene activation via CRISPR/Cas9-mediated trans-epigenetic modulation. *Cell* 171: 1495–1507. e15.

Liu, F., Song, Y., and Liu, D. (1999). Hydrodynamics-based transfection in animals by systemic administration of plasmid DNA. *Gene Ther.* 6: 1258–1266.

Liu, Y., Li, X., He, S. et al. (2020). Efficient generation of mouse models with the prime editing system. *Cell Discov.* 6: 27.

Loonstra, A., Vooijs, M., Beverloo, H.B. et al. (2001). Growth inhibition and DNA damage induced by Cre recombinase in mammalian cells. *Proc. Natl. Acad. Sci. U. S. A.* 98: 9209–9214.

Lundin, A., Porritt, M.J., Jaiswal, H. et al. (2020). ObLiGaRe doxycycline Inducible (ODIn) Cas9 system driving pre-clinical drug discovery, from design to cancer treatment. 2020.04.06.028803. *Nature Communications* 11: 4903.

Luo, X., He, Y., Zhang, C. et al. (2019). Trio deep-sequencing does not reveal unexpected off-target and on-target mutations in Cas9-edited rhesus monkeys. *Nat. Commun.* 10: 5525.

Ma, Y., Ma, J., Zhang, X. et al. (2014). Generation of eGFP and Cre knockin rats by CRISPR/Cas9. *FEBS J.* 281: 3779–3790.

Macleod, R.S., Cawley, K.M., Gubrij, I. et al. (2019). Effective CRISPR interference of an endogenous gene via a single transgene in mice. *Sci. Rep.* 9: 17312.

Maddalo, D., Manchado, E., Concepcion, C.P. et al. (2014). in vivo engineering of oncogenic chromosomal rearrangements with the CRISPR/Cas9 system. *Nature* 516: 423–427.

Maeder, M.L., Linder, S.J., Cascio, V.M. et al. (2013). CRISPR RNA-guided activation of endogenous human genes. *Nat. Methods* 10: 977–979.

Makarova, K.S., Haft, D.H., Barrangou, R. et al. (2011). Evolution and classification of the CRISPR-Cas systems. *Nat. Rev. Microbiol.* 9: 467–477.

Makarova, K.S., Wolf, Y.I., Iranzo, J. et al. (2020). Evolutionary classification of CRISPR-Cas systems: a burst of class 2 and derived variants. *Nat. Rev. Microbiol.* 18: 67–83.

Mali, P., Aach, J., Stranges, P.B. et al. (2013). CAS9 transcriptional activators for target specificity screening and paired nickases for cooperative genome engineering. *Nat. Biotechnol.* 31: 833–838.

Manguso, R.T., Pope, H.W., Zimmer, M.D. et al. (2017). in vivo CRISPR screening identifies Ptpn2 as a cancer immunotherapy target. *Nature* 547: 413–418.

Maresca, M., Lin, V.G., Guo, N., and Yang, Y. (2013). Obligate ligation-gated recombination (ObLiGaRe): custom-designed nuclease-mediated targeted integration through nonhomologous end joining. *Genome Res.* 23: 539–546.

Matharu, N., Rattanasopha, S., Tamura, S. et al. (2019). CRISPR-mediated activation of a promoter or enhancer rescues obesity caused by haploinsufficiency. *Science* 363: 1–11.

Mccarron, A., Cmielewski, P., Reyne, N. et al. (2020). Phenotypic characterization and comparison of cystic fibrosis rat models generated using CRISPR/Cas9 gene editing. *Am. J. Pathol.* 190: 977–993.

Menchaca, A., Dos Santos-Neto, P.C., Souza-Neves, M. et al. (2020). Otoferlin gene editing in sheep via CRISPR-assisted ssODN-mediated homology directed repair. *Sci. Rep.* 10: 5995.

Meshalkina, D.A., Glushchenko, A.S., Kysil, E.V. et al. (2020). SpCas9- and LbCas12a mediated DNA editing produce different gene knockout outcomes in Zebrafish Embryos. *Genes (Basel)* 11.

Mirgayazova, R., Khadiullina, R., Chasov, V. et al. (2020). Therapeutic editing of the TP53 gene: Is CRISPR/Cas9 an option? *Genes (Basel)* 11.

Moretti, A., Fonteyne, L., Giesert, F. et al. (2020). Somatic gene editing ameliorates skeletal and cardiac muscle failure in pig and human models of Duchenne muscular dystrophy. *Nat. Med.* 26: 207–214.

Nakamura, M., Srinivasan, P., Chavez, M. et al. (2019). Anti-CRISPR-mediated control of gene editing and synthetic circuits in eukaryotic cells. *Nat. Commun.* 10: 194.

Nelson, C.E., Hakim, C.H., Ousterout, D.G. et al. (2016). in vivo genome editing improves muscle function in a mouse model of Duchenne muscular dystrophy. *Science* 351: 403–407.

Nissim, L., Perli, S.D., Fridkin, A. et al. (2014). Multiplexed and programmable regulation of gene networks with an integrated RNA and CRISPR/Cas toolkit in human cells. *Mol. Cell* 54: 698–710.

Paix, A., Wang, Y., Smith, H.E. et al. (2014). Scalable and versatile genome editing using linear DNAs with microhomology to Cas9 Sites in Caenorhabditis elegans. *Genetics* 198: 1347–1356.

Parker, B.C., Annala, M.J., Cogdell, D.E. et al. (2013). The tumorigenic FGFR3-TACC3 gene fusion escapes miR-99a regulation in glioblastoma. *J. Clin. Invest.* 123: 855–865.

Patel, A., Zhao, J., Duan, D., and Lai, Y. (2019). Design of AAV vectors for delivery of large or multiple transgenes. *Methods Mol. Biol.* 1950: 19–33.

Perez-Pinera, P., Kocak, D.D., Vockley, C.M. et al. (2013). RNA-guided gene activation by CRISPR-Cas9-based transcription factors. *Nat. Methods* 10: 973–976.

Pettibone, J.R., Yu, J.Y., Derman, R.C. et al. (2019). Knock-in rat lines with Cre recombinase at the dopamine D1 and adenosine 2a receptor loci. *eNeuro* 6.

Platt, R.J., Chen, S., Zhou, Y. et al. (2014). CRISPR-Cas9 knockin mice for genome editing and cancer modeling. *Cell* 159: 440–455.

Port, F., Strein, C., Stricker, M. et al. (2020). A large-scale resource for tissue-specific CRISPR mutagenesis in Drosophila. *elife* 9.

Ran, F.A., Cong, L., Yan, W.X. et al. (2015). in vivo genome editing using *Staphylococcus aureus* Cas9. *Nature* 520 (7546): 186–191.

Rabinowitz, J., Chan, Y.K., and Samulski, R.J. (2019). Adeno-Associated Virus (AAV) versus immune response. *Viruses* 11.

Ran, F.A., Hsu, P.D., Lin, C.Y. et al. (2013a). Double nicking by RNA-guided CRISPR Cas9 for enhanced genome editing specificity. *Cell* 154: 1380–1389.

Ran, F.A., Hsu, P.D., Wright, J. et al. (2013b). Genome engineering using the CRISPR-Cas9 system. *Nat. Protoc.* 8: 2281–2308.

Ren, X., Sun, J., Housden, B.E. et al. (2013). Optimized gene editing technology for Drosophila melanogaster using germ line-specific Cas9. *Proc. Natl. Acad. Sci. U. S. A.* 110: 19012–19017.

Rubin, G.M. and Lewis, E.B. (2000). A brief history of Drosophila's contributions to genome research. *Science* 287: 2216–2218.

Santry, L.A., Ingrao, J.C., Yu, D.L. et al. (2017). AAV vector distribution in the mouse respiratory tract following four different methods of administration. *BMC Biotechnol.* 17: 43.

Schafer, W.R. (2005). Deciphering the neural and molecular mechanisms of C. elegans behavior. *Curr. Biol.* 15: R723–R729.

Shechner, D.M., Hacisuleyman, E., Younger, S.T., and Rinn, J.L. (2015). Multiplexable, locus-specific targeting of long RNAs with CRISPR-Display. *Nat. Methods* 12: 664–670.

Shen, M.W., Arbab, M., Hsu, J.Y. et al. (2018). Predictable and precise template-free CRISPR editing of pathogenic variants. *Nature* 563: 646–651.

Shen, Z., Zhang, X., Chai, Y. et al. (2014). Conditional knockouts generated by engineered CRISPR-Cas9 endonuclease reveal the roles of coronin in C. elegans neural development. *Dev. Cell* 30: 625–636.

Shi, B., Keough, E., Matter, A. et al. (2011). Biodistribution of small interfering RNA at the organ and cellular levels after lipid nanoparticle-mediated delivery. *J. Histochem. Cytochem.* 59: 727–740.

Shirley, J.L., De Jong, Y.P., Terhorst, C., and Herzog, R.W. (2020). Immune responses to viral gene therapy vectors. *Mol. Ther.* 28: 709–722.

Slaymaker, I.M., Gao, L., Zetsche, B. et al. (2016). Rationally engineered Cas9 nucleases with improved specificity. *Science* 351: 84–88.

Soda, M., Choi, Y.L., Enomoto, M. et al. (2007). Identification of the transforming EML4-ALK fusion gene in non-small-cell lung cancer. *Nature* 448: 561–566.

Srivastava, A. (2016). in vivo tissue-tropism of adeno-associated viral vectors. *Curr. Opin. Virol.* 21: 75–80.

Staahl, B.T., Benekareddy, M., Coulon-Bainier, C. et al. (2017). Efficient genome editing in the mouse brain by local delivery of engineered Cas9 ribonucleoprotein complexes. *Nat. Biotechnol.* 35: 431–434.

Suda, T., Gao, X., Stolz, D.B., and Liu, D. (2007). Structural impact of hydrodynamic injection on mouse liver. *Gene Ther.* 14: 129–137.

Suda, T. and Liu, D. (2007). Hydrodynamic gene delivery: its principles and applications. *Mol. Ther.* 15: 2063–2069.

Sun, L., Wu, J., Du, F. et al. (2013). Cyclic GMP-AMP synthase is a cytosolic DNA sensor that activates the type I interferon pathway. *Science* 339: 786–791.

Sun, Y., Chen, X., and Xiao, D. (2007). Tetracycline-inducible expression systems: new strategies and practices in the transgenic mouse modeling. *Acta Biochim. Biophys. Sin. Shanghai* 39: 235–246.

Taheri-Ghahfarokhi, A., Taylor, B.J.M., Nitsch, R. et al. (2018). Decoding non-random mutational signatures at Cas9 targeted sites. *Nucleic Acids Res.* 46: 8417–8434.

Tan, Y., Chu, A.H.Y., Bao, S. et al. (2019). Rationally engineered Staphylococcus aureus Cas9 nucleases with high genome-wide specificity. *Proc. Natl. Acad. Sci. U. S. A.* 116: 20969–20976.

Tanenbaum, M.E., Gilbert, L.A., Qi, L.S. et al. (2014). A protein-tagging system for signal amplification in gene expression and fluorescence imaging. *Cell* 159: 635–646.

Tanihara, F., Hirata, M., Nguyen, N.T. et al. (2018). Generation of a TP53-modified porcine cancer model by CRISPR/Cas9-mediated gene modification in porcine zygotes via electroporation. *PLoS One* 13: e0206360.

Thakore, P.I., Kwon, J.B., Nelson, C.E. et al. (2018). RNA-guided transcriptional silencing in vivo with S. aureus CRISPR-Cas9 repressors. *Nat. Commun.* 9: 1674.

Unniyampurath, U., Pilankatta, R., and Krishnan, M.N. (2016). RNA interference in the age of CRISPR: will CRISPR interfere with RNAi? *Int. J. Mol. Sci.* 17: 291.

Wang, D., Mou, H., Li, S. et al. (2015). Adenovirus-mediated somatic genome editing of pten by CRISPR/Cas9 in mouse liver in spite of Cas9-specific immune responses. *Hum. Gene Ther.* 26: 432–442.

Wang, K., Jin, Q., Ruan, D. et al. (2017). Cre-dependent Cas9-expressing pigs enable efficient in vivo genome editing. *Genome Res.* 27: 2061–2071.

Wang, X., Niu, Y., Zhou, J. et al. (2016). Multiplex gene editing via CRISPR/Cas9 exhibits desirable muscle hypertrophy without detectable off-target effects in sheep. *Sci. Rep.* 6: 32271.

Wangensteen, K.J., Wang, Y.J., Dou, Z. et al. (2018). Combinatorial genetics in liver repopulation and carcinogenesis with a in vivo CRISPR activation platform. *Hepatology* 68: 663–676.

Weber, J., Braun, C.J., Saur, D., and Rad, R. (2020). in vivo functional screening for systems-level integrative cancer genomics. *Nat. Rev. Cancer* 20: 573–593.

Weber, J., Ollinger, R., Friedrich, M. et al. (2015). CRISPR/Cas9 somatic multiplex-mutagenesis for high-throughput functional cancer genomics in mice. *Proc. Natl. Acad. Sci. U. S. A.* 112: 13982–13987.

Wei, T., Cheng, Q., Min, Y.L. et al. (2020). Systemic nanoparticle delivery of CRISPR-Cas9 ribonucleoproteins for effective tissue specific genome editing. *Nat. Commun.* 11: 3232.

Wells, D.J. (2018). Tracking progress: an update on animal models for Duchenne muscular dystrophy. *Dis. Model. Mech.* 11.

Wertz, M.H., Mitchem, M.R., Pineda, S.S. et al. (2020). Genome-wide in vivo CNS screening identifies genes that modify CNS neuronal survival and mHTT toxicity. *Neuron* 106: 76–89. e8.

Wienert, B., Wyman, S.K., Richardson, C.D. et al. (2019). Unbiased detection of CRISPR off-targets in vivo using DISCOVER-Seq. *Science* 364: 286–289.

Wijshake, T., Baker, D.J., and Van De Sluis, B. (2014). Endonucleases: new tools to edit the mouse genome. *Biochim. Biophys. Acta* 1842: 1942–1950.

Wilson, R.C. and Gilbert, L.A. (2018). The promise and challenge of in vivo delivery for genome therapeutics. *ACS Chem. Biol.* 13: 376–382.

Winslow, M.M., Dayton, T.L., Verhaak, R.G. et al. (2011). Suppression of lung adenocarcinoma progression by Nkx2-1. *Nature* 473: 101–104.

Winters, I.P., Chiou, S.H., Paulk, N.K. et al. (2017). Multiplexed in vivo homology-directed repair and tumor barcoding enables parallel quantification of Kras variant oncogenicity. *Nat. Commun.* 8: 2053.

Wu, M., Wei, C., Lian, Z. et al. (2016). Rosa26-targeted sheep gene knock-in via CRISPR-Cas9 system. *Sci. Rep.* 6: 24360.

Wu, X., Scott, D.A., Kriz, A.J. et al. (2014). Genome-wide binding of the CRISPR endonuclease Cas9 in mammalian cells. *Nat. Biotechnol.* 32: 670–676.

Xie, K., Minkenberg, B., and Yang, Y. (2015). Boosting CRISPR/Cas9 multiplex editing capability with the endogenous tRNA-processing system. *Proc. Natl. Acad. Sci. U. S. A.* 112: 3570–3575.

Xu, X. and Qi, L.S. (2019). A CRISPR-dCas toolbox for genetic engineering and synthetic biology. *J. Mol. Biol.* 431: 34–47.

Xue, W., Chen, S., Yin, H. et al. (2014). CRISPR-mediated direct mutation of cancer genes in the mouse liver. *Nature* 514: 380–384.

Yanez Arteta, M., Kjellman, T., Bartesaghi, S. et al. (2018). Successful reprogramming of cellular protein production through mRNA delivered by functionalized lipid nanoparticles. *Proc. Natl. Acad. Sci. U. S. A.* 115: E3351–E3360.

Yang, D., Xu, J., and Chen, Y.E. (2019). Generation of rabbit models by gene editing nucleases. *Methods Mol. Biol.* 1874: 327–345.

Yang, H., Wang, H., Shivalila, C.S. et al. (2013). One-step generation of mice carrying reporter and conditional alleles by CRISPR/Cas-mediated genome engineering. *Cell* 154: 1370–1379.

Yau, E.H. and Rana, T.M. (2018). Next-generation sequencing of genome-wide CRISPR screens. *Methods Mol. Biol.* 1712: 203–216.

Yin, H., Xue, W., Chen, S. et al. (2014). Genome editing with Cas9 in adult mice corrects a disease mutation and phenotype. *Nat. Biotechnol.* 32: 551–553.

Yoshimi, K., Oka, Y., Miyasaka, Y. et al. (2021). Combi-CRISPR: combination of NHEJ and HDR provides efficient and precise plasmid-based knock-ins in mice and rats. *Hum. Genet.* 140 (2): 277–287.

Zetsche, B., Volz, S.E., and Zhang, F. (2015). A split-Cas9 architecture for inducible genome editing and transcription modulation. *Nat. Biotechnol.* 33: 139–142.

Zheng, Y., Shen, W., Zhang, J. et al. (2018). Author correction: CRISPR interference-based specific and efficient gene inactivation in the brain. *Nat. Neurosci.* 21: 894.

Zhou, H., Liu, J., Zhou, C. et al. (2018). in vivo simultaneous transcriptional activation of multiple genes in the brain using CRISPR-dCas9-activator transgenic mice. *Nat. Neurosci.* 21: 440–446.

Zhu, H., Zhang, L., Tong, S. et al. (2019). Spatial control of in vivo CRISPR-Cas9 genome editing via nanomagnets. *Nat. Biomed Eng.* 3: 126–136.

Zincarelli, C., Soltys, S., Rengo, G., and Rabinowitz, J.E. (2008). Analysis of AAV serotypes 1-9 mediated gene expression and tropism in mice after systemic injection. *Mol. Ther.* 16: 1073–1080.

Part 3

Genome Editing in Target Identification and Validation

One of the major applications of CRISPR Cas is in functional genomics studies used to identify novel drug targets. Genome-wide pooled and arrayed sgRNA libraries are available from addgene and commercial sources and protocols established for any nonexpert lab to initiate/perform such projects. With the discovery of CRISPRi and CRISPRa, such studies can even be performed in a setting that more closely mimics drug treatment. Genome editing when combined with single-cell transcriptomics provides a bigger picture of the effect of gene KO on cellular machinery, indicating potential side effects of a drug against target gene. This part covers technological advances and practical suggestions for performing such experiments as well as discusses selected examples of major discoveries from CRISPR Cas functional genomics studies.

8

Pooled CRISPR KO Screens for Target Identification

Antje Grotz and Sumit Deswal

Genome Engineering, Discovery Sciences, BioPharmaceuticals R&D, AstraZeneca, Gothenburg, Sweden

8.1 Introduction

From rare diseases which are often caused by mutations in a single gene to polygenic and complex traits, characterizing the underlying genetic architecture is central to the functional understanding of disease mechanisms and targeted therapies. More specifically, the mapping of genotype–phenotype associations has shown to be fundamental to generate relevant disease models, develop targeted treatments such as gene therapies, and find promising novel drug targets which are twice as likely to successfully pass clinical trials due to their supporting genetic evidence (Nelson et al. 2015; King et al. 2019). The field of functional genomics addresses this challenge by building on next-generation sequencing technologies to study gene and protein function on a large up to genome-wide scale. CRISPR/Cas9 genome editing technologies have further revolutionized this field and led to the development of pooled CRISPR screens which allow for simultaneous and genome-wide investigation of phenotypic consequences upon gene knockout (Shalem et al. 2014; Wang et al. 2014).

Previous limited screening strategies were largely based on RNA interference (RNAi) knockdown technologies or arrayed screen setups (Boutros et al. 2004; Kittler et al. 2004; Whitehurst et al. 2007). While RNAi-based screens using small interfering RNAs (siRNAs) or short hairpin RNAs (shRNAs) have provided important functional insights into the role of specific genes, they rely on transient and post-transcriptional downregulation of transcripts of interest. The associated low level of residual gene expression might however be masking potential phenotypic consequences. Using CRISPR/Cas9 to introduce permanent knockouts on the other hand can be used to detect phenotypes that require a complete protein depletion. Additionally, siRNA libraries must be screened in an arrayed format where each well contains a different siRNA target that must be assessed individually. While allowing for more complex phenotypic readouts as reviewed in Chapter 9, this screening setup is labor-intensive and expensive, making it less suitable for large-scale and genome-wide libraries as opposed to the more cost- and labor-effective pooled screening approaches.

In this chapter, we will introduce pooled CRISPR screening setups from knockout libraries to analysis strategies and we will further summarize selected pooled CRISPR screening applications in oncology and beyond.

8.2 Pooled CRISPR-Cas Screens

Pooled CRISPR screening strategies make large-scale and simultaneous interrogation of thousands of genotype–phenotype relations in a single pool of cells possible. Briefly, CRISPR/Cas9 libraries are being introduced into a pool of cells via titrated lentivirus, the cells are being enriched for a phenotype of interest, and by using next-generation sequencing, genes that play a role in the specific phenotype are being identified (Shalem et al. 2015; Sanjana 2017). To be able to associate the phenotype and an individual perturbation within a cell, the CRISPR reagents are being delivered through lentivirus which stably integrates into the genome. Targeted downstream sgRNA sequencing of all remaining cells within the population of interest will therefore reveal sgRNAs or gene knockouts that have an effect on the chosen phenotype of interest. To ensure that only one sgRNA or gene knockout is being introduced per cell, the viral transduction of lentiviral CRISPR libraries is performed at a low multiplicity of infection (MOI), resulting in a large majority of cells having only a single viral integration. Each gene within a CRISPR library is targeted by several sgRNAs and each sgRNA is being introduced into several hundreds of cells to increase the statistical power and confidence to be able to identify true hits. Cells

Genome Editing in Drug Discovery, First Edition. Edited by Marcello Maresca and Sumit Deswal.

that have not been transduced with lentivirus will be removed through antibiotic selection. The resulting heterogenous cell pool contains cells with the desired gene knockout in addition to a small proportion of cells containing edits from several sgRNAs, no edits at all or modifications that do not affect the open reading frame of the gene, or do not lead to a functional protein knockout. The pool of cells is then being subjected to a certain phenotypic selection of interest to separate the cells into groups that either contain a gene knockout which shows a distinct phenotype under these conditions or cannot be distinguished from control cells. The sgRNAs associated with this specific phenotype will depending on the nature of the readout, subsequently either be enriched or depleted within the pool of cells. To analyze this sgRNA enrichment or depletion, DNA is being extracted from the cells, sgRNA sequences are being amplified using PCR, followed by in-depth read count analysis through next-generation sequencing. After stringent quality control, gene hits are finally being identified by having several sgRNAs with a consistent effect on the phenotype of interest.

Successfully conducting a pooled CRISPR screens requires a suitable and disease-relevant perturbation system, cellular model, and phenotype (Doench 2018). CRISPR gene KO libraries are most commonly and routinely used as a perturbation system to perform loss-of-function screens and are being described in more detail in Section 8.3.1. The opposite direction of effect to induce transcription can be induced using a gain-of-function CRISPR library (CRISPRa) while gene knockdown as opposed to knockout can be achieved using CRISPRi (Gilbert et al. 2014; Konermann et al. 2015). Both of these strategies are based on catalytically inactive or dead Cas9 (dCas9) and are reviewed in Chapter 10. The CRISPR screening toolbox has even further expanded to also include epigenetic (dCas9 fused to DNA or histone modifier), base editing, or noncoding screens in addition to coupling the sgRNA readout to other single-cell techniques such as RNA-Seq (Hilton et al. 2015; Kearns et al. 2015; Dixit et al. 2016; Kweon et al. 2020). In the following sections, however, we are only going to focus on the widely used CRISPR KO perturbation system.

The cellular model chosen for the CRISPR screen must first of all be relevant to the chosen phenotype of interest but it also has to be amendable to genome editing by lentivirus and must be able to be expanded or obtained at a high quantity, therefore often excluding primary cells as screening models (Section 8.3.2). Each screening step from antibiotic selection to viral transduction efficiencies must be tested and optimized within each cell line proposed for screening, which might take longer than the actual screen.

The phenotypic readout is central to the pooled CRISPR screening setup and includes negative, positive, or fluorescence-activated cell sorting (FACS)-based selection strategies. A negative selection screen aims to identify sgRNAs that have been depleted from the final pool of cells, representing genes that are required for cell survival under the selected screening conditions. This strategy has been widely used to assess gene essentialities in cancer cell lines with the ultimate goal to find cancer dependencies and novel drug targets (Cheung et al. 2011; Hart et al. 2015; Meyers et al. 2017; Behan et al. 2019; Dempster et al. 2019). Positive selection strategies however focus on sgRNAs or gene knockouts that have a growth or survival advantage while a selective pressure from culturing conditions or drugs has been applied (Birsoy et al. 2015; Zhang et al. 2016; Jain et al. 2020; Li et al. 2020). While the majority of cells survive in a negative selection screen and only the small number of depleted sgRNAs is of interest, a positive selection screen focuses on the small proportion of cells that survive after the majority of cells has died. Beyond enriching or depleting sgRNAs based on culturing conditions, a phenotypic selection can be performed by selecting populations of interest based on the expression level of a specific marker as measured by FACS (DeJesus et al. 2016; Park et al. 2019; Cortez et al. 2020). After transduction, antibiotic selection, and a desired number of passages, the CRISPR screening cells are being stained for a relevant intra- or extracellular protein of interest through an antibody coupled to a fluorescence marker. Subsequent FACS sorting can sort the pool of cells into several groups based on their level of fluorescence intensity which is representative of the specific protein level. This type of screen can deliver powerful insights by directly probing a disease-relevant protein instead of only measuring survival or growth. The screening setup however is much more complex, requires extensive optimization, and can only be applied to cellular proteins that can be specifically targeted by antibodies (Figures 8.1 and 8.2).

8.3 Reagents

Prior to performing any CRISPR-Cas9 screen, we need to consider the question under investigation and establish the reagents accordingly. Some of the questions to consider would be: Do I need to perform the screen *in vitro* or *in vivo*?, do I need to use human or mouse cell lines (or a different model organism)? or do I need to use a genome-wide library or focus on a smaller subset of genes? Over the last years, genome-wide and focused libraries for both human and mouse have been established. There are also several freely available online algorithms to custom design guide RNAs (sgRNA).

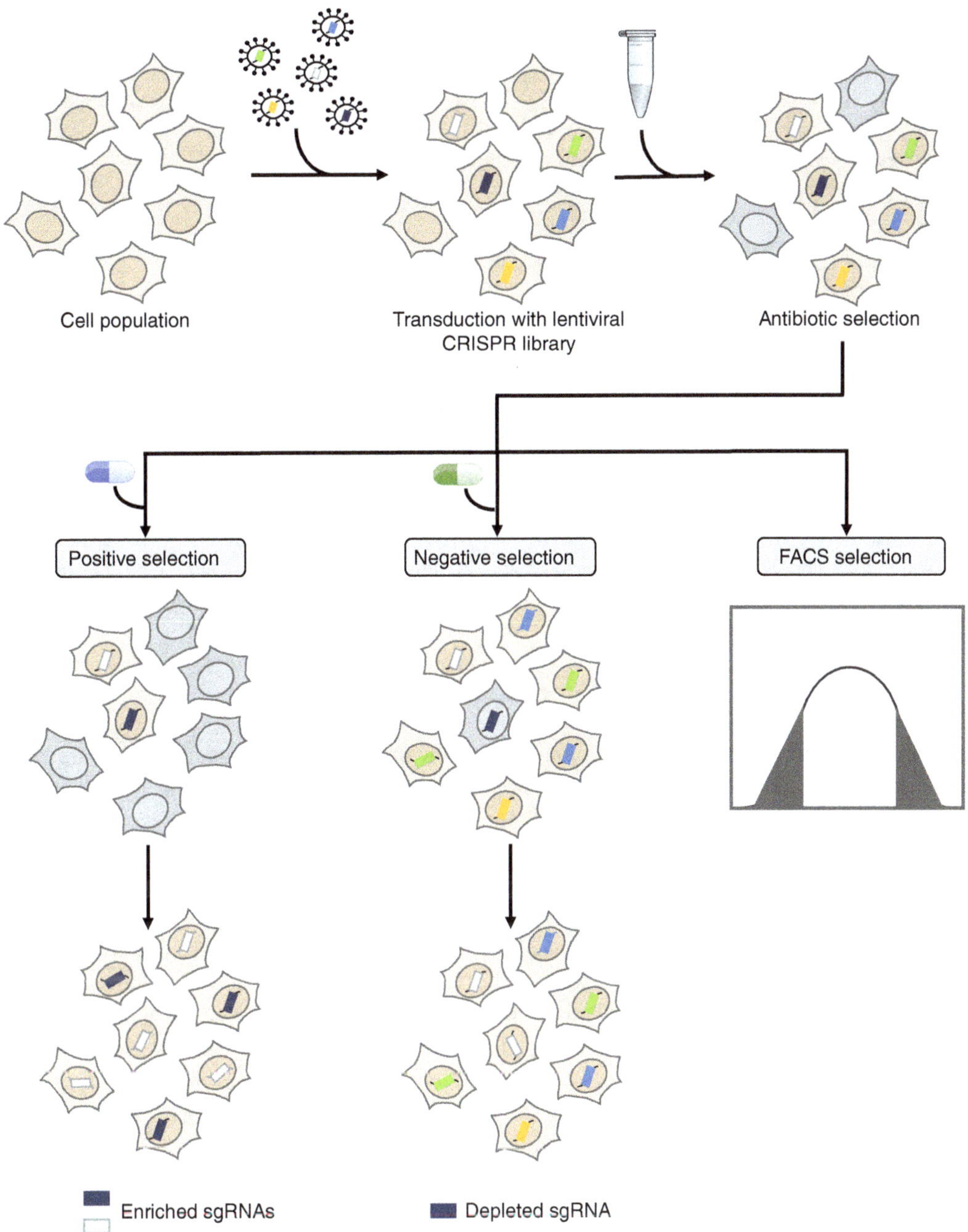

Figure 8.1 CRISPR screening approaches. To introduce CRISPR edits, a cell population of interest is being transduced with a (genome-wide) lentiviral CRISPR library, followed by antibiotic selection to remove untransduced cells (top). A selection of interest is then applied to select for certain cells of interest, for example, through incubation with a specific drug to identify drug-resistant mutants (middle). In a positive selection screen, the majority of cells or sgRNAs are being depleted and the enriched final sgRNAs are of interest (bottom left). In a negative selection screen, a small proportion of cells or sgRNAs is being depleted and are of interest (bottom middle). In addition, cells of interest can also be chosen using FACS selection for a phenotypic marker such as, for example, the level of a surface receptor (bottom right).

8.3.1 Library Design and Synthesis

Many computational tools are available to design single-gene sgRNAs. These are covered extensively in Chapter 5. Some of the sgRNA library design aspects are discussed in this section.

The first genome-wide human (Shalem et al. 2014; Wang et al. 2014) and mouse (Koike-Yusa et al. 2014) sgRNA libraries were published in early 2014. Following these initial reports, many improvements in the sgRNA targeting, tracer sequence, and plasmid architecture have been

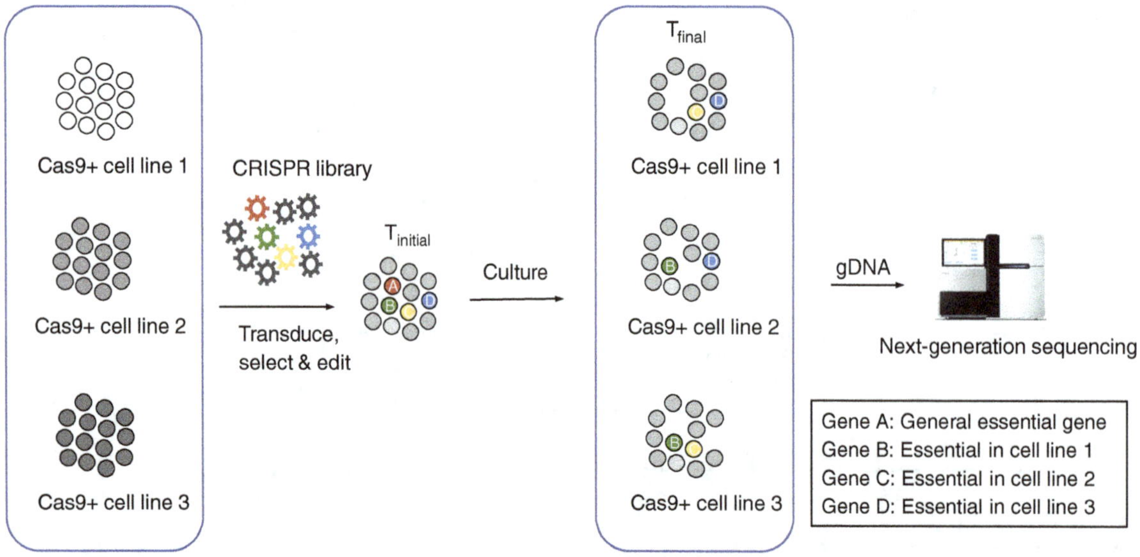

Figure 8.2 Synthetic lethality and lineage dependencies discovery using pooled CRISPR-Cas screens. Essentiality screens can be used to discover tissue-lineage or genotype-specific dependencies. For this, parallel screens need to be performed in cell lines that are derived from different lineages or belong to the same tissue type but vary by their genotype (isogenic cell lines).

reported and improved genome-wide libraries have been published subsequently. To improve sgRNA-mediated knockout efficiency, sgRNAs have been designed to target essential protein domains (Shi et al. 2015). However, generating these libraries requires previous knowledge of functionally relevant protein domains, which is not available genome-wide. Genome-wide sgRNA predictions for human, mouse, and other model organisms can be downloaded from VBC-Score (https://www.vbc-score.org/). These sgRNA predictions took into consideration amino acid composition and conservation resulting in much higher knockout efficiencies (Michlits et al. 2020).

The libraries are being synthesized on arrays, e.g. Agilent Technologies, and the subsequent synthesized oligonucleotides are then cloned in an appropriate library vector. Initial screens used an all-in-one construct, delivering both Cas9 and sgRNA; however, we have found that establishing Cas9 expressing clones of cells and testing them thoroughly followed by a separate delivery of sgRNA library results in higher editing efficiency. It has been shown that in a bulk population of Cas9 expressing cells, not all individual clones of cells show good efficiency of gene targeting (Michlits et al. 2020). Hence, thorough testing and selection of working Cas9 positive clones is recommended for successful CRISPR screening.

Although almost all the libraries displayed below are genome-wide, there are now various libraries available that target subsets of genes, focusing on gene families or biological pathways – for example, libraries targeting kinases (Wang et al. 2019b), transcription factors, epigenetic genes, surface proteins, druggable genes (proteins for which a small molecule inhibitor is available), etc. By reducing the number of genes to study, the number of sgRNAs per gene can be increased to improve statistical confidence in hits. Also studying a focused group of genes reduces the overall cost of screening such as for cell culture and sequencing. Most of the published genome-wide and focused libraries are easily available for the screening end user through plasmid repositories (Table 8.1).

8.3.2 Biological Systems for Screening

A balance must be achieved between relevance of a biological system and ease of utilizing it for experiments. In many disease contexts, an *in vivo* screen is much more relevant to human disease; however, performing *in vivo* CRISPR screens is not always a possibility. Factors concerning *in vivo* screens are discussed later in this chapter (Section 8.6). In this section, we focus on *in vitro* screens that utilize human/mouse cell lines or primary cells. Some of the considerations for a successful CRISPR screen using cell lines is that the target cell line has to be amendable to genome editing, viral transduction, antibiotic selection, and can be expanded as required. In screens using a two plasmid system, one should validate the Cas9 expressing cell line (preferably generate the Cas9 clones and test them thoroughly for genome editing to choose a highly efficient clone).

Table 8.1 Some of the available academic and commercial human and mouse pooled knockout libraries.

Library	Focus	Species	Number of sgRNAs	Number of genes	Reference
Gecko_v1	Genome-wide	Human	64 751	18 080	Shalem et al. (2014)
Gecko_v2	Genome-wide	Human	123 411	19 050	Sanjana et al. (2014)
Wang et al.	Candidate genes	Human	73 151	7114	Wang et al. (2014)
Yusa	Genome-wide	Mouse	87 897	19 150	Koike-Yusa et al. (2014)
Domain library	Epigenetics domains	Mouse	933	192 domains	Shi et al. (2015)
Toronto Knockout (TKOv3) CRISPR Library	Genome-wide	Human	70 948	18 053	Hart et al. (2017)
Brunello	Genome-wide	Human, mouse	77 441	~19 000	Doench et al. (2016)
CRISPR-based double knockout library (CDKO)	Druggable genome	Human	490 000 double-sgRNA combinations	207 genes (3 sgRNAs per gene)	Han et al. (2017)
Avana	Genome-wide	Human	76 106	17 670	Meyers et al. (2017)
UMI	Small pool	Mouse	26 514	6560	Michlits et al. (2017)
VBC	Genome-wide Subpools	Human, mouse	114 584 122 423	18 659 and 19 248	Michlits et al. (2020)
CELLECTA	Genome-wide	Human, mouse	~80 000	~20 000	Cellecta (2020)
Agilent-SureGuide	Genome-wide and Subpools	Human, mouse	Gecko_v2 and custom designs		Agilent (2020)
Thermo Fischer Scientific	Pooled genome-wide and many Subpools	Human	~80 000	~20 000	Thermofisher (2020)

Another important factor, particularly for phenotypic screens, could be the essentiality of the genes. To avoid essential genes dropping out before the phenotypic readout time-point is reached, it is recommended to use inducible expression of Cas9 (e.g. doxycycline-inducible expression). In this system, sgRNA library-infected cells can first be selected under the initial antibiotic selection followed by induced Cas9 expression to analyze a specific time point for the phenotype.

8.4 Library Transduction, Maintenance, and Next-Generation Sequencing

General schemes for performing a pooled CRISPR screen are outlined in Section 8.2. In this section, we discuss practical considerations for a successful screen.

8.4.1 Virus Production and Library Infection

Lentiviral gene delivery has emerged as the primary method for delivering the sgRNA library to cells. Lentiviruses are being integrated in the host genome, enabling easy identification of the sgRNA by deep sequencing. The genetic capacity of lentiviruses is high so that an antibiotic selection marker or even Cas9 itself as part of a single-vector system can be easily incorporated into the sgRNA plasmid. Virus can be produced by standard transient transfection of packaging cell lines, e.g. Lenti-X cells. Virus supernatant is usually concentrated from multiple harvests. A test infection of the target cells under screening conditions or viral titer is recommended to ensure low MOI in the final infection to avoid multiple virus integrations. Maintaining high representation at every step of the screen and mixing of cells from each cell plate within a replicate at each split is recommended to avoid any plate to plate variation within a replicate. Required representation may vary for the screening approaches; however, at least 300× representation is highly recommended. An even higher, up to 1000× representation is recommended for improved statistical power; however, it comes at a cost of higher cell numbers to maintain at each split.

8.4.2 Replicates

Using three biological replicates is a common practice. High Pearson correlation between sgRNA log-fold changes of replicates in a negative selection screen indicates a good screen. For a positive selection screen, most genes do not change their abundance, hence correlation is generally low. Using multiple sgRNAs for each gene (generally 4–6) in the screen provides another level of confidence and some CRISPR screening analysis programs like MAGeCK utilize this factor for statistical significance calculation (Li et al. 2014).

8.4.3 Positive vs Negative Selection Screen

In a negative selection screen such as cancer cells essentiality screens, neutral genes (which form the bulk of the population) remain unaltered and for a fraction of essential genes, the read count drops. Degree of drop in reads (fold change) normally reflects the degree of essentiality of the genes. However, an efficient negative selection requires a near-saturated generation of homozygous knock-out alleles. In contrast to this, homozygous knock-out in a low fraction of cells suffices for achieving strong positive selection effects (Figure 8.3).

8.5 Screen to Target Selection

8.5.1 Statistical Analysis of CRISPR Screens

NGS analysis of the PCR-amplified sgRNA counts results in a large dataset of reads per sgRNA per condition per replicate. Converting these large sheets into list of top statistically significant hits is a challenge and several software programs have been developed for this including MAGECK (Model-based Analysis of Gene Essentiality) (Li et al. 2014), MAGECKflute (Wang et al. 2019a), EdgeR (Dai et al. 2014), caRpools (CRISPR Analyzer for Pooled Screens) (Winter et al. 2016), BAGEL (Bayesian Analysis of Gene Essentiality) (Hart and Moffat 2016), and CRISPRcloud (Jeong et al. 2017) (Figure 8.4).

The first step in the analysis pipeline is deconvolution of FASTQ files, i.e. extraction of the guide sequence and aligning of the extracted guide sequences to the library to generate a read count table for each sgRNA in the library. As a quality control at this step, it is important to check the percentage of mapped reads and the correlation between replicates. Following basic quality controls, sgRNA level analysis is performed. For this, fold change is calculated by comparing the sgRNA read counts within the screening sample to read counts of a control condition or to an earlier control time point, e.g. in depletion screens. Performance of positive and negative control sgRNAs in the library is a good indicator of the screen performance. Several of the analysis tools mentioned earlier also calculate the false discovery rate for sgRNAs. The next step is to identify significant hits at the level of genes. For a gene to be significant and for a consistent effect, multiple sgRNAs need to score. For a positive selection screen with a good sgRNA library, the hit list would not vary much with a change in the analysis tool. For negative selection screens, MAGeCK is one of the most commonly used programs with several options including the visualizing of results.

8.5.2 Gene List to Target Selection

Upon having a list of statistically significant genes in a screen, one needs to perform integrative analysis, combining existing biological knowledge with new results.

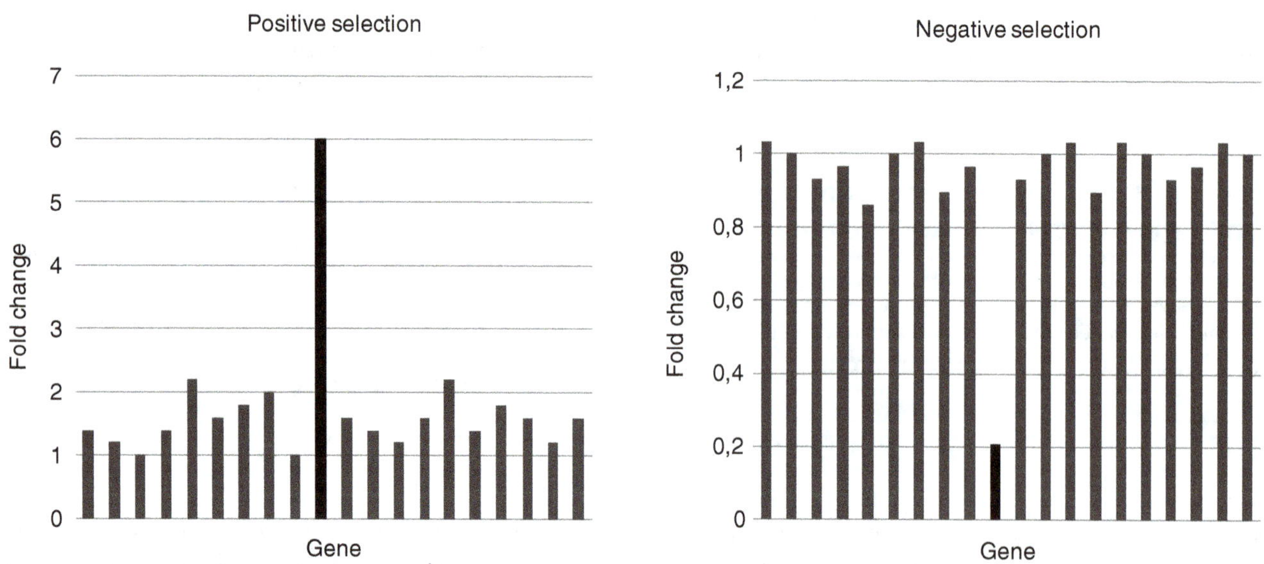

Figure 8.3 Positive selection vs a negative selection pooled screen.

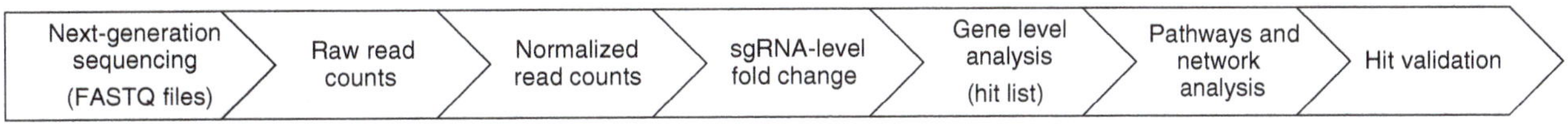

Figure 8.4 Workflow of pooled CRISPR-Cas screen analysis pipeline.

For example, if several genes belong to the same pathway among the top hits, that pathway could be important for the phenotype under investigation. This type of analysis can be done with programs like GSEA (Subramanian et al. 2005), MsigDB (Liberzon et al. 2011), DAVID (Huang da et al. 2009), OncoBase (Li et al. 2019), and DepMap (https://depmap.org/portal/). In the last couple of years, there have been many large-scale studies to characterize human cell lines with respect to their genetic profile, protein abundance, post-translational modifications, functional genomics, and drug-sensitivity profiles (Barretina et al. 2012; Basu et al. 2013; Klijn et al. 2015; Ghandi et al. 2019). Additionally, more than 11 000 cancer patients were characterized by The Cancer Genome Atlas (TCGA) and the International Cancer Genome Consortium (ICGC) for mutations in genes as well as their transcriptomic and epigenetic profil (Liu et al. 2018). Using simple statistical or more advanced machine learning methods, one can explore the molecular signatures that define dependency of a particular cell line/cancer on a gene. Most importantly, further experimental validation of the hits, either by performing a secondary CRISPR screen of a subset of genes (using independent sgRNAs) or follow-up experiments on individual genes, is crucial.

8.6 *In vivo* CRISPR Screens

In vitro screening studies have provided useful information on cell intrinsic properties of cancer cells. However, they do not address the real-world patient condition *in vivo*. In the body, cells are involved in complex interactions with multiple cell types and tumor microenvironment is known to play a major role in the success or failure of cancer therapy. To address this issue, *in vivo* screens are needed. However, performing CRISPR screens *in vivo* requires several technical challenges to be overcome. To date, a large number of *in vivo* screens were performed in a two-step process: cells were transduced with a sgRNA library *in vitro* followed by transplantation into mice to assay the *in vivo* phenotype (Chen et al. 2015). Although compared with purely *in vitro* screens, these transplant-based *in vivo* screens recapitulate some features of carcinogenesis/tumor microenvironment, they also have several limitations. Often, such transplant-based screens are performed in immunocompromised mice, limiting the study of tumor–immune cells interaction. In many cases, cells are transplanted subcutaneously, not resembling the original site of tumor. Such limitations can be overcome by directly mutagenizing the tissue of interest *in vivo*, i.e. performing a truly *in vivo* screen. However, not all organs are easily accessible for transduction. *In vivo* screens were performed in liver by hydrodynamic injection of sgRNA plasmids in tail vein of a Cas9 expressing mouse (Xue et al. 2014). In lungs, CRISPR mutagenesis can be achieved by intratracheal delivery (Sanchez-Rivera et al. 2014).

In vivo screens also have their share of limitations. Factors such as library size, infection efficiency, and library coverage can easily be controlled *in vitro* but are much more difficult to control *in vivo*. As only a limited number of cells can be infected *in vivo*, library size needs to be dramatically reduced. If a large library is used for infection, random sampling errors will lead to many false positive and negative hits. So far in most direct *in vivo* screens, sgRNAs in the range of 10–30 were used (Rogers et al. 2017). Hence, direct *in vivo* screens are more useful for the validation of primary hits coming from either patient sequencing or a primary *in vitro* CRISPR screen, etc.

8.7 Advanced Functional Genomics Screens

Recently, functional genomics screens have been combined with advanced NGS approaches such as single-cell RNA sequencing. Examples of such approaches include Perturb-seq, CROP-seq, and TAP-seq (Dixit et al. 2016; Datlinger et al. 2017). Using these approaches, the effect of each gene on transcriptional regulation can be studied in high-throughput manner. Some of these approaches are discussed in Chapter 12 in this book. Additionally, unique molecular identifier (UMI)-based CRISPR screens were developed to trace single cells to account for cell heterogeneity (Michlits et al. 2017; Schmierer et al. 2017). Methodical details of these screens can be accessed from the respective publications.

8.8 Selected Applications of Pooled CRISPR-Cas Screens

8.8.1 Applications in Oncology

8.8.1.1 Gene Essentiality Screens

An overarching aim in oncology research is to functionally characterize and understand the implications of genetic variation in cancer to develop novel and more efficient treatment strategies and guide future therapies. Currently, cancer drugs have the highest dropout rate in clinical development with on average only one drug achieving final FDA approval out of 15 that initially enter phase 1 clinical trials (Hay et al. 2014). The majority of those studies were suspended due to a lack of efficacy. Identifying the genetic basis of tumor vulnerabilities could reveal potential drug targets with supporting genetic evidence and a defined molecular mechanism which would increase the success rate in clinical development and ultimately cancer therapy prospects. Genome-wide CRISPR/Cas9 screens can be applied to comprehensively identify essential genes in cancer cell lines which are defined as genes that are required for cell survival or proliferation. Targeting these genes or proteins with specific drugs would therefore lead to reduced tumor viability. So far as of 2020, more than 800 cancer cell lines from various cancer types have been used for large-scale CRISPR screening (Meyers et al. 2017; Tsherniak et al. 2017; Behan et al. 2019). The resulting gene hits are either genes which are essential across all cell lines or genes which have only shown to be essential in certain cancer cell lines or in a specific context. Genes that are required for survival in the majority of all tested cancer cell lines or so-called core essential genes are likely involved in essential processes within all cell types and their targeting might be associated with high toxicity in healthy cells as well. Context-specific essential genes on the other hand are only required for cell viability in cancer cell lines with a certain genetic, cell type, or tissue background and represent favorable drug targets with a higher specificity and lower likelihood to induce toxic side effects in healthy cells. Based on 339 cancer cell lines and 18 009 gene targets, Behan et al. identified a median total of 1459 fitness genes in each cell line with a median of 866 cancer-type-specific gene hits and 553 core fitness genes in one of the largest single screening efforts so far (Behan et al. 2019). Together with results from small molecule screens and additional genomic datasets such as transcriptomics and proteomics, the data from these genome-wide CRISPR screens in cancer cell lines have been integrated and made openly accessible as part of an effort to build a Cancer Dependency Map (https://depmap.org/portal/ 2020).

8.8.1.2 Synthetic Lethal Screens

A particularly relevant example of context-specific CRISPR screening approaches in oncology are synthetic lethality screens. Synthetic lethal interactions between genes are defined as a loss of cellular fitness that only occurs upon co-targeting of both genes; knocking out one of the genes individually has no effect on cell survival or growth. This strategy can identify innovative cancer treatment approaches by greatly expanding the range of potential drug targets and by specifically targeting tumor cells. The side effects on healthy cells can be reduced as these cells do not possess the required mutation that leads to a synthetic lethal interaction with the drug target. Cancer cells however have often acquired unique mutations associated with tumor suppressors or oncogenes. Further, if the cancer-associated genes have been classified as undruggable and directly targeting them is not a feasible treatment option, focusing on secondary genes associated with synthetic lethality offers an alternative route to treatment success. CRISPR screens can identify synthetic lethal mutations by performing a deletion screen in a cancer cell line that already contains a primary mutation in a specific gene of interest. The identified genes will therefore be essential genes only in the context of the specific mutation. One of the many conducted synthetic lethality CRISPR screens was performed in several acute myeloid leukemia (AML) cell lines to identify synthetic lethal interactions with the common Ras oncogenes *NRAS* and *KRAS* (Wang et al. 2017). Wang et al. discovered several genes that were only essential in the context of a primary Ras mutation by performing two independent screening approaches, first by comparing Ras mutation harboring AML cell lines with AML cell lines that do not contain Ras mutations and second, by pursuing an isogenic screening strategy using an engineered pro B-cell line (Wang et al. 2017). Similarly, Steinhart et al. performed a genome-wide CRISPR screen to uncover synthetic lethal gene interactions for urgently needed novel therapeutic approaches in pancreatic cancer (Steinhart et al. 2017). In the context of *RNF43* containing pancreatic adenocarcinoma cell lines, the study discovered a synthetic lethal interaction with the Wnt ligand receptor *FZD5*. The authors further demonstrated a potential therapeutic strategy by inhibiting the proliferation of a *RNF43* variant containing patient pancreatic adenocarcinoma cell line with FZD5 targeted antibodies, highlighting how insights generated from CRISPR screens can be directly translated into treatment approaches to ultimately benefit patients.

8.8.1.3 Drug Resistance and Synergy Screens

Resistance to cancer drugs is a common phenomenon due to the fast-evolving nature of the cancer or the presence of resistant cells from the beginning due to tumor heterogeneity. Study of the resistance mechanisms in relevant disease models and hence finding novel drug targets are crucially needed. In many cases, single gene targeting may not work, hence exploring combinatorial therapy options is a good choice. Pooled CRISPR-Cas screening has made it easy to study the resistance mechanisms of drugs as well as identification of synergistic drug combinations for combinatorial cancer therapy (Han et al. 2017). This can be achieved by either performing a conventional CRISPR drop-out screen in the presence and absence of a drug or alternatively by using a double-sgRNA library targeting a pair of genes in a combinatorial manner (Shen et al. 2017). The first approach is technically easier as it only requires performing a regular screen in the presence or absence of a drug; the second approach however additionally requires special cloning of the double-sgRNA library. The library size increases exponentially in the double-sgRNA approach with every additional gene to be included in the library. Hence, this approach requires a preselection of target genes to achieve a manageable screening size.

8.8.1.4 Immuno-Oncology Screens for Cancer Immunotherapy

With the reporting of promising results for immune-checkpoint blocking therapy for cancer treatment (Hodi et al. 2010; Robert et al. 2015), immuno-oncology has been one of the very active areas of research recently. There are many ongoing efforts to enhance antitumor immunity by knocking out immune-suppressing molecules in cancer cells as well as immune cells. CRISPR-Cas screens have also focused on the identification of such molecules (Figure 8.5). For example, CMTM6 was recently identified as a regulator of PD-L1 (a known immune-checkpoint molecule) stability in cancer cells (Burr et al. 2017).

8.8.2 Applications Beyond Oncology

8.8.2.1 Phenotypic Screens to Understand Gene Function and Regulation

Several screens have been performed based on flow cytometry and sorting of cells with a desired phenotype into populations of interest that can be compared. The cells can either be sorted based on antibodies that are targeting an endogenous protein or through the expression of a fluorescently tagged protein. FACS-based screening approaches are highly complex, require specific antibodies and extensive pre-screening optimization that often take longer than the screen itself. Examples of such a screening approach include a screen in dendritic cells to find the regulators of tumor necrosis factor (*TNF*) expression upon bacterial lipopolysaccharide (LPS) stimulation (Parnas et al. 2015), a screen to find regulators of PD-L1 in cancer cells upon stimulation with IFNg (Burr et al. 2017) and a screen to find regulators of ATP level in human cells (Mendelsohn et al. 2018). Stem cell research is another area where CRISPR-Cas screens are making a big impact. Factors involved in specific lineage differentiation and factors for iPSC induction can be discovered with great ease using pooled CRISPR screens. In one such screen, factors for the ability to self-renew and/or survive, suppressors of dissociation-induced cell death, and the regulators of capacity to differentiate were explored in hPSCs (Ihry et al. 2019).

8.8.2.2 Toxicology Applications

CRISPR screens can easily be adapted to study toxicological mechanisms for different toxins. In such a setting, after infection of the cells with a genome-wide sgRNA library, cells are divided into two pools; one pool of cells treated with the toxin and the other left untreated or only treated with a control compound. By comparison, sgRNA that provide a selective survival advantage in the presence of the toxin can be identified. In one such recent study, factors for antibody–drug conjugate toxicity were identified (Tsui et al. 2019). In another such study, *SLC39A9*, a zinc transporter was identified as host factor for Subtilase cytotoxin (Yamaji et al. 2019).

8.8.2.3 Microbiology Applications

Infectious diseases remain a challenge even in the twenty-first century due to the development of antibiotic resistance in bacteria and the fast-evolving nature of the viruses, exemplified by the devastating COVID-19 pandemic around the world in 2020. Pooled CRISPR-Cas screens are being used to address infectious diseases in several ways. One such way is to find regulators of microbial invasion in human cells. The SWI/SNF chromatin remodeling complex and key components of the TGF-β signaling pathway were identified as host factors for SARS-CoV-2 in a genome-wide CRISPR screen for SARS-CoV-2 host factors (Daniloski et al. 2021; Wei et al. 2021). Similarly, the role of *COPI* in Chlamydia trachomatis invasion (a major cause of ocular and genital tract infection in humans) was recently discovered in a genome-wide CRISPR screen in human cells (Park et al. 2019). In addition, screens have also been performed in the infecting agents themselves, meaning transduction of a sgRNA library in the infecting agent instead of human cells. In one such screen, essential genes in toxoplasma were identified (Sidik et al. 2016).

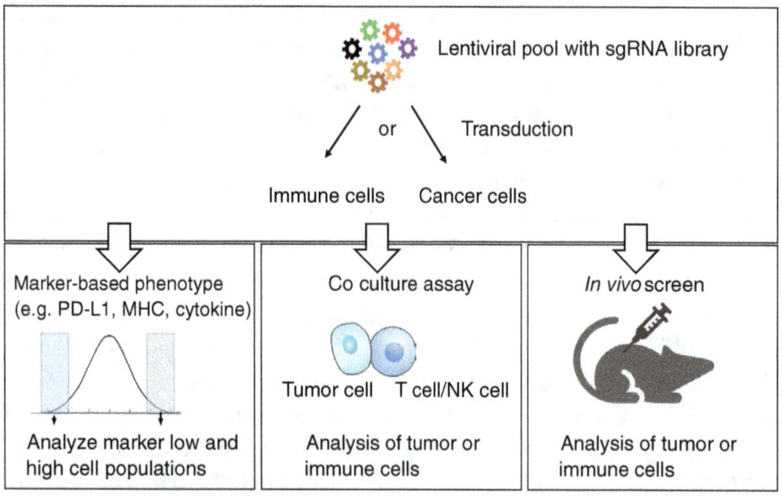

Figure 8.5 A multitude of choices to be made for planning cancer-immunotherapy functional genomics screen. In some screens, cancer cells were infected with a sgRNA library (Manguso et al. 2017; Patel et al. 2017; Pech et al. 2019) while in others, immune cells (CD8 T-cells or NK-cells) were infected (Dong et al. 2019). In phenotypic screens, assay readout is up- or downregulation of a molecule (Burr et al. 2017) while in others, a direct killing of cancer cells by immune cells is tested (Pech et al. 2019). Killing of cancer cells can be tested in a coculture-based assay (Patel et al. 2017; Pech et al. 2019) or directly in a mouse model (Manguso et al. 2017).

8.9 Outlook

Target identification is one of the early and critical decisions that needs to be made in a drug discovery project. Selection of an appropriate target can reduce the chances of failure at a later stage, saving high investments involved in the experiments. Functional genomics screens play an important role in studying genes at large scale. CRISPR-Cas screens have overcome several of the drawbacks associated with RNAi approaches used in the past. This includes higher consistency in knockout generation, high sensitivity for detecting lowly transcribed genes, and lower cell to cell heterogeneity in the phenotypes (Hart et al. 2015). Overall, genome-wide CRISPR screenings have been performed very successfully in human and mouse cancer cell lines. For *in vivo* screens and screens in primary cells, focusing on a subset of genes is ideal to have a good representation and hence lowering of the false discovery rate. With the improvement in sgRNA design algorithms, we expect a lower number of sgRNAs per gene (2–3 instead of 5–6 which is currently used) can be used to make it possible to study more genes in *in vivo* screens as well. A lot of computational tools have been developed to better analyze the large screening data; however, functional follow-up studies to validate individual hits is crucial for all screens.

References

Agilent. 2020. *Crispr libraries*. https://www.agilent.com/en/product/crispr-cas/crispr-libraries (accessed 13 August 2020).

Barretina, J., Caponigro, G., Stransky, N. et al. (2012). The cancer cell line encyclopedia enables predictive modelling of anticancer drug sensitivity. *Nature*, 483, 603–7.

Basu, A., Bodycombe, N.E., Cheah, J.H. et al. (2013). An interactive resource to identify cancer genetic and lineage dependencies targeted by small molecules. *Cell* 154: 1151–1161.

Behan, F.M., Iorio, F., Picco, G. et al. (2019). Prioritization of cancer therapeutic targets using CRISPR-Cas9 screens. *Nature* 568: 511–516.

Birsoy, K., Wang, T., Chen, W.W. et al. (2015). An essential role of the mitochondrial electron transport chain in cell proliferation is to enable aspartate synthesis. *Cell* 162: 540–551.

Boutros, M., Kiger, A.A., Armknecht, S. et al. (2004). Genome-wide RNAi analysis of growth and viability in Drosophila cells. *Science* 303: 832–835.

Burr, M.L., Sparbier, C.E., Chan, Y.C. et al. (2017). CMTM6 maintains the expression of PD-L1 and regulates anti-tumour immunity. *Nature* 549: 101–105.

Cellecta. 2020. *Crispr knockout pooled lentiviral sgrna libraries*. https://cellecta.com/collections/

crispr-knockout-pooled-lentiviral-sgrna-libraries (accessed 13 August 2020).

Chen, S., Sanjana, N.E., Zheng, K. et al. (2015). Genome-wide CRISPR screen in a mouse model of tumor growth and metastasis. *Cell* 160: 1246–1260.

Cheung, H.W., Cowley, G.S., Weir, B.A. et al. (2011). Systematic investigation of genetic vulnerabilities across cancer cell lines reveals lineage-specific dependencies in ovarian cancer. *Proc. Natl. Acad. Sci. U. S. A.* 108: 12372–12377.

Cortez, J.T., Montauti, E., Shifrut, E. et al. (2020). CRISPR screen in regulatory T cells reveals modulators of Foxp3. *Nature* 582: 416–420.

Dai, Z., Sheridan, J.M., Gearing, L.J. et al. (2014). edgeR: a versatile tool for the analysis of shRNA-seq and CRISPR-Cas9 genetic screens. *F1000Res* 3: 95.

Daniloski, Z., Jordan, T.X., Wessels, H.H. et al. (2021). Identification of required host factors for SARS-CoV-2 infection in human cells. *Cell* 184 (1): 92–105.

Datlinger, P., Rendeiro, A.F., Schmidl, C. et al. (2017). Pooled CRISPR screening with single-cell transcriptome readout. *Nat. Methods* 14: 297–301.

Dejesus, R., Moretti, F., Mcallister, G. et al. (2016). Functional CRISPR screening identifies the ufmylation pathway as a regulator of SQSTM1/p62. *elife* 5.

Dempster, J.M., Pacini, C., Pantel, S. et al. (2019). Agreement between two large pan-cancer CRISPR-Cas9 gene dependency data sets. *Nat. Commun.* 10: 5817.

DepMap Consortium, 2020. *DepMap portal*. https://depmap.org/portal (accessed 07 December 2020).

Dixit, A., Parnas, O., Li, B. et al. (2016). Perturb-Seq: dissecting molecular circuits with scalable single-cell RNA profiling of pooled genetic screens. *Cell* 167: 1853–1866. e17.

Doench, J.G. (2018). Am i ready for CRISPR? A user's guide to genetic screens. *Nat. Rev. Genet.* 19: 67–80.

Doench, J.G., Fusi, N., Sullender, M. et al. (2016). Optimized sgRNA design to maximize activity and minimize off-target effects of CRISPR-Cas9. *Nat. Biotechnol.* 34: 184–191.

Dong, M.B., Wang, G., Chow, R.D. et al. (2019). Systematic immunotherapy target discovery using genome-scale in vivo CRISPR screens in CD8 T cells. *Cell* 178: 1189–1204. e23.

Ghandi, M., Huang, F.W., JANE-Valbuena, J. et al. (2019). Next-generation characterization of the cancer cell line encyclopedia. *Nature* 569: 503–508.

Gilbert, L.A., Horlbeck, M.A., Adamson, B. et al. (2014). Genome-scale CRISPR-mediated control of gene repression and activation. *Cell* 159: 647–661.

Han, K., Jeng, E.E., Hess, G.T. et al. (2017). Synergistic drug combinations for cancer identified in a CRISPR screen for pairwise genetic interactions. *Nat. Biotechnol.* 35: 463–474.

Hart, T., Chandrashekhar, M., Aregger, M. et al. (2015). High-resolution CRISPR screens reveal fitness genes and genotype-specific cancer liabilities. *Cell* 163: 1515–1526.

Hart, T. and Moffat, J. (2016). BAGEL: a computational framework for identifying essential genes from pooled library screens. *BMC Bioinformatics* 17: 164.

Hart, T., Tong, A.H.Y., Chan, K. et al. (2017). Evaluation and design of genome-wide CRISPR/SpCas9 knockout screens. *G3 Genes, Genomes, Genetics* 7: 2719–2727.

Hay, M., Thomas, D.W., Craighead, J.L. et al. (2014). Clinical development success rates for investigational drugs. *Nat. Biotechnol.* 32: 40–51.

Hilton, I.B., D'ippolito, A.M., Vockley, C.M. et al. (2015). Epigenome editing by a CRISPR-Cas9-based acetyltransferase activates genes from promoters and enhancers. *Nat. Biotechnol.* 33: 510–517.

Hodi, F.S., O'day, S.J., Mcdermott, D.F. et al. (2010). Improved survival with ipilimumab in patients with metastatic melanoma. *N. Engl. J. Med.* 363: 711–723.

Huang Da, W., Sherman, B.T., and Lempicki, R.A. (2009). Systematic and integrative analysis of large gene lists using DAVID bioinformatics resources. *Nat. Protoc.* 4: 44–57.

Ihry, R.J., Salick, M.R., Ho, D.J. et al. (2019). Genome-scale CRISPR screens identify human pluripotency-specific genes. *Cell Rep.* 27: 616–630. e6.

Jain, I.H., Calvo, S.E., Markhard, A.L. et al. (2020). Genetic screen for cell fitness in high or low oxygen highlights mitochondrial and lipid metabolism. *Cell* 181: 716–727. e11.

Jeong, H.H., Kim, S.Y., Rousseaux, M.W.C. et al. (2017). CRISPRcloud: a secure cloud-based pipeline for CRISPR pooled screen deconvolution. *Bioinformatics* 33: 2963–2965.

Kearns, N.A., Pham, H., Tabak, B. et al. (2015). Functional annotation of native enhancers with a Cas9-histone demethylase fusion. *Nat. Methods* 12: 401–403.

King, E.A., Davis, J.W., and Degner, J.F. (2019). Are drug targets with genetic support twice as likely to be approved? Revised estimates of the impact of genetic support for drug mechanisms on the probability of drug approval. *PLoS Genet.* 15: e1008489.

Kittler, R., Putz, G., Pelletier, L. et al. (2004). An endoribonuclease-prepared siRNA screen in human cells identifies genes essential for cell division. *Nature* 432: 1036–1040.

Klijn, C., Durinck, S., Stawiski, E.W. et al. (2015). A comprehensive transcriptional portrait of human cancer cell lines. *Nat. Biotechnol.* 33: 306–312.

Koike-Yusa, H., Li, Y., Tan, E.P. et al. (2014). Genome-wide recessive genetic screening in mammalian cells with a lentiviral CRISPR-guide RNA library. *Nat. Biotechnol.* 32: 267–273.

Konermann, S., Brigham, M.D., Trevino, A.E. et al. (2015). Genome-scale transcriptional activation by an engineered CRISPR-Cas9 complex. *Nature* 517: 583–588.

Kweon, J., Jang, A.H., Shin, H.R. et al. (2020). A CRISPR-based base-editing screen for the functional assessment of BRCA1 variants. *Oncogene* 39: 30–35.

Li, B., Clohisey, S.M., Chia, B.S. et al. (2020). Genome-wide CRISPR screen identifies host dependency factors for influenza a virus infection. *Nat. Commun.* 11: 164.

Li, W., Xu, H., Xiao, T. et al. (2014). MAGeCK enables robust identification of essential genes from genome-scale CRISPR/Cas9 knockout screens. *Genome Biol.* 15: 554.

Li, X., Shi, L., Wang, Y. et al. (2019). OncoBase: a platform for decoding regulatory somatic mutations in human cancers. *Nucleic Acids Res.* 47: D1044–D1055.

Liberzon, A., Subramanian, A., Pinchback, R. et al. (2011). Molecular signatures database (MSigDB) 3.0. *Bioinformatics* 27: 1739–1740.

Liu, J., Lichtenberg, T., Hoadley, K.A. et al. (2018). An integrated TCGA pan-cancer clinical data resource to drive high-quality survival outcome analytics. *Cell* 173: 400–416. e11.

Manguso, R.T., Pope, H.W., Zimmer, M.D. et al. (2017). in vivo CRISPR screening identifies Ptpn2 as a cancer immunotherapy target. *Nature* 547: 413–418.

Mendelsohn, B.A., Bennett, N.K., Darch, M.A. et al. (2018). A high-throughput screen of real-time ATP levels in individual cells reveals mechanisms of energy failure. *PLoS Biol.* 16: e2004624.

Meyers, R.M., Bryan, J.G., Mcfarland, J.M. et al. (2017). Computational correction of copy number effect improves specificity of CRISPR-Cas9 essentiality screens in cancer cells. *Nat. Genet.* 49: 1779–1784.

Michlits, G., Hubmann, M., Wu, S.H. et al. (2017). CRISPR-UMI: single-cell lineage tracing of pooled CRISPR-Cas9 screens. *Nat. Methods* 14: 1191–1197.

Michlits, G., Jude, J., Hinterndorfer, M. et al. (2020). Multilayered VBC score predicts sgRNAs that efficiently generate loss-of-function alleles. *Nat. Methods* 17: 708–716.

Nelson, M.R., Tipney, H., Painter, J.L. et al. (2015). The support of human genetic evidence for approved drug indications. *Nat. Genet.* 47: 856–860.

Park, J.S., Helble, J.D., Lazarus, J.E. et al. (2019). A FACS-based genome-wide CRISPR screen reveals a requirement for COPI in chlamydia trachomatis invasion. *iScience* 11: 71–84.

Parnas, O., Jovanovic, M., Eisenhaure, T.M. et al. (2015). A genome-wide CRISPR screen in primary immune cells to dissect regulatory networks. *Cell* 162: 675–686.

Patel, S.J., Sanjana, N.E., Kishton, R.J. et al. (2017). Identification of essential genes for cancer immunotherapy. *Nature* 548: 537–542.

Pech, M.F., Fong, L.E., Villalta, J.E. et al. (2019). Systematic identification of cancer cell vulnerabilities to natural killer cell-mediated immune surveillance. *elife* 8.

Robert, C., Long, G.V., Brady, B. et al. (2015). Nivolumab in previously untreated melanoma without BRAF mutation. *N. Engl. J. Med.* 372: 320–330.

Rogers, Z.N., Mcfarland, C.D., Winters, I.P. et al. (2017). A quantitative and multiplexed approach to uncover the fitness landscape of tumor suppression in vivo. *Nat. Methods* 14: 737–742.

Sanchez-Rivera, F.J., Papagiannakopoulos, T., Romero, R. et al. (2014). Rapid modelling of cooperating genetic events in cancer through somatic genome editing. *Nature* 516: 428–431.

Sanjana, N.E. (2017). Genome-scale CRISPR pooled screens. *Anal. Biochem.* 532: 95–99.

Sanjana, N.E., Shalem, O., and Zhang, F. (2014). Improved vectors and genome-wide libraries for CRISPR screening. *Nat. Methods* 11: 783–784.

Schmierer, B., Botla, S.K., Zhang, J. et al. (2017). CRISPR/Cas9 screening using unique molecular identifiers. *Mol. Syst. Biol.* 13: 945.

Shalem, O., Sanjana, N.E., Hartenian, E. et al. (2014). Genome-scale CRISPR-Cas9 knockout screening in human cells. *Science* 343: 84–87.

Shalem, O., Sanjana, N.E., and Zhang, F. (2015). High-throughput functional genomics using CRISPR-Cas9. *Nat. Rev. Genet.* 16: 299–311.

Shen, J.P., Zhao, D., Sasik, R. et al. (2017). Combinatorial CRISPR-Cas9 screens for de novo mapping of genetic interactions. *Nat. Methods* 14: 573–576.

Shi, J., Wang, E., Milazzo, J.P. et al. (2015). Discovery of cancer drug targets by CRISPR-Cas9 screening of protein domains. *Nat. Biotechnol.* 33: 661–667.

Sidik, S.M., Huet, D., Ganesan, S.M. et al. (2016). A genome-wide CRISPR screen in toxoplasma identifies essential apicomplexan genes. *Cell* 166: 1423–1435. e12.

Steinhart, Z., Pavlovic, Z., Chandrashekhar, M. et al. (2017). Corrigendum: genome-wide CRISPR screens reveal a Wnt-FZD5 signaling circuit as a druggable vulnerability of RNF43-mutant pancreatic tumors. *Nat. Med.* 23: 1384.

Subramanian, A., Tamayo, P., Mootha, V.K. et al. (2005). Gene set enrichment analysis: a knowledge-based approach for interpreting genome-wide expression profiles. *Proc. Natl. Acad. Sci. U. S. A.* 102: 15545–15550.

Thermofisher. 2020. *CRISPR Libraries*. https://www.thermofisher.com/se/en/home/life-science/genome-editing/geneart-crispr/crispr-libraries.html (accessed 13 August 2020).

Tsherniak, A., Vazquez, F., Montgomery, P.G. et al. (2017). Defining a cancer dependency map. *Cell* 170: 564–576. e16.

Tsui, C.K., Barfield, R.M., Fischer, C.R. et al. (2019). CRISPR-Cas9 screens identify regulators of antibody-drug conjugate toxicity. *Nat. Chem. Biol.* 15: 949–958.

Wang, B., Wang, M., Zhang, W. et al. (2019a). Integrative analysis of pooled CRISPR genetic screens using MAGeCKFlute. *Nat. Protoc.* 14: 756–780.

Wang, C., Vegna, S., Jin, H. et al. (2019b). Inducing and exploiting vulnerabilities for the treatment of liver cancer. *Nature* 574: 268–272.

Wang, T., Wei, J.J., Sabatini, D.M., and Lander, E.S. (2014). Genetic screens in human cells using the CRISPR-Cas9 system. *Science* 343: 80–84.

Wang, T., Yu, H., Hughes, N.W. et al. (2017). Gene essentiality profiling reveals gene networks and synthetic lethal interactions with oncogenic Ras. *Cell* 168: 890–903. e15.

Wei, J., Alfajaro, M.M., Deweirdt, P.C. et al. (2021). Genome-wide CRISPR screens reveal host factors critical for SARS-CoV-2 infection. *Cell* 184 (1): 76–91.

Whitehurst, A.W., Bodemann, B.O., Cardenas, J. et al. (2007). Synthetic lethal screen identification of chemosensitizer loci in cancer cells. *Nature* 446: 815–819.

Winter, J., Breinig, M., Heigwer, F. et al. (2016). caRpools: an R package for exploratory data analysis and documentation of pooled CRISPR/Cas9 screens. *Bioinformatics* 32: 632–634.

Xue, W., Chen, S., Yin, H. et al. (2014). CRISPR-mediated direct mutation of cancer genes in the mouse liver. *Nature* 514: 380–384.

Yamaji, T., Hanamatsu, H., Sekizuka, T. et al. (2019). A CRISPR screen using subtilase cytotoxin identifies SLC39A9 as a glycan-regulating factor. *iScience* 15: 407–420.

Zhang, R., Miner, J.J., Gorman, M.J. et al. (2016). A CRISPR screen defines a signal peptide processing pathway required by flaviviruses. *Nature* 535: 164–168.

9

Functional Genomics: Arrayed CRISPR KO Screens

Davide Gianni and Leire Escudero-Ibarz

Discovery Biology, BioPharmaceuticals R&D, AstraZeneca, Cambridge, UK

9.1 Introduction

The costs of developing new drugs are continuing to rise and one of the main factors for this is the lack of clinical efficacy of new therapeutic candidates. These failures are expensive and indicative of our inability to fully characterize the efficacy of candidate compounds or drug targets early on in the discovery phase. A study in 2015 demonstrated that the majority of projects failed due to "target linkage to disease not established" or "no validated models available" (Morgan et al. 2018). Moreover, lack of clinical efficacy can also be linked to the fact that most companies are working on the same pool of targets, as they rely on literature reports and external evidence from published data. Therefore, a major challenge in biomedical research and drug discovery is the elucidation of gene function and identification of novel drug targets with a strong genetic link to disease.

Functional genomics is the discipline in molecular and cellular biology field that studies how genes and pathways contribute to a particular disease phenotype. Functional genomic screens are key in target discovery and hold the promise to enable the identification of drug targets with improved clinical attrition rate. The long-term success of target discovery and validation is currently based on four key pillar areas that will be reviewed in this chapter, including: (i) the creation and implementation of successful array screening modalities to perturb individual gene and protein functions in disease relevant models, (ii) the use of translatable and physiologically relevant models that mimic a particular disease phenotype and can be adapted to the screening throughput, (iii) the integration of complex assay readouts to interpret the biological outputs of cellular models upon perturbation, and (iv) the establishment of bioinformatic pipelines to enable quantitative analysis of phenotypic readouts resulting from genetic and non-genetic perturbations.

9.2 Array Format Technologies

To date, multiple tools for phenotypic perturbation have been deployed to identify novel genes and pathways associated with biological processes and disease. Genes or proteins that when enhanced or inactivated cause a change in a specific phenotypic endpoint could be used as effective targets for pharmacological modulation.

In recent years, researchers in the functional genomics field have been able to create and implement new perturbation modalities in phenotypic screening and target identification and validation. Genetic and non-genetic agents are able to perturb the function of relevant and translatable disease models. These modalities include siRNA libraries, CRISPR libraries both in pooled and arrayed format, as well as protein libraries such as the secretome. In addition, a new screening modality has been recently described using fully functionalized fragment (FFF) libraries, which streamlines the process from phenotypic screening to target identification (Parker et al. 2017).

9.2.1 Genetic Perturbation Libraries

Genome-wide and/or bespoke loss-of-function techniques have been widely used to modulate targets in high-throughput experiments. Before the advent of CRISPR technologies, RNA interference libraries (siRNA or shRNA) have been one of the first modalities to explore the "druggable" genome and identify new targets by transcriptional silencing or knock-down. However, while this method

Genome Editing in Drug Discovery, First Edition. Edited by Marcello Maresca and Sumit Deswal.

initially provided valuable insights on biological pathways, it has also shown limitations such as off-target effects, incomplete loss of function, and partial inactivation of the target of interest (Kaelin 2012). Therefore, some of the data might not be representing real biological and translational impact. In contrast, CRISPR-Cas9 techniques have been able to overcome the limitations encountered with siRNA modalities, as this powerful gene-editing technique allows the modulation of the endogenous genome with high efficiency, minimal off-target effects, and higher specificity. Moreover, CRISPR screening allows for immediate association between the target being disrupted and the phenotypic outcome without the need of complex target deconvolution. Recent advances in the pharmaceutical sector have benefited from pre-existing expertise in precise genome editing and automated liquid-handling to enable the expansion of genome-wide CRISPR capabilities, which has accelerated the search for novel drug targets that could have been missed when using siRNA approaches.

Functional genomic screens by CRISPR have revolutionized the field and it is progressively becoming the method of choice for novel target identification. These screens systematically knockout a large number of genes and can be classified into pooled and arrayed screens. As mentioned in the previous chapter, pooled CRISPR screens are based on the delivery of CRISPR reagents using lentivirus-based gRNA libraries in pools of cells. Target deconvolution is performed by next-generation sequencing (NGS), which limits the phenotypic readout to cell proliferation by identifying genetic perturbants that have been either enriched or depleted in the cell population. Additionally, fluorescent reporters or tagged proteins can be also used to sort and enrich cell populations by fluorescence-activated cell sorting (FACS). Ongoing efforts in the field are trying to improve the existing gRNA libraries by optimizing library size and reducing the number of cells per screen. This would reduce resources and costs, and enable large screens in more complex and physiologically relevant systems including primary cells, co-culture, and *in vivo* models.

In contrast, arrayed CRISPR screens incorporate a single perturbation or gRNA per well in a multiwell format, bringing the advantage of multiple and more complex phenotypic readouts, including high-content imaging, enzyme activity, and cytokine release among others (see Section 9.5). However, it is important to highlight that the array format is generally more labor-intensive, more expensive, and requires a more sophisticated automated liquid handling infrastructure. As shown in Figure 9.1, in the array format, typically CRISPR reagents are delivered as synthetic gRNAs in Cas9-expressing cells using lipid transfection methods, as this is the most compatible system with the current high-throughput screening platforms. A recent publication epitomizes this approach with the establishment of a fully integrated CRISPR arrayed platform using synthetic gRNA libraries aimed at identifying novel targets that modulate productive delivery of lipid nanoparticle (LNP) encapsulated mRNA. Using this approach, the authors elegantly validate two genes (UDP-Glucose Ceramide Glucosyltransferase – UGCG and V-type proton ATPase – ATP6V) that significantly modulate the productive delivery of LNP-mRNA. These findings show for the first time how arrayed CRISPR screening technology can elucidate our appreciation of molecular pathways otherwise less understood (Ross-Thriepland et al. 2020). However, one limitation of this workflow using synthetic gRNA is that it is only adoptable for cell lines that can be transfected using lipid-based transfection methods.

When using hard to transfect models, such as primary cells or 3D models, alternative workflows need to be adopted. One possibility is the use of arrayed lentivirus libraries, where individual gRNAs and Cas9 constructs are delivered using lentiviral vectors (Figure 9.2). A recent publication has described the establishment of a lentiviral-based arrayed CRISPR kinome screening platform capable

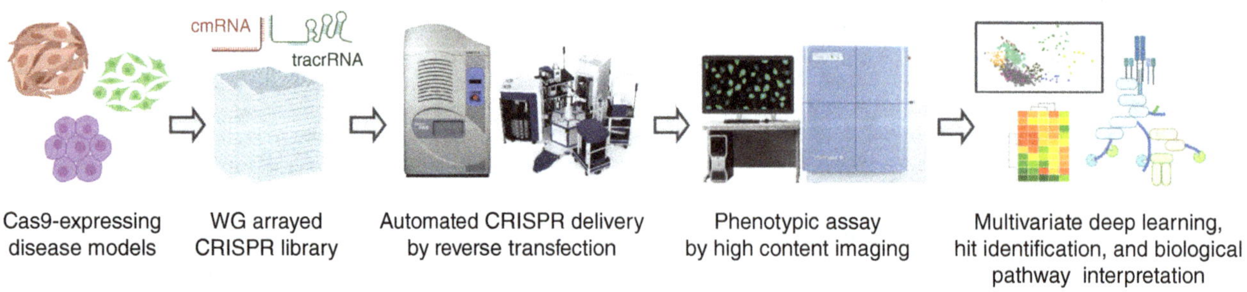

Figure 9.1 **Genome-wide arrayed CRISPR screening platform.** Overview of a standard workflow for whole-genome (WG) arrayed CRISPR screening by lipid transfection. Synthetic gRNA library containing 4 gRNAs per gene/well is delivered to Cas9-expressing cells by lipid reverse transfection using high-throughput automation. CRISPR perturbation is assessed by high-content imaging and a final list of genes associated with a particular phenotypic response are identified and ranked by bioinformatics. *Image created with BioRender.com.*

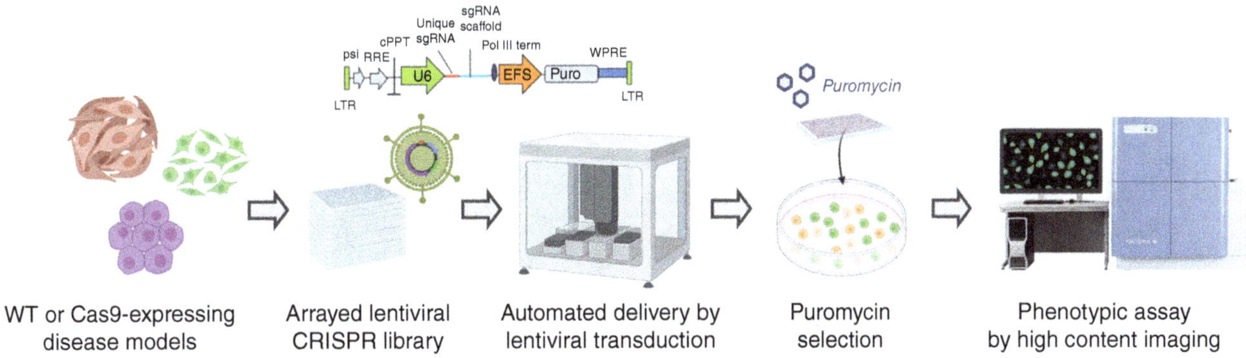

Figure 9.2 Arrayed lentiviral CRISPR screening platform. Overview of a standard workflow for lentiviral CRISPR delivery in array format. Ready-to-use lentiviral particles containing gRNA constructs targeting one gene per well are delivered by transduction in Cas9-expressing cells. Alternatively, WT cells can also be transduced with Cas9 lentivirus prior to gRNA delivery. Successfully transduced cells are selected with puromycin to enrich gRNA-positive populations before assessing CRISPR perturbation by high-content imaging and other relevant phenotypic readouts. *Image created with BioRender.com.*

of identifying a number of canonical modulators of the NF-κB signaling pathway. Using this workflow, the authors identified the majority of kinase genes with known regulatory roles in TNF α-mediated NF-κB signaling at a higher success rate than observed with previous RNAi-based studies (O'Shea et al. 2020). Despite these promising results, this method also presents some limitations, such as potential variable lentiviral titers in commercially available libraries and the need to select out non-transduced cells. Finally, another workflow that holds significant promise and could circumvent some of the previously described limitations is an approach in which the CRISPR reagents are delivered as ribonucleoprotein or RNP (Cas9 protein + synthetic gRNA) via electroporation in arrayed format (Shifrut et al. 2018). This setup could be potentially game changing, especially in an historical moment when academic and industry groups are heavily investing in generation of more translatable models, such as primary cells and iPSC-derived cellular models, which cannot be easily manipulated using lipid-based transfection methods or viral delivery. Recent work from the Marson's lab has shown the value of arrayed CRISPR Cas9 RNP delivery at smaller scale to validate hits from genome-wide pooled knockout screens in CD8 + T cells. This method was used to further characterize the effects of individual perturbations for T cell stimulation and enhanced cancer cell killing. They found that ablation of at least four targets (SOCS1, TCEB2, RASA2, and CBLB) in human T cells enhanced both proliferation and *in vitro* anti-tumor effects (Shifrut et al. 2018). In addition, this delivery method has now been further developed and adapted to perform arrayed *in vitro* validation of pooled knock in screen hits, which has also enabled the discovery of novel chimeric receptors enhancing primary human T cell fitness and tumor clearance (Roth et al. 2020).

In addition to using CRISPR Cas9 nuclease to generate target knockouts (CRISPRn), CRISPR activation (CRISPRa) and CRISPR inactivation (CRISPRi) systems are being adapted to high-throughput screening to yield information about target activation or inhibition and increase confidence in the phenotypic outcomes observed by CRISPRn. These systems use a catalytically inactive Cas9 (nuclease-dead dCas9) fused to DNA effectors to a specific promoter sequence (activator or repressor) to study the upregulation (CRISPRa) and inhibition (CRISPRi) of target genes without the permanent introduction of mutation (see Chapter 10).

To date, functional genomic screens by CRISPR have focused on loss of function or gene activation libraries. However, CRISPR technologies are evolving at a fast pace and it is important to consider the integration of next-generation libraries to expand phenotypic screening toolbox available to researchers in this area. For instance, dCas9 variants can also be fused to DNA methyltransferases allowing screens that target methylation (Liu et al. 2016; Stricker et al. 2016; Klann et al. 2017). Also, dCas9 can be linked to base modification enzymes to enable specific knock-in of point mutations by direct DNA base conversion, which would enable the study of disease caused by single base modifications (Rees and Liu 2018; Anzalone et al. 2019). Additionally, base editing libraries would not only allow for disease-specific single point mutations and drug resistance studies, but would also enable specific protein truncation by introducing STOP codons (Billon et al. 2017) (see Chapter 14). A more recent development is the concept of noncoding RNA CRISPR libraries. Up to now, functional genomic screens have mainly focused on interrogating coding regions of the human genome, despite the fact that protein coding genes only comprise 1% of the human genome. Therefore, being able to query noncoding regions would provide researchers with the unprecedented ability to interrogate much larger sections of the human genome. Of note, there is increasing evidence indicating that noncoding genetic variants contribute significantly to disease mechanisms. Certainly, the ability to study the functional impact of targeting these noncoding polymorphisms would uncover new biology and therapeutic approaches (Papanicolaou and Bonetti 2020). Several groups have already explored this transformative technology, such as studies performing genome-wide screens of noncoding regions using gRNAs to target splice sites, achieving exon skipping or intron retention (Liu et al. 2018; Shukla and Huangfu 2018). In the future, one would therefore expect an increasing demand for noncoding RNA libraries in phenotypic screening, including enhancers regulating gene expression, long noncoding RNAs (lncRNAs), and microRNAs (miRNAs).

9.2.2 Non-Genetic Perturbation Libraries

An alternative screening approach to libraries that affect gene function at the DNA level is the use of non-genetic perturbation libraries for target discovery. One of the most prominent examples of non-genetic perturbation libraries is the secretome. Collaborations between industry and academia, such as AstraZeneca and KTH Royal Institute of Technology in Stockholm, have enabled the implementation of secretome-based phenotypic screening in drug discovery, which has facilitated new drug target identification and the discovery of new biology and signaling pathways in disease (Ding et al. 2020). Institutes such as KTH bring their expertise in secretome-wide scale protein expression and purification, which gives access to a library of thousands of secreted peptides and proteins to screen in a wide range of cell-based phenotypic assays. Secreted proteins in the cell microenvironment and bloodstream are key in regulating cellular phenotypes by transducing signals and modulating cell-to-cell interactions. A main advantage of using secretome libraries is that it allows screening in disease models that are difficult to genetically manipulate by other modalities, such as CRISPR. This is exemplified by successful secretome library screens in primary human lung fibroblasts to study lung senescence, pancreatic beta cells for the treatment of type II diabetes, and cardiac progenitor cells for cardiac regeneration after heart attack (Jennbacken et al. 2019), which are all using physiologically relevant primary models. Also, screening with secretome libraries allows for direct identification of active proteins leading to a specific phenotype, and in some instances also allow for identification of its receptor, expanding the understanding of pathway signaling and disease biology.

Screening library approaches discussed so far rely on the manipulation of genes, or modulation of gene expression and protein function. While these approaches allow researchers to identify targets of interest fairly quickly with "lighter" deconvolution activities, these methods still require identification of small molecule inhibitors to pharmacologically modulate the function of the gene of interest. For this reason, small molecule-based phenotypic drug discovery is a valuable modality that should be deployed together with other screening libraries to achieve successful phenotypic drug discovery. This is particularly important when screening complex disease models, which are not always amenable to genetic manipulation at scale. Moreover, similarly to secretome libraries, small molecule screening brings unique advantages compared with genetic libraries, such as tunable phenotypic effects by using different concentrations and treatment duration of the compound. Also, this screening modality is important when modulating proteins that are essential and their complete knockout would lead to cell lethality. However, the target deconvolution of small molecule hits from small molecule-based phenotypic discovery campaigns is not a straightforward or a quick process, especially for hits with low potency, which require more extensive chemistry optimization. An approach that promises to address this challenge is a phenotypic screening modality that employs FFF libraries, which combines small molecule approaches with chemoproteomics to streamline the process of target identification from phenotypic screening hit molecules. The Cravatt's group has pioneered this approach in a report from 2017 where they have merged fragment-based ligand discovery with quantitative chemical proteomics, allowing the mapping of reversible small molecule–protein interactions directly in cellular models (Parker et al. 2017). Despite the fact that these libraries contain less than 1000 compounds, they still cover a broad chemical space. By embedding photoreactive and clickable reporter groups into bioactive small molecules, it is possible to perform chemical proteomic analysis of protein targets in cells. FFF probes consist of a variable, fragment-like recognition element that confers specificity toward target proteins. This variable region is linked to a photoreactive diazirine group that enables the photoactivated capture (UV light-induced crosslinking) of probes to their interacting proteins directly in living cells. In addition, it also contains an alkyne handle that enables subsequent copper-catalyzed alkine-azide cycloaddition to reporter tags (e.g. biotin, fluorophores) ("clicking" elements) for enrichment and identification of probe-bound proteins by mass spectrometry. Using this library and combining it to screening of physiologically relevant model systems, the authors identify small molecules that promote adipocyte differentiation by engaging the poorly characterized membrane protein PGRMC2. Future studies should assess the applicability of this modality of screening to a set of more diverse physiologically relevant models including cell models that are known to be more scarce and challenging to grow at scale (e.g. iPSC-derived models as well as primary immune cells).

9.3 CRISPR Reagent Delivery Systems

Gene editing success by CRISPR-Cas9 can be only achieved if the delivery of gRNA and Cas9 components into the cells is successful. Genetic modification in several immortalized cell lines has been successfully achieved; however, delivery can become challenging when using hard-to-transfect cells, such as primary cells, pluripotent stem cells, or more advanced models. For these cell types, further optimization is needed to achieve successful delivery conditions.

CRISPR delivery can be divided into two main categories: cargo and delivery methods. CRISPR/Cas9 cargo comes in multiple flavors: (i) DNA plasmid encoding both the Cas9 protein and gRNA, (ii) mRNA for Cas9 translation along with gRNA, (iii) Cas9 protein pre-complexed with gRNA to form RNP, and (iv) gRNA alone in a Cas9 stable line. The delivery methods will depend on the cell model used and the type of cargo. These methods can be divided into (i) physical methods (microinjection, electroporation), (ii) viral methods (adeno-associated virus (AAV), full-sized adenovirus, and lentivirus), and (iii) nonviral methods (lipofection) (Table 9.1). In addition to identifying the optimum delivery method, it is also important to optimize the Cas9-gRNA ratio and the total amount being delivered. These factors are variable among different cell types and could greatly influence successful gene editing.

The standard workflow for pooled CRISPR screening is the delivery of gRNAs by viral transduction in a Cas9-expressing cell line, from either a pool or a single clone. However, stable Cas9 cell generation is not possible for all cell types, such as primary cells, and this approach has mostly been successfully implemented in immortalized cells with replicative capacity.

In the context of high-throughput arrayed CRISPR screening, lipofection is the preferred system as this is amenable to the current liquid handling automation in the pharmaceutical industry. This method is feasible when using easy-to-transfect cells that already express Cas9 stably or inducibly (Maresca et al. 2013), because the subsequent transient delivery of synthetic gRNAs is usually a more efficient step. Moreover, the use of fluorescently tagged Cas9 reporters enable single cell gating on Cas9 expression during high-content image analysis, which undoubtedly provides the advantage to more robustly identify cells with a successful gene editing event. In addition, the use of fluorescently tagged Cas9 reporters allows researchers to screen pool cell populations rather than clones, which also adds the benefit of screening heterogeneous cell populations limiting the possibility of generating artifactual results from the screen of single cell clones. This approach is compatible with plate-based automation platforms and gRNAs can be prepared by acoustic dispensing before adding lipofection reagents and cells by reverse transfection (Ross-Thriepland et al. 2020).

There are now multiple companies commercializing off-the-shelf genome-wide libraries, which are compatible with high-throughput screening and generally show high editing efficiencies and high reproducibility. These libraries are provided with gRNAs that are chemically synthesized to allow the incorporation of chemical modification

Table 9.1 CRISPR reagent delivery methods. General advantages and limitations encountered when delivering CRISPR reagents by viral transduction, electroporation, or lipid transfection.

Delivery system	Advantages	Limitations
Viral	• High transfection efficiency • Low cell toxicity • Amenable to primary or difficult-to-transfect cells	• Time-consuming • Low–medium throughput • Expensive reagents • Additional safety measures
Electroporation	• Medium–high transfection efficiency • Amenable to primary or difficult-to-transfect cells	• Toxic for certain cells types • Low–medium throughput • Expensive reagents
Lipid transfection	• Medium–high transfection efficiency • Low cell toxicity • High-throughput • Cost-effective	• Most primary cells are refractory to lipid transfection

that improve editing efficiencies, reduce degradation and immune activation (Hendel et al. 2015). Synthetic gRNAs can be formatted as dual CRISPR RNA (crRNA) and trans-activating crRNA (tracrRNA) (crRNA:tracrRNA) or as chimeric single-guide RNA (sgRNA). In these libraries, each gene is targeted by multiple gRNAs, rather than relying on one unique gRNA to induce a functional knockout. Normally, the multiguide format is designed using optimized algorithms to predict efficient deletion of large exonic fragments of target genes and increase the likelihood of loss-of-function mutations and significant phenotypic changes. The knockout effect needs to be robust to obtain reliable screening data as single shots and minimize false negatives. Examples of commercial suppliers are Dharmacon and Synthego and they differ in their proprietary multiguide design and synthetic sgRNA chemistry and algorithm designs. Synthego's algorithm combines three synthetic sgRNAs per gene in close proximity to generate large fragment deletions that result in a functional knockout, whereas Dharmacon' s algorithm combines four gRNAs in distant loci and exons for each gene. Moreover, in addition to whole-genome libraries, smaller predesigned libraries are available, including the druggable genome, GPCRs, kinases, epigenetic panel, among others, which allows for screening in cell models that are more challenging to be grown at scale. Also, cherry-pick libraries with a selected set of genes are available to address more bespoke biological questions.

An important factor to take into consideration is the effect of constitutive Cas9 protein expression in cell fitness, currently addressed by using inducible systems as mentioned above. Moreover, when cell function is affected by Cas9 expression or stable Cas9 integration is not feasible (e.g. in difficult-to-transfect models), transient Cas9 expression methods might be advantageous. Also, these approaches might be beneficial to reduce DNA transfection toxicity and the time spent in stable Cas9 cell generation. Transient Cas9 expression also minimizes the time the nuclease remains in cells, potentially reducing off-target effects. Transient delivery of Cas9 can be achieved by mRNA Cas9 transfection or Cas9 protein in RNP complexes. These approaches have proven to be successful not only in immortalized cell lines, but also in patient derived or primary cultures, including primary lung fibroblasts and primary T cells (Seki and Rutz 2018). Most of the time, delivery of RNP complexes in primary cultures require electroporation-based nucleofection methods, which enable direct reagent transfer into the nucleus independently of cell division and achieve high transfection efficiencies in nondividing primary cells. However, unlike lipid transfection methods, the electroporation process is not currently amenable to the automation required for large-scale array CRISPR screens. Recent progress has been made toward optimizing protocols for electroporation in multiwell format, but there are still some improvements to be made to embed the 384 HT electroporator equipment to the existing automation workflows. This will enable the delivery of CRISPR reagents in difficult-to-transfect cells and uncover new targets in more physiologically relevant models. For instance, a recent study has been able to perform whole genome CRISPR pool screens in T cells by co-delivering a pooled sgRNA library with subsequent Cas9 protein delivery by electroporation (Shifrut et al. 2018). This study also showed successful delivery of Cas9 RNPs in T cells by arrayed electroporation, but the scale of this approach in multiwell format is still a bottleneck (Chen et al. 2018; Shifrut et al. 2018; Hultquist et al. 2019; Cortez et al. 2020). Therefore, despite significant advances to implement arrayed screening in translational models, there is still further work to be done to increase scale and throughput for CRISPR delivery in these complex models.

9.4 PreClinical Models in Array Screening

The development and optimization of more translatable *in vitro* models is key to closely mimic disease biology and identify disease-relevant phenotypes. As mentioned before, the use of physiologically relevant models in functional genomics screens is likely to increase the success rate in the identification of novel drug targets and decrease clinical attrition. For this reason, there is an increasing desire to screen in more complex models, such as primary cells, iPSC-derived lineages, co-culture systems, and 3D models, including organoids and Organ-Chip. However, the main challenge in functional genomic screens is to find the right balance between "screenability" and translatability. It is therefore important to consider the feasibility to screen these models at scale and still achieve high perturbation efficiencies.

To date, the most commonly used models for genome-wide genetic screens have been monoculture or 2D cell systems, particularly immortalized cancer cell lines, which are not restricted to cell expansion and manipulation. In this context, the risk of identifying targets that are not relevant to disease is reduced by subsequent validation of a smaller number of identified targets in primary models, when the throughput is no longer a limitation. However, in the future, one would expect the establishment of workflows enabling primary screenings in more advanced cellular models.

For instance, high-throughput screening in primary cells using fibrosis as the main phenotypic readout has recently delivered impactful data in several disease areas, including cardiac, lung, and kidney fibrosis (Sun et al. 2016; Schafer et al. 2017; Palano et al. 2020). Moreover, CRISPR delivery methods in human-derived fibroblasts have been recently optimized with high editing efficiencies in arrayed format; however, the main bottleneck is to reach whole-genome scale screens by embedding the electroporation systems to the existing automation workflows. Alternative methods to overcome the scalability issues have been developed by using lentiviral gRNA systems. A recent publication describes the first whole-genome arrayed CRISPR screen by gRNA lentiviral delivery in a high-content imaging fibrosis model using primary human kidney fibroblasts. Orthogonal approaches to validate hits coming out of the screen were used to identify novel targets linked to kidney fibrosis (Turner et al. 2020).

The advent of cancer immunotherapy in recent years has also led to an increasing demand of more complex models that mimic the tumor microenvironment and mechanisms involved in immune cell recognition of tumor antigens and cancer cell killing. These models include co-culture of immune and cancer cells or organoids in both 2D and 3D format. The ability to deliver CRISPR reagents in immune cells in high-throughput using a co-culture system of immune cells and cancer cells has the potential to unveil new relevant targets to armor T cells for enhanced tumor cell killing. Recent work has demonstrated the feasibility to screen 2D co-culture models in high-throughput by using imaging, impedance, and cytokine release readouts to measure cytotoxicity and exhaustion of T cells (Dufva et al. 2020; Gee et al. 2020). For instance, the manuscript from Gee et al. illustrates efforts to develop a co-culture system utilizing primary T-cells and tumor cells to identify immuno-oncology targets important for tumor cell evasion of immune cell killing. The authors demonstrate that this co-culture system is fit for purpose in arrayed CRISPR screening using a focused synthetic gRNA library. Their approach identifies the gene ICAM1, whose knockout enhances resistance to T-cell killing, highlighting the promise of tumor/immune co-culture systems for the identification of important mechanisms and demonstrating the value of further investigation of different co-culture models that can provide more robust phenotypic output.

Furthermore, 3D co-culture models are not only able to measure T cell killing, but also assess immune cell infiltration in the tumor site, and associated components of the tumor microenvironment, including fibroblasts and the extracellular matrix (Di Modugno et al. 2019; Osswald et al. 2019). A different approach to 3D models is the generation of organoid and tumoroid models, which originate from tissue-derived adult stem cells or patient-derived tumor tissue, and can be grown in self-organized organoid structures. Surprisingly, these models can be genetically modified, expanded, and cryopreserved, although it might be still challenging to perform genetic manipulation at scale (Drost and Clevers 2018). A recent study described a scalable cancer-spheroid model and performed genome-wide CRISPR screens in 2D monolayer and 3D lung-cancer spheroids, demonstrating that CRISPR phenotypes in 3D more accurately recapitulated those of *in vivo* tumors (Han et al. 2020). Of note, this study overcomes the scalability restrictions by generating 3D spheroids from lung cancer cells, instead of patient-derived tissue. Going forward, a number of collaborations between academic and industry labs will enable the development of screening workflows in organoid models in oncology, respiratory, and metabolic disease, by using established organoid cultures from primary tissue (colon, breast, prostate, among others) and fetal and/or iPSC-derived models (alveolar lung organoids).

Increasing complexity can be also achieved by using *in vivo* models for target discovery, being the most challenging system to adapt to the throughput required in functional genomics. Several studies have demonstrated the feasibility of genetic screens in mouse models, although limited to the oncology field (Manguso et al. 2017; Van Der Weyden et al. 2017; Dong et al. 2019). Traditionally, mouse models have been the *in vivo* method of choice in functional genomics; however, alternative models such as zebrafish might be more suitable to the scale in phenotypic screening. Zebrafish represents a middle ground between the more simplistic *in vitro* models and the most complex *in vivo* mouse models. Zebrafish is not only valuable in basic research and developmental biology, but in combination with CRISPR genome editing tools, also offers the advantage of human disease modeling for its use in target discovery and safety profiling (Rubbini et al. 2020). Injection of CRISPR gRNA reagents in zebrafish embryos has achieved up to 100% editing efficiency rates, enabling the generation of homozygous knockout mutants in the first larvae generation. In comparison with genetically modified mouse models, the high rate of egg and larvae production in zebrafish models circumvents the long waiting time of multiple generations required before obtaining bi-allelic mutants. Zebrafish would therefore streamline the processes required to bring new targets to the pipeline, with reduced costs and time compared with preclinical studies in mouse. Beyond its use in genetic screening and mechanism of action studies, zebrafish embryos are also amenable to phenotypic screening approaches using high content and high-throughput imaging, which would enable large-scale compound screening as well as genetic modeling (Gehrig et al. 2018; Westhoff et al. 2020).

Overall, zebrafish translatability and feasibility to medium/high-throughput screening makes it an attractive model for target validation in multiple disease areas.

9.5 Phenotypic Screening Readouts

Emerging methods in phenotypic screening require advanced readout technologies. In order to observe the phenotypic response to biological perturbations in disease models, researchers need to build plate-based and multiparametric readouts that offer the throughput, resolution, and quick turnaround times in phenotypic screening. In the context of array screening, several cellular technologies and readouts have been optimized in target identification and validation, including high-throughput imaging, cytokine release, viability assays, and transcriptomics.

One major readout in target identification and validation is high-content imaging, as this method has evolved to become an important part of high-throughput screening. This method enables to capture detailed cellular phenotypic modulation, including protein expression, cellular localization, motility, cell cycle, proliferation, and morphological changes. Imaging methods not only enable the assessment of phenotypic effects of screening libraries, but also provide valuable information around cytotoxicity and mechanism of action. Therefore, high-content imaging facilitates the interrogation of complex and multivariate parameters, which can also capture single cell level and spatial features in heterogeneous populations. A recent development in the imaging field is the use of "Cell Painting" assays for compound profiling, as this method allows the extraction of thousands of cellular features (Bray et al. 2016). The main limitation in immunofluorescence imaging is the need for specific and sensitive antibodies for accurate target protein detection. Therefore, extensive assay development is required prior to phenotypic screening with an imaging readout. On the other hand, the main advantage for this method is the ability to multiplex several antibodies, allowing for protein quantification of multiple targets, as well as measuring intensity, cell morphology, and texture features. During functional genomics screening campaigns, the generation of high volume and complex datasets also requires the use of sophisticated image analysis software and algorithms. Therefore, the development of robust computational tools to QC and analysis is even more critical in arrayed screening and is key to deliver meaningful screening outputs. Future technologies will enable the adaptation of imaging readouts to more physiologically relevant models, which face challenges associated with the distinction of multiple cell types in co-culture systems, and detection of targets at the core of 3D structures or infiltrating cells. Moreover, emerging advancements around assay miniaturization would benefit for the execution of large-scale screens with minimal use of reagents and cell material. In this context, collaborations with academic organizations and industry partners will be crucial to investigate novel techniques of 3D imaging as well as develop new miniaturized screening workflows.

An additional readout that is gaining traction in the phenotypic screening field is the combination of single-cell transcriptomics with imaging readouts. This method allows for spatial transcriptomic profiling within individual cells and tissues, building expression profiles and spatial distribution of RNA in the cell native environment (Xia et al. 2019). This approach includes multiplexed error-robust fluorescence in situ hybridization (MERFISH), achieving multiplex single-molecule FISH measurements. By increasing the throughput in array format, it is possible to simultaneously detect RNA transcript abundance of thousands of genes in single cells and with subcellular structure resolution. The ability to perform genome-wide RNA profiling with single cell granularity would enable the interrogation of new biology that has not been yet explored. Besides detecting coding RNA transcripts, this method is also able to detect noncoding RNA, which combined with noncoding RNA CRISPR screening, could represent a complete new screening modality for phenotypic screening. Nonetheless, further work is needed to adapt this technology to existing automation and reduce the demanding imaging time. Also, MERFISH is a biased method and still limited by the choice of a small subset of transcripts. While RNAseq would overcome this issue, this technology is not currently amenable to the high-throughput screening. However, emerging transcriptomics-based technologies of lower resolution are being explored as an alternative to RNAseq in target discovery, such as DRUGseq (Ye et al. 2018; Yang et al. 2020).

Another major readout in phenotypic screening is the assessment of cytokine release in high-throughput, which is also amenable to more complex models. Increasing demand for this method has been observed with the advent of the immuno-oncology and cell therapy. Over recent years, the sensitivity, affinity, and specificity of cytokine detection methods have greatly improved and has allowed detection of low-abundant secreted molecules in minimal cell culture and extracellular supernatant volumes. Cytokine release methods include flow cytometry, ELISA, and bead-based assay readouts. These methods have been also adapted to high-throughput screening, with the use of AlphaLisa Immunoassay kits and iQue Flow Cytometry systems. AlphaLisa methods allow for multiplex cytokine detection by using luminescence plate readers in 96- or 384-well plate format. The iQue system offers the advantage of single cell resolution by the use of fluorescently tagged antibodies and flow cytometry in 96-well plate format. Cytokine release assays coupled with phenotypic screening libraries have accelerated target discovery in the cell therapy field. Recent work has been done in cancer cells-T cells co-culture systems that rely not only on cell viability readouts, but also measure cytotoxic molecule release as a surrogate of T cell killing. Moreover, T cell exhaustion is one of the main limitations for efficacious T cell therapy, and being able to measure exhaustion markers in high-throughput will likely enable the discovery of targets whose modulation could reverse the exhaustion phenotype. One of the main limitations for cytokine release assays is the potential degradation of the target molecule in solution, as well as measuring extracellular (as opposed to intracellular) cytokine production. In addition, reagent costs might be a constraint when performing high-throughput screening, although this can be reduced by designing multiplexed assays and optimizing minimal assay volumes. Therefore, thorough optimization of assay parameters needs to be performed prior to start the phenotypic screen that employs cytokine release as a read out.

9.6 Bioinformatic Pipeline

Despite the clear advantages that functional genomic screens bring to the field, data interpretation can be one of the main bottlenecks. The large datasets generated from phenotypic screening provide valuable insights in novel biology, identification of new therapeutic targets, and potential resistance/MoA of existing compounds. However, data mining, curation, and interpretation can be one of the main bottlenecks in this process, and it is important to manage data outputs following the FAIR guiding principle: data are findable, accessible, interoperable, and reusable (Wilkinson et al. 2016). Collaborative workstreams between cell biology and bioinformatic teams are crucial to build internal and external pipelines to support the computational analysis and interpretation of large-scale and genome-wide phenotypic screening, in order to increase the efficiency in the drug discovery process. A recent study from industry-based scientists describes the sequential workflow to interpret large datasets originating from CRISPR arrayed screens (Guerriero et al. 2020). This computational platform integrates (i) the evaluation of gRNA library quality, (ii) processing of data outputs, analysis, and QC, (iii) hit identification and scoring, and (iv) data visualization and biological insights. Moreover, this pipeline includes deep learning approaches to extract more relevant information from imaging-based screening outputs, including cell segmentation from label-free digital phase contrast (DPC) images. This optimized computational workflow has been successfully established for imaging outputs from CRISPR-based screening libraries; however, parallel efforts should be done to expand this capability to (i) additional phenotypic screening modalities, such as secretome, (ii) different screening readouts, including cytokine release and spatial transcriptomics, and (iii) emerging disease areas and technology, such as cell therapy and more complex co-culture models.

References

Anzalone, A.V., Randolph, P.B., Davis, J.R. et al. (2019). 'Search-and-replace genome editing without double-strand breaks or donor DNA', *Nature*. *Nature Research* 576 (7785): 149–157. https://doi.org/10.1038/s41586-019-1711-4.

Billon, P., Bryant, E.E., Joseph, S.A. et al. (2017). CRISPR-mediated base editing enables efficient disruption of eukaryotic genes through induction of STOP codons. *Mol. Cell*. Cell Press 67 (6): 1068–1079. e4. doi: https://doi.org/10.1016/j.molcel.2017.08.008.

Bray, M.A., Singh, S., Han, H. et al. (2016). Cell painting, a high-content image-based assay for morphological profiling using multiplexed fluorescent dyes. *Nat. Protoc.* 11 (9) https://doi.org/10.1038/nprot.2016.105.

Chen, X., Kozhaya, L., Tastan, C. et al. (2018). Functional interrogation of primary human T cells via CRISPR genetic editing. *J. Immunol.* The American Association of Immunologists 201 (5): 1586–1598. https://doi.org/10.4049/jimmunol.1701616.

Cortez, J.T., Montauti, E., Shifrut, E. et al. (2020). CRISPR screen in regulatory T cells reveals ubiquitination modulators of Foxp3. *bioRxiv*. Cold Spring Harbor Laboratory: 2020. 02.26.966911. doi: https://doi.org/10.1101/2020.02.26.966911.

Ding, M., Tegel, H., Sivertsson, Å. et al. (2020). Secretome-based screening in target discovery. *SLAS DISCOVERY: Advancing the Science of Drug Discovery*. SLAS Discov: 247255522091711. https://doi.org/10.1177/2472555220917113.

Dong, M.B., Wang, G., Chow, R.D. et al. (2019). Systematic immunotherapy target discovery using genome-scale in vivo CRISPR screens in CD8 T cells. *Cell*. Elsevier Inc. 178 (5): 1189–1204. e23. doi: https://doi.org/10.1016/j.cell.2019.07.044.

Drost, J. and Clevers, H. (2018). Organoids in cancer research. *Nat. Rev. Cancer*. Nature Publishing Group: 407–418. https://doi.org/10.1038/s41568-018-0007-6.

Dufva, O., Koski, J., Maliniemi, P. et al. (2020). Integrated drug profiling and CRISPR screening identify essential pathways for CAR T-cell cytotoxicity. *Blood*. NLM (Medline) 135 (9): 597–609. https://doi.org/10.1182/blood.2019002121.

Gee, S., Nelson, N., Bornot, A. et al. (2020). Developing an arrayed CRISPR-Cas9 Co-culture screen for immuno-oncology target ID. *SLAS Discov*. SAGE Publications Inc. 25 (6): 581–590. https://doi.org/10.1177/2472555220916457.

Gehrig, J., Pandey, G., and Westhoff, J.H. (2018). Zebrafish as a model for drug screening in genetic kidney diseases. *Front. Pediatr.* Frontiers Media S.A.: 183. https://doi.org/10.3389/fped.2018.00183.

Guerriero, M.L., Corrigan, A., Bornot, A. et al. (2020). Delivering robust candidates to the drug pipeline through computational analysis of arrayed CRISPR screens. *SLAS DISCOVERY: Advancing the Science of Drug Discovery*. SAGE Publications: 247255522092113. https://doi.org/10.1177/2472555220921132.

Han, K., Pierce, S.E., Li, A. et al. (2020). CRISPR screens in cancer spheroids identify 3D growth-specific vulnerabilities. *Nature*. Nature Research 580 (7801): 136–141. https://doi.org/10.1038/s41586-020-2099-x.

Hendel, A., Bak, R.O., Clark, J.T. et al. (2015). Chemically modified guide RNAs enhance CRISPR-Cas genome editing in human primary cells'. *Nat. Biotechnol*. Nature Publishing Group 33 (9): 985–989. https://doi.org/10.1038/nbt.3290.

Hultquist, J.F., Hiatt, J., Schumann, K. et al. (2019). CRISPR–Cas9 genome engineering of primary CD4 + T cells for the interrogation of HIV–host factor interactions. *Nat. Protoc.* Nature Publishing Group 14 (1): 1–27. https://doi.org/10.1038/s41596-018-0069-7.

Jennbacken, K., Wågberg, F., Karlsson, U. et al. (2019). Phenotypic screen with the human secretome identifies FGF16 as inducing proliferation of iPSC-derived cardiac progenitor cells. *Int. J. Mol. Sci.* MDPI AG 20 (23): 6037. https://doi.org/10.3390/ijms20236037.

Kaelin, W.G. (2012). Use and abuse of RNAi to study mammalian gene function. *Science*. American Association for the Advancement of Science: 421–422. https://doi.org/10.1126/science.1225787.

Klann, T.S., Black, J.B., Chellappan, M. et al. (2017). CRISPR-Cas9 epigenome editing enables high-throughput screening for functional regulatory elements in the human genome. *Nat. Biotechnol*. Nature Publishing Group 35 (6): 561–568. https://doi.org/10.1038/nbt.3853.

Liu, X. S., Wu, H., Ji, X., Stelzer, Y., Wu, X., Czauderna, S., Shu, J., Dadon, D., Young, R.A., Jaenisch, R., (2016) 'Editing DNA methylation in the mammalian genome', *Cell*. Cell Press, 167(1), pp. 233–247.e17. doi: https://doi.org/10.1016/j.cell.2016.08.056.

Liu, Y., Cao, Z., Wang, Y. et al. (2018). Genome-wide screening for functional long noncoding RNAs in human cells by Cas9 targeting of splice sites. *Nat. Biotechnol.* Nature Publishing Group 36 (12): 1203–1210. https://doi.org/10.1038/nbt.4283.

Manguso, R.T., Pope, H.W., Zimmer, M.D. et al. (2017). In vivo CRISPR screening identifies Ptpn2 as a cancer immunotherapy target. *Nature*. Nature Publishing Group 547 (7664): 413–418. https://doi.org/10.1038/nature23270.

Maresca, M., Lin, V.G., Guo, N., and Yang, Y. (2013). Obligate ligation-gated recombination (ObLiGaRe): Custom-designed nuclease-mediated targeted integration through nonhomologous end joining. *Genome Res.* 23 (3): 539–546. https://doi.org/10.1101/gr.145441.112.

Di Modugno, F., Colosi, C., Trono, P. et al. (2019). 3D models in the new era of immune oncology: focus on T cells, CAF and ECM. *J. Exp. Clin. Cancer Res.* BioMed Central Ltd. doi: https://doi.org/10.1186/s13046-019-1086-2.

Morgan, P., Brown, D.G., Lennard, S. et al. (2018). Impact of a five-dimensional framework on R&D productivity at AstraZeneca. *Nat. Publ. Group*. Nature Publishing Group 17 (3): 167–181. https://doi.org/10.1038/nrd.2017.244.

O'Shea, P., Wildenhain, J., Leveridge, M. et al. (2020). A novel screening approach for the dissection of cellular regulatory networks of NF-κB using arrayed CRISPR gRNA libraries. *SLAS Discov*. SAGE Publications Inc. 25 (6): 618–633. https://doi.org/10.1177/2472555220926160.

Osswald, A., Hedrich, V., and Sommergruber, W. (2019). 3D-3 tumor models in drug discovery for analysis of immune cell infiltration. *Methods Mol. Biol.* Humana Press Inc.: 151–162. https://doi.org/10.1007/978-1-4939-9145-7_10.

Palano, G., Jansson, M., Backmark, A. et al. (2020). A high-content, in vitro cardiac fibrosis assay for high-throughput, phenotypic identification of compounds with anti-fibrotic activity. *J. Mol. Cell. Cardiol.* Academic Press 142: 105–117. https://doi.org/10.1016/j.yjmcc.2020.04.002.

Papanicolaou, N. and Bonetti, A. (2020). The new frontier of functional genomics: from chromatin architecture and noncoding RNAs to therapeutic targets. *SLAS Discov.* SAGE Publications Inc. 25 (6): 568–580. https://doi.org/10.1177/2472555220926158.

Parker, C.G., Galmozzi, A., Wang, Y. et al. (2017). Ligand and target discovery by fragment-based screening in human cells. *Cell* 168 (3): 527–541. e29. doi: https://doi.org/10.1016/j.cell.2016.12.029.

Rees, H.A. and Liu, D.R. (2018). Base editing: precision chemistry on the genome and transcriptome of living cells. *Nat. Rev. Genet.* Springer US 19 (12): 770–788. https://doi.org/10.1038/s41576-018-0059-1.

Ross-Thriepland, D., Bornot, A., Butler, L. et al. (2020). Arrayed CRISPR screening identifies novel targets that enhance the productive delivery of mRNA by MC3-based lipid nanoparticles. *SLAS discovery: advancing life sciences R & D.* SAGE Publications Inc.: 2472555220925770. https://doi.org/10.1177/2472555220925770.

Roth, T.L., Li, P.J., Blaeschke, F. et al. (2020). Pooled knockin targeting for genome engineering of cellular immunotherapies. *Cell* Elsevier BV. doi: https://doi.org/10.1016/j.cell.2020.03.039.

Rubbini, D., Cornet, C., Terriente, J., and Di Donato, V. (2020). CRISPR meets Zebrafish: accelerating the discovery of new therapeutic targets. *SLAS Discov.* SAGE Publications Inc.: 552–567. https://doi.org/10.1177/2472555220926920.

Schafer, S., Viswanathan, S., Widjaja, A.A. et al. (2017). IL-11 is a crucial determinant of cardiovascular fibrosis HHS Public Access. *Nature* 552 (7683): 110–115. https://doi.org/10.1038/nature24676.

Seki, A. and Rutz, S. (2018). Optimized RNP transfection for highly efficient CRISPR/Cas9-mediated gene knockout in primary T cells. *J. Exp. Med.* 215 (3): 985–997. https://doi.org/10.1084/jem.20171626.

Shifrut, E., Carnevale, J., Tobin, V. et al. (2018). Genome-wide CRISPR screens in primary human T cells reveal key regulators of immune function. *Cell.* Elsevier Inc. 175 (7): 1958–1971. e15. doi: https://doi.org/10.1016/j.cell.2018.10.024.

Shukla, A. and Huangfu, D. (2018). Decoding the noncoding genome via large-scale CRISPR screens. *Curr. Opin. Genet. Dev.* Elsevier Ltd: 70–76. https://doi.org/10.1016/j.gde.2018.06.001.

Stricker, S.H., Köferle, A., and Beck, S. (2016). From profiles to function in epigenomics. *Nat. Rev. Genet.* Nature Publishing Group: 51–66. https://doi.org/10.1038/nrg.2016.138.

Sun, K.H., Chang, Y., Reed, N.I., and Sheppard, D. (2016). α-smooth muscle actin is an inconsistent marker of fibroblasts responsible for force-dependent TGFβ activation or collagen production across multiple models of organ fibrosis. *Am. J. Physiol. Lung Cell. Mol. Physiol.* 310 (9): L824–L836. https://doi.org/10.1152/ajplung.00350.2015.

Turner, R.J., Golz, S., Wollnik, C. et al. (2020). A whole genome-wide arrayed CRISPR screen in primary organ fibroblasts to identify regulators of kidney fibrosis. *SLAS Discov.* SAGE Publications Inc. 25 (6): 591–604. https://doi.org/10.1177/2472555220915851.

Westhoff, J., Steenbergen, P., Thomas, L. et al. (2020). In vivo high-content screening in zebrafish for developmental nephrotoxicity of approved drugs. *bioRxiv* https://doi.org/10.1101/2020.04.21.052688.

Van Der Weyden, L., Arends, M.J., Campbell, A.D. et al. (2017). Genome-wide in vivo screen identifies novel host regulators of metastatic colonization. *Nature.* Nature Publishing Group 541 (7636): 233–236. https://doi.org/10.1038/nature20792.

Wilkinson, M.D., Dumontier, M., Aalbersberg, I.J. et al. (2016). Comment: The FAIR guiding principles for scientific data management and stewardship. *Sci. Data.* Nature Publishing Groups 3 (1): 1–9. https://doi.org/10.1038/sdata.2016.18.

Xia, C., Fan, J., Emanuel, G. et al. (2019). Spatial transcriptome profiling by MERFISH reveals subcellular RNA compartmentalization and cell cycle-dependent gene expression. *Proc. Natl. Acad. Sci.* National Academy of Sciences 116 (39): 19490–19499. https://doi.org/10.1073/PNAS.1912459116.

Yang, X., Kui, L., Tang, M. et al. (2020). High-throughput transcriptome profiling in drug and biomarker discovery. *Front. Genet.* Frontiers Media S.A.: 19. https://doi.org/10.3389/fgene.2020.00019.

Ye, C., Ho, D.J., Neri, M. et al. (2018). DRUG-seq for miniaturized high-throughput transcriptome profiling in drug discovery. *Nat. Commun.* Nature Publishing Group 9 (1): 1–9. https://doi.org/10.1038/s41467-018-06500-x.

10

Applications of CRISPRi and CRISPRa in Drug Discovery

Luke A. Gilbert

Departments of Urology and Cellular & Molecular Pharmacology, and UCSF Helen Diller Family Comprehensive Cancer Center, University of California San Francisco, San Francisco, CA, USA

10.1 Introduction

One of the grand challenges in human biology is to understand the function of each gene encoded by the human genome. One method for understanding the function of a gene is to experimentally perturb its activity and then to measure a phenotype of interest. A key requirement for approaching this grand challenge is that methods for perturbing gene function must be robust, precise, scalable, and multiplexable. Understanding gene function and dysregulation in disease is a key to understanding human health and enables drug discovery. This book chapter will discuss how CRISPR systems that enable us to control the expression of human genes can be used to understand gene function and to advance drug discovery efforts.

Broadly, drug discovery efforts are made possible by any experimental approach that identifies genes that drive disease. Drug discovery efforts also necessitate elucidating drug mechanism of action, identification of biomarkers associated with drug response, as well as strategies for drug repurposing or development of innovative combination therapies. It is also worth noting that although we conventionally think of drug discovery efforts in the context of druggable genes that can be targeted by small molecules or proteins and undruggable genes that cannot, this paradigm is outdated. New breakthroughs in small molecule drug design have rendered previously undruggable proteins, such as KRAS, druggable (McCormick 2020; Ostrem et al. 2013). Perhaps more importantly, CRISPR genome and epigenome editing have enabled us to consider a future in which any gene is in essence druggable through programmed genetic or epigenetic perturbations (Doudna and Charpentier 2014; Pickar-Oliver and Gersbach 2019; Wang et al. 2016). This book chapter will cover how CRISPR has enabled each of these facets of drug discovery and point to future directions enabled by CRISPR systems.

10.2 Retooling CRISPR to Repress Gene Expression in Human Cells

Repurposed CRISPR systems can robustly and precisely edit the human genome, epigenome, and transcriptome, and these technologies have greatly enabled our capacity to experimentally search for genes that modulate disease phenotypes. Programmable epigenetic editing is enabled by coupling nuclease inactive Cas9 (dCas9) or Cas12 (dCas12) directly to enzymes or to protein scaffold domains that recruit enzymes that edit the epigenetic code to activate or repress transcription (Pickar-Oliver and Gersbach 2019; Pulecio et al. 2017) (Figure 10.1). Epigenetic editing enables perturbation of protein coding genes, nonprotein coding genes, and also genomic regulatory elements such as enhancers. Together these tools, described in more detail below, have enabled us to turn human gene expression up and down to probe gene function in human disease.

Human gene expression can be turned off or down by a method known as CRISPR interference (CRISPRi) (Gilbert et al. 2013; Qi et al. 2013). The CRISPRi approach most commonly consists of dCas9 fused to a KRAB domain although alternate approaches have used additional silencing domains such as LSD1, SID, FOG1, EZH2, and MECP2 (Alerasool et al. 2020; Kearns et al. 2015; O'Geen et al. 2017, 2019; Pickar-Oliver and Gersbach 2019; Yeo et al. 2018). KRAB domains are thought to silence gene expression by first binding a second scaffold protein, KAP1, ultimately forming a protein complex with chromatin regulators such as CBX5/HP1α and SETDB1 (Lupo et al. 2013;

Genome Editing in Drug Discovery, First Edition. Edited by Marcello Maresca and Sumit Deswal.

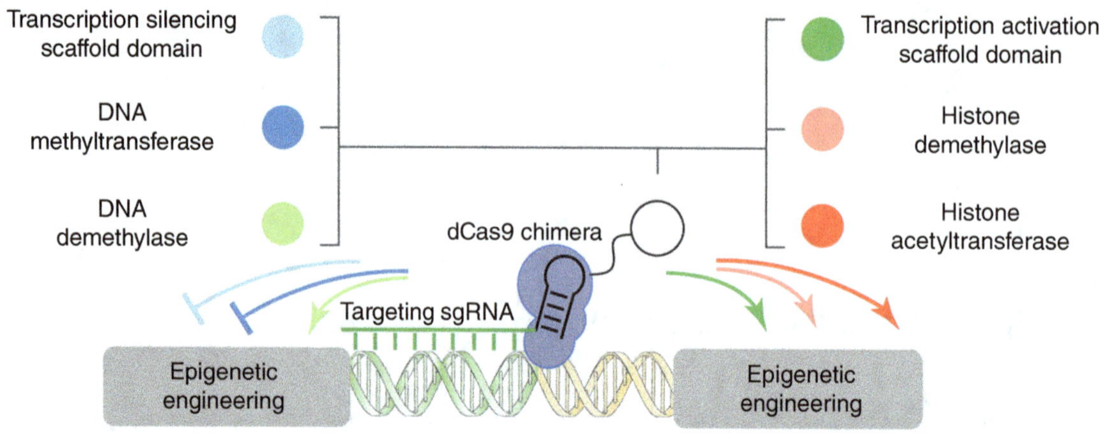

Figure 10.1 dCas9 is a programmable RNA-guided DNA-binding platform that can be used to recruit protein domains to specific genomic locations to edit the epigenome or control transcription in human cells. Our first-generation epigenome editors use protein scaffolds to control transcription or edit the epigenome. These protein scaffolds function by recruiting DNA and histone-modifying enzymes (CRISPRi/a) and transcription machinery (CRISPRa).

Zhu et al. 2020). KRAB domains are present naturally on human Zinc Finger transcription factors and are thought to be important for silencing repetitive elements encoded by the human genome. Fused to dCas9, KRAB domains function as short-range transcriptional repressors that act in part by locally inducing deposition of H3K9Me3 and loss of H3K27Ac (Gilbert et al. 2014; Liu et al. 2017; Yeo et al. 2018). CRISPRi repression of target genes can be highly specific without significant off-target activity (Gilbert et al. 2013; Liu et al. 2017).

Other epigenetic editing approaches have silenced human gene expression by editing DNA methylation (Liu et al. 2016; Pflueger et al. 2018; Stepper et al. 2017; Vojta et al. 2016). Epigenetic editing domains can also be recruited in combination, resulting in emergent properties such as epigenetic memory (Amabile et al. 2016; O'Geen et al. 2019). In one example, transient co-recruitment of KRAB, DNMT3A, and DNMT3L to a locus writes an epigenetic memory that represses gene expression heritably over many cell divisions in a manner maintained by the host cell (Amabile et al. 2016).

The CRISPRi dCas9-KRAB epigenetic editor has been used to repress gene expression in a wide variety of cell types including pluripotent stem cells and their differentiated derivatives such as neurons and cardiomyocytes (Liu et al. 2017; Mandegar et al. 2016; Tian et al. 2019). It is worth noting that unlike CRISPR/Cas9, CRISPRi epigenetic editing enables loss-of-function genetic experiments without induction of DNA double-stranded breaks (DSBs), a property which could be especially useful in stem cell applications (Ihry et al. 2018), as DSBs can result in TP53-dependent and -independent cell death. Together these CRISPR tools enable us to robustly and specifically turn off or down the expression of protein coding genes to identify how individual genes regulate biological processes and disease phenotypes.

Preclinical evidence has demonstrated AAV-mediated in vivo delivery of CRISPRi can modulate disease phenotypes (Thakore et al. 2018). In this example, a single dose of AAV encoding *Staphylococcus aureus* Cas9-based repressor (dSaCas9KRAB) resulted in long-term gene silencing of Pcsk9, reduced serum Pcsk9, and lower cholesterol levels. This establishes CRISPRi epigenetic editing as a potential therapeutic for long-term gene repression in post-mitotic tissues. Epigenetic editing clinical trials have also been proposed for the treatment of neurodegeneration (Zeitler et al. 2019).

Epigenetic editing can also perturb the function of non-coding genomic elements such as enhancers and insulators to modulate gene expression (Fulco et al. 2016; Gasperini et al. 2019; Hilton et al. 2015; Kearns et al. 2015; Li et al. 2020; Pickar-Oliver and Gersbach 2019; Stricker et al. 2017; Tarjan et al. 2019). An early example demonstrated that dCas9-mediated recruitment of KRAB or LSD1 to the promoter or distal and proximal enhancers of the Oct4 gene differentially modulates Oct4 gene expression in mouse embryonic stem cells (Kearns et al. 2015). This illustrates that epigenetic editing may enable us to move beyond correlative measurements of transcription and epigenetic marks and instead allow us to write and remove epigenetic marks on histones and DNA to elucidate their function directly (Stricker et al. 2017). This approach is relevant to drug discovery in that if we are able to identify developmental stage-specific, lineage-restricted enhancers, then we can target gene expression in specific cell types by disrupting enhancer sequences via genome editing. In a hallmark example of this approach, it has been demonstrated that disruption of an erythroid-specific enhancer of BCL11A expression results in decreased expression of BCL11A in erythroid cells and concomitant derepression of fetal hemoglobin expression which could alleviate β-hemoglobinopathy phenotypes (Bauer et al. 2013). Current clinical trials are testing this therapeutic genome engineering approach for the β-hemoglobinopathies (Zipkin 2019).

It is worth noting that epigenetic editor fusion protein design and activity can be unpredictable. Differential protein domain placement and peptide linker design could influence fusion protein folding, recruitment of binding partners, or enzyme accessibility to DNA or histones. For example, placement of different KRAB domains on the amino versus carboxy terminus of dCas9 greatly impacts CRISPRi activity (Alerasool et al. 2020). Any level of cleavage within the linker peptide will also impact epigenetic editing on- and off-target activity. New epigenetic editor proteins should be tested carefully and optimized as much as possible using positive control gene targets. Ideally, future advances in protein engineering will enable systematic evaluation of large numbers of protein variants or improved prediction of protein properties. These considerations will become ever more important as human therapeutic epigenetic editing enters the clinic.

10.3 Retooling CRISPR to Activate Gene Expression in Human Cells

Transcription of human genes can also be turned up or turned on using a method known as CRISPR activation (CRISPRa) (Cheng et al. 2013; Gilbert et al. 2013, 2014; Konermann et al. 2013; Maeder et al. 2013a; Mali et al. 2013; Pickar-Oliver and Gersbach 2019). CRISPRa most commonly consists of dCas9 fused to one or more transactivation domains. The most commonly used transactivation domains are derived from human and viral proteins including VP16, Rta, NFκB- p65, and HSF1 (Chavez et al. 2015, 2016; Konermann et al. 2015). These protein domains are thought to act as scaffolds recruiting basal Pol II transcription factors, the Mediator coactivator as well proteins such as CBP/p300 that act in a coordinated manner to activate transcription (Hall and Struhl 2002; Lee and Young 2013; Vojnic et al. 2011). A general theme of transactivation domains is that co-recruitment of multiple domains in a homo or hetero-typic manner results in an additive or synergistic activation of transcription (Chavez et al. 2015; Konermann et al. 2015; Maeder et al. 2013b; Tanenbaum et al. 2014). Transcription can also be activated by recruiting enzymes such as DNA demethylases or histone acetyltransferases (Choudhury et al. 2016; Halmai et al. 2020; Hilton et al. 2015; Josipović et al. 2019; Liu et al. 2016; Maeder et al. 2013c; Morita et al. 2016, 2020; Xu et al. 2016). CRISPRa has been applied in a variety of cell models, including in pluripotent stem cells, as well as in vivo settings (Alda-Catalinas et al. 2020; Braun et al. 2016; Liu et al. 2018b; Matharu et al. 2019; Wangensteen et al. 2018; Yang et al. 2019; Zhou et al. 2018). CRISPRa can also modulate the function of noncoding enhancer elements. In an early example, it was shown that recruitment of the p300 histone acetyltransferase domain to the HS2 hemoglobin enhancer element upregulates hemoglobin gene expression (Hilton et al. 2015).

Preclinical evidence suggests CRISPRa could be used to rescue diseases of haploinsufficiency by upregulating gene expression from the wild-type allele of a gene. This has been reduced to practice in a mouse model of obesity driven by Sim1 haploinsufficiency where upregulation of Sim1 by CRISPRa rescued obesity in vivo (Matharu et al. 2019). CRISPR DNA methylation editing was also used to activate expression of FMR1 and rescue Fragile X syndrome phenotypes (Liu et al. 2018a). Lastly, CRISPRa was used in a model of epilepsy to upregulate Kcna1 in mouse hippocampal excitatory neurons resulting in decreased seizures and rescue of cognitive deficits (Colasante et al. 2020). These studies demonstrate CRISPRa can be delivered via AAV into a tissue to modulate gene expression and disease phenotypes in vivo.

10.4 Multiplexed CRISPRi/a Genetic Perturbations

Beyond single gene perturbations, pairs or sets of genes can be turned up and down by CRISPRi/a in a multiplexed manner by expressing more than one sgRNA per cell targeting different genes (Adamson et al. 2016; Cheng et al. 2013; Du et al. 2017; Horlbeck et al. 2018; McCarty et al. 2020; Norman et al. 2019; Zhou et al. 2018). This should greatly enable drug target identification as, for example, we can model how inhibition or activation of multiple kinases influences a phenotype of interest. In one CRISPRa example, 10 genes were co-activated within the brain of a mouse (Zhou et al. 2018). This type of experiment should enable us to reveal how genes work together to control complex phenotypes as discussed at the end of this chapter. It has also been demonstrated that CRISPRi/a can be deployed to inactivate one gene and activate a second gene within the same cell (Boettcher et al. 2018; Zalatan et al. 2015). This enables one to probe how a flux of products in multi-output pathways is modulated, for example, in metabolism. Lastly, using both CRISPRi and CRISPRa, one can titrate the level of gene repression or activation by modulating sgRNA sequence or spacing from the transcription start site of a gene (Jost et al. 2020a). For CRISPRa, one can also express two or more sgRNAs targeting the same gene to additively or synergistically increase gene expression.

Multiplexed CRISPRi/a genetic perturbations may enable innovation in complex engineered cell therapies such as universal CAR-T. In one example, generation of optimized universal CAR-T cells required mutating three genes simultaneously which has been shown to result in DNA damage phenotypes and a high rate of chromosomal translocations (Bothmer et al. 2020; Ren et al. 2017). In principle, optimized universal CAR-T cells could be generated without DSB by multiplexed hit-and-run heritable epigenetic editing (Amabile et al. 2016).

10.5 CRISPRi/a Functional Genomics as a Discovery Modality

Large-scale CRISPR functional genomics experiments use massively parallel perturbation of gene function to systematically identify genes that modulate biological processes associated with disease phenotypes. Epigenetic editing is easily scalable and can be highly specific enabling genome-scale functional genomics experiments with a relatively low false-positive hit rate (Figure 10.2).

The first genome-scale loss-of-function CRISPRi functional genomics experiments targeted dCas9-KRAB to repress expression of protein coding genes (Gilbert et al. 2014). These experiments revealed expected common essential hits such as genes encoding protein components of the ribosome, spliceosome, and DNA replication machinery that are generally required for cell proliferation or survival. This first CRISPRi genome-scale screen also identified conditionally essential drivers of cell state such as the transcription factor *GATA1*. Additional experiments demonstrated CRISPRi screens can elucidate determinants of cellular response to protein toxins and small molecules (Gilbert et al. 2014; Jost et al. 2017; Lou et al. 2019; le Sage et al. 2017). CRISPRi functional genomics experiments have also been used to identify nonprotein coding genes such as long noncoding RNAs (lncRNAs) that modulate cell proliferation or survival (Liu et al. 2017, 2020a,b). In one example, a CRISPRi screen identified several hundred lncRNA genes that modulate stem cell self-renewal, proliferation, or survival in induced pluripotent stem cells (iPSCs) (Liu et al. 2017). Lastly, Perturb-seq pairs single-cell RNA-sequencing with CRISPRi-based screens connecting individual genetic perturbations to its transcriptional consequences. Perturb-seq has been used to map genes that regulate cellular response to small molecules revealing mechanisms of cancer cell death (Adamson et al. 2016; Replogle et al. 2020).

CRISPRi functional genomics experiments can also systematically map enhancer–promoter function across relatively large regions of the genome (Klann et al. 2017; Stricker et al. 2017). In a hallmark example, a custom CRISPRi sgRNA library was used to target >1 megabase region of the genome around the genes *MYC* and *GATA1* revealing 9 distal enhancers that modulate gene expression (Fulco et al. 2016). Because CRISPRi has been shown to be a short-range epigenetic editor that locally perturbs the epigenome, it is possible to design custom sgRNA libraries that perturb large swaths of the human genome (Tycko et al. 2019). Early sgRNA tiling screens suggested that CRISPRi acts to disrupt promoter function within approximately ~500–1000 nucleotides of an sgRNA binding site (Gilbert et al. 2014). If we take this to be true for the entire genome, then to functionally profile a 100 megabase nucleotide stretch of a human chromosome, it may be possible to pick 2–3 highly active and specific sgRNAs per 1000 nucleotides resulting in an sgRNA library of only 200–300 000 elements, which would represent a large but manageable

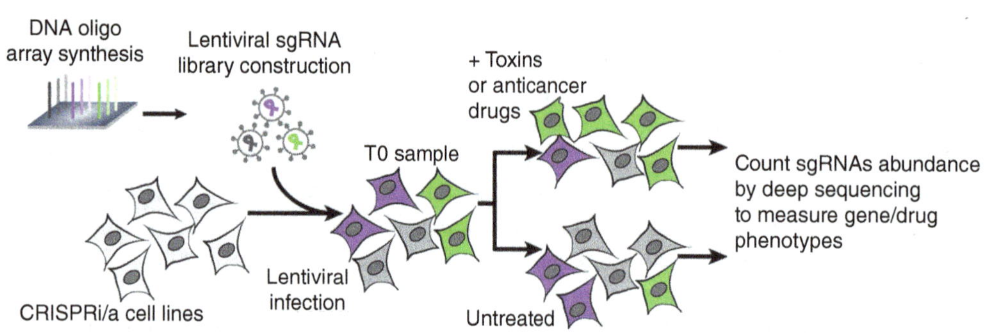

Figure 10.2 An approach for pooled CRISPRi/a screens in human cells. Custom CRISPRi/a libraries contain 1000–1 000 000 sgRNAs. Genetic interaction mapping libraries utilize a modified lentiviral construct that encodes two linked sgRNAs.

sgRNA library. Such a sparse sgRNA tiling library design incorporates a substantial number of unknowns and many sgRNA tiling libraries have instead chosen to pick all possible sgRNAs that satisfy off-target criteria. In contrast, to profile the genome at nucleotide resolution using Cas9 or a base editor would require a much larger sgRNA library and more problematically will require relaxation of sgRNA off-target activity scores which can induce artifactual results (Tycko et al. 2019). Future improvements in epigenetic editing may derive longer range epigenetic editors that could profile ever greater swaths of the human genome with relatively small sgRNA libraries to identify cell type-specific enhancers that regulate the expression of key disease genes.

The first genome-scale gain-of-function CRISPRa functional genomics experiments revealed that the top hits in a cell proliferation and survival screen are tumor suppressor genes, transcription factors that regulate cell state, and regulators of the cell cycle apparatus (Gilbert et al. 2014). Tumor suppressor genes should serve as expected positive control genes for CRISPRa screens performed in proliferating cells. Upregulated expression of tumor suppressor genes can block cell cycle progression or induce cell death resulting in depletion of sgRNAs targeting tumor suppressor genes within a CRISPRa functional genomics screen. Additional experiments demonstrated CRISPRa genome-scale screens can identify genes that regulate cellular response to protein toxins and anticancer drugs that are not identified by CRISPRi or CRISPR/Cas9 screens (Gilbert et al. 2014; Konermann et al. 2015). Large-scale CRISPRa functional genomics experiments have also identified lncRNAs that modulate cancer cell response to anticancer drugs (Bester et al. 2018; Joung et al. 2017). CRISPRa functional genomics approaches can also be used to identify functional regulatory elements (Klann et al. 2017; Tycko et al. 2019). CRISPRa perturb-seq experiments enable mapping of cell state transitions driven by transcription factors at single-cell resolution which may be useful for designing cell therapies or optimizing differentiation protocols (Norman et al. 2019).

Currently, multiple CRISPRi and CRISPRa genome-scale sgRNA libraries exist (Gilbert et al. 2014; Horlbeck et al. 2016; Konermann et al. 2015; Sanson et al. 2018). Newer CRISPRi/a sgRNA libraries that incorporate improved algorithms for predicting sgRNA activity generally identify more hits in an experiment and thus a more complete picture of the genes that modulate a phenotype. Some CRISPRi/a sgRNA libraries incorporate large sets of negative control sgRNAs which enables additional statistical analysis of screen data quality (Horlbeck et al. 2016). CRISPRi libraries have been informatically compared through precision recall analysis to CRISPR/Cas9 libraries and generally both approaches reveal the expected set of gold standard essential human genes required for proliferation and survival of human cells suggesting these sgRNA libraries are highly active and silence or mutate the majority of genes targeted (Horlbeck et al. 2016; Sanson et al. 2018). CRISPRi may reveal hits missed in CRISPR/Cas9 screens for genes that are essential for cell proliferation or viability through partial gene knockdown (le Sage et al. 2017). For CRISPRa, it is largely unknown what fraction of genes encoded by the genome can be activated with different types of epigenetic editors as we do not have a large set of expected positive control genes in most experiments. It is possible that future experiments will empirically validate the activity of all sgRNAs included in CRISPRi/a genome-scale libraries using newer functional genomics screening modalities such as perturb-seq that can directly define sgRNA activity by measuring target gene expression (Adamson et al. 2016; Norman et al. 2019; Replogle et al. 2020).

10.6 Identifying Gene Targets for the Treatment of Disease Using CRISPRi/a

CRISPRi/a functional genomics experiments can identify genes that modulate disease phenotypes which is the first step in many drug development efforts. In theory, these approaches could be applied to any disease; however, in practice many efforts are limited by not having an appropriate disease model for such studies. In one aspect, complex diseases associated with incompletely penetrant genetic variants identified by genome-wide association studies could be difficult to model via cellular models. For example, it is currently unclear how one would dissect the biology of human addiction, anxiety, or autism via functional genomics. In other aspects, disease-relevant cell types may not be easy to manipulate via CRISPRi/a in vitro or in vivo. Some diseases may require in vivo analysis. Despite this, there are many current and future questions in disease biology that can be answered by CRISPRi/a functional genomics experiments.

Perhaps the most important first choice in identifying gene targets for the treatment of disease is choosing the disease model that is most relevant to the question at hand. Can you identify genes that regulate a phenotype of interest in vitro in a cell culture model or does the biology require in vivo models? For in vitro applications, can you use a simple two dimensional cell culture model or do you need to use a more complicated three-dimensional cell model such as an organoid model? Should you use a mouse or a human cellular model? Can you use a human cell model implanted into a mouse? There are no right or wrong answers but rather this is where a deep understanding of the biology question you are asking is key to setting up a successful search for gene targets. For example, it may not be possible to perform a genome-scale screen in vivo in hematopoietic stem cells as this cell type is quite rare while it has been possible to perform genome-scale screens in developing skin epithelium (Beronja et al. 2013; Lu et al. 2011). Directed differentiation of human stem cells may enable functional genomics experiments in diverse cell types, such as neurons, or even in organoid models composed of multiple cell types (Tian et al. 2019).

Where possible, I always start to plan a CRISPRi/a functional genomics screen by first identifying a set of positive controls genes from the literature and then try to do small-scale experiments that demonstrate single gene perturbations result in a phenotype that is of interest and can be measured reproducibly. The exact dynamic range you need to implicate a gene as contributing to a phenotype depends on the assay being used for the screen (e.g. cell proliferation vs fluorescent reporter) as well as the reproducibility of the assay. Generally, screens that represent each sgRNA within the CRISPRi/a library with more cells (e.g. 500–1000 cells per sgRNA) generate higher quality data. Negative control sgRNA distributions can be helpful for calling hit genes. These general topics have been excellently reviewed recently (Doench 2018; Hanna and Doench 2020).

Perhaps the simplest phenotype in functional genomics is cell proliferation or survival. Proliferation and survival phenotypes identify genes of diverse functions into a broad catch-all assay that is easy to perform and usually fairly quantitative. For example, a proliferation or survival screen could identify the following types of genes which have very different functions: (i) transcription factors that regulate cell state such as terminal differentiation into functional cell types, (ii) common essential genes required for cell replication such as DNA polymerase in all human cells, and (iii) genes that modulate various forms of cell health such as autophagy and apoptosis. A proliferation and survival phenotype is appropriate for screens that seek genes required for survival, such as, for example, a synthetic lethal screen in KRAS-mutant cancer models, or that rescue a fitness defect, for example, genes that rescue neurons from mitochondrial dysfunction. For these two examples, one can imagine both CRISPRi and CRISPRa being informative and it is likely for loss- and gain-of-function approaches to reveal different biology. The key to using the results from any cell proliferation or survival screen is applying informatic filters to narrow in on genes that are selectively relevant to the biology that is being queried. For example, genome-scale CRISPRi screens performed in cancer cells reveal on average more than 1000 genes required for proliferation or survival (Horlbeck et al. 2016). The vast majority of these genes are not likely to be high-priority drug targets as most of these genes are not selectively essential for a specific cancer cell state but rather are generally essential for maintaining the viability of proliferating human cells (Hart et al. 2015). Robust strategies for comparative analysis of genomics and functional genomics data to prioritize drug targets are key challenges.

To prioritize new oncology drug targets, CRISPR/Cas9 cell viability screens have been performed in more than 700 cancer cell models (Behan et al. 2019; Meyers et al. 2017). This effort has become known as the Cancer Dependency Map (DepMap). This effort has revealed genes that are broadly required for human cell viability as well as genes that are selective regulators of survival or proliferation in a specific cancer cell state. The DepMap CRISPR functional genomics data can be integrated with the genomic, transcriptomic, and metabolomic data from the Cancer Cell Line Encyclopedia project or from cancer genomics efforts such as TCGA and together these resources represent an enormously useful dataset for a broad community (Ding et al. 2018; Ghandi et al. 2019; Li et al. 2019). These data enabled an oncology drug target score to prioritize new drug targets for cancer therapy (Behan et al. 2019). For example, by combining genomic and CRISPR functional genomic data, one can search for mutation-specific synthetic lethal genetic dependencies. A strong synthetic lethal relationship between DNA mismatch repair mutant cancer cells and the gene WRN was identified which has resulted in new drug discovery efforts (Behan et al. 2019; Chan et al. 2019). The DepMap drug prioritization score weights genes by a target tractability metric resulting in emphasis on drug targets that are currently thought to be targetable by small molecules and therapeutic extracellular proteins. Importantly, cross-platform analysis that integrates CRISPRi functional genomics data with DepMap data enables interrogation of cancer cell vulnerabilities associated with specific mutations suggesting this is a resource that should be considered and queried by any laboratory generating functional genomics data in bespoke cell models or assays (Lou et al. 2019). Looking forward, it is clear that we must move beyond identification of genes required for cancer cell viability in vitro to identify drug targets that are important in vivo (Braun et al. 2016; Chen et al. 2015; Dong et al. 2019; Jin et al. 2019; O'Loughlin and Gilbert 2019; Xue et al. 2014; Yin et al. 2014; Zhu et al. 2021).

Beyond cancer biology, the DepMap data has also enabled co-essentiality analysis of gene pairs across many cell types which can identify genetic interactions and protein complexes (Bayraktar et al. 2020; Wainberg et al. 2021; Wang et al. 2017). In an unlimited research environment, a CRISPRa DepMap effort would reveal key tumor suppressors and transcription factors that regulate cancer cell state. Alternately, a CRISPRi DepMap targeting noncoding genomic elements such as enhancer and insulators as well as lncRNAs would profile how the noncoding genome regulates cancer cell viability. Lastly, an effort to profile genetic dependencies across many types of primary human cells, organoids, or iPSC-derived cell types would be of broad interest and enable disease research beyond oncology.

10.7 Identification of Mechanisms of Response and Resistance to Drugs by CRISPRi and CRISPRa

CRISPRi/a genome-scale screens allow us to turn on or off all genes encoded by the human genome and ask how this modulates cellular response to small molecules (Jost and Weissman 2018; Kampmann 2017). This approach has been used to dissect mechanisms of response and resistance to anticancer therapies targeting microtubules, HSP70, protein translation, BRAF, and KRAS (Jost et al. 2017; Konermann et al. 2015; Liaud et al. 2019; Lou et al. 2019; le Sage et al. 2017; Shao et al. 2018). The results from CRISPRi/a drug screen inform our understanding of drug mechanism of action, identify potential biomarkers, and enable innovative combination therapies. For example, KRAS inhibitors could be combined with SHP2, CDK4/6, or EGFR inhibitors to drive deeper remissions or broaden the cohort of patients who respond to KRAS inhibitors (Lou et al. 2019).

Comparative analysis of CRISPRi to CRISPRa genome-scale cell proliferation screens revealed that loss- and gain-of-function experiments reveal nonoverlapping sets of genes that regulate cancer cell proliferation, survival, or cell state (Gilbert et al. 2014). This principle that loss- and gain-of-function screens combined yield a more complete view of a biological process was extended by a study that used genome-scale CRISPRi/a and CRISPR/Cas9 to identify genes that regulate response to vemurafenib, a BRAF inhibitor, in a melanoma cancer model (le Sage et al. 2017). Together these examples motivate the use of both CRISPRi and CRISPRa in drug discovery as gain- and loss-of function experiments together reveal the network of genes that modulate complex biological processes.

It has also been shown that together CRISPRi/a experiments can identify genes that have inverse cellular phenotypes upon activation or repression (Gilbert et al. 2014). For example, for certain genes it has been observed that decreasing gene expression promotes cellular sensitivity to a perturbation while increasing gene expression promotes resistance. Genes that have inverse phenotypes in loss- and gain-of-function genetic experiments are often the direct molecular target of a small molecules or protein toxins (Jost and Weissman 2018). This genetic principle has also been established by yeast loss- and gain-of-function functional genomics studies (Hoon et al. 2008). For example, combined CRISPRi/a genome-scale experiments profiling genes that regulate Cholera toxin biology demonstrated that together these data pinpointed the Cholera glycosphingolipid receptor molecule, Gm1a (Gilbert et al. 2014). CRISPRa turned up levels of Gm1a resulting in cellular sensitivity to Cholera toxin while CRISPRi turned down the levels of cellular Gm1a resulting in resistance. This same genetic principle has been utilized to understand the drug mechanism of action and to pinpoint the direct molecular target of small molecules. A combined CRISPRi/a approach was used to investigate, rigosertib, a poorly characterized drug with multiple proposed protein targets (Jost et al. 2017, 2020b) revealing that rigosertib is a microtubule-destabilizing agent. This hypothesis generated by CRISPRi/a genome-scale screens was then supported by a variety of cell biology, biochemical, and structural biology experiments culminating in structure-guided mutation in the rigosertib binding pocket on beta tubulin which rescued rigosertib toxicity.

10.8 CRISPRi/a Genetic Interaction Mapping for Drug Discovery

Epigenetic editing can also be multiplexed to perturb more than one gene per cell for combinatorial searches for targets that modulate disease biology. Genetic interaction maps measure the cellular consequences of perturbing hundreds of thousands of gene pairs to systematically identify synthetic and buffering genetic relationships (Costanzo et al. 2019; Domingo et al. 2019). In drug discovery, mapping genetic interactions can reveal synthetic lethal vulnerabilities in tumors or identify genes that act to suppress genetic disorders. CRISPRi and CRISPRa genetic interaction maps have enabled unbiased characterization of gene function by using each gene's unique signature of genetic interactions within an experiment to cluster genes by their function (Horlbeck et al. 2018; Norman et al. 2019). For example, a CRISPRi genetic interaction map genetically identified TMEM261, a poorly characterized gene, as required for the function of Electron Transport Chain Complex 1 (Horlbeck et al. 2018). CRISPRi maps reveal strong synthetic lethal genetic interactions that nominate new anticancer strategies (Du et al. 2017; Horlbeck et al. 2018). Practically, small-scale genetic interaction maps comprised of hit genes identified in primary CRISPRi/a screens should enable classification of hit gene mechanisms of action and this is likely to be the most widespread utility of this approach. Genetic interaction maps also reveal functional buffering between paralogs (Dede et al. 2020). Genetic interactions can also be mapped at single-cell resolution by perturb-seq which should enable dissection of heterogenous cellular responses to drugs or enable innovative therapeutic approaches to suppress drug resistance (Norman et al. 2019). Lastly, expression of two or more genes can be turned off and on within the same cell revealing directional genetic interactions allowing for a more complete understanding of gene regulatory networks in disease (Boettcher et al. 2018). These features of CRISPRi and CRISPRa genetic interaction maps enable us to systematically search for combinations of genes that are drug targets, drug response modifier genes, or biomarkers (Ramkumar and Kampmann 2018).

More broadly, a global genetic interaction map in yeast revealed principles of cellular organization and signaling (Costanzo et al. 2016). Problematically, human biology encompasses many more genes and also at least several hundred cell types making it a vast challenge to systematically map genetic interactions in human development, health, and disease. Pairwise genetic interaction maps of n genes require libraries of size $n^2/2$. As such, it may be that a hybrid approach of iteratively measuring and predicting genetic interactions will enable systematic understanding of human genetic interactions (Norman et al. 2019). In this study, a combination of genetic interaction map measurements and single-cell RNA-sequencing was used to predict genetic interactions but additional approaches are likely to be useful. Beyond pairwise genetic interactions, we know that higher order genetic interactions regulate complex phenotypes. For example, a genetic interaction between Oct4, Sox2, c-Myc, and Klf4 reprograms fibroblasts to pluripotent stem cells (Takahashi and Yamanaka 2006). Trigenic genetic interactions have been mapped at large scale in yeast but measuring all trigenic or higher order genetic interactions is unlikely in yeast or human cells, thus necessitating prediction approaches (Kuzmin et al. 2018).

10.9 Conclusion

The ability to robustly and precisely manipulate gene expression by editing the epigenome enables fundamental and translational research which should drive drug discovery applications. CRISPRi/a functional genomics approaches identify new drug targets, biomarkers, and combination therapies. Epigenetic editing also promises to be a new therapeutic approach for treatment of disease via engineered cells or in vivo therapies.

References

Adamson, B., Norman, T.M., Jost, M. et al. (2016). A multiplexed single-cell CRISPR screening platform enables systematic dissection of the unfolded protein response. *Cell 167*: 1867–1882.e21.

Alda-Catalinas, C., Bredikhin, D., Hernando-Herraez, I. et al. (2020). A single-cell transcriptomics CRISPR-activation screen identifies epigenetic regulators of the zygotic genome activation program. *Cell Syst.* 11: 25–41.e9.

Alerasool, N., Segal, D., Lee, H., and Taipale, M. (2020). An efficient KRAB domain for CRISPRi applications in human cells. *Nat. Methods 17*: 1093–1096.

Amabile, A., Migliara, A., Capasso, P. et al. (2016). Inheritable silencing of endogenous genes by hit-and-run targeted epigenetic editing. *Cell 167*: 219–232.e14. DOI: 10.1016/j.cell.2021.03.025

Bauer, D.E., Kamran, S.C., Lessard, S. et al. (2013). An erythroid enhancer of BCL11A subject to genetic variation determines fetal hemoglobin level. *Science 342*: 253–257.

Bayraktar, E.C., La, K., Karpman, K. et al. (2020). Metabolic coessentiality mapping identifies C12orf49 as a regulator of SREBP processing and cholesterol metabolism. *Nat. Metab. 2*: 487–498.

Behan, F.M., Iorio, F., Picco, G. et al. (2019). Prioritization of cancer therapeutic targets using CRISPR–Cas9 screens. *Nature 568*: 511–516.

Beronja, S., Janki, P., Heller, E. et al. (2013). RNAi screens in mice identify physiological regulators of oncogenic growth. *Nature 501*: 185–190.

Bester, A.C., Lee, J.D., Chavez, A. et al. (2018). An integrated genome-wide CRISPRa approach to functionalize lncRNAs in drug resistance. *Cell 173*: 649–664.e20.

Boettcher, M., Tian, R., Blau, J.A. et al. (2018). Dual gene activation and knockout screen reveals directional dependencies in genetic networks. *Nat. Biotechnol. 36*: 170–178.

Bothmer, A., Gareau, K.W., Abdulkerim, H.S. et al. (2020). Detection and modulation of DNA translocations during multi-gene genome editing in T cells. *CRISPR J.* 3: 177–187.

Braun, C.J., Bruno, P.M., Horlbeck, M.A. et al. (2016). Versatile in vivo regulation of tumor phenotypes by dCas9-mediated transcriptional perturbation. *Proc. Natl. Acad. Sci. U. S. A. 113*: E3892–E3900.

Chan, E.M., Shibue, T., McFarland, J.M. et al. (2019). WRN helicase is a synthetic lethal target in microsatellite unstable cancers. *Nature 568*: 551–556.

Chavez, A., Scheiman, J., Vora, S. et al. (2015). Highly-efficient Cas9-mediated transcriptional programming. *Nat. Methods 12*: 326–328.

Chavez, A., Tuttle, M., Pruitt, B.W. et al. (2016). Comparison of Cas9 activators in multiple species. *Nat. Methods 13*: 563–567.

Chen, S., Sanjana, N.E., Zheng, K. et al. (2015). Genome-wide CRISPR screen in a mouse model of tumor growth and metastasis. *Cell 160*: 1246–1260.

Cheng, A.W., Wang, H., Yang, H. et al. (2013). Multiplexed activation of endogenous genes by CRISPR-on, an RNA-guided transcriptional activator system. *Cell Res. 23*: 1163–1171.

Choudhury, S.R., Cui, Y., Lubecka, K. et al. (2016). CRISPR-dCas9 mediated TET1 targeting for selective DNA demethylation at BRCA1 promoter. *Oncotarget 7*: 46545–46556.

Colasante, G., Qiu, Y., Massimino, L. et al. (2020). In vivo CRISPRa decreases seizures and rescues cognitive deficits in a rodent model of epilepsy. *Brain 143*: 891–905.

Costanzo, M., VanderSluis, B., Koch, E.N. et al. (2016). A global genetic interaction network maps a wiring diagram of cellular function. *Science 353*: aaf1420-1–aaf1420-14.

Costanzo, M., Kuzmin, E., van Leeuwen, J. et al. (2019). Global genetic networks and the genotype-to-phenotype relationship. *Cell 177*: 85–100.

Dede, M., McLaughlin, M., Kim, E., and Hart, T. (2020). Multiplex enCas12a screens detect functional buffering among paralogs otherwise masked in monogenic Cas9 knockout screens. *Genome Biol. 21*: 262.

Ding, L., Bailey, M.H., Porta-Pardo, E. et al. (2018). Perspective on oncogenic processes at the end of the beginning of cancer genomics. *Cell 173*: 305–320.c10.

Doench, J.G. (2018). Am I ready for CRISPR? A user's guide to genetic screens. *Nat. Rev. Genet. 19*: 67–80.

Domingo, J., Baeza-Centurion, P., and Lehner, B. (2019). The causes and consequences of genetic interactions (epistasis). *Annu. Rev. Genomics Hum. Genet. 20*: 433–460.

Dong, M.B., Wang, G., Chow, R.D. et al. (2019). Systematic immunotherapy target discovery using genome-scale in vivo CRISPR screens in CD8 T cells. *Cell 178*: 1189–1204.e23.

Doudna, J.A. and Charpentier, E. (2014). Genome editing. The new frontier of genome engineering with CRISPR-Cas9. *Science 346*: 1258096.

Du, D., Roguev, A., Gordon, D.E. et al. (2017). Genetic interaction mapping in mammalian cells using CRISPR interference. *Nat. Methods 14*: 577–580.

Fulco, C.P., Munschauer, M., Anyoha, R. et al. (2016). Systematic mapping of functional enhancer–promoter connections with CRISPR interference. *Science 354*: 769–773.

Gasperini, M., Hill, A.J., McFaline-Figueroa, J.L. et al. (2019). A genome-wide framework for mapping gene regulation via cellular genetic screens. *Cell* 176: 377–390.e19.

Ghandi, M., Huang, F.W., Jané-Valbuena, J. et al. (2019). Next-generation characterization of the cancer cell line encyclopedia. *Nature 569*: 503–508.

Gilbert, L.A., Larson, M.H., Morsut, L. et al. (2013). CRISPR-mediated modular RNA-guided regulation of transcription in eukaryotes. *Cell 154*: 442–451.

Gilbert, L.A., Horlbeck, M.A., Adamson, B. et al. (2014). Genome-scale CRISPR-mediated control of gene repression and activation. *Cell 159*: 647–661.

Hall, D.B. and Struhl, K. (2002). The VP16 activation domain interacts with multiple transcriptional components as determined by protein-protein cross-linking In vivo. *J. Biol. Chem. 277*: 46043–46050.

Halmai, J.A.N.M., Deng, P., Gonzalez, C.E. et al. (2020). Artificial escape from XCI by DNA methylation editing of the CDKL5 gene. *Nucleic Acids Res. 48*: 2372–2387.

Hanna, R.E. and Doench, J.G. (2020). Design and analysis of CRISPR–Cas experiments. *Nat. Biotechnol. 38*: 813–823.

Hart, T., Chandrashekhar, M., Aregger, M. et al. (2015). High-resolution CRISPR screens reveal fitness genes and genotype-specific cancer liabilities. *Cell 163*: 1515–1526.

Hilton, I.B., D'Ippolito, A.M., Vockley, C.M. et al. (2015). Epigenome editing by a CRISPR-Cas9-based acetyltransferase activates genes from promoters and enhancers. *Nat. Biotechnol. 33*: 510–517.

Hoon, S., Smith, A.M., Wallace, I.M. et al. (2008). An integrated platform of genomic assays reveals small-molecule bioactivities. *Nat. Chem. Biol. 4*: 498–506.

Horlbeck, M.A., Gilbert, L.A., Villalta, J.E. et al. (2016). Compact and highly active next-generation libraries for CRISPR-mediated gene repression and activation. *ELife 5*.

Horlbeck, M.A., Xu, A., Wang, M. et al. (2018). Mapping the genetic landscape of human cells. *Cell* 174: 953–967.e22.

Ihry, R.J., Worringer, K.A., Salick, M.R. et al. (2018). p53 inhibits CRISPR-Cas9 engineering in human pluripotent stem cells. *Nat. Med. 24*: 939–946.

Jin, X., Simmons, S.K., Guo, A.X. et al. (2019). In vivo perturb-seq reveals neuronal and glial abnormalities associated with autism risk genes. *BioRxiv*: 791525.

Josipović, G., Tadić, V., Klasić, M. et al. (2019). Antagonistic and synergistic epigenetic modulation using orthologous CRISPR/dCas9-based modular system. *Nucleic Acids Res. 47*: 9637–9657.

Jost, M. and Weissman, J.S. (2018). CRISPR approaches to small molecule target identification. *ACS Chem. Biol. 13*: 366–375.

Jost, M., Chen, Y., Gilbert, L.A. et al. (2017). Combined CRISPRi/a-based chemical genetic screens reveal that rigosertib is a microtubule-destabilizing agent. *Mol. Cell 68*: 210–223.e6.

Jost, M., Santos, D.A., Saunders, R.A. et al. (2020a). Titrating gene expression using libraries of systematically attenuated CRISPR guide RNAs. *Nat. Biotechnol. 38*: 355–364.

Jost, M., Chen, Y., Gilbert, L.A. et al. (2020b). Pharmaceutical-grade rigosertib is a microtubule-destabilizing agent. *Mol. Cell* 79: 191–198.e3.

Joung, J., Engreitz, J.M., Konermann, S. et al. (2017). Genome-scale activation screen identifies a lncRNA locus regulating a gene neighbourhood. *Nature 548*: 343–346.

Kampmann, M. (2017). Elucidating drug targets and mechanisms of action by genetic screens in mammalian cells. *Chem. Commun. Camb. Engl. 53*: 7162–7167.

Kearns, N.A., Pham, H., Tabak, B. et al. (2015). Functional annotation of native enhancers with a Cas9-histone demethylase fusion. *Nat. Methods 12*: 401–403.

Klann, T.S., Black, J.B., Chellappan, M. et al. (2017). CRISPR-Cas9 epigenome editing enables high-throughput screening for functional regulatory elements in the human genome. *Nat. Biotechnol. 35*: 561–568.

Konermann, S., Brigham, M.D., Trevino, A. et al. (2013). Optical control of mammalian endogenous transcription and epigenetic states. *Nature 500*: 472–476.

Konermann, S., Brigham, M.D., Trevino, A.E. et al. (2015). Genome-scale transcriptional activation by an engineered CRISPR-Cas9 complex. *Nature 517*: 583–588.

Kuzmin, E., VanderSluis, B., Wang, W. et al. (2018). Systematic analysis of complex genetic interactions. *Science 360*: 1–9.

Lee, T.I. and Young, R.A. (2013). Transcriptional regulation and its misregulation in disease. *Cell 152*: 1237–1251.

Li, H., Ning, S., Ghandi, M. et al. (2019). The landscape of cancer cell line metabolism. *Nat. Med. 25*: 850–860.

Li, K., Liu, Y., Cao, H. et al. (2020). Interrogation of enhancer function by enhancer-targeting CRISPR epigenetic editing. *Nat. Commun. 11*: 485.

Liaud, N., Horlbeck, M.A., Gilbert, L.A. et al. (2019). Cellular response to small molecules that selectively stall protein synthesis by the ribosome. *PLoS Genet. 15*: e1008057.

Liu, X.S., Wu, H., Ji, X. et al. (2016). Editing DNA methylation in the mammalian genome. *Cell 167*: 233–247.e17.

Liu, S.J., Horlbeck, M.A., Cho, S.W. et al. (2017). CRISPRi-based genome-scale identification of functional long noncoding RNA loci in human cells. *Science 355*: 1–9.

Liu, X.S., Wu, H., Krzisch, M. et al. (2018a). Rescue of fragile X syndrome neurons by DNA methylation editing of the FMR1 gene. *Cell 172*: 979–992.e6.

Liu, Y., Yu, C., Daley, T.P. et al. (2018b). CRISPR activation screens systematically identify factors that drive neuronal fate and reprogramming. *Cell Stem Cell 23*: 758–771.e8.

Liu, S., Harmston, N., Glaser, T.L. et al. (2020a). Wnt-regulated lncRNA discovery enhanced by in vivo identification and CRISPRi functional validation. *Genome Med. 12*: 89.

Liu, S.J., Malatesta, M., Lien, B.V. et al. (2020b). CRISPRi-based radiation modifier screen identifies long non-coding RNA therapeutic targets in glioma. *Genome Biol. 21*: 1–18.

Lou, K., Steri, V., Ge, A.Y. et al. (2019). KRASG12C inhibition produces a driver-limited state revealing collateral dependencies. *Sci. Signal. 12*: 1–15.

Lu, R., Neff, N.F., Quake, S.R., and Weissman, I.L. (2011). Tracking single hematopoietic stem cells in vivo using high-throughput sequencing in conjunction with viral genetic barcoding. *Nat. Biotechnol. 29*: 928–933.

Lupo, A., Cesaro, E., Montano, G. et al. (2013). KRAB-zinc finger proteins: a repressor family displaying multiple biological functions. *Curr. Genomics 14*: 268–278.

Maeder, M.L., Linder, S.J., Cascio, V.M. et al. (2013a). CRISPR RNA-guided activation of endogenous human genes. *Nat. Methods 10*: 977–979.

Maeder, M.L., Linder, S.J., Reyon, D. et al. (2013b). Robust, synergistic regulation of human gene expression using TALE activators. *Nat. Methods 10*: 243–245.

Maeder, M.L., Angstman, J.F., Richardson, M.E. et al. (2013c). Targeted DNA demethylation and activation of endogenous genes using programmable TALE-TET1 fusion proteins. *Nat. Biotechnol. 31*: 1137–1142.

Mali, P., Aach, J., Stranges, P.B. et al. (2013). CAS9 transcriptional activators for target specificity screening and paired nickases for cooperative genome engineering. *Nat. Biotechnol. 31*: 833–838.

Mandegar, M.A., Huebsch, N., Frolov, E.B. et al. (2016). CRISPR interference efficiently induces specific and reversible gene silencing in human iPSCs. *Cell Stem Cell 18*: 541–553.

Matharu, N., Rattanasopha, S., Tamura, S. et al. (2019). CRISPR-mediated activation of a promoter or enhancer rescues obesity caused by haploinsufficiency. *Science 363*: eaau0629.

McCarty, N.S., Graham, A.E., Studená, L., and Ledesma-Amaro, R. (2020). Multiplexed CRISPR technologies for gene editing and transcriptional regulation. *Nat. Commun. 11*: 1281.

McCormick, F. (2020). Sticking it to KRAS: covalent inhibitors enter the clinic. *Cancer Cell 37*: 3–4.

Meyers, R.M., Bryan, J.G., McFarland, J.M. et al. (2017). Computational correction of copy number effect improves specificity of CRISPR-Cas9 essentiality screens in cancer cells. *Nat. Genet. 49*: 1779–1784.

Morita, S., Noguchi, H., Horii, T. et al. (2016). Targeted DNA demethylation in vivo using dCas9-peptide repeat and scFv-TET1 catalytic domain fusions. *Nat. Biotechnol. 34*: 1060–1065.

Morita, S., Horii, T., Kimura, M., and Hatada, I. (2020). Synergistic upregulation of target genes by TET1 and VP64 in the dCas9-SunTag platform. *Int. J. Mol. Sci. 21*.

Norman, T.M., Horlbeck, M.A., Replogle, J.M. et al. (2019). Exploring genetic interaction manifolds constructed from rich single-cell phenotypes. *Science* 365: 786–793.

O'Geen, H., Ren, C., Nicolet, C.M. et al. (2017). dCas9-based epigenome editing suggests acquisition of histone methylation is not sufficient for target gene repression. *Nucleic Acids Res. 45*: 9901–9916.

O'Geen, H., Bates, S.L., Carter, S.S. et al. (2019). Ezh2-dCas9 and KRAB-dCas9 enable engineering of epigenetic memory in a context-dependent manner. *Epigenetics Chromatin 12*: 26. DOI: 10.1016/j.cell.2021.03.025

O'Loughlin, T.A. and Gilbert, L.A. (2019). Functional genomics for cancer research: applications in vivo and in vitro. *Annu. Rev. Cancer Biol. 3*: 345–363.

Ostrem, J.M., Peters, U., Sos, M.L. et al. (2013). K-Ras(G12C) inhibitors allosterically control GTP affinity and effector interactions. *Nature 503*: 548–551.

Pflueger, C., Tan, D., Swain, T. et al. (2018). A modular dCas9-SunTag DNMT3A epigenome editing system overcomes pervasive off-target activity of direct fusion dCas9-DNMT3A constructs. *Genome Res. 28*: 1193–1206.

Pickar-Oliver, A. and Gersbach, C.A. (2019). The next generation of CRISPR–Cas technologies and applications. *Nat. Rev. Mol. Cell Biol. 20*: 490–507.

Pulecio, J., Verma, N., Mejía-Ramírez, E. et al. (2017). CRISPR/Cas9-based engineering of the epigenome. *Cell Stem Cell 21*: 431–447.

Qi, L.S., Larson, M.H., Gilbert, L.A. et al. (2013). Repurposing CRISPR as an RNA-guided platform for sequence-specific control of gene expression. *Cell 152*: 1173–1183.

Ramkumar, P. and Kampmann, M. (2018). CRISPR-based genetic interaction maps inform therapeutic strategies in cancer. *Transl. Cancer Res. 7*: S61–S67.

Ren, J., Liu, X., Fang, C. et al. (2017). Multiplex genome editing to generate universal CAR T cells resistant to PD1 inhibition. *Clin. Cancer Res. 23*: 2255–2266.

Replogle, J.M., Norman, T.M., Xu, A. et al. (2020). Combinatorial single-cell CRISPR screens by direct guide RNA capture and targeted sequencing. *Nat. Biotechnol. 38*: 954–961.

le Sage, C., Lawo, S., Panicker, P. et al. (2017). Dual direction CRISPR transcriptional regulation screening uncovers gene networks driving drug resistance. *Sci. Rep. 7*: 17693.

Sanson, K.R., Hanna, R.E., Hegde, M. et al. (2018). Optimized libraries for CRISPR-Cas9 genetic screens with multiple modalities. *Nat. Commun. 9*: 5416.

Shao, H., Li, X., Moses, M.A. et al. (2018). Exploration of benzothiazole rhodacyanines as allosteric inhibitors of protein–protein interactions with Heat Shock Protein 70 (Hsp70). *J. Med. Chem. 61*: 6163–6177.

Stepper, P., Kungulovski, G., Jurkowska, R.Z. et al. (2017). Efficient targeted DNA methylation with chimeric dCas9–Dnmt3a–Dnmt3L methyltransferase. *Nucleic Acids Res. 45*: 1703–1713.

Stricker, S.H., Köferle, A., and Beck, S. (2017). From profiles to function in epigenomics. *Nat. Rev. Genet. 18*: 51–66.

Takahashi, K. and Yamanaka, S. (2006). Induction of pluripotent stem cells from mouse embryonic and adult fibroblast cultures by defined factors. *Cell 126*: 663–676.

Tanenbaum, M.E., Gilbert, L.A., Qi, L.S. et al. (2014). A protein-tagging system for signal amplification in gene expression and fluorescence imaging. *Cell 159*: 635–646.

Tarjan, D.R., Flavahan, W.A., and Bernstein, B.E. (2019). Epigenome editing strategies for the functional annotation of CTCF insulators. *Nat. Commun. 10*: 4258.

Thakore, P.I., Kwon, J.B., Nelson, C.E. et al. (2018). RNA-guided transcriptional silencing in vivo with S. aureus CRISPR-Cas9 repressors. *Nat. Commun. 9*: 1674.

Tian, R., Gachechiladze, M.A., Ludwig, C.H. et al. (2019). CRISPR interference-based platform for multimodal genetic screens in human iPSC-derived neurons. *Neuron 104*: 239–255.e12.

Tycko, J., Wainberg, M., Marinov, G.K. et al. (2019). Mitigation of off-target toxicity in CRISPR-Cas9 screens for essential non-coding elements. *Nat. Commun. 10*: 4063.

Vojnic, E., Mourão, A., Seizl, M. et al. (2011). Structure and VP16 binding of the mediator Med25 activator interaction domain. *Nat. Struct. Mol. Biol. 18*: 404–409.

Vojta, A., Dobrinić, P., Tadić, V. et al. (2016). Repurposing the CRISPR-Cas9 system for targeted DNA methylation. *Nucleic Acids Res. 44*: 5615–5628.

Wainberg, M., Kamber, R.A., Balsubramani, A. et al. (2021). A genome-wide almanac of co-essential modules assigns function to uncharacterized genes. *Nat. Genet.* 53 (5): 638–649.

Wang, H., La Russa, M., and Qi, L.S. (2016). CRISPR/Cas9 in genome editing and beyond. *Annu. Rev. Biochem. 85*: 227–264.

Wang, T., Yu, H., Hughes, N.W. et al. (2017). Gene essentiality profiling reveals gene networks and synthetic lethal interactions with oncogenic ras. *Cell 168*: 890–903.e15.

Wangensteen, K.J., Wang, Y.J., Dou, Z. et al. (2018). Combinatorial genetics in liver repopulation and carcinogenesis with a in vivo CRISPR activation platform. *Hepatology 68*: 663–676.

Xu, X., Tao, Y., Gao, X. et al. (2016). A CRISPR-based approach for targeted DNA demethylation. *Cell Discov. 2*: 1–12.

Xue, W., Chen, S., Yin, H. et al. (2014). CRISPR-mediated direct mutation of cancer genes in the mouse liver. *Nature 514*: 380–384.

Yang, J., Rajan, S.S., Friedrich, M.J. et al. (2019). Genome-scale CRISPRa screen identifies novel factors for cellular reprogramming. *Stem Cell Rep. 12*: 757–771.

Yeo, N.C., Chavez, A., Lance-Byrne, A. et al. (2018). An enhanced CRISPR repressor for targeted mammalian gene regulation. *Nat. Methods 15*: 611–616.

Yin, H., Xue, W., Chen, S. et al. (2014). Genome editing with Cas9 in adult mice corrects a disease mutation and phenotype. *Nat. Biotechnol. 32*: 551–553.

Zalatan, J.G., Lee, M.E., Almeida, R. et al. (2015). Engineering complex synthetic transcriptional programs with CRISPR RNA scaffolds. *Cell 160*: 339–350.

Zeitler, B., Froelich, S., Marlen, K. et al. (2019). Allele-selective transcriptional repression of mutant HTT for the treatment of Huntington's disease. *Nat. Med. 25*: 1131–1142.

Zhou, H., Liu, J., Zhou, C. et al. (2018). In vivo simultaneous transcriptional activation of multiple genes in the brain using CRISPR-dCas9-activator transgenic mice. *Nat. Neurosci. 21*: 440–446.

Zhu, Y., Sun, D., Jakovcevski, M., and Jiang, Y. (2020). Epigenetic mechanism of SETDB1 in brain: implications for neuropsychiatric disorders. *Transl. Psychiatry 10*: 1–8.

Zhu, X.G., Chudnovskiy, A., Baudrier, L. et al. (2021). Functional genomics in vivo reveal metabolic dependencies of pancreatic cancer cells. *Cell Metab.* 33 (1): 211–221.

Zipkin, M. (2019). CRISPR's "magnificent moment" in the clinic. *Nat. Biotechnol.* https://doi.org/10.1038/d41587-019-00035-2.

11

Sequence Diversification Screens with CRISPR-Cas9-Guided Base Editors

Jenna Persson and Bernhard Schmierer

CRISPR Functional Genomics, Karolinska Institutet and SciLifeLab, Stockholm, Sweden

11.1 Introduction

Perturbing a system to understand its function underpins modern biology. We owe much of our understanding of genes and genomes to mutant phenotypes, both natural and induced. These efforts continue today, with siRNA and CRISPR technology allowing genome-scale functional genomics studies. Many questions cannot be answered by perturbation of entire genes or regulatory elements, however. Understanding which precise amino acid residues are involved in protein–protein or protein–small molecule interactions, for example, would require the interrogation of thousands of sequence variants, as does directed evolution. Mapping which specific sequence motifs are of regulatory importance for genes of interest requires similar sequence diversification. There is considerable demand from the pharmaceutical industry for new methods for mapping the interactions between drug targets and bioactive molecules, both to better understand mechanisms of action and drug specificity and to predict resistance mutations. New CRISPR-based treatments may also be able to alter gene expression by editing regulatory sequences if these sequences can be mapped well enough. Generating these variants is not as straightforward as creating loss-of-function mutations, and has been the focus of considerable recent method development.

An ideal method for sequence diversification should be specific, endogenous, and affordable. Early forward genetic screens relied on random chemical mutagenesis or irradiation (e.g. Lewis and Bacher 1968, Nüsslein-Volhard and Wieschaus 1980; Taylor and Trotter 1967). Obtaining saturated diversification via random mutagenesis requires prohibitively large mutant libraries, however, as few mutations will be in the gene of interest. More targeted approaches are thus desirable. Mutagenesis specific to the element of interest has been achieved *in vitro* by error-prone PCR (Arnold 1993; Leung et al. 1989), by gene shuffling (Stemmer 1994), or by gene synthesis, followed by overexpression in cells. Variants frequently can only be expressed in less-relevant model systems, however, or in the system of interest but at nonphysiological levels. This can lead to the formation of erroneous protein complexes and results that do not translate to the physiological context in the system of interest. In the specific case of cis-regulatory elements, the endogenous chromatin context is essential to maintain functional long-range interactions, and meaningful results are difficult to obtain in exogenous model systems. The level of control provided by gene synthesis allows for semi-rational design and reduced screen scale, but is still very expensive. A relatively new approach uses the precise genomic targeting afforded by CRISPR-Cas systems to recruit mutagenizing DNA base deaminases to genes of interest (Anzalone et al. 2020; Rees and Liu 2018). Two 2016 studies, from the Bassik and Chang labs, have applied these base editor tools to high-throughput screening (Hess et al. 2016; Ma et al. 2016). While the method has yet to be widely adopted, base-editor screening has the potential to offer specific, endogenous, and rapid sequence diversification at an affordable cost.

This chapter explains the principles of genetic screening using CRISPR by first focusing on conventional CRISPR screens that interrogate gene function at the level of the

Genome Editing in Drug Discovery, First Edition. Edited by Marcello Maresca and Sumit Deswal.

entire transcriptional unit – either using wild-type Cas9, or nuclease dead Cas9 (dCas9) fusions to transcriptional inhibitors or activators. We then give a short overview of the rapidly developing base-editing field, and finally review the recent developments that have harnessed base editing for sequence diversification in a high-throughput screening setting.

11.2 CRISPR as a Genetic Screening Method

For most of the twentieth century, forward genetic screens to identify genes responsible for particular phenotypes were limited to genetically tractable model organisms such as yeast, *Drosophila melanogaster*, and *Caenorhabditis elegans*. The discovery of RNA interference (RNAi) around the turn of the century finally allowed genetic screens to be performed in mammalian cells. Together with the completion of the human genome sequence, RNAi revolutionized our understanding of gene function and provided important insights into the wiring of genetic networks. In the drug discovery field, RNAi quickly became essential as a high-throughput screening tool for target identification and deconvolution, and as a tool for target validation.

With the advent of CRISPR a decade later, a more versatile option for genetic screening has become available. Many obstacles from proof-of-principle to practical application of CRISPR in genetic screens had already been overcome during the heyday of RNAi, and it took less than two years from the first-ever applications of CRISPR in human cells to the first reports of high-throughput, genome-wide CRISPR screens (Koike-Yusa et al. 2014; Shalem et al. 2014; Wang et al. 2014; Zhou et al. 2014; Figure 11.1). Pooled screening technology was then further improved by the inclusion of unique molecular identifiers, which allow more precise hit calling (Michlits et al. 2017; Schmierer et al. 2017).

Like RNAi screens, CRISPR screens can be performed in one of two formats: arrayed or pooled. Irrespective of the format, a high-throughput CRISPR screen requires three key components: a method to deliver Cas9 together with one specific guide into cells to create the desired perturbations at scale; a method to read out the phenotype; and a method to assign the phenotype back to the perturbation. All three components differ between the arrayed and the pooled format (Table 11.1).

The obvious advantages of arrayed screens are the flexibility and complexity of the phenotypic readout; their equally obvious disadvantages are high cost and the need for automation. Pooled screens by contrast are relatively cheap and do not require either miniaturization or automation, but are limited to physically separable phenotypes. This disadvantage of pooled screens can often be overcome by creative assay design or prior creation of suitable reporter cell lines. Recently, the strict distinction between pooled and arrayed screens has been increasingly blurred by emerging

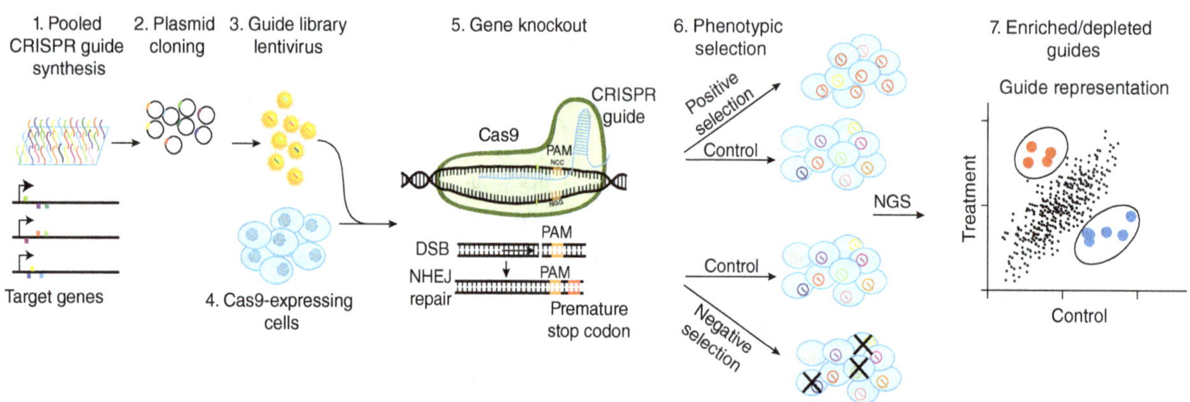

Figure 11.1 General principle of a pooled CRISPR screen. The example shown is for CRISPR knock-out screens. Guide sequences are synthesized as DNA oligos on commercially available oligo arrays (Step 1). Commonly, at least four guides per gene are included, and a genome-wide human or mouse library thus contains around 80 000 guides. The oligos are cloned in pool into a lentiviral plasmid downstream of an RNA Pol III responsive promoter (most commonly U6, Step 2). The resulting, highly complex plasmid library is then sequenced for quality control, ascertaining full library coverage. The plasmid library containing the guide sequences is packaged into lentiviral particles (Step 3), and transduced into a cell line stably expressing Cas9 (Step 4). This is done at a low multiplicity of infection (MOI = 0.3), and untransduced cells are killed off by antibiotic selection. Low MOI ensures that the vast majority of cells obtain one and only one guide. Editing occurs (Step 5), and a pool of cells is created in which each cell has a specific gene targeted – which gene this is, is flagged by the stable integrated guide sequence each cell harbors. Importantly, in a pooled screen, isolation of phenotypes of interest is required (Step 6). This can either be live/dead screening, or sorting of phenotypes by flow cytometry. An amplicon containing the guide expression cassette is then subjected to next-generation sequencing, and guides that are significantly enriched or depleted in the population of interest are identified (Step 7).

Table 11.1 Arrayed versus pooled CRISPR screens.

	Arrayed CRISPR screen	Pooled CRISPR screen
Format	Microtiter plates. High throughput requires miniaturization and automation.	Large-scale cell culture. No automation required.
Delivery method	Any delivery method (plasmids, ribonucleoproteins, virus, RNA, . . .).	Viral delivery – cells need a stably integrated marker flagging the perturbation (normally the guide sequence).
Phenotyping method	Virtually any assay is possible. No hard limits on the complexity of a phenotype that can be read out.	Requires physical separation of phenotypes into subpopulations: live/dead, competition, phenotypes that are sortable by flow cytometry.
Phenotype–perturbation assignment	Each microtiter well contains one perturbation. Phenotype–perturbation assignment is trivial.	Each cell contains one perturbation. A) Subpopulations are analyzed by NGS: Which guide sequences are enriched/depleted in the subpopulations? B) Single cells are analyzed for phenotype and perturbation by either microscopy/in situ guide sequencing, or by single-cell transcriptomics / guide-RNA sequencing.

single-cell technologies, which combine the advantages of pooled perturbation and arrayed readout (Datlinger et al. 2017; Dixit et al. 2016; Feldman et al., 2019; Jaitin et al. 2016; Replogle et al. 2020).

In contrast to RNAi, which has only a single mode of action (i.e. the destabilization of RNA) and is applicable only in loss-of-function screens, CRISPR can be flexibly engineered to perturb gene expression, and indeed genetic information in general, in more than just a single way.

The CRISPR-Cas system is covered extensively elsewhere in this volume. In brief, a short RNA guides a Cas endonuclease to bind and cleave homologous nucleic acids, either RNA or DNA depending on the system. Originally evolved as a form of bacterial immunity targeting invading phages (Mojica and Rodriguez-Valera 2016), CRISPR systems are now widespread laboratory tools that allow genome engineering with unprecedented speed and precision. The *Streptococcus pyogenes* Cas9 nuclease was among the first to be characterized (Jinek et al. 2012) and continues to dominate laboratory use due to its amenability as a tool. *Sp*Cas9 targets DNA, which it cleaves efficiently in all studied exogenous systems. It is also able to function without additional proteins, a prerequisite for ease of use. Finally, *Sp*Cas9 has a short and common (5'-NGG-3') *protospacer adjacent motif* or PAM, a sequence bordering the target site without which most Cas proteins cannot recognize their target. Together, the Cas9 nuclease and the guide RNA introduce a double-strand break at the target site, which is then repaired by the cell's DNA repair pathways. If a repair template is provided, there is a chance that the template sequence will replace the endogenous sequence, allowing for precise genome editing. In the majority of cases, however, repair is untemplated and results in short insertion/deletion mutations.

11.3 Conventional Genetic Loss- and Gain-of-Function Screens Using CRISPR

11.3.1 CRISPRko Screens

For loss-of-function screens, wild-type *Sp*Cas9 is most commonly used. In this approach (CRISPRko), the Cas9 enzyme is targeted by suitable guide sequences to exons of genes of interest. Resulting DNA double-strand breaks are repaired by non-homologous end joining (NHEJ), which is error-prone, and several cycles of cut and repair quickly lead to small insertions or deletions. These so-called "indel" mutations cause a frameshift mutation and loss of expression of a functional protein in the majority of cases.

11.3.2 CRISPRi and CRISPRa Screens

It is also possible to inactivate both of Cas9's nuclease domains with single point mutations, resulting in a targetable platform to deliver additional effector proteins to genomic loci of interest. This strategy is used in a different type of loss-of-function screen, CRISPR inhibition (CRISPRi) screens (Qi et al. 2013). Nuclease dCas9 is fused to an inhibitory domain such as a KRAB (Bellefroid et al. 1991) or SID domain (Ayer et al. 1996), and this fusion protein is expressed together with guide sequences targeting the transcription start sites of genes of interest. Guiding the dCas9-fusion protein to the transcription start site of a gene prevents gene expression through recruitment of corepressors, such as TIF1-beta (TRIM28) to a dCas9-fused KRAB domain, or the mSin3 corepressor complex to a dCas9-fused SID domain.

Both loss-of-function approaches, CRISPRko and CRISPRi, have their own strengths and weaknesses. DNA double-strand breaks introduced by wild-type Cas9 can evoke a DNA-damage response (Haapaniemi et al. 2018; Ihry et al. 2018), which exerts an additional, guide-sequence-independent selection pressure. Complete disruption of protein expression, while commonly observed, is not guaranteed by the introduction of small indels: In-frame repair can maintain protein function in full or in part. Similarly, other compensation mechanisms such as internal translation initiation or exon skipping during splicing can lead to expression of a partially functional, or even hyperfunctional, protein (Smits et al. 2019). Finally, gene knock-out by nonsense mutation is restricted to protein coding genes, and does not allow interrogation of noncoding RNAs. CRISPRi with dCas9 by contrast always acts through knock-down of genes rather than complete knock-out, and is thus more similar to RNAi-based approaches. Because CRISPRi is based on transcriptional repression, it is suitable for the interrogation of both protein-coding and noncoding genes. However, transcription start site positions need to be accurately defined to determine sgRNA position, making successful CRISPRi experiments dependent on cap-analysis gene expression (CAGE) data – currently the most reliable method of transcription start site determination.

Gain-of-function screens are an important complement to loss-of-function approaches, particularly in mammalian systems where functional redundancy is common. Gain-of-function manipulation of endogenous genes can be achieved with CRISPR activation (CRISPRa), which uses a similar strategy to the CRISPRi approach to achieve the opposite result. In CRISPRa, transcriptional activators such as a VP64 domain are recruited via a dCas9-guide complex to the transcription start sites of genes of interest, which can strongly induce expression of the targeted gene. Several methods have been developed, which differ mainly in the exact means of coactivator recruitment. How strongly target gene expression can be activated is difficult to predict and depends on factors including the coactivators used, the basal expression levels (repressed or active state) of the targeted gene, and cell type, and is difficult to predict. As with CRISPRi, CRISPRa can also be used to perturb expression of nonprotein coding genes.

11.4 Sequence Diversification Screens Using CRISPR Base Editing

11.4.1 A Short History of Base Editors

In spite of increasing excitement about the potential of CRISPR-based treatments to cure genetic diseases, the CRISPR-Cas9 system based on wild-type, active Cas9 nuclease is much better at gene knock-out than precision repair of mutations. While treatments based on gene knock-out for cancer and monogenic blood disorders are under development, the number of human genetic diseases that can be cured by gene deletion is limited. In fact, the majority of pathogenic genetic variants are small point mutations (Rees and Liu 2018 and references therein). The efficiency of correcting such mutations by homologous recombination is typically below the threshold of clinical effect, necessitating a new approach. Base editing was first published by the Liu lab (Komor et al. 2016). Here, the CRISPR-Cas9 system serves as a platform to direct DNA base deaminases to a small target region of the genome, achieving high rates of conversion from one base to another without accompanying indel mutations.

In its first incarnation, Base Editor 1 (BE1) consisted of a nuclease dCas9 protein fused to the rat APOBEC1 cytidine deaminase. Deamination converts cytosine (C) to uracil (U), which is read as thymine (T) during DNA replication or repair, leading to $C \rightarrow T$ mutations. Cytidine deaminases target RNA or single-stranded DNA, and are integral to cellular processes from immunity to metabolism. While capable of editing 40–50% of target cytosines *in vitro,* BE1 efficiency plummeted *in vivo*: active uracil DNA glycosylase present in mammalian cells removes the created uracil and triggers base excision repair (BER), reverting the mutation to wild-type in all but a few percent of cases. To block uracil DNA glycosylase activity, the Liu lab added a uracil glycosylase inhibitor (UGI) from the PBS1 *Bacillus subtilis* bacteriophage to create BE2. They then restored DNA nicking activity at one of the Cas9 nuclease domains (HNH, H840 restored) such that the non-edited strand would be nicked. This results in the edited strand being used as the template, raising editing efficiency to 60–70% *in vivo* (BE3), though also elevating indel mutation frequency.

In parallel, the Kondo lab (Nishida et al. 2016) developed the TARGET-AID system for sequence diversification. Rather than an APOBEC family deaminase, this system uses the sea lamprey protein PmCDA1, a homolog of the human activation-induced cytidine deaminase AID. Nishida et al. use Cas9-CDA1 fusions to mutagenize endogenous budding yeast genes, achieving 40–50% base conversion with Cas9 nickase fusions. They reach similar levels of conversion in mammalian cells with the additional fusion of a UGI to their editor, resulting in a similar tool to that described in Komor et al. (2016). The activity windows of the two editors are slightly different, and so could be used to target different bases. Interestingly, PmCDA1 is most active at cooler temperatures, and Nishida et al. follow transfection with a 25 °C incubation. This base editor's reduced activity under standard cell culture conditions could be used to limit off-target editing.

In a major breakthrough, 2017 saw the addition of adenine base editors (ABE) to the toolbox (Gaudelli et al. 2017). Following the same principle as cytosine base editors (CBE), ABEs use nickase Cas9 to direct a modified TadA protein to target adenine bases. Because there are no known DNA adenine deaminases, Komor et al. subjected an RNA adenine deaminase, the *Escherichia coli* TadA protein, to repeated rounds of directed evolution until it was able to efficiently edit DNA in mammalian cells. The adenines are deaminated to inosine, which is interpreted by the cell as guanine, resulting in A → G mutations. This is of particular interest because 48% of pathogenic human SNPs can be potentially corrected by an A → G transition (Gaudelli et al. 2017).

Codon optimization, improved nuclear localization signals, and ancestral sequence reconstruction led to increased editing efficiency for both CBEs and ABEs (Koblan et al. 2018; Zafra et al. 2018), and many groups have worked to improve the base editors' versatility by reducing stringency of PAM requirements (Kim et al. 2017) or changing the editing window (Huang et al. 2019), incorporating different Cas proteins (Kleinstiver et al. 2019; Richter et al. 2020), and reducing off-target and RNA editing (Grünewald et al. 2019). Most recently, the Liu and Doudna labs collaborated to address a weakness of the ABEs: even with PAM requirements relaxed to 5'-NG-3', the majority of disease-associated C to T transitions are inaccessible to ABE activity. In addition, ABEs proved relatively inefficient, making them dependent on the long residency time of *Sp*Cas9 and precluding ABE use with other Cas proteins, a strategy possible for CBEs. Several additional rounds of PACE (Esvelt et al. 2011) evolution resulted in ABE8e, which is several hundred-fold more efficient than previously published ABEs and is consequently able to edit during the shorter residencies of, for instance, Cas12a variants (Richter et al. 2020).

All of the above base editor tools principally cause transition mutations, either C to T or A to G. The most recent addition to the base editor tool kit, a C-to-G base editor (CGBE), breaks this pattern (Kurt et al. 2020). Where other base editor strategies seek to control the outcome of editing by inhibiting uracil *N*-glycosylase (UNG) activity and blocking BER, CGBE exploits this pathway. Composed of the BE4max editor (Komor et al. 2017) with uracil glycosylase inhibitors replaced by an *E. coli* UNG fusion, CGBE was able to convert C to G with up to 63% efficiency. Typical conversion rates were lower, however, and generated a minor additional C to A transversion product. Although control of outcome is weaker than with CBE and ABE tools, CGBE will useful for generating mutations previously accessible only by homologous recombination.

The obvious utility of base editing for the introduction of precise mutations has resulted in widespread adaptation to fit diverse applications. Many of these developments were driven by the need to edit inaccessible bases, resulting in base editor fusions with alternative Cas proteins or retooling of the original Cas9 fusions (Thuronyi et al. 2019). The Joung lab made Cpf1/Cas12a CBEs (Kleinstiver et al. 2019), which are easier to target to AT-rich regions due to their PAM requirements. The Bock lab introduced more rigid linker sequences of varying length to the CBE fusions, resulting in more control of the editing window. They further increased precision by replacing the APOBEC cytidine deaminase favored by the Liu lab with truncations of the sea lamprey AID homolog CDA1 (Tan et al. 2019). Base editors have been adapted to a wide range of systems, from important crop species (Wu et al. 2020) to human organoids (Geurts et al. 2020), and are now reaching the clinic, where the stakes for precision editing are the highest.

11.4.2 Base-Editing Sequence Diversification Screens – CRISPR-X and CRISPR-TAM

CRISPRko, CRISPRi, and CRISPRa are powerful screening tools, but all act at the level of the entire gene, and are thus limited to interrogation of entire genes. In drug discovery, these methods can be applied for target identification but are not suitable for studying drug–target interactions (DTI). Particularly in the case of essential genes, drug resistance typically develops through the acquisition of missense mutations rather than wholesale loss of a gene product. Furthermore, once a drug target has been identified, the ability to focus further investigation on smaller sequence motifs, such as functional domains and active sites, is crucial to learn more about the drug's mode of binding and mechanism of action.

Base editing has been optimized commonly for precision and specificity, with the aim of correcting disease-causing mutations in therapeutic applications. However, the approach can also be used to create localized sequence diversity by designing the base-editing complex such that it can modify a broader window of nucleotides around the guide-binding position (Figure 11.2, upper panel). Combined with a "tiling" design, where a region of interest is densely covered with guide sequences, this approach is an effective means of sequence diversification in a targeted and localized manner. The process is somewhat reminiscent of naturally occurring somatic hypermutation during antibody maturation. In 2016, two papers introduced this approach for sequence diversification, CRISPR-X (Hess et al. 2016) and CRISPR-TAM (Ma et al. 2016). These controlled, traceable, and

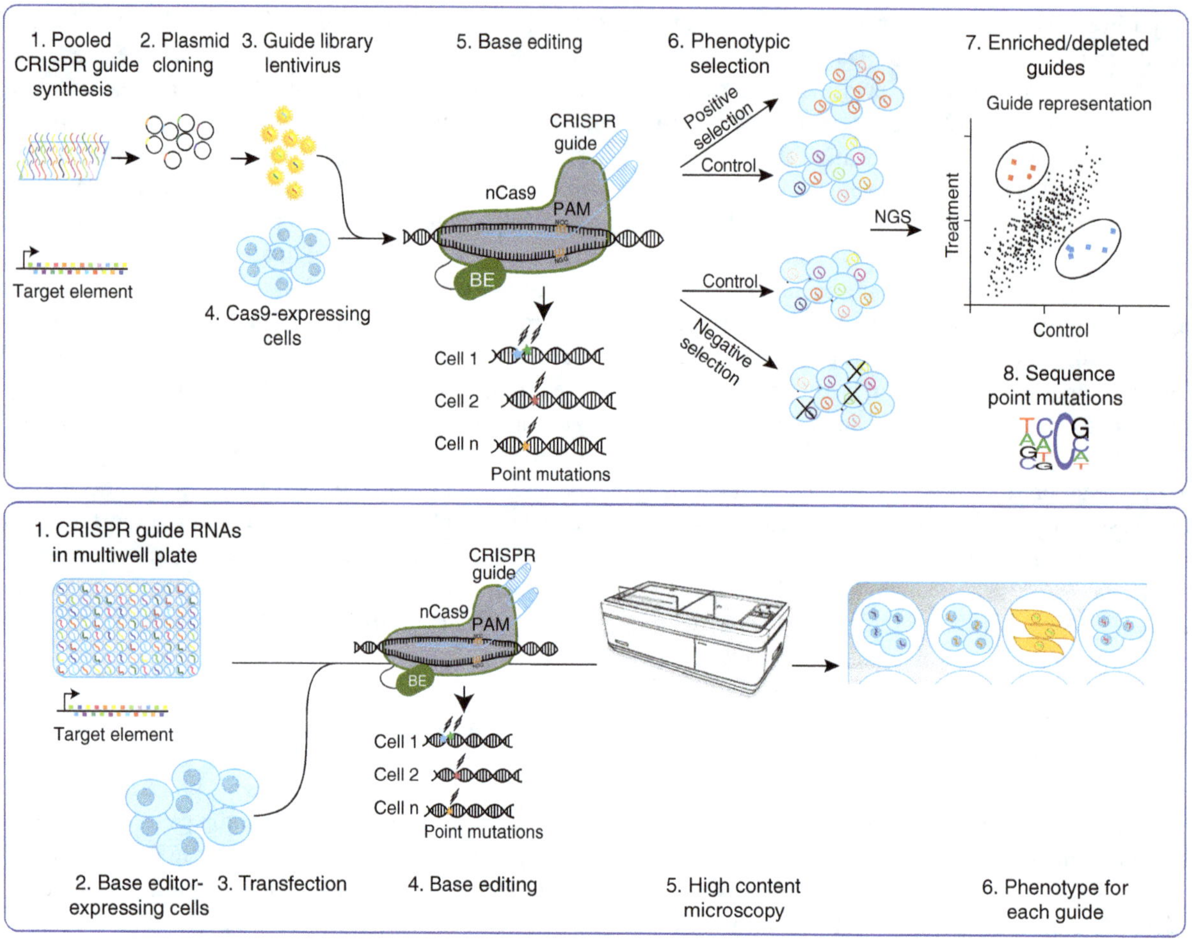

Figure 11.2 **Top. Workflow of a pooled sequence diversification screen using base editors.** Guide sequences densely tiling one or several regions of interest are synthesized as DNA oligos, cloned into a lentiviral plasmid, and transduced into cells stably expressing Cas9 fused to a base editor (BE). Just as in a conventional pooled screen, library transduction is performed at a low multiplicity of infection, so the vast majority of cells obtain a single guide. Base editing happens in close vicinity to the target sequence of the sgRNA, and cells obtain point mutations in the region specified by the guide sequence they contain. The phenotype of interest is isolated (for instance, cells that survive treatment with a cytotoxic drug), and compared with an untreated control by assessing the guide frequencies in both populations. Guides that are enriched in the surviving population are likely to have mediated creation of a resistance mutation. The guide sequence itself informs on the exact region where this mutation happened, which can then be amplified and amplicon sequenced to determine the exact nucleotide substitution responsible for the resistance. **Bottom. Workflow of an arrayed sequence diversification screen using base editors.** In an arrayed format, base-editor expressing cells are transfected with guide RNAs in a multiwell format. After base editing has happened, high-content microscopy or other phenotypic readouts can be performed to evaluate the effects of each guide. Amplicon sequencing of interesting phenotypes is then used to identify the precise nucleotide substitutions causing the phenotype.

cost-effective methods to explore mutational space have significant advantages over prior approaches such as random mutagenesis or DNA synthesis.

Although the principle of both methods is the same, there are some differences in the design. The CRISPR-TAM system initially fused human AID to the Cas9 C-terminus. This editor was able to correct a premature stop codon in a GFP, albeit at rather low frequencies. The system was then improved by truncating AID to remove its C-terminus, which contains a nuclear export signal. Efficiency of the correction of the premature stop codon could be doubled, and overall a mutation frequency of more than 20% was observed across the targeted GFP locus. The base edits introduced yielded all possible transitions and transversions from cytosine and guanine, with the expected C→T and G→A transitions being slightly favored. UGI co-expression limited the observed base substitutions to C→T transitions, with the preferred position of editing in a narrow region -12 to -16 bases upstream of the PAM. The CRISPR-X system recruits a hyperactive version of AID containing the three point mutations K10E, T82I, and E156G. As in the

CRISPR-TAM sequence, the nuclear export signal of AID was removed. In this approach, however, the hyperactive AID was not fused directly to Cas9, but rather to an MS2 protein and expressed separately. MS2-binding aptamers included in the tracrRNA recruit this MS2-AID fusion to a dCas9-guide complex, an approach pioneered for CRISPRa in Feng Zhang's lab (Konermann et al. 2015). As with CRISPR-TAM, CRISPR-X yielded all three possible transitions and transversion from the original C or G, with only a slight bias toward C→T. Mutations occurred in a +12 to +32 bp window downstream of the PAM when using C-terminally truncated AID. Introduction of the abovementioned three point mutations increased efficiency, and extended the window to −50 to +50 with respect to the PAM. These findings demonstrate that both targeting window and efficiency are tunable, even though optimization into a tool with predictable behavior may be challenging. In summary, these methods allow introduction of point mutations with a low rate of indels, an important feature for diversification of protein coding sequences, where indel formation would lead to premature stop codons and loss of full-length proteins.

Like any other kind of CRISPR-based screen, base editing screens can be conducted in arrayed format as well as in pool (Figure 11.2, lower panel). With each guide confined to an individual cell culture well, it is possible to link complex phenotypes to a given guide. Arrayed screening requires advanced robotics and high content imaging capacity, however.

11.4.3 Recent Developments in Base-Editing Sequence Diversification

Several recent publications build on the CRISPR-X and CRISPR-TAM proofs of principle. Després et al. used Target-AID to mutate 17 000 sites in 1500 essential budding yeast genes (Després et al. 2020). They found hundreds of sites with significant fitness contribution, particularly at conserved and structurally important residues. This clearly shows the potential of base-editor screening to explore protein structure and map functional amino acids. Seeking to adapt base-editor screening to more high-throughput applications than those seen in the original CRISPR-X and CRISPR-TAM publications, Després et al. avoid guide RNAs with multiple cytosines in the peak editing window of Target-AID. They then rely on computational models to predict the editing outcomes of significantly depleted guides, which they argue eliminates the need to sequence mutagenized sites. It remains to be seen whether the streamlining and cost effectiveness of this approach will be considered worth the tradeoff in specific mutagenesis profile. A compromise would be to first determine active guides based on abundance in the post-competition cell pool, and then only sequence the genomic regions targeted by a few interesting guides. It would also be possible to determine which specific mutations are functional in the essential validation step that follows any pooled screen.

11.4.4 Single-Cell Transcriptomic Read Out of Sequence Diversification Screens

Screening with base editors was also recently coupled to a single-cell transcriptomic readout. The A375 melanoma cell line harbors the BRAF V600E mutation, rendering it sensitive to MAPK pathway inhibition by e.g. vemurafenib or selumetinib. This allows positive selection for mutations that cause resistance to either vemurafenib or selumetinib via a genomic pooled CRISPR screen (Doench et al. 2016). Jun et al. made use of the A375/vemurafenib system, tiling the coding regions of three genes associated with vemurafenib resistance with guides (Jun et al. 2020). In this study, the BE3 editor (APOBEC1-nCas9-UGI; Komor et al. 2016) was used to generate point mutations, and the transcriptomes of single vemurafenib-resistant cells were then sequenced using the CROP-seq method (Datlinger et al. 2017). Approximately 10% base substitution was observed in the absence of selective pressure after 25 days in culture. Vemurafenib-treated cells showed enrichment of guides targeting a clinically validated residue. The authors were also able to characterize the resulting perturbations to the transcriptome. This pipeline of base-editor screening followed by single-cell transcriptomics is likely to become more popular as the cost of single-cell sequencing decreases.

11.4.5 Combining ABE and CBE Editors

To maximize sequence diversification, it is appealing to try to combine cytosine and ABE. Li et al. experiment with a number of different nCas9 fusions to APOBEC3A and TadA, developing tools for simultaneous C→T and A→G editing that they name saturated targeted endogenous mutagenesis editors (STEMEs) (Li et al. 2020a). They apply STEME-1 to directed evolution in rice protoplasts, mutating the gene for acetyl-coenzyme A carboxylase, which is targeted by the herbicide haloxyfop. STEME-1 mutagenesis yields rice seedlings carrying different resistance mutations. The majority of these are C→T transitions, probably indicative of the kinetic advantages of APOBEC3A over the evolved TadA version used in this construct (TadA7.10, Koblan et al. 2018). STEME-1 outperforms the simultaneous targeting of separate CBE and ABE constructs with the same guide for maximum mutagenesis. This finding is corroborated in a recent paper (Sakata et al. 2020), in which AID and ABE10.7 are both fused to the same nickase for targeted mutagenesis in human cell lines. This study finds that the ABE10.7 construct in particular does not edit well when targeted with the same guide as Target-AID. Nevertheless, simultaneous targeting of plant-optimized nCas9-AID*D (Ren et al. 2018) and nCas9-TadA (Yan et al. 2018) by the same guide RNA is used successfully in another recent directed evolution mutagenesis screen in rice (Kuang et al. 2020). Using their own base-editing-mediated gene evolution (BEMGE) method, Kuang et al. find novel mutations conferring resistance to the herbicide bispyribac-sodium.

One obvious drawback of base editing for sequence diversification is its bias toward transition mutations (C→T, A→G and vice versa). APOBEC/AID/CDA1-based editors preferentially convert cytosine to thymine, while the TadA component of ABS can only convert adenine to guanine. Although different systems permit varying levels of BER, which can result in any base being substituted for the excised base, transversions are relatively rare. There is also a tradeoff between efficiency of editing and diversity of outcome. Uracil glycosylase inhibitors protect C to T transitions, preventing the removal of DNA-uracil and improving editing efficiency by several-fold *in vivo* (Komor et al. 2016). Without the formation of an apurinic/apyrimidinic site, however, editing outcome is limited to C→T transition. The new CGBE tool (Kurt et al. 2020) harnesses BER, and may point the way for future transversion editors. Gajula suggests adding translesion synthesis polymerases to fusion editor complexes (Gajula 2019). This could recapitulate the situation in B cells undergoing somatic hypermutation to create antibody diversity, with the editing result depending on the polymerase used. In addition to this interesting speculation, a few studies have explored alternative methods to leverage CRISPR systems for sequence diversification.

11.4.6 Other CRISPR-Based Sequence Diversification Methods

Shortly after the publication of CRISPR-X and CRISPR-TAM, the Doench lab also published an exploration of phenotypic screening of sequence variation, though with conventional active Cas9 (Donovan et al. 2017). Pooled CRISPR screens typically include several guides against each gene, with concordance between guides being the key criterion for calling hits. In the course of a loss-of-function screen in the BRAF V600E/MAPK inhibitor system (Doench et al. 2016), Donovan et al. observed one guide behaving differently from the other three guides targeting MAP2K1 under MAPK inhibition selection. MAP2K1 codes for MEK1, which positively regulates the MAPK pathway. MAP2K1 knock-out should be selected against during MAPK inhibition, and the majority of guides reflected this. A minority of guides resulted instead in strong resistance phenotypes, leading the authors to suspect the creation of resistant protein variants. After confirming this, the group tiled guides along MAP2K to find drug-resistant protein variants. This demonstrates that in the case of powerful positive selection, screening with nuclease-active Cas9 can reveal functionally distinct sequence variants. The missense mutants of interest are a byproduct and minority outcome in this type of screen, however, and Donovan et al. discuss the possible advantages of using base editors in the future.

More recently, Halperin et al. (2018) explored fusing an error-prone nick-translating polymerase to Cas9. This class of polymerases synthesize from a single-strand break, or nick, in the DNA, displacing and cleaving the downstream strand. Halperin et al. chose *E. coli* PolI as a starting point, introducing three known fidelity-reducing mutations for a processive sequence diversification tool they named EvolvR. Using enhanced Cas9 (eCas9, Slaymaker et al. 2016), with its longer target residency time, increased mutation rate, while inserting a thioredoxin binding domain into the polymerase generated a more processive, though less mutagenic, EvolvR tool. At its most processive, the EvolvR mutational window is about 50bp from the Cas9-induced nick. Although Halperin et al. only use their EvolvR tool in *E. coli*, this is an exciting proof of concept. The polymerase incorporates all four bases, though with a bias toward A and T. It would be very interesting to see whether directed evolution could, with the EvolvR tools as a starting point, generate diversifiers with a more frequent and random incorporation.

11.5 Applications for Base-Editor Screening

11.5.1 Studying Drug–Target Interactions (DTI)

Identification of targets for novel small molecules (SM) is challenging, and best approached by a combination of chemical, proteomic, and genetic methods. The genetic methods of choice are genome-wide CRISPRko, CRISPRi, and CRISPRa screens, which can help pinpoint protein targets and target pathways. Once target hypotheses have been generated, sequence diversification screening by base editing can help deconvolute possible targets and verify that a small molecule of interest indeed exerts its physiological effect through interaction with these targets. The use of base editors for sequence diversification in coding genes, as first demonstrated in the CRISPR-X and

CRISPR-TAM approaches, has enormous potential for the characterization of DTI (Figure 11.3a). Determining the specific residues required for drug function, as can be achieved with a base editing screen, will have several positive effects. Drug candidates frequently fail in clinical trials, after years of investment, when the desired phenotype is found to be mediated by pathways other than the presumed target protein (Lin et al. 2019). The creation and identification of a mutation in a hypothetical target that confers resistance to the effect of the small compound is perhaps the most compelling evidence for a small molecule–target interaction. Such information could prevent the costly pursuit of off-target "mechanisms of action."

Better DTI mapping could also contribute to personalized medicine, ensuring that only patients with a vulnerable protein sequence receive a given treatment. In the same way, base-editor screening could predict potential resistance mutations to monitor for.

Beyond DTI exploration, sequence diversification with base editors could be used to map essential domains for protein function (Figure 11.3a) or for protein–protein interaction (Figure 11.3b). This method could also reveal *de novo* protein functions that arise from SNVs, as opposed to the majority of other CRISPR-based screening techniques, which operate at the level of the entire gene or element.

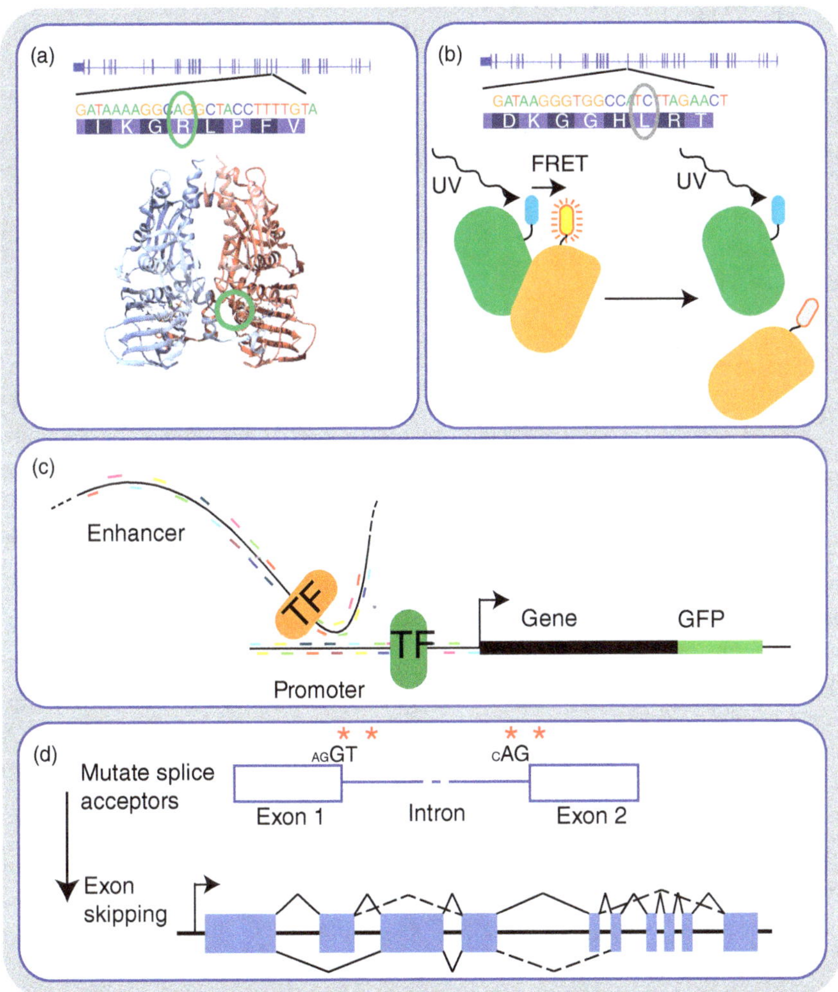

Figure 11.3 **Applications of sequence diversification screens.** (a) The major strength of sequence diversification screens lies in the identification of important amino acid residues, for protein function, drug–target interaction, protein–protein interaction, etc. (b) In a pooled scenario, FRET sensors provide a sortable phenotype to read out mutations that disrupt protein–protein interaction. (c) Tiling enhancer and/or promoter regions with mutagenic guides can identify small sequence motifs important for the regulation of gene expression. (d) Libraries targeting splice sites to induce exon skipping can help elucidate the functional importance of exons and splice variants.

11.5.2 Regulatory Elements

There is growing interest in using the CRISPR system to alter gene regulatory sequences for clinical applications. For example, the β-hemoglobinopathies sickle cell disease and beta thalassemia are ameliorated by hereditary persistence of fetal hemoglobin (HPFH), a condition in which fetal hemoglobin (HbF; $\alpha2\gamma2$) is expressed in adults. This has inspired numerous attempts to mimic HPFH as a treatment by means of genetic engineering or bioactive chemicals (Demirci et al. 2020). One approach now in multiple clinical trials achieves constitutive HbF expression by using CRISPR-Cas9 editing to knock out BCL11A, which normally begins to repress fetal (γ) hemoglobin shortly after birth (Chang et al. 2017). Mutating the binding sites of repressors BCL11A or LRF in the γ-globin promoter also results in γ-globin upregulation, and has been achieved both by conventional CRISPR mutagenesis (Martyn et al. 2018) and base editing (Wang et al. 2020).

A similar avenue of treatment has been proposed for the *TERT* gene (Li et al. 2020b). *TERT* codes for telomerase, which extends telomere ends, allowing highly proliferating cancer cells to escape senescence. *TERT* promoter mutations that create *de novo* binding sites for ETS-family transcription factors, thus elevating gene expression, are common in human somatic and germline tumors, particularly glioblastomas. Destroying these expression-enhancing transcription factor binding sites using a *Campylobacter jejuni* (*Cj*) nickase Cas9-ABE strongly reduced both ETS-family transcription factor binding and *TERT* expression. Glioblastoma cell line telomere length was consequently reduced, inhibiting proliferation and inducing senescence (Li et al. 2020b).

As a general mechanism, blocking transcriptional repressor binding is likely to have fewer off-target effects than knocking out the repressor itself. Using base editors to target transcription factor binding sites is thus emerging as a clinically important field. While it is easy to envision correcting known disease-linked SNVs, as in the case of *TERT*, or generating ameliorating mutations as in the case of HbF, altering healthy promoters or enhancers to compensate for distal disease phenotypes will be less straightforward. Screens for functional point mutations with disruptive regulatory effects could reveal a wealth of new candidates for future therapies (Figure 11.3c).

11.5.3 Alternative Splicing

The vast majority of human genes give rise to multiple protein isoforms via alternative splicing (Wang et al. 2008). Misregulation of splicing plays an increasingly recognized role in cancer and other diseases, with elevated levels of aberrantly spliced mRNAs detected in a wide range of tumors (Kahles et al. 2018; Singh and Cooper 2012; Supek et al. 2014). Base editing tools have been applied to the regulation of splicing in a variety of configurations (Figure 11.3d). In CRISPR-SKIP, Gapinske et al. use BE3 (Komor et al. 2016) to convert a highly conserved splice acceptor guanine to an adenine, resulting in exon skipping (Gapinske et al. 2018). In the same year, the CRISPR-TAM base editor (Ma et al. 2016) was applied to control exon inclusion or skipping (Yuan et al. 2018), and to restore the function of a mutant dystrophin gene in patient-derived iPSCs. More recently, the Liu lab ABE (Koblan et al. 2018; Richter et al. 2020) was used to correct a splicing defect in the survival motor neuron (SMN) gene of spinal muscular atrophy (SMA) patient-derived iPSCs (Lin et al. 2020). Editing splice sites can also effectively knock out a gene by inducing early intron inclusion. Kluesner et al. (2020) conclude that intron inclusion is more effective for loss-of-function screening than creating a premature stop codon (CRISPR-STOP, Kuscu et al. 2017). Although not the primary focus of the work, the SpliceR web tool introduced here could be used to easily generate libraries for high-throughput interrogation of alternative splicing.

11.5.4 Limitations

Genetic screening with base editor tools is not without limitations. As with any pooled genetic screen, pooled base-editor screening requires a separable phenotypic readout. This is usually achievable with reporter engineering or, in some cases, using single-cell transcriptomes as the screen readout. Pooled screening requires large numbers of cells to be kept in culture to maintain library coverage. This may be even more true in the case of a base editor screen, where the possible editing outcomes for every guide are more diverse than in a canonical CRISPR screen with catalytically active Cas9. In the case of very efficient editing, the guide will eventually cease to recognize the target DNA sequence. If base editing is to be used for advanced *in vitro* evolution applications, where several simultaneous mutations are often required for significant improvement, target recognition could be an obstacle. Finally, transversion mutations are currently more difficult to achieve than transitions. BER introduces all bases to sites of mismatch, but with possible tradeoffs in editing efficiency. The recent addition of the C to G base editor CGBE (Kurt et al. 2020) depends on the creation of an abasic site, tethering a uracil glycosylase (UNG) to the APOBEC cytidine editor. In spite of these considerations, pooled screening with base editors has enormous potential to explore sequence space for a variety of applications.

11.6 Conclusion

Developing bioactive SM into suitable drug candidates requires an understanding of their mechanism of action and their biological selectivity. Validated mechanism of action information is critical: Currently, 97% of cancer drugs never make it into the clinic, often because the proposed mechanism of action is incorrect and observed drug effects are caused by unspecific, off-target toxicity (Lin et al. 2019). In-depth DTI characterization early in the drug discovery process is thus important to reduce attrition rate and avoid costly and time-consuming follow-up of poor leads. Base-editor screening has the potential for specific, endogenous, and rapid sequence diversification at an affordable cost. As a genetic method, base-editor screening is complementary to chemical and proteomic DTI studies, and is thus a valuable addition to the current drug discovery toolbox.

At the dawn of the genome era, the world was surprised to learn that humans have only twice as many genes as fruit flies, and somewhat fewer than mice (Lander et al. 2001; Venter et al. 2001). We have begun to understand the regulatory complexity that makes this possible, encoded in what was once considered "junk DNA." CRISPR screening has revolutionized the functional genomics of coding sequences. Base-editor screening will be a crucial tool in efforts to improve our characterization of regulatory sequences and alternative splicing, as well as functional mutational spaces.

Acknowledgements

CRISPR Functional Genomics is a Swedish national research facility funded by Science for Life Laboratory (SciLifeLab) and Karolinska Institutet. The authors thank Dr Allegra Lord for critical review of the manuscript.

References

Anzalone, A.V., Koblan, L.W., and Liu, D.R. (2020). Genome editing with CRISPR-Cas nucleases, base editors, transposases and prime editors. *Nat. Biotechnol.* 38 (7): 824–844. https://doi.org/10.1038/s41587-020-0561-9.

Arnold, F.H. (1993). Protein engineering for unusual environments. *Curr. Opin. Biotechnol.* 4 (4): 450–466. https://doi.org/10.1016/0958-1669(93)90011-k.

Ayer, D.E., Laherty, C.D., Lawrence, Q.A. et al. (1996). Mad proteins contain a dominant transcription repression domain. *Mol. Cell. Biol.* 16: 5772–5781.

Bellefroid, E.J., Poncelet, D.A., Lecocq, P.J. et al. (1991). The evolutionarily conserved Krüppel-associated box domain defines a subfamily of eukaryotic multifingered proteins. *Proc. Natl. Acad. Sci. U. S. A.* 88: 3608–3612.

Chang, K.H., Smith, S.E., Sullivan, T. et al. (2017). Long-term engraftment and fetal globin induction upon BCL11A gene editing in bone-marrow-derived CD34+ hematopoietic stem and progenitor cells. *Mol. Ther. Methods Clin. Dev.* 4: 137–148. Published 2017 Jan 11. doi:https://doi.org/10.1016/j.omtm.2016.12.009.

Datlinger, P., Rendeiro, A.F., Schmidl, C. et al. (2017). Pooled CRISPR screening with single-cell transcriptome readout. *Nat. Methods* 14: 297–301.

Demirci, S., Leonard, A., and Tisdale, J.F. (2020). Genome editing strategies for fetal hemoglobin induction in beta-hemoglobinopathies [published online ahead of print, 2020 May 14]. *Hum. Mol. Genet.*: ddaa088. https://doi.org/10.1093/hmg/ddaa088.

Després, P.C., Dubé, A.K., Seki, M. et al. (2020). Perturbing proteomes at single residue resolution using base editing. *Nat. Commun.* 11 (1): 1871. Published 2020 Apr 20. doi:https://doi.org/10.1038/s41467-020-15796-7.

Dixit, A., Parnas, O., Li, B. et al. (2016). Perturb-seq: dissecting molecular circuits with scalable single-cell RNA profiling of pooled genetic screens. *Cell* 167: 1853–1866.e17.

Doench, J.G., Fusi, N., Sullender, M. et al. (2016). Optimized sgRNA design to maximize activity and minimize off-target effects of CRISPR-Cas9. *Nat. Biotechnol.* 34 (2): 184–191. https://doi.org/10.1038/nbt.3437.

Donovan, K.F., Hegde, M., Sullender, M. et al. (2017). Creation of novel protein variants with CRISPR-Cas9-mediated mutagenesis: turning a screening by-product into a discovery tool. *PLoS One* 12 (1): e0170445. Published 2017 Jan 24. doi:https://doi.org/10.1371/journal.pone.0170445.

Esvelt, K.M., Carlson, J.C., and Liu, D.R. (2011). A system for the continuous directed evolution of biomolecules. *Nature* 472 (7344): 499–503. https://doi.org/10.1038/nature09929.

Feldman D, Singh A, Schmid-Burgk JL et al. (2019). Optical Pooled Screens in Human Cells. *Cell* 179(3):787–799.e17. https://doi.org/10.1016/j.cell.2019.09.016

Gajula, K.S. (2019). Designing an elusive C•G→G•C CRISPR base editor. *Trends Biochem. Sci.* 44 (2): 91–94. https://doi.org/10.1016/j.tibs.2018.10.004.

Gapinske, M., Luu, A., Winter, J. et al. (2018). CRISPR-SKIP: programmable gene splicing with single base editors.

Genome Biol. 19 (1): 107. Published 2018 Aug 15. doi:https://doi.org/10.1186/s13059-018-1482-5.

Gaudelli, N.M., Komor, A.C., Rees, H.A. et al. (2017). Programmable base editing of A•T to G•C in genomic DNA without DNA cleavage [published correction appears in Nature. 2018 May 2;]. *Nature* 551 (7681): 464–471. https://doi.org/10.1038/nature24644.

Geurts, M.H., de Poel, E., Amatngalim, G.D. et al. (2020). CRISPR-based adenine editors correct nonsense mutations in a cystic fibrosis organoid biobank. *Cell Stem Cell* 26 (4): 503–510.e7. https://doi.org/10.1016/j.stem.2020.01.019.

Grünewald, J., Zhou, R., Iyer, S. et al. (2019). CRISPR DNA base editors with reduced RNA off-target and self-editing activities. *Nat. Biotechnol.* 37 (9): 1041–1048. https://doi.org/10.1038/s41587-019-0236-6.

Haapaniemi, E., Botla, S., Persson, J. et al. (2018). CRISPR-Cas9 genome editing induces a p53-mediated DNA damage response. *Nat. Med.* 24 (7): 927–930. https://doi.org/10.1038/s41591-018-0049-z.

Halperin, S.O., Tou, C.J., Wong, E.B. et al. (2018). CRISPR-guided DNA polymerases enable diversification of all nucleotides in a tunable window. *Nature* 560 (7717): 248–252. https://doi.org/10.1038/s41586-018-0384-8.

Hess, G.T., Frésard, L., Han, K. et al. (2016). Directed evolution using dCas9-targeted somatic hypermutation in mammalian cells. *Nat. Methods* 13 (12): 1036–1042. https://doi.org/10.1038/nmeth.4038.

Huang, T.P., Zhao, K.T., Miller, S.M. et al. (2019). Circularly permuted and PAM-modified Cas9 variants broaden the targeting scope of base editors [published correction appears in Nat Biotechnol. 2019 Jun 10;]. *Nat. Biotechnol.* 37 (6): 626–631. https://doi.org/10.1038/s41587-019-0134-y.

Ihry, R.J., Worringer, K.A., Salick, M.R. et al. (2018). p53 inhibits CRISPR-Cas9 engineering in human pluripotent stem cells. *Nat. Med.* 24 (7): 939–946. https://doi.org/10.1038/s41591-018-0050-6.

Jaitin, D.A., Weiner, A., Yofe, I. et al. (2016). Dissecting immune circuits by linking CRISPR-pooled screens with single-cell RNA-seq. *Cell* 167 (7): 1883–1896.e15. doi:https://doi.org/10.1016/j.cell.2016.11.039.

Jinek, M., Chylinski, K., Fonfara, I. et al. (2012). A programmable dual-RNA-guided DNA endonuclease in adaptive bacterial immunity. *Science* 337 (6096): 816–821. https://doi.org/10.1126/science.1225829.

Jun, S., Lim, H., Chun, H. et al. (2020). Single-cell analysis of a mutant library generated using CRISPR-guided deaminase in human melanoma cells. *Commun Biol.* 3 (1): 154. Published 2020 Apr 2. doi:https://doi.org/10.1038/s42003-020-0888-2.

Kahles, A., Lehmann, K.V., Toussaint, N.C. et al. (2018). Comprehensive analysis of alternative splicing across tumors from 8,705 patients. *Cancer Cell* 34 (2): 211–224.e6. doi:https://doi.org/10.1016/j.ccell.2018.07.001.

Kim, Y.B., Komor, A.C., Levy, J.M. et al. (2017). Increasing the genome-targeting scope and precision of base editing with engineered Cas9-cytidine deaminase fusions. *Nat. Biotechnol.* 35 (4): 371–376. https://doi.org/10.1038/nbt.3803.

Kleinstiver, B.P., Sousa, A.A., Walton, R.T. et al. (2019). Engineered CRISPR-Cas12a variants with increased activities and improved targeting ranges for gene, epigenetic and base editing [published correction appears in Nat Biotechnol. 2020 Jul;38(7):901]. *Nat. Biotechnol.* 37 (3): 276–282. https://doi.org/10.1038/s41587-018-0011-0.

Kluesner, M.G., Lahr, W.S., Lonetree, C.L. et al. (2020). CRISPR-Cas9 cytidine and adenosine base editing of splice-sites mediates highly-efficient disruption of proteins in primary cells. *Nat Commun.* 12 (1): 2437. https://doi.org/10.1038/s41467-021-22009-2.

Koblan, L.W., Doman, J.L., Wilson, C. et al. (2018). Improving cytidine and adenine base editors by expression optimization and ancestral reconstruction. *Nat. Biotechnol.* 36 (9): 843–846. https://doi.org/10.1038/nbt.4172.

Koike-Yusa, H., Li, Y., Tan, E.P. et al. (2014). Genome-wide recessive genetic screening in mammalian cells with a lentiviral CRISPR-guide RNA library. *Nat. Biotechnol.* 32 (3): 267–273. https://doi.org/10.1038/nbt.2800.

Komor, A.C., Kim, Y.B., Packer, M.S. et al. (2016). Programmable editing of a target base in genomic DNA without double-stranded DNA cleavage. *Nature* 533 (7603): 420–424. https://doi.org/10.1038/nature17946. Epub 2016 Apr 20.

Komor, A.C., Zhao, K.T., Packer, M.S. et al. (2017). Improved base excision repair inhibition and bacteriophage Mu Gam protein yields C:G-to-T:A base editors with higher efficiency and product purity. *Sci. Adv.* 3 (8): eaao4774.

Konermann, S., Brigham, M.D., Trevino, A.E. et al. (2015). Genome-scale transcriptional activation by an engineered CRISPR-Cas9 complex. *Nature* 517 (7536): 583–588. https://doi.org/10.1038/nature14136.

Kuang, Y., Li, S., Ren, B. et al. (2020). Base-editing-mediated artificial evolution of OsALS1 in planta to develop novel herbicide-tolerant rice germplasms. *Mol. Plant* 13 (4): 565–572. https://doi.org/10.1016/j.molp.2020.01.010.

Kurt, I.C., Zhou, R., Iyer, S. et al. (2020). CRISPR C-to-G base editors for inducing targeted DNA transversions in human cells [published online ahead of print, 2020 Jul 20]. *Nat. Biotechnol.* https://doi.org/10.1038/s41587-020-0609-x.

Kuscu, C., Parlak, M., Tufan, T. et al. (2017). CRISPR-STOP: gene silencing through base-editing-induced nonsense mutations. *Nat. Methods* 14 (7): 710–712. https://doi.org/10.1038/nmeth.4327.

Lander, E., Linton, L., Birren, B. et al. (2001). Initial sequencing and analysis of the human genome. *Nature* 409: 860–921.

Leung, D.W., Chen, E., and Goeddel, D.V. (1989). A method for random mutagenesis of a defined DNA segment using a modified polymerase chain reaction. *Technique* 1: 11–15.

Lewis, E.B. and Bacher, F. (1968). Methods of feeding ethyl methane sulphonate (EMS) to Drosophila males. *Drosoph. Inf. Serv.* 43: 193.

Li, C., Zhang, R., Meng, X. et al. (2020a). Targeted, random mutagenesis of plant genes with dual cytosine and adenine base editors. *Nat. Biotechnol.* 38 (7): 875–882. https://doi.org/10.1038/s41587-019-0393-7.

Li, X., Qian, X., Wang, B. et al. (2020b). Programmable base editing of mutated TERT promoter inhibits brain tumour growth. *Nat. Cell Biol.* 22 (3): 282–288. https://doi.org/10.1038/s41556-020-0471-6.

Lin, A., Giuliano, C.J., Palladino, A. et al. (2019). Off-target toxicity is a common mechanism of action of cancer drugs undergoing clinical trials. *Sci. Transl. Med.* 11 (509): eaaw8412. https://doi.org/10.1126/scitranslmed.aaw8412.

Lin, X., Chen, H., Lu, Y.Q. et al. (2020). Base editing-mediated splicing correction therapy for spinal muscular atrophy. *Cell Res.* 30 (6): 548–550. https://doi.org/10.1038/s41422-020-0304-y.

Ma, Y., Zhang, J., Yin, W. et al. (2016). Targeted AID-mediated mutagenesis (TAM) enables efficient genomic diversification in mammalian cells. *Nat. Methods* 13: 1029–1035. https://doi.org/10.1038/nmeth.4027.

Martyn, G.E., Wienert, B., Yang, L. et al. (2018). Natural regulatory mutations elevate the fetal globin gene via disruption of BCL11A or ZBTB7A binding. *Nat. Genet.* 50 (4): 498–503. https://doi.org/10.1038/s41588-018-0085-0.

Michlits, G., Hubmann, M., Wu, S.H. et al. (2017). CRISPR-UMI: single-cell lineage tracing of pooled CRISPR-Cas9 screens. *Nat. Methods* 14 (12): 1191–1197. https://doi.org/10.1038/nmeth.4466.

Mojica, F.J. and Rodriguez-Valera, F. (2016). The discovery of CRISPR in archaea and bacteria. *FEBS J.* 283 (17): 3162–3169. https://doi.org/10.1111/febs.13766.

Nishida, K., Arazoe, T., Yachie, N. et al. (2016). Targeted nucleotide editing using hybrid prokaryotic and vertebrate adaptive immune systems. *Science* 353 (6305): aaf8729. https://doi.org/10.1126/science.aaf8729.

Nüsslein-Volhard, C. and Wieschaus, E. (1980). Mutations affecting segment number and polarity in Drosophila. *Nature* 287: 795–801.

Qi, L.S., Larson, M.H., Gilbert, L.A. et al. (2013). Repurposing CRISPR as an RNA-guided platform for sequence-specific control of gene expression. *Cell* 152 (5): 1173–1183. https://doi.org/10.1016/j.cell.2013.02.022.

Rees, H.A. and Liu, D.R. (2018). Base editing: precision chemistry on the genome and transcriptome of living cells. *Nat. Rev. Genet.* 19 (12): 770–788. https://doi.org/10.1038/s41576-018-0059-1.

Ren, B., Yan, F., Kuang, Y. et al. (2018). Improved base editor for efficiently inducing genetic variations in rice with CRISPR-Cas9-guided hyperactive hAID mutant. *Mol. Plant* 11 (4): 623–626. https://doi.org/10.1016/j.molp.2018.01.005.

Replogle, J.M., Norman, T.M., Xu, A. et al. (2020). Combinatorial single-cell CRISPR screens by direct guide RNA capture and targeted sequencing. *Nat. Biotechnol.* 38 (8): 954–961. https://doi.org/10.1038/s41587-020-0470-y.

Richter, M.F., Zhao, K.T., Eton, E. et al. (2020). Phage-assisted evolution of an adenine base editor with improved Cas domain compatibility and activity [published correction appears in Nat Biotechnol. 2020 May 20;]. *Nat. Biotechnol.* 38 (7): 883–891. https://doi.org/10.1038/s41587-020-0453-z.

Sakata, R.C., Ishiguro, S., Mori, H. et al. (2020). Base editors for simultaneous introduction of C-to-T and A-to-G mutations [published correction appears in Nat Biotechnol. 2020 Jun 5;]. *Nat. Biotechnol.* 38 (7): 865–869. https://doi.org/10.1038/s41587-020-0509-0.

Schmierer, B., Botla, S.K., Zhang, J. et al. (2017). CRISPR/Cas9 screening using unique molecular identifiers. *Mol. Syst. Biol.* 13 (10): 945. Published 2017 Oct 9. doi:https://doi.org/10.15252/msb.20177834.

Shalem, O., Sanjana, N.E., Hartenian, E. et al. (2014). Genome-scale CRISPR-Cas9 knockout screening in human cells. *Science* 343 (6166): 84–87. https://doi.org/10.1126/science.1247005.

Singh, R.K. and Cooper, T.A. (2012). Pre-mRNA splicing in disease and therapeutics. *Trends Mol. Med.* 18 (8): 472–482. https://doi.org/10.1016/j.molmed.2012.06.006.

Slaymaker, I.M., Gao, L., Zetsche, B. et al. (2016). Rationally engineered Cas9 nucleases with improved specificity. *Science* 351 (6268): 84–88. https://doi.org/10.1126/science.aad5227.

Smits, A.H., Ziebell, F., Joberty, G. et al. (2019). Biological plasticity rescues target activity in CRISPR knock outs. *Nat. Methods* 16 (11): 1087–1093. https://doi.org/10.1038/s41592-019-0614-5.

Stemmer, W.P. (1994). Rapid evolution of a protein in vitro by DNA shuffling. *Nature* 370 (6488): 389–391. https://doi.org/10.1038/370389a0.

Supek, F., Miñana, B., Valcárcel, J. et al. (2014). Synonymous mutations frequently act as driver mutations in human cancers. *Cell* 156 (6): 1324–1335. https://doi.org/10.1016/j.cell.2014.01.051.

Tan, J., Zhang, F., Karcher, D., and Bock, R. (2019). Engineering of high-precision base editors for site-specific single nucleotide replacement [published correction appears in Nat Commun. 2019 May 1;10(1):2019]. *Nat. Commun.* 10 (1): 439. Published 2019 Jan 25. doi:https://doi.org/10.1038/s41467-018-08034-8.

Taylor, A.L. and Trotter, C.D. (1967). Revised linkage map of *Escherichia coli*. *Bacteriol. Rev.* 31 (4): 332–353.

Thuronyi, B.W., Koblan, L.W., Levy, J.M. et al. (2019). Continuous evolution of base editors with expanded target compatibility and improved activity [published correction appears in Nat Biotechnol. 2019 Aug 12;]. *Nat. Biotechnol.* 37 (9): 1070–1079. https://doi.org/10.1038/s41587-019-0193-0.

Venter, C.J., Adams, M.D., Myers, E.W. et al. (2001). The sequence of the human genome. *Science* 291 (5507): 1304–1135.

Wang, E.T., Sandberg, R., Luo, S. et al. (2008). Alternative isoform regulation in human tissue transcriptomes. *Nature* 456 (7221): 470–476. https://doi.org/10.1038/nature07509.

Wang, T., Wei, J.J., Sabatini, D.M., and Lander, E.S. (2014). Genetic screens in human cells using the CRISPR-Cas9 system. *Science* 343 (6166): 80–84. https://doi.org/10.1126/science.1246981.

Wang, L., Li, L., Ma, Y. et al. (2020). Reactivation of γ-globin expression through Cas9 or base editor to treat β-hemoglobinopathies. *Cell Res.* 30 (3): 276–278. https://doi.org/10.1038/s41422-019-0267-z.

Wu, J., Chen, C., Xian, G. et al. (2020). Engineering herbicide-resistant oilseed rape by CRISPR/Cas9-mediated cytosine base-editing. *Plant Biotechnol. J.* 18 (9): 1857–1859. https://doi.org/10.1111/pbi.13368.

Yan, F., Kuang, Y., Ren, B. et al. (2018). Highly efficient A·T to G·C base editing by Cas9n-guided tRNA adenosine deaminase in rice. *Mol. Plant* 11 (4): 631–634. https://doi.org/10.1016/j.molp.2018.02.008.

Yuan, J., Ma, Y., Huang, T. et al. (2018). Genetic modulation of RNA splicing with a CRISPR-guided cytidine deaminase. *Mol. Cell* 72 (2): 380–394.e7. doi:https://doi.org/10.1016/j.molcel.2018.09.002.

Zafra, M.P., Schatoff, E.M., Katti, A. et al. (2018). Optimized base editors enable efficient editing in cells, organoids and mice. *Nat. Biotechnol.* 36 (9): 888–893. https://doi.org/10.1038/nbt.4194.

Zhou, Y., Zhu, S., Cai, C. et al. (2014). High-throughput screening of a CRISPR-Cas9 library for functional genomics in human cells. *Nature* 509 (7501): 487–491. https://doi.org/10.1038/nature1316.

12

Single-Cell Transcriptomics and Epigenomics for CRISPR-Mediated Perturbation Studies

Saumyaa Saumyaa, Ramy Elgendy, and Alessandro Bonetti

Translational Genomics, Discovery Sciences, BioPharmaceuticals R&D, AstraZeneca, Molndal, Sweden

12.1 Introduction

The development of transcriptomics and epigenomics has allowed rapid advances in our understanding of biological processes. We can now dissect cell subtypes and states, and characterize gene expression regulatory programs at an unparalleled depth, yielding key details on cell identity and the molecular mechanisms governing it. However, conventional bulk analysis dilutes cell-specific data by providing only average information, thereby limiting the resolution of transcriptomic and functional genomic studies (Gawad et al. 2016). The importance of cell-to-cell variability has been the object of several studies that have uncovered the cellular and regulatory heterogeneity within tissues, in development and disease (Jaitin et al. 2014).

Single-cell transcriptomics allow the profiling of thousands of individual cells and the identification of cell types at an unprecedented resolution. Recent methodological advances provide cost-efficient and scalable approaches that can be used to identify gene expression signatures of different cell types and generate atlases of cellular identities across tissues. The advent of single-cell epigenomics methodologies has further progressed the knowledge of cell type-specific regulatory programs that control gene expression. However, a major limitation of single-cell technologies has been the descriptive nature of these methodologies and therefore their inability to elucidate causality unless combined with perturbation approaches (Junker and van Oudenaarden 2014; Moignard and Göttgens 2014).

After the initial demonstration of programmable DNA cleavage by Cas nuclease, Clustered Regularly Interspaced Short Palindromic Repeats (CRISPR)-based technologies have propelled our capacity to elucidate genetic function with an exponential growth for discovery, engineering, and application of CRISPR-Cas genome editing tools. Apart from the conventional indels generated by the original CRISPR/Cas9, nowadays scientists can achieve an impressive array of perturbations by employing different proteins or domains fused with the Cas9 protein. Different types of transcriptional regulations can be obtained through CRISPR interference (CRISPRi) or activation (CRISPRa) to inhibit or activate gene expression, respectively (Lau 2014). Combined with large-scale genetic screening approaches, we now have the possibility to investigate the function of several genes in a single experiment.

Recent advances in the ability to measure transcriptomes and epigenomes in single cells have enabled the development of high-content genetic screens where high-throughput CRISPR perturbations can be coupled with single-cell technologies. Furthermore, single-cell approaches enable the use of CRISPR-based perturbations *in vivo* to examine their effect on heterogeneous populations of cells within a tissue.

In this chapter, we discuss the application of single-cell transcriptomics and epigenomics methodologies for the phenotypic characterization of CRISPR-Cas9-mediated genome editing screens. We focus solely on the combination of CRISPR perturbations and single-cell technologies that use next-generation sequencing readouts. For readers interested in emerging spatial transcriptomics technologies with single-cell resolution, we suggest the review by Waylen and colleagues (Waylen et al. 2020). Finally, we

Genome Editing in Drug Discovery, First Edition. Edited by Marcello Maresca and Sumit Deswal.

provide future perspectives regarding the use of single-cell technologies in combination with CRISPR-based perturbations to unveil the function of genomic regulatory elements that control cell identity.

12.2 CRISPR-Based Genetic Screens with Single-Cell Transcriptomics Readout

Conventional methods for large-scale genetic screens to investigate gene function are usually divided into pooled or arrayed CRISPR screens. In pooled CRISPR screens, multiple perturbations can be simultaneously applied to a bulk population of cells in a single experiment, making them a suitable tool for performing genome-wide screens. However, they are limited to easily measurable phenotypes (e.g. cell viability, reporter gene expression, or drug resistance) and are unpowered to distinguish between different mechanisms that produce similar phenotypes. In arrayed CRISPR screens, a single perturbation is applied to a bulk cell population, one experiment at a time. Although they allow more complex phenotypic and molecular readouts at the bulk population level (e.g. secreted protein or transcriptional profile), they fall behind on throughput and cost-effectiveness offered by pooled screens.

To design methods that can yield rich readouts while maintaining the throughput, several groups developed new CRISPR-based pooled screening approaches that employ high-throughput scRNA-seq technology to enable the measurement of more complex phenotypes such as transcriptional responses (Figure 12.1). Integration of gene editing and single-cell transcriptomics enables simultaneous measurement of the repression or knock-out of a target gene and its related changes on the transcriptome, thereby considerably increasing the number of potential readouts.

The design for any of the above-described genetic screens relies on the delivery of lentiviral libraries carrying different single-guide RNAs (sgRNAs) into a cell population that expresses CRISPR-Cas9. All these newly developed single-cell-based CRISPR approaches share the key technical innovation of enabling the identification of the sgRNA from each cell during scRNA-seq via addition of a polyadenylation signal to the sgRNA identifier. All the 3'-based scRNA-seq technologies adopt a universal approach to make use of a poly(T) primer to capture the polyadenylated transcripts. Making use of this strategy, a polyadenylated sgRNA identifier gets captured alongside cellular transcripts, facilitating the detection of sgRA in scRNA-seq experiments and linking it to respective single-cell transcriptome profiles. With this strategy, it is now possible to identify the perturbed gene without sequencing the sgRNA cassette from genomic DNA as done in previous studies. The vector design for CRISP-seq, Perturb-seq, and Mosaic-seq generates an RNA polymerase II-driven polyadenylated guide barcode (gBC) transcript, with each unique gBC associated to a unique sgRNA, to indirectly identify the sgRNAs during scRNA-seq (Dixit et al. 2016; Jaitin et al. 2016; Xie et al. 2017). CROP-seq vector design is different as it generates an RNA polymerase II transcribed polyadenylated sgRNA transcript, which is directly detected during scRNA-seq (Datlinger et al. 2017).

As for the strategy of choice to partition single cells and generate scRNA-seq libraries, there are two main options: plate-based or droplet-based methods. Plate-based methods such as MARS-seq (Jaitin et al. 2014) and Smart-seq 2 (Picelli et al. 2014) involve physical separation of single cells into individual plate wells via flow cytometry, making them lower throughput and more labor intensive. Droplet-based methods, such as Drop-seq (Macosko et al. 2015), use microfluidic devices for partitioning of single cells into individual droplets, where transcripts from each cell acquire a unique barcode. Thus, these methods have higher throughput, allowing the investigation of thousands of cells in a single experiment. Each of the single-cell CRISPR screen approach employed a different scRNA-seq technology, CRISP-seq used MARS-seq (Jaitin et al. 2016), Perturb-seq used 10x Genomics platform (Dixit et al. 2016), and CROP-seq and MOSAIC-seq used Drop-seq (Datlinger et al. 2017; Xie et al. 2017).

ScRNA-seq technologies provide high-dimensional data but it is also characterized by signal sparsity where most genes are not measured due to technology of choice, sequencing depth, or expression level of the RNA. In order to perform computational analyses, several pipelines have been developed. As we will not cover the computational methods developed for the analysis of scRNA-seq data, we direct the reader to recently published reviews for more detailed discussion (Hie et al. 2020; Lähnemann et al. 2020).

Herein, we discuss in greater detail each strategy, their inherent advantages, limitations, and applications (Table 12.1).

12.2.1 CRISP-seq

CRISP-seq has been one of the first approaches that integrates pooled CRISPR screens with scRNA-seq technology (Jaitin et al. 2016). Like other similar techniques, CRISP-seq delivers CRISPR-Cas9 components to cells via lentiviral transduction. The vector used for this approach consists of a sgRNA expression cassette transcribed by RNA polymerase III (to cause perturbations), a fluorescent selection marker that enables the enrichment of transduced cells by flow cytometry and an RNA polymerase II-transcribed polyadenylated gBC that serves as a proxy for sgRNA identification. As the gBCs are located several kilobases away from

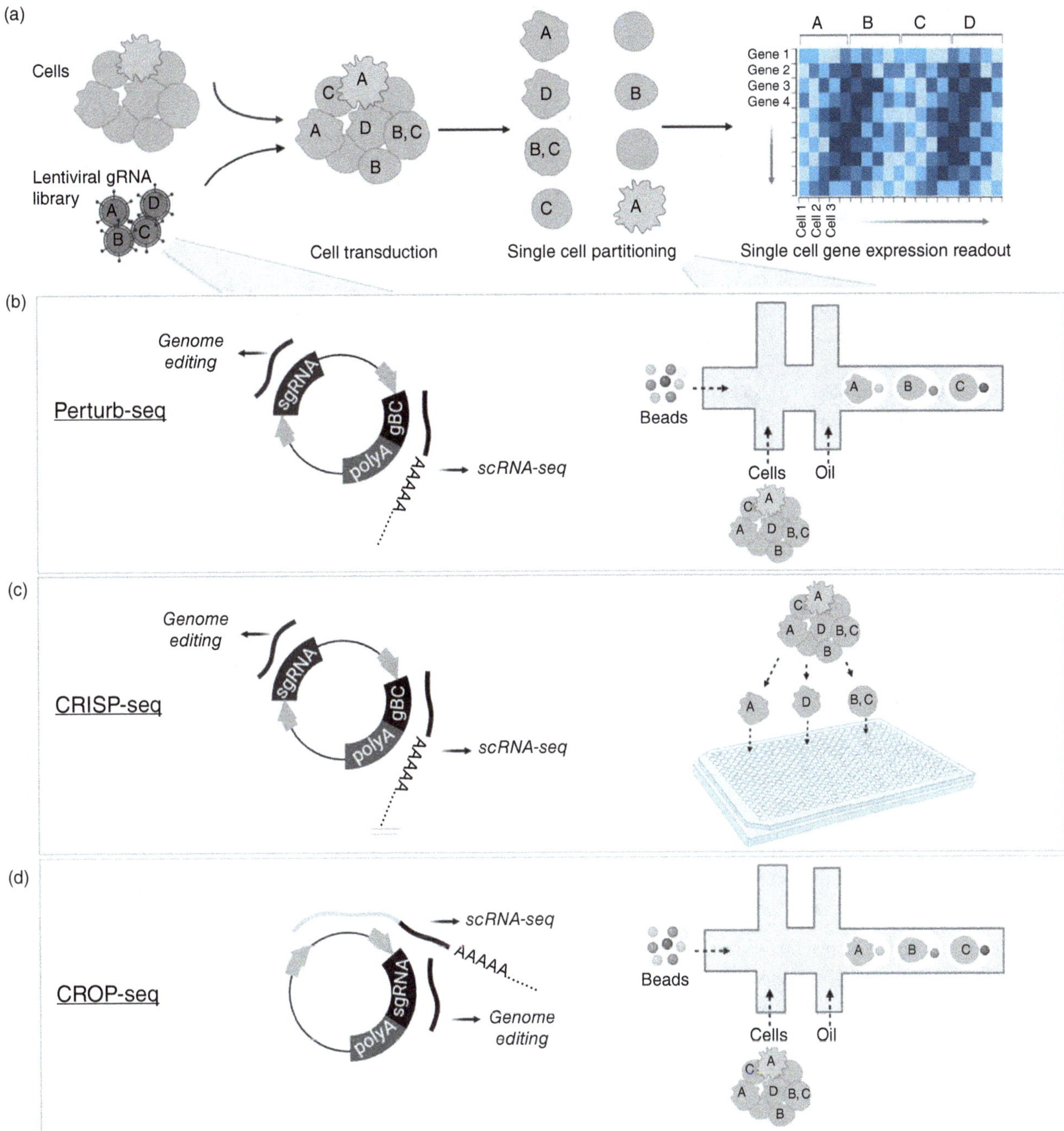

Figure 12.1 Schematic representation of scRNA-seq approaches for evaluating CRISPR-mediated perturbations. (a) Cells are transduced with lentiviral CRISPR sgRNA libraries, followed by partitioning into single cells. mRNA from each single cell along with the polyA-tailed sgRNA or its associated guide barcode (gBC) transcript is captured via a poly(T)-based scRNA-seq technology. The readout is each cell's transcriptional profile in response to the specific perturbation. Various approaches have been developed, each with its own choice of vector design and scRNA-seq technology. (b) Perturb-seq lentiviral vector consists of a gRNA cassette which expresses an RNA Polymerase III transcribed functional sgRNA transcript for genome editing, and a gBC cassette which expresses an RNA polymerase II transcribed polyadenylated gBC transcript captured during scRNA-seq to identify the sgRNA. It employs a droplet based scRNA-seq technology (now commercially available from 10x Genomics). (c) CRISP-seq lentiviral vector also consists of a gRNA cassette which expresses an RNA polymerase III transcribed functional sgRNA transcript for genome editing, and a gBC cassette which expresses an RNA polymerase-II transcribed polyadenylated gBC transcript as sgRNA identifier for scRNA-seq. It uses a plate-based scRNA-seq approach as in MARS-seq. (d) CROP-seq lentiviral vector consists of a gRNA cassette which gets duplicated during viral integration. It expresses an RNA Polymerase III transcribed functional sgRNA transcript for genome editing, and another RNA polymerase-II transcribed polyadenylated sgRNA transcript which gets sequenced during scRNA-seq. The scRNA method employed is a droplet-based microfluidic separation approach as in Drop-seq.

Table 12.1 Summary of pooled genetic screens that combine CRISPR perturbations with scRNA-seq.

Technology	Application	Advantage	Limitation	Reference
Perturb-seq	Identification of regulators of complex cellular phenotypes	No risk of barcode swapping with arrayed lentiviral production, sgRNA enrichment by amplification	No direct link between gRNA expression and observed phenotype; Limited scalability with arrayed lentiviral production	Dixit et al. (2016) and Adamson et al. (2016)
CRISP-seq	Identification of regulators of myeloid differentiation	Scalability due to pooled lentiviral production	Barcode swapping; Not ideal for higher throughput	Jaitin et al. (2016)
CROP-seq	Identification of key mediators of T-cell receptor induction	Improved vector design, scalable pooled lentiviral production, no risk of barcode swapping	Detection of sgRNA in only 40–60% of single cells; Larger size of the vector limits the inclusion of multiple sgRNAs	Datlinger et al. (2017)
Mosaic-seq	Identification of functional enhancer regions	Silencing of enhancers (rather than mutagenesis) abrogates enhancer functionality	Each sgRNA needs to infect hundreds of cells to gain statistical power; Barcode swapping	Xie et al. (2017)

their corresponding sgRNAs, viral recombination can generate barcode swapping among different sgRNAs, resulting in loss of accuracy and sensitivity (Hill et al. 2018). Another limitation of CRISP-seq is the use of a plate-based approach where single cells are sorted by flow cytometry into individual wells, followed by the preparation of scRNA-seq libraries. This strategy significantly limits the throughput of the approach.

This method has been applied both in *in-vitro* models and in mouse models to identify regulators of myeloid differentiation, hematopoiesis, and antiviral immune response pathways (Giladi et al. 2018; Jaitin et al. 2016).

12.2.2 Perturb-seq

Perturb-seq has been developed by two independent groups that combined different CRISPR-based perturbations with scRNA-seq technology. Dixit and colleagues employed the classical CRISPR methodology to knock-out specific target genes whereas Adamson et al. used CRISPRi technique to downregulate the transcripts of interest (Adamson et al. 2016; Dixit et al. 2016). As with CRISP-seq and Mosaic-seq, Perturb-seq lentiviral vector design includes an RNA polymerase III transcribed sgRNA expression cassette (to cause perturbations), a fluorescent selection marker that enables the enrichment of transduced cells by flow cytometry, and an RNA polymerase II-transcribed gBC that allows the identification of the sgRNA from scRNA-seq data. The cDNA obtained from gBC during the scRNA-seq library preparation is amplified using an enrichment PCR for improved detection of gBCs. Unlike CRISP-seq and Mosaic-seq that contemplate a pooled lentiviral production, Perturb-seq protocol uses an arrayed generation of lentiviral vectors, hence reducing the risks of barcode swapping. Perturb-seq differs from CRISP-seq also with respect to the preparation of scRNA-seq libraries by the use of droplet-based microfluidics (i.e. 10x Genomics platform) that increases its throughput.

Dixit and colleagues demonstrated the power of Perturb-seq by analyzing 200 000 cells in immune cells and cell lines, focusing on the transcription factors (TFs) regulating the response of dendritic cells to lipopolysaccharide (Dixit et al. 2016). The study from Adamson et al. investigated the unfolded protein response and uncovered different pathways that respond to specific stresses (Adamson et al. 2016).

More recently, Perturb-seq was applied *in vivo* to assess the effects of genetic disruption for genes associated with the risk of developing autism spectrum disorder or neurodevelopmental delay (Jin et al. 2020). Using a mice strain that constitutively expresses CRISPR-Cas9, the developing cortex *in utero* was transduced with lentiviral constructs targeting these genes, followed by single-cell transcriptomic analysis of perturbed cells from the early postnatal brain. Out of 35 targeted genes, 9 had a detectable effect at the transcriptomic level in different cell types, identifying lineage-specific transcriptional programs and highlighting the complexity of these pathologies (Jin et al. 2020). However, the remaining 26 gene ablations did not have any effect in the scRNA-seq data, raising the possibility that some gene deletions with no effect are not effectively mutated. This points out the difficult interpretation of negative results for all pooled genetic screenings that use scRNA-seq as phenotypic readout.

Direct-capture Perturb-seq is a recently developed method which directly captures and amplifies the expressed sgRNAs alongside single-cell transcriptomes, and does not rely on gBC transcripts for indirect indexing of sgRNAs (Replogle et al. 2020). This method involves a modification of the design of scRNA-seq-compatible sgRNAs. The beads that capture and prime the RT of cellular mRNA transcripts through a complementary poly(T) oligo in the droplet-based 3' scRNA-seq technology have additional oligos with a specific capture sequence. A complementary capture sequence is added to the constant region of the sgRNA, enabling the direct capture and RT priming of sgRNA transcripts. 3' direct-capture Perturb-seq is compatible with approaches aimed at lineage tracing (Chan et al. 2019).

This method also combines single-cell CRISPR screens with 5' scRNA-seq technology, whereas all the above-described methods only employ 3'- scRNA-seq technology. In 5' scRNA-seq, the mRNA transcripts are first reverse transcribed using a poly(T) primer, before their 5' end capture and cell barcoding happens. Similar to their 3' approach, the sgRNA transcripts for 5' assay also contained a capture sequence (different from polyA sequence) in their constant region. They are then first reverse transcribed using a primer complementary to that capture sequence, followed by bead capture from 5' end and cell barcoding alongside mRNA. A similar approach for 5' assay called ECCITE-seq was published around the same time (Mimitou et al. 2019). 5' direct-capture Perturb-seq enables the analysis of V(D)J clonotypes in response to genetic perturbations.

Major advantages of this approach are the direct identification of sgRNA and its design flexibility that provides the possibility of working with both 5' and 3' scRNA-seq.

12.2.3 CROP-seq

CROP-seq (CRISPR droplet sequencing) is a complementary approach that integrates pooled CRISPR screening and scRNA-seq technology with some key differences (Datlinger et al. 2017). Unlike Perturb-Seq and CRISP-seq that rely on a cloning strategy that indirectly links each sgRNA to a specific barcoded transcript detected by scRNA-seq, CROP-seq detects the sgRNA directly by scRNA-seq, without the use of barcodes. The lentiviral vector also includes an RNA polymerase III transcribed functional sgRNA transcript that causes perturbation, like the above-described methods. But its vector design positions the sgRNA cassette such that the RNA polymerase II also generates a copy of the sgRNA transcript and add a poly(A) tail to it, which makes it detectable by scRNA-seq technology together with the cellular mRNAs. Furthermore, during reverse transcription and viral integration, the functional Pol III transcribed sgRNA cassette is duplicated, therefore providing additional copies of sgRNAs for the gene repression (Datlinger et al. 2017). Additionally, this strategy simplifies the construction of large-scale sgRNA libraries since barcodes do not need to be anymore included.

Similarly to Perturb-seq and Mosaic-seq, CROP-seq employs a droplet-based microfluidic approach to generate scRNA libraries to maximize the throughput. By using CROP-seq, Datlinger and colleagues identified proteins involved in the signaling cascade upon T cell receptor activation (Datlinger et al. 2017).

However, CROP-seq presents some limitations. Due to the constraints on the cassette size, CROP-seq is likely incompatible with the delivery of multiple sgRNAs with the same vector. Also, only about half of the sgRNAs are recovered from the scRNA-seq data, leading to limited sensitivity of the generated single-cell transcriptomes (Datlinger et al. 2017; Hill et al. 2018). Hill and colleagues have proposed an improved approach that includes a targeted amplification of the sgRNA cassette during scRNA-seq preparation (Hill et al. 2018). This additional step significantly increases the assignment rate of guides to cells from the 42–47% of the original CROP-seq publication to 94% (Hill et al. 2018).

An extended application of CROP-seq, expression CROP-seq, provides an experimental approach to screen single nucleotide polymorphisms (SNPs) that are known to likely influence disease risk based on the statistical inferences from genome-wide association studies (Pan et al. 2020). Candidate SNPs are targeted by gRNAs for mutations, micro-deletion or micro-insertions at or near the SNP site. scRNA-seq then provides a readout of the change in expression of the transcripts linked to that SNP. To validate this approach, they screened several SNPs located in regulatory genomic intervals of three genes previously associated with inflammatory bowel disease. Although promising as a tool for fine-mapping of causal variants, this approach is still limited by the occurrence of gRNA recognition sequences (protospacer adjacent motif – PAM sequences) encompassing the SNP of interest and the reduced power for lowly expressed genes (Pan et al. 2020).

A recently developed modification of CROP-seq combines targeted scRNAseq (instead of the whole transcriptome assay used in rest of the methods) with pooled genetic screens. This new method, targeted Perturb-seq (TAP-seq), addresses some limitations of traditional methods concerning the inefficient measurement of lowly expressed genes and the prohibitive cost of a genome-scale screen (Schraivogel et al. 2020). Instead of analyzing the expression levels of all genes from single cells, TAP-seq focuses sequencing coverage on a preselected set of genes that are relevant to a specific biological question. By amplifying genes of interest with target-specific nested primers, TAP-seq identifies changes in gene expression with higher sensitivity compared with the original Perturb-seq, while lowering the sequencing requirements up to 50-fold (Schraivogel et al. 2020). The authors used TAP-seq to generate high-resolution, function-based enhancer–promoter interaction maps for two large genomic regions, identifying regulatory regions for a much larger fraction of candidate genes (24–32%) as opposed to only 3.5–6% identified by CROP-seq in a previous study (Gasperini et al. 2019).

12.2.4 Mosaic-seq

Mosaic-seq combines CRISPRi and scRNA-seq, allowing the functional characterization of the noncoding regulatory elements, such as enhancer regions in their physiological state (Xie et al. 2017). In this approach, the enhancer regions are repressed rather than mutated to ensure that the regulatory element is functionally altered. In Mosaic-seq protocol, the first step is to establish a cell population that stably expresses the potent repressor domain KRAB fused to a catalytically inactive Cas9 (dCas9). A library of vectors that share similar design with CRISP-seq and Perturb-seq, expressing an sgRNA targeting a specific enhancer and a gBC for identifying the linked sgRNA, is introduced in the cells via a lentiviral transduction. Finally, scRNA-seq libraries are prepared from the cell pool by following a similar approach as in Drop-seq (Macosko et al. 2015). By using Mosaic-seq, Xie and colleagues dissected the relative contribution of each enhancer element clustered into specific regulatory regions called superenhancers (Xie et al. 2017). The high resolution of the single-cell phenotypic readout combined with the powerful genome-editing approach identified the surprising cell-to-cell variability in enhancer usage that controls the expression of specific genes. Apart from the potential issues with barcode swapping (as in CRISP-seq), one limitation of Mosaic-seq is the requirement of a robust number of infected cells with each specific sgRNA to gain statistical power during the analysis of scRNA-seq data.

12.3 CRISPR-Based Genetic Screens with Single-Cell Epigenomics Readout

CRISPR-mediated perturbation experiments can also induce changes at the level of the epigenome. With the term epigenomics, we include the study of a wide class of modifications that affect gene expression without changing the underlying DNA sequence. These include chromatin accessibility, DNA methylation, histone modifications, and transcription factors occupancy. Historically, technological advances in the epigenomic field have followed methodological progresses achieved in the transcriptomic field and only recently researchers have started to investigate state- or condition-specific epigenomic signatures at the single-cell level (Clark et al. 2016).

Compared with transcriptomics-based approaches, single-cell epigenomic assays have the potential to uncover mechanisms that govern gene regulation by identifying mediators of chromatin structure, transcription initiation, or elongation. Therefore, single-cell epigenomic techniques can be employed to assess both cellular diversity and discover gene regulatory mechanisms. Single-cell epigenomic technologies targeting chromatin accessibility or histone modifications can identify cis-regulatory elements (e.g. enhancers) that control the expression of nearby genes from trans-acting factors such as TFs involved in the activation of specific loci.

Transcription activity is tightly regulated by the coordinated action of chromatin modifiers, TFs, and general accessibility of the genome regulatory regions. Although cell states defined by single-cell transcriptomics should correlate with cell states detected by single-cell epigenomics, some biological processes can induce significant mismatches between epigenomic and transcriptomic landscapes. For example, promoter regions can be classified as repressed, poised, or active according to their histone modifications profile. Although repressed and poised promoters are characterized by the absence of transcriptional output, their epigenetic states are different, thus enabling a more granular classification that would not be possible using scRNA-seq. Conversely, ligand-activated responses can induce fast changes at the transcriptional level without exhibiting significant differences at the chromatin level.

As epigenomic technologies can detect different layers of gene regulation, here we give a brief overview of different approaches that can be used to assess changes in the epigenetic status at the single-cell level following CRISPR perturbations (Table 12.2).

Chromatin accessibility defines genomic regions that are not occupied by nucleosomes and therefore available for interaction with TFs and the transcription machinery. Therefore, open chromatin regions include actively transcribed segments and active regulatory elements that impact gene expression. ATAC-seq (Assay for Transposase-Accessible Chromatin) and DNase-seq represent most commonly used technologies that identify genomic regions in open chromatin configuration both in bulk and single-cell populations (Buenrostro et al. 2015; Jin et al. 2015).

Recent technological improvements of already established technologies have enabled the identification of histone modifications and TFs binding sites at the single-cell level. Chromatin immunoprecipitation sequencing (ChIP-seq) and Cut&Tag represent two separate approaches that interrogate mechanisms responsible for the regulation of genetic programs underlying distinct cell states (Rotem et al. 2015; Wu et al. 2020).

Table 12.2 Epigenomic technologies that have been developed for single-cell resolution.

Technology	Approach	Use	Reference
DNase-seq	Use of DNase I to detect open chromatin regions	Genome-wide identification of chromatin accessibility	Jin et al. (2015)
ATAC-seq	Use of hyperactive Tn5 to detect open chromatin regions	Genome-wide identification of chromatin accessibility	Buenrostro et al. (2015)
ChIP-seq	Immunoprecipitation of proteins of interest using a specific antibody followed by sequencing of bound genomic regions	Identification of histone modifications and TFs binding sites	Rotem et al. (2015)
Cut&Tag	Antibody-specific insertion of adapter sequences nearby regions bound by protein of interest using a proteinA-Tn5 fusion protein	Identification of histone modifications and TFs binding sites	Wu et al. (2020)
DNA methylation (several technologies)	Bisulfite and bisulfite-free approaches	Identification of repressed genomic regions	Karemaker and Vermeulen (2018)

Another important layer of epigenomic regulation is the methylation state of specific genomic regions. Addition of methyl marks to DNA intervals causes compaction of chromatin and repression of gene expression. During the recent years, several methods have been developed to identify genomic regions that are silenced in specific cell populations or cell states (Karemaker and Vermeulen 2018).

Recent studies have shown that several genes are involved in the establishment or maintenance of epigenetic modifications (Cavalli and Heard 2019). Therefore, changes in specific epigenomic marks following CRISPR perturbations can confirm the role of known chromatin regulators or reveal novel players involved in gene regulatory networks. The only study that has so far used CRISPR-based perturbation employing single-cell epigenomics measurements is perturbation-indexed single-cell ATAC-seq (Perturb-ATAC) (Rubin et al. 2019). Perturb-ATAC experimental strategy consists of transducing a cell population expressing CRISPR-Cas9 with a pool of lentiviral vectors containing different sgRNAs. As in Perturb-seq, Perturb-ATAC vector design includes an RNA polymerase II-driven marker gene containing a gBC to enrich for transduced cells and indirectly identify the sgRNA, respectively. To confirm the wide applicability of the approach, the authors performed CRISPRi and CRISPR Knockout (KO) technologies separately. After CRISPR-based perturbation, the cells are captured in single-cell chambers using a microfluidic platform, lysed and subjected to scATAC-seq protocol with the Tn5 transposase to identify regions of open chromatin. After DNA transposition, the transcribed barcodes that identify the sgRNAs are reverse transcribed using target-specific primers. Transposed genomic DNA and gBC cDNA are then PCR amplified with single-cell-specific barcoded primers, pooled and sequenced (Rubin et al. 2019). By integrating CRISPR perturbations on TFs, chromatin-modifying factors, and noncoding RNAs with scATAC-seq, Perturb-ATAC identified regulators of chromatin accessibility, a hierarchy of TFs that control specific cell states and epistatic interactions between trans-acting factors. For future studies, the authors envisioned small modifications of the design strategy to increase the number of targets while decreasing the number of cells infected with the same sgRNAs. A current limitation is the relatively low-throughput that can be improved in the following iterations by using other approaches for sorting single cells (e.g. FACS).

A very recent preprint seems to address some of these limitations by proposing a much higher throughput droplet-based method named SPEAR-ATAC (Single-cell perturbations with an accessibility read-out using scATAC-seq). It offers an additional advantage of directly reading integrated sgRNA from the genome, as opposed to relying on sgRNA identifying transcripts (Pierce et al. 2020).

12.4 Future Perspectives

Recent advances in pooled genetic screenings that integrate CRISPR-mediated genome editing with single-cell phenotypic readouts have enabled researchers to elucidate gene function, characterize regulatory networks, and identify genetic interactions. Future improvements in the fields of genome editing and single-cell methodologies will likely have a major impact in our understanding of the precise mechanisms that govern gene expression and cell identity. Ongoing efforts to bring the cost per cell down and to increase the throughput further will make these approaches more accessible and scalable. A possible approach to reduce the sequencing cost is to perform enrichment of targets and depletion of low interest, high-abundance transcripts. Another area for development is handling the high-dimensional and sparse data generated by these methods. Parallel advances in data analysis, especially with increase in the scale of such studies, will be a pressing requirement.

One current limitation of these approaches is that the CRISPR-induced mutations introduced into the analyzed cells are not assessed directly in the data but are linked to the detection of the sgRNA or its associated gBC. Furthermore, gene ablation is not measured efficiently for each target gene, leaving the possibility that a KO with no effect is not perturbed in reality, resulting in important genes not being identified in the screens.

Another important area of improvement will be identification of off-target genome editing. Due to nonspecific binding of the gRNA, editing of undesired genomic locations can confound the downstream analyses of on-target perturbations or limit therapeutic applications. Future progress that increases sequencing reads coverage for each single cell will enable the identification of off-target events in individual cells. Designing experiments where multiple gRNAs targeting the same gene in a significant number of transduced cells will help researchers to dissect on-target responses (significantly affected transcripts shared between the two gRNAs) from off-targets (significantly altered transcripts that are not shared between the two gRNAs).

The very recent development of new technological approaches that pair chromatin accessibility with whole transcriptome measurements at the single-cell level will provide valuable tools for the identification of key regulators of gene expression as well as their mechanism of action (Chen et al. 2019).

Another emerging theme is the role of genome organization in regulation of gene expression. Although few publications have shown the application of chromatin conformation capture technologies at the single nuclei level, major drawbacks are the limited amount of cells that can be processed in a single experiment and the sparse matrices derived from each single cell (Díaz et al. 2018). With future technological improvements, we could envision the integration of CRISPR-screens and single-cell Hi-C sequencing to study chromatin conformation changes at greater resolution and identify its relationship with the regulation of gene expression.

Acknowledgments

Figure 12.1 has been generated with Biorender.

References

Adamson, B., Norman, T.M., Jost, M. et al. (2016). A multiplexed single-cell CRISPR screening platform enables systematic dissection of the unfolded protein response. *Cell* 167: 1867–1882. https://doi.org/10.1016/j.cell.2016.11.048.

Buenrostro, J.D., Wu, B., Litzenburger, U.M., Ruff, D., Gonzales, M.L., Snyder, M.P., Chang, H.Y., Greenleaf, W.J., 2015. Single-cell chromatin accessibility reveals principles of regulatory variation. *Nature*. 523, 486–490. https://doi.org/10.1038/nature14590.

Cavalli, G. and Heard, E. (2019). Advances in epigenetics link genetics to the environment and disease. *Nature* 571: 489–499. https://doi.org/10.1038/s41586-019-1411-0.

Chan, M.M., Smith, Z.D., Grosswendt, S. et al. (2019). Molecular recording of mammalian embryogenesis. *Nature* 570: 77–82. https://doi.org/10.1038/s41586-019-1184-5.

Chen, S., Lake, B.B., and Zhang, K. (2019). High-throughput sequencing of the transcriptome and chromatin accessibility in the same cell. *Nat. Biotechnol.* 37: 1452–1457. https://doi.org/10.1038/s41587-019-0290-0.

Clark, S.J., Lee, H.J., Smallwood, S.A. et al. (2016). Single-cell epigenomics: powerful new methods for understanding gene regulation and cell identity. *Genome Biol.* 17: 72. https://doi.org/10.1186/s13059-016-0944-x.

Datlinger, P., Rendeiro, A.F., Schmidl, C. et al. (2017). Pooled CRISPR screening with single-cell transcriptome readout. *Nat. Methods* 14: 297–301. https://doi.org/10.1038/nmeth.4177.

Díaz, N., Kruse, K., Erdmann, T. et al. (2018). Chromatin conformation analysis of primary patient tissue using a low input Hi-C method. *Nat. Commun.* 9: 4938. https://doi.org/10.1038/s41467-018-06961-0.

Dixit, A., Parnas, O., Li, B. et al. (2016). Perturb-seq: dissecting molecular circuits with scalable single-cell RNA profiling of pooled genetic screens. *Cell* 167: 1853–1866.e17. https://doi.org/10.1016/j.cell.2016.11.038.

Gasperini, M., Hill, A.J., McFaline-Figueroa, J.L. et al. (2019). A genome-wide framework for mapping gene regulation via cellular genetic screens. *Cell* 176: 377–390. https://doi.org/10.1016/j.cell.2018.11.029.

Gawad, C., Koh, W., and Quake, S.R. (2016). Single-cell genome sequencing: current state of the science. *Nat. Rev. Genet.* 17: 175–188. https://doi.org/10.1038/nrg.2015.16.

Giladi, A., Paul, F., Herzog, Y. et al. (2018). Single-cell characterization of haematopoietic progenitors and their trajectories in homeostasis and perturbed haematopoiesis. *Nat. Cell Biol.* 20: 836–846. https://doi.org/10.1038/s41556-018-0121-4.

Hie, B., Peters, J., Nyquist, S.K. et al. (2020). Computational methods for single-cell RNA sequencing. *Annu. Rev. Biomed. Data Sci.* 3: 339–364. https://doi.org/10.1146/annurev-biodatasci-012220-100601.

Hill, A.J., McFaline-Figueroa, J.L., Starita, L.M. et al. (2018). On the design of CRISPR-based single-cell molecular screens. *Nat. Methods* 15: 271–274. https://doi.org/10.1038/nmeth.4604.

Jaitin, D.A., Kenigsberg, E., Keren-Shaul, H. et al. (2014). Massively parallel single-cell RNA-seq for marker-free decomposition of tissues into cell types. *Science* 343: 776–779. https://doi.org/10.1126/science.1247651.

Jaitin, D.A., Weiner, A., Yofe, I. et al. (2016). Dissecting immune circuits by linking CRISPR-pooled screens with single-cell RNA-seq. *Cell* 167: 1883–1896.e15. https://doi.org/10.1016/j.cell.2016.11.039.

Jin, W., Tang, Q., Wan, M. et al. (2015). Genome-wide detection of DNase I hypersensitive sites in single cells and FFPE tissue samples. *Nature* 528: 142–146. https://doi.org/10.1038/nature15740.

Jin, X., Simmons, S.K., Guo, A. et al. (2020). In vivo perturb-seq reveals neuronal and glial abnormalities associated with autism risk genes. *Science* 370: eaaz6063. https://doi.org/10.1126/science.aaz6063.

Junker, J.P. and van Oudenaarden, A. (2014). Every cell is special: genome-wide studies add a new dimension to single-cell biology. *Cell* 157: 8–11. https://doi.org/10.1016/j.cell.2014.02.010.

Karemaker, I.D. and Vermeulen, M. (2018). Single-cell DNA methylation profiling: technologies and biological applications. *Trends Biotechnol.* 36: 952–965. https://doi.org/10.1016/j.tibtech.2018.04.002.

Lähnemann, D., Köster, J., Szczurek, E. et al. (2020). Eleven grand challenges in single-cell data science. *Genome Biol.* 21: 31. https://doi.org/10.1186/s13059-020-1926-6.

Lau, E. (2014). CRISPR screening from both ways. *Nat. Rev. Genet.* 15: 778–779. https://doi.org/10.1038/nrg3850.

Macosko, E.Z., Basu, A., Satija, R. et al. (2015). Highly parallel genome-wide expression profiling of individual cells using nanoliter droplets. *Cell* 161: 1202–1214. https://doi.org/10.1016/j.cell.2015.05.002.

Mimitou, E.P., Cheng, A., Montalbano, A. et al. (2019). Multiplexed detection of proteins, transcriptomes, clonotypes and CRISPR perturbations in single cells. *Nat. Methods* 16: 409–412. https://doi.org/. doi: 10.1038/s41592-019-0392-0.

Moignard, V. and Göttgens, B. (2014). Transcriptional mechanisms of cell fate decisions revealed by single cell expression profiling. *Bioessays* 36: 419–426. https://doi.org/10.1002/bies.201300102.

Pan, Y., Tian, R., Lee, C., Bao, G., Gibson, G., 2020. Fine-mapping within eQTL credible intervals by expression CROP-seq. *Biol. Methods Protoc.* 5, bpaa008. https://doi.org/10.1093/biomethods/bpaa008.

Picelli, S., Faridani, O.R., Björklund, A.K. et al. (2014). Full-length RNA-seq from single cells using smart-seq2. *Nat. Protoc.* 9: 171–181. https://doi.org/10.1038/nprot.2014.006.

Pierce, S.E., Granja, J.M., and Greenleaf, W.J. (2020). High-throughput single-cell chromatin accessibility CRISPR screens enable unbiased identification of regulatory networks in cancer. *BioRxiv* https://doi.org/10.1101/2020.11.02.364265.

Replogle, J.M., Norman, T.M., Xu, A. et al. (2020). Combinatorial single-cell CRISPR screens by direct guide RNA capture and targeted sequencing. *Nat. Biotechnol.* 38: 954–961. https://doi.org/10.1038/s41587-020-0470-y.

Rotem, A., Ram, O., Shoresh, N. et al. (2015). Single-cell ChIP-seq reveals cell subpopulations defined by chromatin state. *Nat. Biotechnol.* 33: 1165–1172. https://doi.org/10.1038/nbt.3383.

Rubin, A.J., Parker, K.R., Satpathy, A.T. et al. (2019). Coupled single-cell CRISPR screening and epigenomic profiling reveals causal gene regulatory networks. *Cell* 176: 361–376. https://doi.org/10.1016/j.cell.2018.11.022.

Schraivogel, D., Gschwind, A.R., Milbank, J.H. et al. (2020). Targeted perturb-seq enables genome-scale genetic screens in single cells. *Nat. Methods* 17: 629–635. https://doi.org/10.1038/s41592-020-0837-5.

Waylen, N.L., Nim, H.T., Martellotto, L.G., and Ramialison, M. (2020). From whole-mount to single-cell spatial assessment of gene expression in 3D. *Commun. Biol.* 3: 602. https://doi.org/10.1038/s42003-020-01341-1.

Wu, S.J., Furlan, S.N., Mihalas, A.B. et al. (2020). Single-cell analysis of chromatin silencing programs in developmental and tumor progression. *Biorxiv* https://doi.org/10.1101/2020.09.04.282418.

Xie, S., Duan, J., Li, B. et al. (2017). Multiplexed engineering and analysis of combinatorial enhancer activity in single cells. *Mol. Cell* 66: 285–299.e5. https://doi.org/10.1016/j.molcel.2017.03.007.

Part 4

Therapeutic Genome Editing

Translation of genome editing technology to the clinic is underway, with announced or impending clinical trials utilizing *ex vivo* strategies for anticancer immunotherapy (for example, CAR T cells) or correction of hemoglobinopathies. This section covers the various aspects of therapeutic genome editing with examples of base editors, RNA editing, applications for T cell therapy, editing in stem cells, and formulation challenges for genome editing.

13

DNA Repair Pathways in the Context of Therapeutic Genome Editing

Sandra Wimberger[1], *Nina Akrap*[1], *and Amir Taheri-Ghahfarokhi*[2]

[1] *Discovery Biology, Discovery Sciences, BioPharmaceuticals R&D, AstraZeneca, Gothenburg, Sweden*
[2] *Quantitative Biology, Discovery Sciences, BioPharmaceuticals R&D, AstraZeneca, Gothenburg, Sweden*

13.1 Reprogrammable Nucleases

Before the era of genome editing and employment of designer nucleases for targeted gene editing, the gene therapy field utilized random integration of transgenes using viral vectors. In the early 2000s, a clinical study aiming to treat X-linked severe combined immune deficiency (XSCID) in children (Cavazzana-Calvo et al. 2000) exposed a major complication with this approach. Tragically, two of the patients developed leukemia, which could possibly be linked to the integration of the transgene near an oncogene and its subsequent transactivation (Hacein-Bey-Abina et al. 2003). This incident posed a setback on gene therapy programs and highlighted the demand for methods to specifically repair disease-causing mutations in their genomic context in place of viral-mediated random integrations.

The development of programmable nucleases enables researchers to introduce targeted DNA double-strand breaks (DSBs) at the disease-causing genomic site with acceptable efficiency and specificity for clinical applications. Initially, researchers developed meganucleases and zinc-finger nucleases (ZFNs) (Urnov et al. 2010), followed by the expansion of the tool set with transcription activator-like effector nucleases (TALENs) (Christian et al. 2010). ZFNs and TALENs consist of distinct DNA-binding motifs, linked to the cleavage domain of a restriction enzyme called *Fok*I. The *Fok*I cleavage domain requires protein dimerization to cut the DNA, therefore a pair of ZFNs or TALENs must bind to the DNA in favorable proximity and orientation to enable *Fok*I dimerization. This in turn increases the chance of locating a unique site in large genomes. ZFNs and TALENs enable targeted genome editing, however, their construction requires complex protein engineering for each target site, representing a considerable disadvantage.

A major breakthrough in therapeutic genome editing (TGE) was the adaptation of CRISPR-Cas (clustered regularly interspaced short palindromic repeats/CRISPR-associated protein) systems for genome editing purposes in human cells (Cong et al. 2013). CRISPR-centric technologies overcome the limitations of ZFNs and TALENs by providing a simple and rapid method for reprogramming the nuclease, thus most of todays' clinical trials employ CRISPR-related designer nucleases.

Designer nucleases allow researchers to introduce a DSB in a region of interest, but ultimately it is the cells' own DNA repair mechanisms that detect and repair the DNA lesions. Therefore, an understanding of the underlying DNA repair pathways behind the formation of genome editing events is fundamental in designing safe and efficacious gene editing strategies. Genome editing strategies that are currently in ongoing clinical trials utilize either non-homologous end joining (NHEJ), or homology-directed repair (HDR) pathways. In this chapter, we review clinical studies from the last decade using ZFNs, TALENs, meganucleases, and CRISPR-Cas as identified on clinicaltrials.gov in the context of the exploited DNA repair pathway. As of January 2021, there are 56 clinical trials for TGE: nine studies using TALENs, 15 studies with ZFNs, and 31 studies employing CRISPR-Cas systems.

13.2 DNA Double-Strand Break Repair Pathways

DSBs, regardless of the underlying cause, are among the most detrimental assaults to the integrity of a genome. Hence, eukaryotic cells have evolved an array of sophisticated pathways and signaling networks to maintain their genomic integrity. There are four distinct DSB repair pathways in human cells, in order of their dependence on the

Genome Editing in Drug Discovery, First Edition. Edited by Marcello Maresca and Sumit Deswal.

homologous base pairs, non-homologous end joining (NHEJ), microhomology-mediated end joining (MMEJ), single-strand annealing (SSA), and homology-directed repair (HDR) (Figure 13.1). The cells' choice for the repair pathways and the balance between them depends on several factors, *e.g.* characteristics of DNA lesion, cell cycle state, and the DNA repair capacity of the cell (Scully et al. 2019).

To some extent, the choice of DNA repair pathway is determined even before the occurrence of a DSB, because cells keep their genome under persistent surveillance using proteins that sense DSBs and influence the choice of DNA repair pathways. It is remarkable how fast human cells can detect a DSB in their 3 billion base pairs large genome. Phosphorylation of the H2AX (called γH2AX), a sensitive biomarker for DSBs, is detectable within minutes after gamma radiation and maximizes 30 minutes postradiation (Redon et al. 2009; Mah et al. 2010). Sensing DSBs is the specialty of a few proteins: (i) Ku dimer – Ku70 and Ku80 – encoded by *XRCC6* and *XRCC5* genes, respectively, (ii) MRN complex encoded by *MRE11*, *RAD50*, *NBN* genes, (iii) PARP1, and (iv) SIRT6 (Onn et al. 2020). The first proteins that arrive at the break site influence the downstream signaling and choice of repair pathway. For example, Ku dimers and MRN complex are known to be associated with NHEJ and HDR, respectively. A logical implication of DSB sensors' influence on the choice of the DNA repair pathways is that their abundance may correlate with the prevalence of a specific DNA repair pathway. Based on previous comprehensive analyses of gene expression in human tissues (Fagerberg et al. 2014), we could perform a side-by-side comparison for the abundance of the sensing proteins across different tissues to provide a better overview of their expression. Genes that encode DSB sensing proteins are ubiquitously expressed across all human tissues (Figure 13.2). Interestingly, Ku dimers are consistently expressed in all tissues at several folds higher than the MRN complex.

13.2.1 Non-Homologous End Joining (NHEJ)

Ku dimers are the key DSB sensors that promote the repair by NHEJ and are highly abundant in human cells. Consequently, NHEJ is the dominant repair pathway in human cells across different tissues. Ku dimers form a ring-like structure that encircles the broken ends, thus limiting the exposure of DNA ends to the enzymes that catalyze DNA end resection. In addition to the Ku dimers, tumor protein p53 binding protein 1 (53BP1; encoded by the *TP53BP1* gene) is also known as an early NHEJ player with an anti-resection role. In addition to the detection and protection of DNA broken ends, Ku dimers also provide a docking station for other NHEJ components to join and orient the DNA repair. DNAPK catalytic subunit (DNAPKcs, encoded by the *PRKDC* gene) and Ligase 4 (encoded by the *LIG4* gene) are considered as other key NHEJ players (Ceccaldi et al. 2016). Compound-mediated inhibition or genetic loss of Ku70/80, 53BP1, DNAPK, or Ligase 4 shifts DNA repair toward other repair pathways (Yeh et al. 2019).

There are a few aspects about NHEJ that are especially relevant in the context of genome editing. First, the term "non-homologous" does not imply that NHEJ is not capable of using homologies during the repair process. This term has been used historically to emphasize its difference to HDR, which uses several hundreds of homologous base-pairs (bp) for high-fidelity DNA repair. NHEJ has been reported to use microhomologies of 1–4 bp. Second, NHEJ often has been referred to as an "error-prone" pathway, which undermines NHEJs' precision. The term "error-prone" does not imply that NHEJ events always cause errors, rather denoting this possibility besides perfect repair. In contrast, MMEJ is inherently erroneous. Indeed,

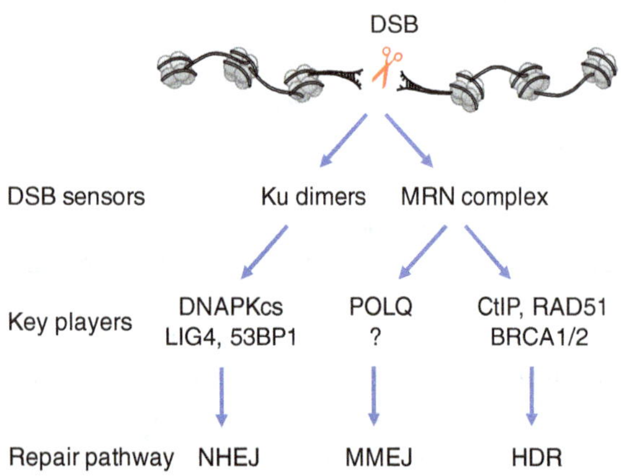

Figure 13.1 Schematic representation of the major DNA double-strand break (DSB) repair pathways in human cells. The first main responders to the DSBs include Ku dimer and MRN complex that orient the repair of the DSBs to either of the homology-independent or homology-dependent repair pathways, respectively. NHEJ: non-homologous end joining, MMEJ: microhomology-mediated end joining, HDR: homology-directed repair.

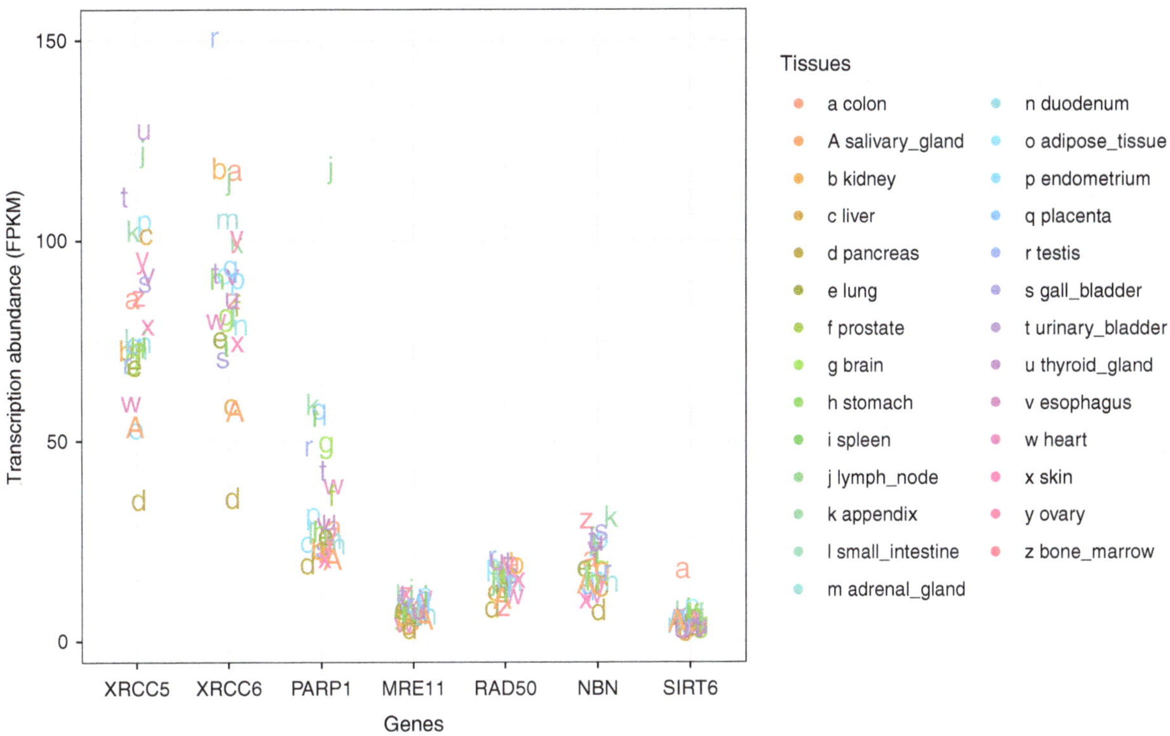

Figure 13.2 Transcript abundance of DSB sensor proteins across normal tissues. Quantification scores for selected genes are represented as FPKM (fragments per kilobase of exon model per million mapped reads) values. *Source*: Data shown in this figure were obtained from a previous quantitative transcriptomics analysis (RNA-Seq) of genes across a set of human organs and tissues (Fagerberg et al. 2014).

several lines of evidence support the precision of NHEJ, *e.g.* transgenes could precisely ligate into the Cas9 cleavage site via NHEJ (Maresca et al. 2013; Suzuki et al. 2016), and elevated levels of mutagenesis have been reported when TREX2, an exonuclease, is coupled with Cas9 transfection (Chari et al. 2015; Taheri-Ghahfarokhi et al. 2018). Third, NHEJ-associated insertions and deletions (indels) are often referred to as "random," while it has been shown that the interplay between Cas9 cuts and repair by NHEJ results in highly predictable outcome (Li et al. 2015; van Overbeek et al. 2016). Cas9 generates a mixture of blunt and single-stranded overhangs in a context-dependent manner. For example, it is more likely that Cas9 generates single nucleotide overhangs at the target sites containing thymidine or guanine at the fourth and third positions before a protospacer adjacent motif (PAM), respectively (Taheri-Ghahfarokhi et al. 2018). This results in highly predictable single nucleotide insertion and deletions at the Cas9 cut sites, of which insertions are indeed duplications of the preceding nucleotides. Fourth, as explained above, NHEJ inherently minimizes end processing of the broken ends. Thus, NHEJ-associated indels are expected to be small.

In TGE, the NHEJ pathway has been harnessed to create gene knockouts. High prevalence of NHEJ reduces the efficacy of HDR-mediated strategies. However, NHEJ-mediated knock-in (KI) approaches would offer an attractive alternative in scenarios where a transgene integration is desirable. NHEJ can be used to efficiently insert transgenes through usage of DNA donor templates without homology arms, a strategy termed homology-independent targeted integration (HITI) (Suzuki et al. 2016). This approach takes advantage of the blunt-end cleavage by Cas9 to insert a blunt-end linear donor and is active in dividing and to some extent in non-dividing cells, both *in vitro* and *in vivo*. Integration of the DNA donor can theoretically appear in forward and reverse orientations. Moreover, previous reports have shown high prevalence of unintended integration of the adeno-associated virus (AAV) genome in muscle, eye, and liver cells (Hanlon et al. 2019; Nelson et al. 2019). Even though this causes serious safety concerns for the use of AAV as a tool for Cas9 and single guide RNA (sgRNA) delivery, it also implies an efficacious route for delivery and integration of therapeutic transgenes intended for site-specific integration. Moreover, for short transgenes which are shorter than half of AAVs' capacity, both strands of the AAV could be used to ensure transgene expression regardless of the orientation the AAV genome integrates into the cleavage site.

13.2.2 Microhomology-Mediated End Joining (MMEJ)

The terms MMEJ and alternative end joining (alt-EJ, a-EJ, also known as alt-NHEJ) have been used interchangeably throughout the literature, all referring to an inherently erroneous pathway that relies on microhomologies (less than 20 bp) to repair broken DNA ends. The term alternative has been used for this pathway because it has been assumed that MMEJ acts in NHEJ-deficient cells; however, several lines of evidence support MMEJ as a major pathway that contributes to the DSB repair even in cells proficient for NHEJ. Thus, in this chapter, we intentionally avoided using the "alt-EJ" term. Moreover, MMEJ represents the nature of this pathway more accurately than alt-EJ.

The first responder to the DSBs that can initialize the repair by the MMEJ pathway is the MRN complex. MRN complex acts in the repair of DSBs in several ways including detection, signaling, and processing of the broken ends. MRN complex can promote the repair by any of the repair pathways; however, it is playing an essential role in 3′–5′ resection of the DNA ends which provides short single-stranded DNA substrates for homology-dependent repair pathways, *i.e.* MMEJ, SSA, and HDR. In addition to the MRN complex, DNA Polymerase Theta (POLQ) plays a key role in the MMEJ pathway (Taheri-Ghahfarokhi et al. 2018). Analysis of Cas9-induced deletions in POLQ knockout human cells revealed a significant reduction in the formation of MMEJ-associated deletions. Moreover, MMEJ has a slower kinetic compared with NHEJ. It has been shown that NHEJ-associated indels were detectable as early as four hours after Cas9 transfection, while MMEJ-associated deletions arise up to 48 hours post transfection (Taheri-Ghahfarokhi et al. 2018).

MMEJ repair anneals broken DNA ends at the short microhomology sequences surrounding the DSBs; hence, MMEJ repair results in the formation of deletions that can be distinguished from randomly generated deletions. MMEJ footprints are also evident in the formation of the deletion variants that are captured in dbSNP and ClinVar databases. For example, 57% of the 3 bp deletions in human genome were associated with perfect 3 bp microhomologies which exceeds the 1.56% probability expected by random base distribution (0.25^3) (Grajcarek et al. 2019).

Exploiting the MMEJ repair pathway has several implications in the context of therapeutic genomic editing. Repair by MMEJ is context-dependent and understanding its mechanism aides the prediction of the editing outcome with a high degree of reliability. This means that the expected deletions and their subsequent effects could be predicted for Cas9 cleavages at different sites and hence designing more efficacious genome editing strategies. For example, disease causing micro-duplications can be precisely removed using a simple Cas9 cleavage and MMEJ repair which eliminates the need of recruiting the less-frequent HDR pathway. However, the size of the MMEJ-associated deletions at the Cas9 cleavage sites and their prevalence has remained enigmatic. A recent analysis of Cas9-induced large deletions revealed high prevalence of microhomologies at breakpoint junctions. Notably, unwanted large deletions have been reported at the Cas9-targeted sites in different cell types at various frequencies. Therefore, a complete understanding of MMEJ-associated deletions is essential to fully anticipate the genome editing outcome. Besides, the MMEJ pathway has been exploited to facilitate integration of donor DNA. A technique called precise integration into target chromosome (PITCh) uses the MMEJ pathway and donor templates with extremely short microhomologies (≤40 bp) for targeted integration (Sakuma et al. 2015). However, like NHEJ-mediated integration and HITI, PITCh may cause indels at the DNA junctions.

Furthermore, MMEJ repair pathway can result in the formation of deleterious translocations. Due to its slow kinetics and its reliance on frequently found microhomologies, MMEJ can potentially even promote translocations between the on-target and/or off-target *loci*, thus serious safety considerations need to be taken into account for TGE projects.

13.2.3 Single-Strand Annealing (SSA)

SSA, like MMEJ, is an inherently erroneous repair process, which results in the loss of several kilobases of the DNA. Homology requirements for SSA are above 20 nucleotides, relying on the presence of homologies longer than what MMEJ requires. A key protein in SSA is RAD52. It is not clear to what extent SSA contributes to the formation of genetic alteration at Cas9-targeted sites in human cells. Such large homologies have not been observed in studies that aimed to analyze large deletions. However, a recent analysis revealed highly frequent SSA-mediated deletions in *Leishmania* (Zhang and Matlashewski 2019).

13.2.4 Homology-Directed Repair (HDR)

HDR and homologous recombination (HR) have been used interchangeably in the context of genome editing, both referring to a seamless repair of the Cas cleavage site. A prerequisite for the HR pathway to repair the DSBs is the extended resection of the broken ends, for which the contribution of

C-terminal binding protein 1 (CtIP) is essential to produce long single-stranded DNA strands. Following the production of the single-stranded DNA, RAD51 recombinase binds and facilitates the invasion of the broken DNA ends to a homologous template, where the lost bases will be added precisely to the broken ends. Sister chromatids represent the naturally occurring templates but can also be provided artificially by using an exogenous DNA donor template. In the former case, the repair could potentially result in the loss of heterozygosity, while the latter enables researchers to precisely introduce the intended changes in between the homologous sequences. Therefore, exploiting the HDR pathway to repair the Cas9 cleavage sites allows for the introduction of DNA edits in a controlled manner. This is a major difference to the repair by NHEJ and MMEJ; although their editing outcome is predictable to some degree, NHEJ and MMEJ produce a range of insertions and deletions that may not have therapeutic value. The challenge with HDR is its low efficacy compared with NHEJ and MMEJ pathways. HDR is restricted to the late S/G2 phases of the cell cycle (Heyer et al. 2010) and it is not active in non-mitotic cells, posing a major challenge for clinical applications. Utilization of HDR are however very promising, as HDR can facilitate the integration of small and large DNA cassettes for correction of, *e.g.* point mutations or even fully functional genes. Despite these obstacles, clinical applications of HDR hold great promise; therefore, numerous multifaceted and elaborate strategies were developed to address the current shortcomings.

13.3 Strategies to Improve Knock-In (KI)

The lower efficacy of HDR is a direct consequence of the presence of competing NHEJ and MMEJ pathways; therefore, various approaches have been developed to either enhance the efficacy of HDR using, *e.g.* compound-mediated inhibition of NHEJ, modified Cas enzymes, timed expression of Cas9, or alternatively utilizing NHEJ/MMEJ pathways for KI purposes. Other strategies applied fusion of proteins to Cas9 which can recruit HDR enzymes close to the cut site. Below we discuss the most commonly applied preclinical techniques to increase HDR efficacies.

13.3.1 Co-Enrichment Strategies

In certain circumstances as is the case for, *e.g. ex vivo* applications methods to enrich for cells with gene editing events could be applied to compensate for low KI efficiency. Collectively these methods rely on the observation that cells proficient for genomic editing at one locus are more likely to be proficient for editing events at another locus (Kim et al. 2011; Ramakrishna et al. 2014). Most of these co-selection methods have in common that they introduce an exogenous selectable marker, such as antibiotic resistance genes or fluorescent markers, which is often not desirable in a therapeutic setting. This constraint may be overcome by using transient selection markers that can be removed via the Cre/loxP system as has been demonstrated in mice and human-induced pluripotent stem cells (Sluch et al. 2018; Steyer et al. 2018; Supharattanasitthi et al. 2019). More recently developed enrichment methods utilize endogenous selection markers for enrichment (Agudelo et al. 2017; Wiebking et al. 2020; Li et al. 2021) and might be better suited for clinical applications, provided the introduced mutation in the selection locus is safe and enrichment does not cause additional unintended side effects.

13.3.2 Compound-mediated Inhibition of Non-Homologous End Joining (NHEJ)

Since the efficiency of HDR-mediated KI is low and constrained by HDR-competing DSB repair pathways, in particular NHEJ, one major strategy is the blockage of NHEJ components to increase HDR. Several small molecule inhibitors for several NHEJ proteins have been identified, including DNA-PKcs (Robert et al. 2015; Shy et al. 2016; Yao et al. 2017; Zhang et al. 2017; Riesenberg and Maricic 2018; Yao et al. 2018; Riesenberg et al. 2019), ATR serine/threonine kinase (Ma et al. 2018), Ku70/80 proteins (Weterings et al. 2016), and Ligase 4 inhibitors (Chu et al. 2015; Maruyama et al. 2015; Singh et al. 2015; Ma et al. 2016; Yang et al. 2016; Li et al. 2017b; Gerlach et al. 2018; Hu et al. 2018), as well as the peptide i53 inhibiting 53BP1 (Canny et al. 2018), which offer an option to increase HDR in *ex vivo* settings. Cell-type-specific effectiveness and toxicity of individual compounds need to be carefully evaluated experimentally as well as CRISPR-related effects on off-targets, translocations, and large deletions.

Alternative methods aim to activate the HDR pathway. Binding of helical RAD51 filaments to ssDNA is essential for effective HDR and overexpression of RAD51 in Chinese hamster ovary fibroblasts and human embryonic stem cells has been shown to promote HDR (Vispe 1998; Takayama 2017). However, effects of the recently identified RAD51 enhancer RS-1 showed mixed results in augmenting HDR-mediated KI (Jayathilaka et al. 2008; Pinder et al. 2015). The more recently identified CRISPY mix that consists of four small molecules has been used to increase HDR efficiencies in human-induced pluripotent stem cells, but had varied responses in some immortalized and primary cells, suggesting that the utilized DNA repair pathways are cell-type specific (Riesenberg and Maricic 2018). The CRISPY mix includes the following components: NU7026 (DNA-PK inhibition), Trichostatin A (activation of ATM-dependent DNA damage signaling pathway), MLN4924 (inhibition of CtIP neddylation and increase of DNA end resection), and NSC 15520 (prevents association of RPA with p53 and RAD9, possibly increasing RPA abundance and increase of HDR). Combined inhibition of several NHEJ proteins or additional DNA repair pathways with different inhibitor cocktails may increase HDR rates further in other cell types.

13.3.3 Cell Cycle Manipulation

In addition to manipulating DNA repair pathways with small molecules to increase HDR-mediated KI, some approaches utilize drugs that cause cell cycle arrest and cell cycle synchronization. After reversible cell cycle synchronization, the inhibitors are removed, and gene editing reagents are delivered into cells in a timed manner. This gauged delivery of CRISPR reagents can augment KI efficiencies. Examples of cell cycle manipulating drugs among others include nocodazole and ABT-751 (Lin et al. 2014; Yang et al. 2016). Likewise, Cas9 enzymatic activity has been linked to proteins that are subjected to tight cell cycle control. One of these proteins, geminin, is a substrate for G1 proteasome degradation and if fused to Cas9 can limit its expression to S, G2, and M phases of the cell cycle. This design can promote HDR events, while reducing unwanted NHEJ events in human immortalized and stem cells (Gutschner et al. 2016; Lomova et al. 2019).

13.3.4 Modifications of sgRNA and HDR Donors

The CRISPR-Cas9 system can be programmed to cut DNA target sequences through rational design of sgRNAs, theoretically permitting to easily target any DNA sequence of interest. Introducing chemical modifications to synthetic sgRNAs, Cas9 mRNA and donor templates provide a variety of protection options for gene editing reagents (Hendel et al. 2015; Rahdar et al. 2015; Li et al. 2017a; Yin et al. 2017; Finn et al. 2018; Li et al. 2018; McMahon et al. 2018). Properties of chemically modified nucleotides were shown to improve nucleic acid stability by resisting nuclease digestion, reduction of off-target effects, and decrease of immunogenic responses and delivery (Kariko et al. 2005; Judge et al. 2006; Cekaite et al. 2007). sgRNA modifications can be classified as modifications of the linkage between nucleotides and modifications of the nucleic acid sugars and nucleobases (Kelley et al. 2016). Modifications of the backbone such as phosphorothioates change the charge of the phosphate backbone and improve delivery and nuclease resistance (Eckstein 2014). Sugar modifications, like 2′-O-methyl, 2′-F, or locked nucleotide acid, improve base pairing and resistance against nucleases (Allerson et al. 2005). In addition, RNA can be modified at the 3′ and 5′ ends with a variety of functional moieties (Kelley et al. 2016). It has been shown that the co-delivery of Cas9 mRNA with sgRNA with 2′-O-methyl-3′-phosphorothioate modifications at both termini improved gene editing efficiency in different human primary cells (Hendel et al. 2015). Apart from chemical modifications, engineering of the sgRNA secondary structure was shown to increase editing specificity (Li et al. 2020). These modifications include sgRNA truncation, partial replacement with DNA, and addition of a 5′ hairpin (Fu et al. 2014; Yin et al. 2018; Kocak et al. 2019).

Besides the various means of sgRNA optimization, numerous methods for HDR donor template were devised, including improved template designs (Paquet et al. 2016; Renaud et al. 2016; Richardson et al. 2016) and chemical donor DNA modifications, co-localization of DNA donor templates to the specific target site of nuclease-induced DSBs (Carlson-Stevermer et al. 2017; Aird et al. 2018; Savic et al. 2018) as well as delivery of the DNA donor template with AAVs (Nishiyama et al. 2017).

13.4 Genome Editing in Clinical Trials

The initial focus of TGE has been set on monogenetic disorders as these provide attractive candidates for gene therapy trials. Monogenetic diseases are caused by inheritance of a mutation in a single gene although the phenotypic manifestation may depend on further factors, such as the presence of additional genetic variants, epigenetic alterations as well as environmental and life-style factors. Genetic diseases are caused by several types of mutations, including missense and nonsense mutations, insertions, deletions, frameshift mutations, and repeated expansions. Depending on the mutation type, distinct genome editing strategies can be applied for correction, engaging distinct cellular DNA repair mechanisms.

Current clinical trials on genome editing primarily focus on treating hematological, hepatological, and ocular monogenetic diseases (Table 13.1) as well as on immuno-oncology applications (Table 13.2). All reported clinical trials are either phase I studies evaluating safety, or phase II studies assessing safety and dosing of the therapeutic agent. These trials will help to pave the way for future applications of designer nucleases for specific and curative treatments of genetic disorders.

13.4.1 Strategies Utilizing Non-Homologous End Joining (NHEJ)

The high activity of NHEJ repair directly after the introduction of a targeted DSB with designer nucleases and its

Table 13.1 Phase I and II clinical trials with applied designer nucleases and the underlying DNA repair pathway.

Area	Condition	DNA repair			Gene	Editor	Method/delivery/cell type	Identifier
Viral diseases	HIV – acquired immunodeficiency syndrome (AIDS)	NHEJ	Inactivation (exon)	Indels	*CCR5*	ZFNs	*Ex vivo*/adenovirus/CD34+ T-cells/CD34+ HSPC	NCT00842634 NCT01044654 NCT01252641 NCT02225665 NCT02388594 NCT02500849 NCT03617198 NCT03666871
				Deletion	*CCR5*	Cas9	*Ex vivo*/RNP/CD34+ HSPC	NCT03164135
	HPV – related cervical cancer	NHEJ	Inactivation (exon)	Indels	HPV16 E6/E7	Cas9	*In vivo*/plasmid/cervical epithelia	NCT02800369
					HPV16 E6/E7	Cas9 & TALENs	*In vivo*/plasmid/cervical epithelia	NCT03057912 NCT03226470
	HSV – refractory keratitis	NHEJ	Inactivation (exon)	Indels	*HSV-1*	Cas9	*In vivo*/BDmRNA viroid	NCT04560790
Metabolic diseases	Mucopolysaccharidosis I & II	HDR/NHEJ	Gene replacement (intron)	Knock-in	*ALB*	ZFNs	*In vivo*/AAV/hepatocytes	NCT02702115 NCT03041324
Hematological diseases	Hemophilia B	HDR/NHEJ	Gene replacement (intron)	Knock-in	*ALB*	ZFNs	*In vivo*/AAV/hepatocytes	NCT02695160
	β-thalassemia	NHEJ	Enhancement (regulatory region)	Indels	*BCL11A*	Cas9 ZFNs	*Ex vivo*/electroporation/ CD34+ HSPCs	NCT03655678 NCT03432364
		HDR	Mutation correction (exon)	Precise	*HBB*	Cas9	*Ex vivo*/autologous iHSC	NCT03728322
	Sickle cell disease	NHEJ	Enhancement (regulatory region)	Indels	*BCL11A*	ZFNs Cas9	*Ex vivo*/electroporation/ CD34+ HSPCs	NCT03653247 NCT03745287 NCT04443907
Eye disease	Leber congenital amaurosis 10	NHEJ	Restoration of splicing (intron)	Deletion	*CEP290*	Cas9	*In vivo*/AAV/retinal epithelial cells	NCT03872479
Neuro-degenerative diseases	Hereditary transthyretin amyloidosis	NHEJ	Inactivation (exon)	Indels	*TTR*	Cas9	*In vivo*/LNP /liver (hepatocytes)	NCT04601051

Human immunodeficiency virus (HIV), human papilloma virus (HPV), herpes simplex virus (HSV), non-homologous end joining (NHEJ), homology-directed repair (HDR), zinc-finger nucleases (ZNFs), transcription activator-like effector nucleases (TALENs), CRISPR-associated protein 9 (Cas9), adeno-associated virus (AAV), hematopoietic stem cell (HSPC), induced hematopoietic stem cell (iHSC).

Table 13.2 Phase I and II clinical trials with designer nucleases in immuno-oncology.

DNA repair	Editor	Targeted gene	Identifier
NHEJ – inactivation	TALENs	TRAC or/and TCBR	NCT02746952 NCT02808442 NCT03190278 NCT04106076 NCT04150497 NCT04093596 NCT04142619 NCT03939026 NCT03190278 NCT04106076
		CD52	NCT02746952 NCT02808442 NCT03190278 NCT04106076 NCT04150497 NCT03939026
		SLAM	NCT04142619
	Cas9	PDC1	NCT02863913 NCT02793856 NCT03081715 NCT03399448 NCT03044743 NCT03545815 NCT03747965 NCT04417764
		TRAC or/and TRBC	NCT03166878 NCT03399448 NCT04035434 NCT04637763 NCT04557436 NCT04629729 NCT04244656 NCT04438083 NCT04502446
		β2m	NCT03166878 NCT04035434 NCT04244656 NCT04438083 NCT04502446
		CDC7	NCT03690011
		HPK1	NCT04037566
		CISH	NCT04426669
		CD38	NCT04614636
	Meganucleases	TRAC	NCT04171843 NCT04030195 NCT03666000
HDR – gene insertions	Cas9	TRAC CD19-directed CAR	NCT04035434 NCT04637763 NCT04629729
		TRAC BCMA-directed CAR	NCT04244656
		TRAC CD70-directed CAR	NCT04438083 NCT04502446
	Meganucleases	TRAC BCMA-directed CAR	NCT04171843
		TRAC CD20-directed CAR	NCT04030195
		TRAC CD19-directed CAR	NCT03666000

Non-homologous end joining (NHEJ), homology-directed repair (HDR), zinc-finger nucleases (ZNFs), transcription activator-like effector nucleases (TALENs), CRISPR-associated protein 9 (Cas9).

high prevalence in most cell types are major advantages of the NHEJ pathway for clinical applications. NHEJ is described as error-prone, resulting in random indels; in the absence of exogenous nucleases, however, NHEJ most often leads to accurate ligation of DSBs. In fact, the targeted repeated cutting of the genomic locus by designer nucleases results in error-prone repair of DNA lesions. These mechanisms are used in TGE to either reduce/eliminate or replace/correct a disease-related gene. The NHEJ pathway is harnessed for the following strategies:

1) Inactivation of a gene; generated indels or defined deletions lead to a premature stop codon or frameshift in an endogenous gene.
2) Enhancement or suppression of gene expression by introducing indels or defined deletions in a regulatory region.
3) Facilitation of exon skipping or restoration of correct splicing through the formation of indels or defined deletions in intronic splice sites.
4) Gene replacements by KI of larger transgenes (Figure 13.3).

The following paragraph describes examples for each strategy from ongoing clinical studies as summarized in Table 13.1.

In 2009, only four years after the initial report about ZFN-induced targeted DSBs, the first clinical trial with ZFNs was initiated to treat the human immunodeficiency virus type 1 (HIV-1) infection. A pair of ZFNs is applied to introduce indels in the *CCR5* gene to knockout the CCR5 receptor in human CD4+ T cells (Tebas et al. 2014). HIV preferentially infects CD4+ T cells, where CCR5 functions as a co-receptor of CD4 for viral entry into the cell (Choe et al. 1996). Humans with a naturally occurring homozygous 32 bp deletion in the *CCR5* locus are resistant to HIV-1 infections and show no disease symptoms related to the lack of CCR5 (Hutter et al. 2009; Gupta et al. 2019). The *ex vivo* gene editing strategy generates a DSB in the *CCR5*

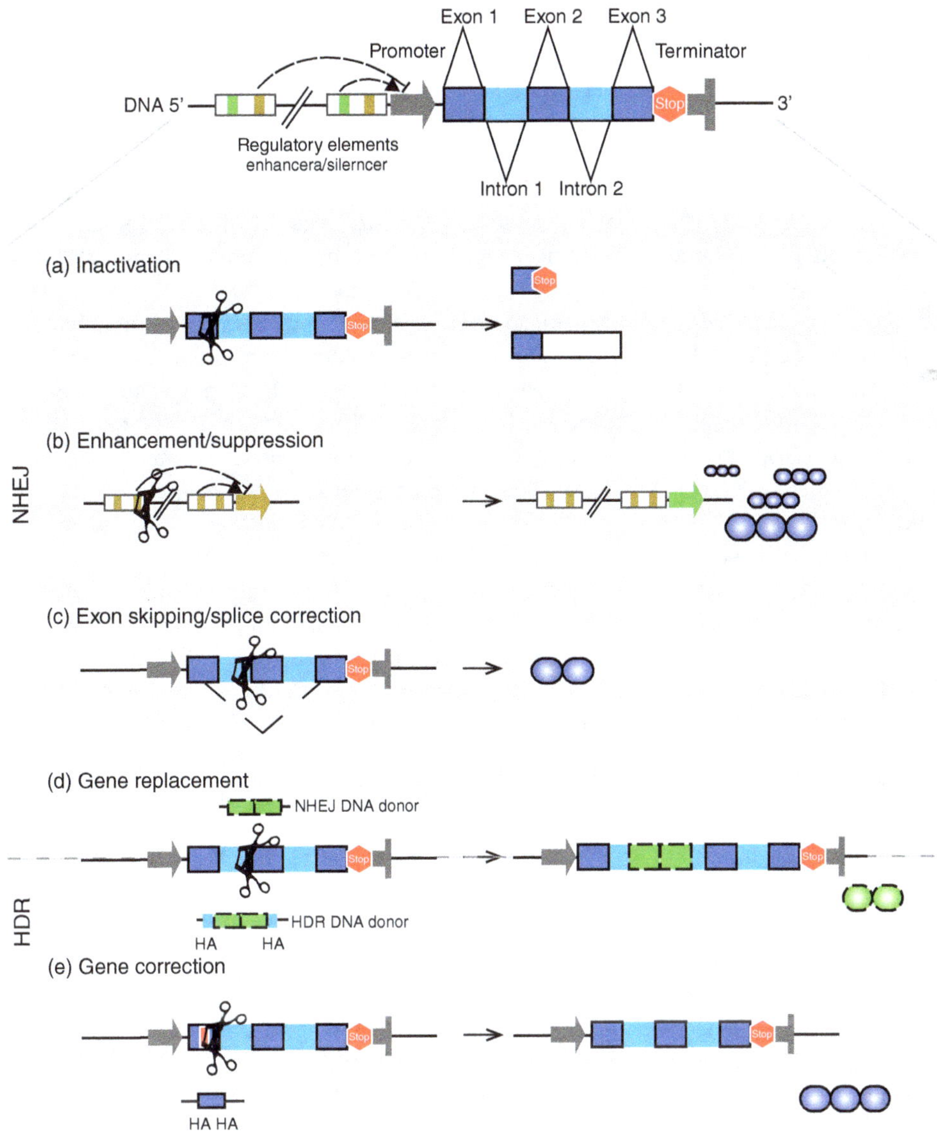

Figure 13.3 Genome editing strategies utilizing non-homologous end joining (NHEJ) or homology-directed repair (HDR) for gene therapy, after introducing targeted DNA double-strand breaks (DSBs) with designer nucleases. The NHEJ-mediated formation of indels or precise deletions are harnessed in diverse fashions depending on the location of the targeted DSB in the genome: (a) Generated mutations in an exon leads to the formation of premature stop codons or frameshifts and subsequent inactivation/knockout of the endogenous gene. (b) DNA manipulations in regulatory regions enhance or suppress gene expression. (c) Indels in specific intronic regions facilitate exon skipping or correction of aberrant splice sites. (d) Gene replacements are either engaging NHEJ- or HDR-mediated repair depending on the type of the co-delivered DNA donor. Double-stranded DNA donors without homology arms (HA) are ligated into the DSB through NHEJ, while single- or double-stranded DNA donors with HA are incorporated via HDR. (e) HDR is utilized to correct a disease-causing mutation precisely.

gene upstream of the delta32 mutation in CD4+ T cells isolated from HIV-positive patients. The activated NHEJ repair pathway produces indels, leading to a nonfunctional or truncated form of the CCR5 receptor, which is absent from the cell surface (NCT00842634) (Perez et al. 2008). This study demonstrated the safety and feasibility of ZFN to install resistance against HIV infections. Encouraged by these results, ZFNs were subsequently applied to knockout the CCR5 receptor in autologous CD34+ hematopoietic stem/progenitor cells (HSPCs), with the intention to generate a long-lasting supply of HIV-1-resistant T helper cells and CD4+ monocytes (NCT02500849). Eight years later, another study evaluated the safety and feasibility of a Cas9-mediated knockout of the CCR5 receptor in HSPCs from healthy donors. The allogenic cell therapy used a dual-guide approach, simultaneously introducing two DSBs in close spatial proximity. The dual cut excises a small DNA fragment, of which the ends are united via NHEJ. In comparison to random indels, this approach aims to install a predictable 35 bp deletion, resulting in knockout through introduction of a frameshift mutation (NCT03164135). The results of the study showed that one year after transplantation, the most abundant mutation type was the desired 35 bp deletion. Yet, the percentage of persistent cells with a CCR5 disruption was low. The limited editing efficiency in combination with modest engraftment of edited cells proved to be insufficient to achieve efficient treatment of HIV. However, the authors revealed that transplantation and long-term engraftment of CRISPR-edited allogeneic HSPCs is indeed possible (Xu et al. 2019).

A persistent infection with human papillomavirus (HPV) types 16 and 18 may lead to precancerous lesions and are correlated with a higher risk of cervical cancer (Burd 2003). Antiviral gene therapy directly targets virus-encoded oncogenes E6 and E7 (NCT02800369 & NCT03057912). NHEJ-mediated indels generated in the viral genes with either ZFNs, CRISPR-Cas9, or TALENs showed a reduction of E6 and E7 expression, inducing apoptosis and inhibiting cell growth *in vitro* (Ding et al. 2014; Hu et al. 2014).

The introduction of a frameshift or stop-codon mutation for gene knockouts is widely used in immuno-oncology clinical applications. These include *ex vivo* genome editing in T cells to ablate different surface proteins such as the checkpoint regulator, programmed cell death protein 1 (PD1), to improve the function and persistence of engineered T cells, the T cell receptor (TCR) locus to prevent graft versus host disease or the major histocompatibility complex (MHC), to avoid rejection of the donor cells by the host. Often, different genes are targeted simultaneously giving rise to additional challenges. A detailed review about immuno-oncology and gene editing in T cells is given in Chapter 16.

The clinical trials described thus far utilized *ex vivo* gene therapy. Clinical protocols usually involve cell isolation from either patients or healthy donors, *ex vivo* genetic modifications, and administration of cells to the patient. Advantages of this strategy involve easier delivery of the gene editing reagents and the possibility to select the cells before transplantation to increase safety and efficiency.

The prospect of *ex vivo* therapy is also advantageous in developing strategies to treat two inherited hematological disorders, sickle cell disease (SCD) and β-thalassemia, with millions of people affected around the world and limited curative treatments (Cao and Galanello 2010; Piel et al. 2017). Both diseases are characterized by mutations in the beta globin gene. SCD is caused by a single point mutation, replacing glutamic acid with valine at position 6, leading to malformation of red blood cells in SCD. In contrast, numerous mutations are correlated with β-thalassemia, which commonly lead to reduced levels or complete absence of hemoglobin β.

Most of the current clinical trials employ a strategy where indels are introduced at the erythroid-specific enhancer region of the transcriptional repressor *BCL11A*, causing reduced expression levels and thus restoring fetal γ-globin expression. Fetal hemoglobin has a higher binding affinity to oxygen than adult hemoglobin to ensure efficient oxygen transport to the developing fetus from the mother's bloodstream. The expression of fetal hemoglobin decreases within the first year after birth (Sankaran et al. 2010). Re-expression of fetal γ-globin is believed to rescue the disease phenotype in SCD and β-thalassemia. In early 2021, extremely promising initial results of two patients, who received cell therapy treatment were reported (Frangoul et al. 2021); one patient was diagnosed with β-thalassemia and the other participant with SCD, both were administered with autologous CRISPR-Cas9 edited CD34+ HSPCs. The two patients showed elevated levels of fetal hemoglobin and good editing in engrafted hematopoietic stem cells, which made the patients independent from regularly required blood transfusions. Studies on long-term effects and inclusion of additional patients in the trial will give a broader perspective on the success of this promising treatment in the future.

One of the first Cas9-based *in vivo* gene editing approaches in trails facilitated NHEJ repair to introduce a defined deletion in an intronic region to cure Leber congenital amaurosis type 10 (LCA10) (NCT03872479). The rare autosomal inherited eye disease is characterized by severe loss of vision or blindness in children, linked to a loss of function mutation in the centrosomal protein 290 (CEP290) gene. The most frequent disease-causing mutation is a point mutation in intron 26, converting an adenine to guanine, which creates a strong splice donor and the inclusion of a cryptic exon, generating a premature stop codon (den Hollander et al. 2006). The therapeutic gene editing strategy involves injections of AAVs loaded with *Staphylococcus aureus* Cas9 nuclease and two guide RNAs into the sub-retina. The dual-guide RNAs, flanking the mutation in intron 26, excise the aberrant splice site and the autonomous DNA ends are either fused back directly, resulting in a deletion, or incorporate the inverted fragment. Both outcomes are expected to be therapeutically beneficial (Maeder et al. 2019). The same strategy can be used to introduce indels close to a splice acceptor of a normal exon to enable exon skipping and restore an aberrant reading frame. The therapeutic relevance has been shown in preclinical studies for Duchenne muscular dystrophy

(DMD) treatments (Ousterout et al. 2013; Amoasii et al. 2017; Amoasii et al. 2018).

The generation of indels and longer deletions produced by the NHEJ pathway is a powerful tool for TGE. In addition, the NHEJ pathway can be utilized for whole-gene or longer sequence insertions in dividing and non-dividing cells. To this end, designer nucleases introduce the DSB at the desired genomic locus and simultaneously generate a linear donor template in the cell, which will be ligated directly into the target-site (Orlando et al. 2010; Maresca et al. 2013; Suzuki et al. 2016). This strategy is evaluated in several clinical trials, such as for treatment of hemophilia B, mucopolysaccharidosis type I (MPS I or Hurler syndrome), and mucopolysaccharidosis type II (MPS II, Hunter syndrome). The bleeding disorder hemophilia B is caused by mutations in the factor IX gene, leading to a deficiency of coagulation factor XI. Children suffering from autosomal recessive inherited Hurler syndrome have a mutation in the gene encoding for alpha-L iduronidase, whereas mutations in the iduronate-2-sulfatase gene causes the x-linked recessive inherited Hunter syndrome. In the absence of the metabolic enzymes alpha-L iduronidase or iduronate-2-sulfatase, patients are unable to break down glycosaminoglycans and accumulate these in lysosomes. This process eventually leads to skeletal, cardiac, respiratory, and neurological abnormalities, atypically enlarged organs and in severe forms, premature death. Hurler syndrome is the most severe form of MPS (Muenzer 2011). The site-specific gene-editing strategy to treat these diseases does not target the mutated locus itself; instead, a corrected copy of the transgene is inserted into intron 1 of the albumin gene. Integration into a considerably transcribed locus refractory to silencing is thought to ensure sufficient protein expression to achieve the desired therapeutic effect.

AAVs are used to deliver the expression cassette, consisting of a splice acceptor followed by a complementary DNA encoding factor IX, alpha-L iduronidase, or iduronate-2-sulfatase, flanked by homology arms and inverted terminal repeats, together with a pair of albumin-specific ZFNs into hepatocytes. The entire donor template can be integrated via NHEJ repair. Alternatively, homology arms can facilitate integration via HDR (Sharma et al. 2015; Laoharawee et al. 2018).

13.4.2 Strategies Engaging Homology-Directed Repair (HDR)

In contrast to NHEJ, HDR is limited to the S/G2 phase and requires a repair template. Replication from externally delivered donor templates enables precise genome editing to introduce defined changes in the DNA. Therefore, engaging HDR and NHEJ repair to deliver the transgenes of factor IX, alpha-L iduronidase, or iduronate-2-sulfatase into the albumin locus of hepatocytes is an excellent strategy to reach both dividing and nondividing cells. In both cases, the transcriptional outcome will be identical due to the donor design and subsequent interaction of splicing donor and acceptor. Intron 1 of the albumin locus is not only an attractive target because of high expression levels, but also for the signal peptide, which is encoded by exon 1 and cleaved from the final protein version of albumin. During transcription, the transgene will be fused to the signal peptide and afterward separated, theoretically allowing for the expression of any secreted protein from the albumin locus as a general approach for protein replacement therapy (Sharma et al. 2015). In 2019, progress reports of the clinical trials to treat Hurler and Hunter syndromes revealed low levels of transgene expression, which will be a major obstacle to tackle in the future (Joseph Muenzer et al. 2019; Paul Harmatz et al. 2019). Apart from inserting a transgene into a locus of interest, HDR can be utilized to directly correct a disease-causing mutation.

In 2019, a clinical trial was initiated to correct mutations in the HBB gene, which leads to β-thalassemia. The strategy involves transplantation of *in vitro* gene-corrected, patient-specific-induced HSCs (NCT03728322). The exact editing strategy has not been revealed, but one potential approach could be the co-delivery of a donor template carrying the disease-correcting mutation together with an endonuclease.

Overall, to accomplish beneficial therapeutic effects through HDR gene modifications, the repair outcomes often have to compete with the error-prone NHEJ, which can be challenging in human cells. With continuous new discoveries and engineering of designer nucleases, creative strategies independent from HDR or NHEJ mechanisms will most likely enter the clinic in the future. In 2020, a pioneer study was launched, exploiting the unique characteristic of the CRISPR-Cas3 system, to introduce large deletions in genomic DNA to fight urinary tract infections caused by *Escherichia coli* (NCT04191148).

13.5 Major Safety Considerations in TGE Clinical Trials

All TGE programs are in their preliminary stages of clinical trials to evaluate their safety and efficacy. However, a myriad of preclinical studies has already identified several potential safety issues. A detailed description of these issues is out of the scope of this chapter; however, here we provide a concise summary of the issues that clinical trials may face.

13.5.1 Activation of p53 upon Introduction of DNA Breaks

CRISPR-induced DSBs can cause cells to initiate apoptosis rather than the intended gene edit to occur (Hu et al. 2014). Further safety concerns were revealed when two studies showed that CRISPR-mediated gene editing had the potential for inadvertent selection of cells with a deregulated p53 pathway (Haapaniemi et al. 2018; Ihry et al. 2018). Hence, effective CRISPR edits are more likely to happen in p53-suppressed cells, which can lead to a selection bias toward oncogenic cell populations. p53 inactivation is associated with uncontrolled cell proliferation and the development of cancer. However, follow-up studies demonstrated that this risk can be mitigated through optimization of laboratory protocols (Schiroli et al. 2019; Wu et al. 2019).

13.5.2 Off-Target Activity, Large Deletions, and Complex Rearrangements

CRISPR off-targets represent a major safety issue in genome editing. A detailed discussion about off-targets and how to mitigate the associated risks can be found in Chapter 20 of this book. In addition to off-targets, large deletions of several kilobases and more complex chromosomal rearrangements have been reported as unintended consequences of CRISPR-mediated gene editing (Kosicki et al. 2018; Cullot et al. 2019). These findings highlighted a major concern for clinical applications of DSB-inducing gene editing therapies. Strategies to reduce these risks include utilization of catalytically inactive Cas9 variants, such as dead Cas9 (dCas9), in which both endonuclease domains are deactivated and do not produce a DSB. dCas9 can be fused to transcriptional activator or repressor domains as well as DNA-binding effectors to transiently modulate gene expression without generating a DSB. Other Cas9 variants such as Cas9 nickases (Cas9n) with only one active endonuclease domain can also be considered. Instead of introducing a DSB, Cas9n generates a DNA single-strand cut. DSBs can be introduced through the use of paired nickases for genome editing (Gilbert et al. 2013; Maeder et al. 2013; Perez-Pinera et al. 2013; Qi et al. 2013). Additional modifications of these Cas9 variants have led to the development of base- and prime-editors (Komor et al. 2016; Gaudelli et al. 2017; Anzalone et al. 2019). Base-editors are composed of a catalytically impaired Cas9 fused to a single-stranded DNA deaminase domain. Today, two main types of base-editors have been developed, *i.e.* cytosine base-editors, catalyzing the conversion of a C·G base pair to a T·A base pair and adenine base-editors, which convert an A·T base pair into a G·C base pair. An extensive overview about base editing is provided in Chapter 14 of this book. Prime-editing is a recently developed technology that allows for the installation of precise point mutations as well as small insertions and deletions. The prime-editing system consists of a Cas9 nickase that is fused to a reverse transcriptase enzyme and a prime editing gRNA (pegRNA). The active endonuclease domain of Cas9 nicks the DNA and the prime-editor then uses the released 3′ DNA end at the target site to prime reverse transcription using the pegRNA as template. The pegRNA contains the desired edit and a region of homology to the target site to enable DNA repair.

13.5.3 Pre-Existing Immunity to Cas9 Variants

Besides technical constraints, gene editing with bacterial-derived designer nucleases raises concerns about immunogenic toxicity. A recent report demonstrates that more than half of the participants of the study had pre-existing antibodies against the most widely used Cas9 orthologs, derived from *Streptococcus pyogenes* and *Staphylococcus aureus* (Charlesworth et al. 2019). Pre-existing adaptive immune responses in humans may be an obstacle for safe and efficacious gene editing treatments or worse trigger significant toxicity in patients. In previous gene therapy trials with viral vectors, no therapeutic benefit was derived from the therapy in study participants that had pre-existing adaptive immune responses against the used viral vector, because either of vector neutralization by antibodies (DePolo et al. 2000; Manno et al. 2006) or clearance of transduced cells by cytotoxic T cells (Mingozzi et al. 2007). Thus, adaptive immune responses in humans can represent a substantial barrier to gene therapy and cannot be adequately evaluated in nonhuman systems (Mingozzi and High 2013). Specific safety concerns associated with viral delivery or lipid nanoparticles are discussed in the respective chapters.

13.6 Conclusion

The efficacy and safety of reprogrammable nucleases that are used in TGE programs is owed to the existence of the highly efficient repair mechanisms in human cells. Designing genuine genome editing strategies to exploit the full capacity of the DNA repair pathways is as important as harnessing new reprogrammable nucleases. Recent developments to incorporate new functionalities to the Cas9 variants, *e.g.* base- and prime-editors, are major steps toward gaining more control of the type of edits at the targeted loci by orienting the repair processes. TGE has an immense potential in shifting the paradigms in medicine, moving from treating to curing a wide range of diseases.

References

Agudelo, D., Duringer, A., Bozoyan, L. et al. (2017). Marker-free coselection for CRISPR-driven genome editing in human cells. *Nat. Methods* 14: 615–620.

Aird, E.J., Lovendahl, K.N., St Martin, A. et al. (2018). Increasing Cas9-mediated homology-directed repair efficiency through covalent tethering of DNA repair template. *Commun. Biol.* 1: 54.

Allerson, C.R., Sioufi, N., Jarres, R. et al. (2005). Fully 2'-modified oligonucleotide duplexes with improved in vitro potency and stability compared to unmodified small interfering RNA. *J. Med. Chem.* 48: 901–904.

Amoasii, L., Long, C., Li, H. et al. (2017). Single-cut genome editing restores dystrophin expression in a new mouse model of muscular dystrophy. *Sci. Transl. Med.* 9: 1–11.

Amoasii, L., Hildyard, J.C.W., Li, H. et al. (2018). Gene editing restores dystrophin expression in a canine model of Duchenne muscular dystrophy. *Science* 362: 86–91.

Anzalone, A.V., Randolph, P.B., Davis, J.R. et al. (2019). Search-and-replace genome editing without double-strand breaks or donor DNA. *Nature* 576: 149–157.

Burd, E.M. (2003). Human papillomavirus and cervical cancer. *Clin. Microbiol. Rev.* 16: 1–17.

Canny, M.D., Moatti, N., Wan, L.C.K. et al. (2018). Inhibition of 53BP1 favors homology-dependent DNA repair and increases CRISPR-Cas9 genome-editing efficiency. *Nat. Biotechnol.* 36: 95–102.

Cao, A. and Galanello, R. (2010). Beta-thalassemia. *Genet. Med.* 12: 61–76.

Carlson-Stevermer, J., Abdeen, A.A., Kohlenberg, L. et al. (2017). Assembly of CRISPR ribonucleoproteins with biotinylated oligonucleotides via an RNA aptamer for precise gene editing. *Nat. Commun.* 8: 1711.

Cavazzana-Calvo, M., Hacein-Bey, S., de Saint Basile, G. et al. (2000). Gene therapy of human severe combined immunodeficiency (SCID)-X1 disease. *Science* 288: 669–672.

Ceccaldi, R., Rondinelli, B., and D'Andrea, A.D. (2016). Repair pathway choices and consequences at the double-strand break. *Trends Cell Biol.* 26: 52–64.

Cekaite, L., Furset, G., Hovig, E., and Sioud, M. (2007). Gene expression analysis in blood cells in response to unmodified and 2'-modified siRNAs reveals TLR-dependent and independent effects. *J. Mol. Biol.* 365: 90–108.

Chari, R., Mali, P., Moosburner, M., and Church, G.M. (2015). Unraveling CRISPR-Cas9 genome engineering parameters via a library-on-library approach. *Nat. Methods* 12: 823–826.

Charlesworth, C.T., Deshpande, P.S., Dever, D.P. et al. (2019). Identification of preexisting adaptive immunity to Cas9 proteins in humans. *Nat. Med.* 25: 249–254.

Choe, H., Farzan, M., Sun, Y. et al. (1996). The beta-chemokine receptors CCR3 and CCR5 facilitate infection by primary HIV-1 isolates. *Cell* 85: 1135–1148.

Christian, M., Cermak, T., Doyle, E.L. et al. (2010). Targeting DNA double-strand breaks with TAL effector nucleases. *Genetics* 186: 757–761.

Chu, V.T., Weber, T., Wefers, B. et al. (2015). Increasing the efficiency of homology-directed repair for CRISPR-Cas9-induced precise gene editing in mammalian cells. *Nat. Biotechnol.* 33: 543–548.

Cong, L., Ran, F.A., Cox, D. et al. (2013). Multiplex genome engineering using CRISPR/Cas systems. *Science* 339: 819–823.

Cullot, G., Boutin, J., Toutain, J. et al. (2019). CRISPR-Cas9 genome editing induces megabase-scale chromosomal truncations. *Nat. Commun.* 10: 1136.

Depolo, N. J., Reed, J. D., Sheridan, P. L., Townsend, K., Sauter, S. L., Jolly, D. J. & Dubensky, T. W., JR. 2000. VSV-G pseudotyped lentiviral vector particles produced in human cells are inactivated by human serum. *Mol. Ther.*, 2, 218–222.

Ding, W., Hu, Z., Zhu, D. et al. (2014). Zinc finger nucleases targeting the human papillomavirus E7 oncogene induce E7 disruption and a transformed phenotype in HPV16/18-positive cervical cancer cells. *Clin. Cancer Res.* 20: 6495–6503.

Eckstein, F. (2014). Phosphorothioates, essential components of therapeutic oligonucleotides. *Nucleic Acid Ther.* 24: 374–387.

Fagerberg, L., Hallstrom, B.M., Oksvold, P. et al. (2014). Analysis of the human tissue-specific expression by genome-wide integration of transcriptomics and antibody-based proteomics. *Mol. Cell. Proteomics* 13: 397–406.

Finn, J.D., Smith, A.R., Patel, M.C. et al. (2018). A single administration of CRISPR/Cas9 lipid nanoparticles achieves robust and persistent in vivo genome editing. *Cell Rep.* 22: 2227–2235.

Frangoul, H., Altshuler, D., Cappellini, M.D. et al. (2021). CRISPR-Cas9 gene editing for sickle cell disease and beta-thalassemia. *N. Engl. J. Med.* 384: 252–260.

Fu, Y., Sander, J.D., Reyon, D. et al. (2014). Improving CRISPR-Cas nuclease specificity using truncated guide RNAs. *Nat. Biotechnol.* 32: 279–284.

GAUDELLI, N.M., KOMOR, A.C., REES, H.A. et al. (2017). Programmable base editing of A*T to G*C in genomic DNA without DNA cleavage. *Nature* 551: 464–471.

Gerlach, M., Kraft, T., Brenner, B. et al. (2018). Efficient knock-in of a point mutation in porcine fibroblasts using the CRISPR/Cas9-GMNN fusion gene. *Genes (Basel)* 9.

Gilbert, L.A., Larson, M.H., Morsut, L. et al. (2013). CRISPR-mediated modular RNA-guided regulation of transcription in eukaryotes. *Cell* 154: 442–451.

Grajcarek, J., Monlong, J., Nishinaka-Arai, Y. et al. (2019). Genome-wide microhomologies enable precise template-free editing of biologically relevant deletion mutations. *Nat. Commun.* 10: 4856.

Gupta, R.K., Abdul-Jawad, S., Mccoy, L.E. et al. (2019). HIV-1 remission following CCR5Delta32/Delta32 haematopoietic stem-cell transplantation. *Nature* 568: 244–248.

Gutschner, T., Haemmerle, M., Genovese, G. et al. (2016). Post-translational regulation of Cas9 during G1 enhances homology-directed repair. *Cell Rep.* 14: 1555–1566.

Haapaniemi, E., Botla, S., Persson, J. et al. (2018). CRISPR-Cas9 genome editing induces a p53-mediated DNA damage response. *Nat. Med.* 24: 927–930.

Hacein-Bey-Abina, S., Von Kalle, C., Schmidt, M. et al. (2003). A serious adverse event after successful gene therapy for X-linked severe combined immunodeficiency. *N. Engl. J. Med.* 348: 255–256.

Hanlon, K.S., Kleinstiver, B.P., Garcia, S.P. et al. (2019). High levels of AAV vector integration into CRISPR-induced DNA breaks. *Nat. Commun.* 10: 4439.

Harmatz, P., Lau, H.A., Heldermon, C., et al. (2019). EMPOWERS: a phase 1/2 clinical trial of SB-318 ZFN-mediated in vivo human genome editing for treatment of MPS I (hurler syndrome). *Mol. Genet. Metab.* 126: S86.

Hendel, A., Bak, R.O., Clark, J.T. et al. (2015). Chemically modified guide RNAs enhance CRISPR-Cas genome editing in human primary cells. *Nat. Biotechnol.* 33: 985–989.

Heyer, W.D., Ehmsen, K.T., and Liu, J. (2010). Regulation of homologous recombination in eukaryotes. *Annu. Rev. Genet.* 44: 113–139.

den Hollander, A.I., Koenekoop, R.K., Yzer, S. et al. (2006). Mutations in the CEP290 (NPHP6) gene are a frequent cause of leber congenital amaurosis. *Am. J. Hum. Genet.* 79: 556–561.

Hu, Z., Yu, L., Zhu, D. et al. (2014). Disruption of HPV16-E7 by CRISPR/Cas system induces apoptosis and growth inhibition in HPV16 positive human cervical cancer cells. *Biomed. Res. Int.* 2014: 612823.

Hu, Z., Shi, Z., Guo, X. et al. (2018). Ligase IV inhibitor SCR7 enhances gene editing directed by CRISPR-Cas9 and ssODN in human cancer cells. *Cell Biosci.* 8: 12.

Hutter, G., Nowak, D., Mossner, M. et al. (2009). Long-term control of HIV by CCR5 delta32/Delta32 stem-cell transplantation. *N. Engl. J. Med.* 360: 692–698.

Ihry, R.J., Worringer, K.A., Salick, M.R. et al. (2018). p53 inhibits CRISPR-Cas9 engineering in human pluripotent stem cells. *Nat. Med.* 24: 939–946.

Jayathilaka, K., Sheridan, S.D., Bold, T.D. et al. (2008). A chemical compound that stimulates the human homologous recombination protein RAD51. *Proc. Natl. Acad. Sci. U. S. A.* 105: 15848–15853.

Judge, A.D., Bola, G., Lee, A.C., and Maclachlan, I. (2006). Design of noninflammatory synthetic siRNA mediating potent gene silencing in vivo. *Mol. Ther.* 13: 494–505.

Kariko, K., Buckstein, M., Ni, H., and Weissman, D. (2005). Suppression of RNA recognition by toll-like receptors: the impact of nucleoside modification and the evolutionary origin of RNA. *Immunity* 23: 165–175.

Kelley, M.L., Strezoska, Z., He, K. et al. (2016). Versatility of chemically synthesized guide RNAs for CRISPR-Cas9 genome editing. *J. Biotechnol.* 233: 74–83.

Kim, H., Um, E., Cho, S.R. et al. (2011). Surrogate reporters for enrichment of cells with nuclease-induced mutations. *Nat. Methods* 8: 941–943.

Kocak, D.D., Josephs, E.A., Bhandarkar, V. et al. (2019). Increasing the specificity of CRISPR systems with engineered RNA secondary structures. *Nat. Biotechnol.* 37: 657–666.

Komor, A.C., Kim, Y.B., Packer, M.S. et al. (2016). Programmable editing of a target base in genomic DNA without double-stranded DNA cleavage. *Nature* 533: 420–424.

Kosicki, M., Tomberg, K., and Bradley, A. (2018). Repair of double-strand breaks induced by CRISPR-Cas9 leads to large deletions and complex rearrangements. *Nat. Biotechnol.* 36: 765–771.

Laoharawee, K., Dekelver, R.C., Podetz-Pedersen, K.M. et al. (2018). Dose-dependent prevention of metabolic and neurologic disease in murine MPS II by ZFN-mediated in vivo genome editing. *Mol. Ther.* 26: 1127–1136.

Li, Y., Park, A.I., Mou, H. et al. (2015). A versatile reporter system for CRISPR-mediated chromosomal rearrangements. *Genome Biol.* 16: 111.

Li, B., Zhao, W., Luo, X. et al. (2017a). Engineering CRISPR-Cpf1 crRNAs and mRNAs to maximize genome editing efficiency. *Nat. Biomed. Eng.* 1.

Li, G., Zhang, X., Zhong, C. et al. (2017b). Small molecules enhance CRISPR/Cas9-mediated homology-directed genome editing in primary cells. *Sci. Rep.* 7: 8943.

Li, B., Zeng, C., and Dong, Y. (2018). Design and assessment of engineered CRISPR-Cpf1 and its use for genome editing. *Nat. Protoc.* 13: 899–914.

Li, B., Niu, Y., Ji, W., and Dong, Y. (2020). Strategies for the CRISPR-based therapeutics. *Trends Pharmacol. Sci.* 41: 55–65.

Li, S., Akrap, N., Cerboni, S. et al. (2021). Universal toxin-based selection for precise genome engineering in human cells. *Nat. Commun.* 12: 497.

Lin, S., Staahl, B.T., Alla, R.K., and Doudna, J.A. (2014). Enhanced homology-directed human genome engineering by controlled timing of CRISPR/Cas9 delivery. *Elife* 3: e04766.

Lomova, A., Clark, D.N., Campo-Fernandez, B. et al. (2019). Improving gene editing outcomes in human hematopoietic stem and progenitor cells by temporal control of DNA repair. *Stem Cells* 37: 284–294.

Ma, Y., Chen, W., Zhang, X. et al. (2016). Increasing the efficiency of CRISPR/Cas9-mediated precise genome editing in rats by inhibiting NHEJ and using Cas9 protein. *RNA Biol.* 13: 605–612.

Ma, X., Chen, X., Jin, Y. et al. (2018). Small molecules promote CRISPR-Cpf1-mediated genome editing in human pluripotent stem cells. *Nat. Commun.* 9: 1303.

Maeder, M.L., Linder, S.J., Cascio, V.M. et al. (2013). CRISPR RNA-guided activation of endogenous human genes. *Nat. Methods* 10: 977–979.

Maeder, M.L., Stefanidakis, M., Wilson, C.J. et al. (2019). Development of a gene-editing approach to restore vision loss in Leber congenital amaurosis type 10. *Nat. Med.* 25: 229–233.

Mah, L.J., El-Osta, A., and Karagiannis, T.C. (2010). gammaH2AX: a sensitive molecular marker of DNA damage and repair. *Leukemia* 24: 679–686.

Manno, C.S., Pierce, G.F., Arruda, V.R. et al. (2006). Successful transduction of liver in hemophilia by AAV-Factor IX and limitations imposed by the host immune response. *Nat. Med.* 12: 342–347.

Maresca, M., Lin, V.G., Guo, N., and Yang, Y. (2013). Obligate ligation-gated recombination (ObLiGaRe): custom-designed nuclease-mediated targeted integration through nonhomologous end joining. *Genome Res.* 23: 539–546.

Maruyama, T., Dougan, S.K., Truttmann, M.C. et al. (2015). Increasing the efficiency of precise genome editing with CRISPR-Cas9 by inhibition of nonhomologous end joining. *Nat. Biotechnol.* 33: 538–542.

Mcmahon, M.A., Prakash, T.P., Cleveland, D.W. et al. (2018). Chemically modified Cpf1-CRISPR RNAs mediate efficient genome editing in mammalian cells. *Mol. Ther.* 26: 1228–1240.

Mingozzi, F. and High, K.A. (2013). Immune responses to AAV vectors: overcoming barriers to successful gene therapy. *Blood* 122: 23–36.

Mingozzi, F., Maus, M.V., Hui, D.J. et al. (2007). CD8(+) T-cell responses to adeno-associated virus capsid in humans. *Nat. Med.* 13: 419–422.

Muenzer, J. (2011). Overview of the mucopolysaccharidoses. *Rheumatology (Oxford)* 50 (Suppl 5): v4–v12.

Muenzer, J., C.E.P., Burton, B., Lau, H.A. et al. (2019). CHAMPIONS: a phase 1/2 clinical trial with dose escalation of SB-913 ZFN-mediated in vivo human genome editing for treatment of MPS II (hunter syndrome). *Mol. Genet. Metab.* 126: S104.

Nelson, C.E., Wu, Y., Gemberling, M.P. et al. (2019). Long-term evaluation of AAV-CRISPR genome editing for Duchenne muscular dystrophy. *Nat. Med.* 25: 427–432.

Nishiyama, J., Mikuni, T., and Yasuda, R. (2017). Virus-mediated genome editing via homology-directed repair in mitotic and postmitotic cells in mammalian brain. *Neuron* 96: 755–768. e5.

Onn, L., Portillo, M., Ilic, S. et al. (2020). SIRT6 is a DNA double-strand break sensor. *Elife* 9.

Orlando, S.J., Santiago, Y., Dekelver, R.C. et al. (2010). Zinc-finger nuclease-driven targeted integration into mammalian genomes using donors with limited chromosomal homology. *Nucleic Acids Res.* 38: e152.

Ousterout, D.G., Perez-Pinera, P., Thakore, P.I. et al. (2013). Reading frame correction by targeted genome editing restores dystrophin expression in cells from Duchenne muscular dystrophy patients. *Mol. Ther.* 21: 1718–1726.

Paquet, D., Kwart, D., Chen, A. et al. (2016). Efficient introduction of specific homozygous and heterozygous mutations using CRISPR/Cas9. *Nature* 533: 125–129.

Perez, E.E., Wang, J., Miller, J.C. et al. (2008). Establishment of HIV-1 resistance in CD4+ T cells by genome editing using zinc-finger nucleases. *Nat. Biotechnol.* 26: 808–816.

Perez-Pinera, P., Kocak, D.D., Vockley, C.M. et al. (2013). RNA-guided gene activation by CRISPR-Cas9-based transcription factors. *Nat. Methods* 10: 973–976.

Piel, F.B., Steinberg, M.H., and Rees, D.C. (2017). Sickle cell disease. *N. Engl. J. Med.* 376: 1561–1573.

Pinder, J., Salsman, J., and Dellaire, G. (2015). Nuclear domain 'knock-in' screen for the evaluation and identification of small molecule enhancers of CRISPR-based genome editing. *Nucleic Acids Res.* 43: 9379–9392.

Qi, L.S., Larson, M.H., Gilbert, L.A. et al. (2013). Repurposing CRISPR as an RNA-guided platform for sequence-specific control of gene expression. *Cell* 152: 1173–1183.

Rahdar, M., Mcmahon, M.A., Prakash, T.P. et al. (2015). Synthetic CRISPR RNA-Cas9-guided genome editing in human cells. *Proc. Natl. Acad. Sci. U. S. A.* 112: E7110–E7117.

Ramakrishna, S., Cho, S.W., Kim, S. et al. (2014). Surrogate reporter-based enrichment of cells containing RNA-guided Cas9 nuclease-induced mutations. *Nat. Commun.* 5: 3378.

Redon, C.E., Dickey, J.S., Bonner, W.M., and Sedelnikova, O.A. (2009). gamma-H2AX as a biomarker of DNA damage induced by ionizing radiation in human peripheral blood lymphocytes and artificial skin. *Adv. Space Res.* 43: 1171–1178.

Renaud, J.B., Boix, C., Charpentier, M. et al. (2016). Improved genome editing efficiency and flexibility using modified oligonucleotides with TALEN and CRISPR-Cas9 nucleases. *Cell Rep.* 14: 2263–2272.

Richardson, C.D., Ray, G.J., Dewitt, M.A. et al. (2016). Enhancing homology-directed genome editing by catalytically active and inactive CRISPR-Cas9 using asymmetric donor DNA. *Nat. Biotechnol.* 34: 339–344.

Riesenberg, S. and Maricic, T. (2018). Targeting repair pathways with small molecules increases precise genome editing in pluripotent stem cells. *Nat. Commun.* 9: 2164.

Riesenberg, S., Chintalapati, M., Macak, D. et al. (2019). Simultaneous precise editing of multiple genes in human cells. *Nucleic Acids Res.* 47: e116.

Robert, F., Barbeau, M., Ethier, S. et al. (2015). Pharmacological inhibition of DNA-PK stimulates Cas9-mediated genome editing. *Genome Med.* 7: 93.

Sakuma, T., Takenaga, M., Kawabe, Y. et al. (2015). Homologous recombination-independent large gene cassette knock-in in CHO cells using TALEN and MMEJ-directed donor plasmids. *Int. J. Mol. Sci.* 16: 23849–23866.

Sankaran, V.G., Xu, J., and Orkin, S.H. (2010). Advances in the understanding of haemoglobin switching. *Br. J. Haematol.* 149: 181–194.

Savic, N., Ringnalda, F.C., Lindsay, H. et al. (2018). Covalent linkage of the DNA repair template to the CRISPR-Cas9 nuclease enhances homology-directed repair. *Elife* 7.

Schiroli, G., Conti, A., Ferrari, S. et al. (2019). Precise gene editing preserves hematopoietic stem cell function following transient p53-mediated DNA damage response. *Cell Stem Cell* 24: 551–565. e8.

Scully, R., Panday, A., Elango, R., and Willis, N.A. (2019). DNA double-strand break repair-pathway choice in somatic mammalian cells. *Nat. Rev. Mol. Cell Biol.* 20: 698–714.

Sharma, R., Anguela, X.M., Doyon, Y. et al. (2015). In vivo genome editing of the albumin locus as a platform for protein replacement therapy. *Blood* 126: 1777–1784.

Shy, B.R., Macdougall, M.S., Clarke, R., and Merrill, B.J. (2016). Co-incident insertion enables high efficiency genome engineering in mouse embryonic stem cells. *Nucleic Acids Res.* 44: 7997–8010.

Singh, P., Schimenti, J.C., and Bolcun-Filas, E. (2015). A mouse geneticist's practical guide to CRISPR applications. *Genetics* 199: 1–15.

Sluch, V.M., Chamling, X., Wenger, C. et al. (2018). Highly efficient scarless knock-in of reporter genes into human and mouse pluripotent stem cells via transient antibiotic selection. *PLoS One* 13: e0201683.

Steyer, B., Bu, Q., Cory, E. et al. (2018). Scarless genome editing of human pluripotent stem cells via transient puromycin selection. *Stem Cell Rep.* 10: 642–654.

Supharattanasitthi, W., Carlsson, E., Sharif, U., and Paraoan, L. (2019). CRISPR/Cas9-mediated one step bi-allelic change of genomic DNA in iPSCs and human RPE cells in vitro with dual antibiotic selection. *Sci. Rep.* 9: 174.

Suzuki, K., Tsunekawa, Y., Hernandez-Benitez, R. et al. (2016). In vivo genome editing via CRISPR/Cas9 mediated homology-independent targeted integration. *Nature* 540: 144–149.

Taheri-Ghahfarokhi, A., Taylor, B.J.M., Nitsch, R. et al. (2018). Decoding non-random mutational signatures at Cas9 targeted sites. *Nucleic Acids Res.* 46: 8417–8434.

Takayama, K., et al. (2017). "Highly efficient biallelic genome editing of human ES/iPS cells using a CRISPR/Cas9 or TALEN system." *Nucleic Acids Res* 45(9): 5198–5207.

Tebas, P., Stein, D., Tang, W.W. et al. (2014). Gene editing of CCR5 in autologous CD4 T cells of persons infected with HIV. *N. Engl. J. Med.* 370: 901–190.

Urnov, F.D., Rebar, E.J., Holmes, M.C. et al. (2010). Genome editing with engineered zinc finger nucleases. *Nat. Rev. Genet.* 11: 636–646.

Van Overbeek, M., Capurso, D., Carter, M.M. et al. (2016). DNA repair profiling reveals nonrandom outcomes at Cas9-mediated breaks. *Mol. Cell* 63: 633–646.

Vispe, S., et al. (1998). "Overexpression of Rad51 protein stimulates homologous recombination and increases resistance of mammalian cells to ionizing radiation." *Nucleic Acids Res* 26(12): 2859–2864.

Weterings, E., Gallegos, A.C., Dominick, L.N. et al. (2016). A novel small molecule inhibitor of the DNA repair protein Ku70/80. *DNA Repair (Amst)* 43: 98–106.

Wiebking, V., Patterson, J.O., Martin, R. et al. (2020). Metabolic engineering generates a transgene-free safety switch for cell therapy. *Nat. Biotechnol.* 38: 1441–1450.

Wu, Y., Zeng, J., Roscoe, B.P. et al. (2019). Highly efficient therapeutic gene editing of human hematopoietic stem cells. *Nat. Med.* 25: 776–783.

Xu, L., Wang, J., Liu, Y. et al. (2019). CRISPR-edited stem cells in a patient with HIV and acute lymphocytic leukemia. *N. Engl. J. Med.* 381: 1240–1247.

Yang, D., Scavuzzo, M.A., Chmielowiec, J. et al. (2016). Enrichment of G2/M cell cycle phase in human pluripotent stem cells enhances HDR-mediated gene repair with customizable endonucleases. *Sci. Rep.* 6: 21264.

Yao, X., Wang, X., Hu, X. et al. (2017). Homology-mediated end joining-based targeted integration using CRISPR/Cas9. *Cell Res.* 27: 801–814.

Yao, X., Zhang, M., Wang, X. et al. (2018). Tild-CRISPR allows for efficient and precise gene knockin in mouse and human cells. *Dev. Cell* 45: 526–536. e5.

Yeh, C.D., Richardson, C.D., and Corn, J.E. (2019). Advances in genome editing through control of DNA repair pathways. *Nat. Cell Biol.* 21: 1468–1478.

Yin, H., Song, C.Q., Suresh, S. et al. (2017). Structure-guided chemical modification of guide RNA enables potent non-viral in vivo genome editing. *Nat. Biotechnol.* 35: 1179–1187.

Yin, H., Song, C.Q., Suresh, S. et al. (2018). Partial DNA-guided Cas9 enables genome editing with reduced off-target activity. *Nat. Chem. Biol.* 14: 311–316.

Zhang, W.W. and Matlashewski, G. (2019). Single-strand annealing plays a major role in double-strand DNA break repair following CRISPR-Cas9 cleavage in leishmania. *mSphere* 4.

Zhang, J.P., Li, X.L., Li, G.H. et al. (2017). Efficient precise knockin with a double cut HDR donor after CRISPR/Cas9-mediated double-stranded DNA cleavage. *Genome Biol.* 18: 35.

14

DNA Base Editing Strategies for Genome Editing

Matthew Coelho[1], *Songyuan Li*[2], *and Benjamin J.M. Taylor*[3]

[1] *Wellcome Sanger Institute, Cambridge, UK*
[2] *Genome Engineering, Discovery Sciences, BioPharmaceuticals R&D, AstraZeneca, Gothenburg, Sweden*
[3] *Discovery Sciences, BioPharmaceuticals R&D, AstraZeneca, Cambridge, UK*

14.1 Introduction

Since the human genome was first published in 2001, our understanding of pathogenic genetic variants has increased at an accelerated speed. Among all known pathogenic mutations, single-nucleotide polymorphisms (SNPs) represent the largest class by far (Landrum et al. 2014, 2016). Methods using CRISPR-associated nucleases to precisely reverse pathogenic mutations rely on homology-directed repair (HDR). The efficiency of HDR is often restricted by cell cycle stage and is generally low-to-modest. Moreover, double-strand breaks (DSBs) generated by nucleases have been shown to induce chromosome rearrangements and large deletions, posing safety concerns for therapeutic applications (Adikusuma et al. 2018; Kosicki et al. 2018). In order to reverse pathogenic SNPs safely and efficiently, base editors were developed to install point mutations with minimal double-strand break formation.

The first versions of base editors reported were designed for enzymatic conversion of cytosine to thymine, through a uracil intermediate (Komor et al. 2016), so-called cytosine base editors (CBEs). Base editor iteration 3 (BE3) includes a rat APOBEC1 as the cytidine deaminase domain, as this was found to be most active *in vitro* (Figure 14.1a). However, other deaminase domains have since been used. Shortly after the description of BE3, sea lamprey *Pm*CDA1 (an ortholog of activation-induced cytidine deaminase, AID) was used in Target-AID (Nishida et al. 2016). In addition to using different cytidine deaminases, these early versions of base editors had either an N-terminal fusion (BE3) or a C-terminal fusion (Target-AID) to Cas9-D10A and different linkers. Collectively, these differences conferred different editing activity windows (activity window refers to the nucleotide position within the protospacer sequence where the deaminase domain of the base editor is predicted to have maximum editing activity; the protospacer adjacent motif (PAM) is counted as positions 21–23). BE3 is most active at positions 4–8, and Target-AID mostly edits between positions 1 and 4. Notably, Target-AID seems to work suboptimally at mammalian body temperature, probably because sea lamprey *Pm*CDA1 has adapted to work at lower temperatures. Currently, BE3 seems to be used more widely, partly owing to the continuous refinement of this architecture in recent times. BE3 typically gives 15–75% DNA editing in a bulk population of cells, with <1% indels. However, this is dependent on the genomic locus, partly owing to BE3 partially retaining the APOBEC sequence preference for TC, with GC contexts having the poorest editing efficiencies.

In 2017, only a year after the description of the first CBEs, the first adenine base editor (ABE) was reported (Figure 14.1b). ABE converts adenosine to guanine, through an inosine intermediate (Gaudelli et al. 2017). The development of ABE was considerably more technically challenging than for CBE, as there was no known natural deaminase for adenosine that worked on ssDNA. Instead, *Escherichia coli* TadA tRNA adenosine deaminase was used as a starting point for an impressive directed evolution and structure-guided mutagenesis campaign to direct TadA to accept DNA substrates. ABE7.10 emerged after 7 rounds of redesign, giving approximately 50% editing yields in HEK293 cells, with <0.1% indels. The lower rate of indel formation with ABEs compared with CBE is linked to the fact that excision of uracils in mammalian DNA is

Genome Editing in Drug Discovery, First Edition. Edited by Marcello Maresca and Sumit Deswal.

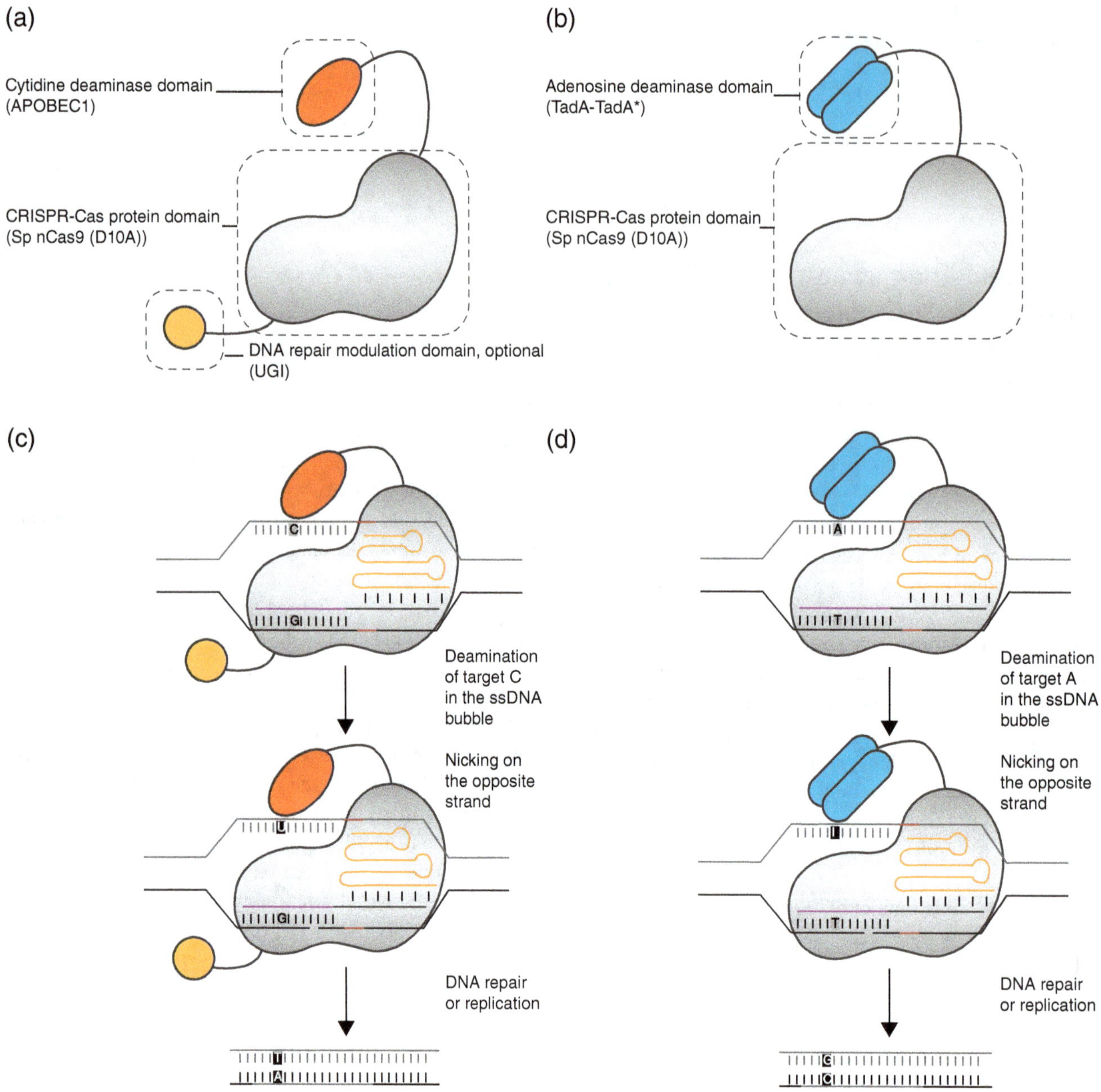

Figure 14.1 Cytosine and adenine base editing in DNA. (a) Architecture of cytosine base editors exemplified by BE3. UGI uracil DNA glycosylase inhibitor. (b) Architecture of adenine base editors exemplified by ABE7.10. (c) Schematic of cytosine base editing in DNA. (d) Schematic of adenine base editing in DNA.

common and not entirely inhibited by the UGI component of BE3, whereas repair of inosine in genomic DNA appears to be far less common. The editing window is between positions 3 and 7 for ABE7.10, and between 4 and 9 for ABE7.9. Collectively, CBEs and ABEs can convert C → T, G → A, A → G, and C → T, considerably widening the scope for correction of genetic disease-causing mutations (Figure 14.1c,d). Since CBE and ABE introduce directly opposing edits, CBE could be used to revert ABE-induced edits and *vice versa*.

So far, we have discussed base editors capable of making irreversible chemical changes to the DNA code. More recently, RNA editing has been explored as an alternative avenue, with potentially wide-ranging applications for research and therapeutics. One advantage of RNA editing is perhaps its safety, as it is transient in nature. Inherently, a disadvantage of this is that one must edit a sufficiently high number of RNA molecules in each cell (not just two DNA strands) to elicit the desired phenotypic effect. RNA editing approaches are discussed in detail in Chapter 15.

14.2 Base Editor Architectures

Several alternative deaminases are available, which all have distinct activity windows and activities. For example, rat APOBEC1 displays a more notable TC sequence context preference than both AID and CDA1. Coelho et al. investigated the use of APOBEC3A and APOBEC3B human deaminases in a BE3 architecture (Coelho et al. 2018). APOBEC3A had a broader editing window, perhaps relating to the relatively open structural conformation of the active site making APOBEC3A more permissive to substrate entry than APOBEC3B. Notably, APOBEC3B had a similar editing profile to rat APOBEC1 in terms of editing window, on-target editing and off-target editing. Joung and colleagues also reported an APBEC3A base editor, which was refined to have a strong TC context preference by introduction of a single amino acid substitution. This resulted in an editor that had very little bystander activity, but is restricted in its editing scope (Gehrke et al. 2018).

Interestingly, changing how the deaminase domain is recruited, or how the deamination event is repaired, can drastically change the window of editing and the diversity of the editing outcome. Both approaches have been used to make base editors for mutagenesis, i.e. introducing genetic diversity over a broad region.

CRISPR-X uses a fusion of hyperactive AID which lacks the nuclear-export signal, to MS2, which in turn can be recruited to dCas9 by gRNAs containing 2x MS2 binding regions (Hess et al. 2016). In principle, multiple AID-MS2 fusions can bind to a single dCas9-gRNA complex, a key difference to classical tethering approaches. This modified architecture gives a wide and shifted editing window, relative to BE3. CRISPR-X has an editing window of approximately 50 nt distal (in both 3′ and 5′ directions) to the gRNA PAM.

A similar mutagenesis approach called Targeted AID-mediated Mutagenesis (TAM) also uses hyperactive AID, but directly tethered to dCas9 (Ma et al. 2016). Both CRISPR-X and TAM deliberately lack the UGI component of BE3, which reduces editing efficiency but increases the diversity of the editing outcomes achieved from cytidine deamination, leading to all possible bases at edited positions; however, C to T editing still predominates. Since CRISPR-X and TAM give significantly reduced editing efficiencies, a strong positive selective pressure is required in functional genomics approaches. For example, these approaches were used to discover mutations conferring resistance to the drugs bortezomib and imatinib.

Finally, Base Editor for Programming Larger C to U (T) Scope, or BE-PLUS, uses an innovative recruitment strategy dependent on the SunTag system (Jiang et al. 2018). Ten copies of a GCN4 peptide were fused to Cas9D10A to recruit a single chain variant fragment (scFV) antibody, which is tethered to APOBEC-UGI-GB1. The GB1 portion is a small soluble tag to reduce protein aggregation of the scFV. BE-PLUS harbors a UGI molecule at the C-terminus, so predominantly gives C → T mutations, but the SunTag system results in a much broader editing window than BE3 (from position 4 to 14). A higher number of uracil lesions in tandem apparently reduces the number of indels and non-C-T mutations from cytidine deamination events. In addition, a higher number of UGI molecules more thoroughly inhibits base excision repair of the uracil lesion (Komor et al. 2017). Together, these factors contribute to BE-PLUS yielding fewer indels and a higher product purity compared with BE3.

14.3 Safety Considerations for Base Editing

A major safety advantage of base editing over traditional Cas9 approaches is the dramatic reduction in DNA DSBs. DSBs can cause the introduction of large deletions after DNA repair (Kosicki et al. 2018). In addition, DSBs from Cas9 nuclease can lead to a p53 response (Haapaniemi et al. 2018; Ihry et al. 2018), and cell cycle arrest, even in cancer cell lines when the targeted gene lies in an amplified region (Aguirre et al. 2016). Moreover, in-frame indels have the potential to produce novel protein domains that may stimulate an immune response or have an altered function in the cell.

Due to the very low rates of DSBs and indel formation with base editing, it is an attractive alternative approach for therapeutic genome editing. However, the safety of base editing technologies is currently under investigation and they are not without their own risks. Broadly, liabilities can be split into two categories: off-target DNA editing and off-target RNA editing.

14.3.1 Off-Target DNA Editing

Many cancers exhibit mutational signatures consistent with elevated levels of DNA cytosine deamination, which has been linked to dysregulation of APOBEC activity (Alexandrov et al. 2013). Deamination events are particularly enriched in TC dinucleotide motifs and in regions where ssDNA is commonly apparent, for example at transcription bubbles (Taylor et al. 2013). Overexpression of APOBECs or UGI as part of base editors could contribute to driving somatic mutagenesis through causing cytosine deamination events and blocking their repair.

Genome-wide, unbiased strategies to identify sites of off-target DNA editing by base editors have been carried out using ingenious modifications to existing methods to capture Cas9 off-target edits through next-generation sequencing. For example, modified versions of Digenome-seq were successfully used to identify off-target sites of CBE and ABE (Kim et al. 2017a, 2019). These studies find some overlap with Cas9 off-targets, suggesting that gRNA binding with mismatches is unsurprisingly still a concern. However, some off-target sites identified are private to base editors, and it is apparent that the rules governing off-target editing for a given gRNA are not the same as for Cas9. Notably, these *in vitro* methods cannot detect events where uracil or inosine base editing DNA intermediates are repaired in the cell, so a technology that could directly detect these in genomic DNA in cells or *in vivo* would be invaluable.

Whole-genome sequencing studies have looked at the end results of off-target DNA editing *in vivo* by CBE and ABE, demonstrating that ABE is much more specific than CBE. Editing in mouse embryos (Zuo et al. 2019) or rice (Jin et al. 2019) yielded off-target deamination events by CBEs even in the absence of gRNAs. The mechanism and extent of Cas9-independent off-target editing from overexpressed deaminase domains in live cells is not yet fully understood.

14.3.2 Off-Target RNA editing

RNA sequencing studies revealed that ABEs and CBEs can cause extensive, transcriptome-wide off-target RNA editing. Apart from the obvious concern that the abundance of RNA off-targets could have deleterious phenotypic effects in the cell, the authors also observed editing of the base editor transcripts themselves, meaning a worrying lack of control of the editing machinery at work (Grunewald et al. 2019b). To address this, Joung and colleagues introduced several mutations into the deaminase domain of rat APOBEC1 to attenuate RNA binding, dramatically reducing unwanted RNA editing (Grunewald et al. 2019a). Selective curbing of unwanted RNA editing (SECURE) variants identified included BE3-R33A/K34A, which reduced RNA editing to approximately background levels and had comparable on-target DNA editing to BE3 for most loci tested. However, the deaminase mutations also caused a more pronounced sequence preference for TC, which reduces the editing window but also reduces the number of possible loci that can be efficiently edited. Notably, the authors analyzed off-target RNA editing in cells expressing high levels of base editor from plasmids, and in some of their experiments, enriched for cells with the top 5% of BE3 expression. The levels of RNA editing in bulk populations of cells, with RNP delivery, for example, is likely to be much lower (Kim et al. 2014).

ABE was initially generated through directed evolution to accept DNA substrates, but the *E. coli* TadA deaminase was not specifically selected to reject its native RNA substrate. Therefore, it is perhaps not surprising that off-target RNA editing is still an issue. However, several ABE variants with mutations in the deamination domains have yielded base editors with greatly reduced RNA editing and similar DNA on-target efficiency (Grunewald et al. 2019b; Rees et al. 2019). Notably, inactivation of the wild-type TadA monomer in the TadA-TadA* heterodimer of ABE7.10, or its complete removal, could reduce RNA editing without having a significant impact on DNA editing. Moreover, mutating the evolved TadA* so that it no longer tolerates the 2′hydroxyl on RNA substrates, greatly reduced residual RNA editing of ABE. The authors recommend using the ABEmaxAW, miniABEmax(V82G), or miniABEmax(K20A/R21A) versions to reduce unwanted RNA editing.

14.4 Improving Precision, Efficiency, and Specificity

For the perfect base editor, three main principles have to be satisfied: precision, efficiency, and specificity. Here, precision refers to the purity of the edit, minimizing bystander mutations, indels, and undesired on-target mutations. Efficiency ensures the desired edit is achieved at high penetrance and in a large proportion of cells, maximizing the chances of success in conferring the desired phenotypic outcome. Finally, the most specific base editors bind and edit only the desired locus, and do not cause collateral editing in the genome or transcriptome, be it through gRNA-dependent or -independent mechanisms.

14.4.1 Precision

Base editing off-targets, base editing bystander mutations, and impure base edits are all issues which have been partly overcome by protein engineering efforts. The addition of a second UGI and increasing the length of both linkers in BE3 led to BE4, which achieved higher editing efficiencies of around 1.5-fold over BE3, with fewer non-C-to-T by-products (Komor et al. 2017). This suggests that cellular uracil DNA glycosylase is still functional and only partly inhibited by the single UGI molecule on BE3. However, massive overexpression of UGI may also increase global deamination rates in the genome. To further reduce indel formation, Liu and colleagues added the bacteriophage dsDNA end-binding protein Gam to BE4 to give rise to BE4-Gam (Komor et al. 2017). YE1-BE3, YE2-BE3, and YEE-BE3 are newer BE3 versions with mutations introduced to restrict the base editing activity window, resulting in fewer bystander mutations, thus increased precision (Kim et al. 2017b). YE1-BE3 is an attractive alternative to BE3 as the editing window is reduced from 5 to as few as 2 bases, and the maximal editing efficiency at these positions is comparable to BE3.

14.4.2 Efficiency

Protein expression and distribution have been critical in optimizing editing efficiencies in mammalian cells. Mislocalized editors are obviously useless, and so the simple addition of an additional nuclear localization sequence (NLS) at the N-terminus increased editing efficiencies (Zafra et al. 2018). The authors also removed possible premature polyadenylation sequences from BE3. Equally, base editors that have been carefully codon optimized for expression in mammalian cells have higher expression levels and thus, activity (Zafra et al. 2018). Codon optimization and redesign of the APOBEC deaminase domain directed by ancestral protein reconstruction led to the generation of BE4max, ABEmax, and AncBE4max, with increased activity in cells (Koblan et al. 2018).

Another limitation of base editor efficiency is the restriction of the deaminase activity window position relative to the PAM. For example, for BE3, the desired mutation must fall within positions 4–8 of the protospacer for optimal editing. By using alternative orthologs to *Sp*Cas9, one can generate base editors that use alternative PAMs, and by mutagenesis of the Cas9 protein itself, one can generate base editor variants that have relaxed, unnatural PAM specificities. For example, *Staphylococcus aureus* Cas9 base editor variants are capable of editing proximal to NNGRRT PAMs (Komor et al. 2017), and an enhanced version of Cas12a-BE can efficiently edit proximal to TTTV PAMs, where V is A, C, or G (Kleinstiver et al. 2019). Mutant *Sp*Cas9 variants adopted for base editing include Cas9-NG (Nishimasu et al. 2018) or xCas9 (Hu et al. 2018), with PAM specificity relaxed from NGG to NG, thus increasing the available PAMs for editing from approximately every eight bases on either strand (NGG PAM for BE3), to one in every two bases on either strand (NG PAM). Cas-NG and associated base editor variants seem to outperform xCas9 versions in the vast majority of genomic loci. More recently, SpRY Cas9 has been developed, which is essentially PAM-less (albeit with lower efficiency at NYN PAMs, i.e. NCN and NTN), dramatically widening the base editing space to almost every adenine and cytosine base in the genome (Walton et al. 2020). This will be invaluable for semi-saturation mutagenesis screens using base editors.

Finally, when other components of efficiency have been satisfied, co-selection strategies can be used to enrich for edited populations of cells (Coelho et al. 2018).

14.4.3 Specificity

One of the most simple and effective ways to reduce off-target base editing is to deliver RNP formulations rather than plasmid, to minimize the amount of base editor and exposure time to the bare minimum required for on-target editing (Kim et al. 2014). In addition to this, high-fidelity Cas9 variants generally contain mutations that help discriminate between perfectly matched gRNAs and mismatched gRNA sequences when it comes to activating nuclease activity. However, many high-fidelity variants of Cas9 also have reduced on-target editing efficiencies. Of note, the *Sp*Cas9 variant harboring a single amino acid substitution (R691A), HiFi-Cas9, is widely used for its exceptional specificity, and minimal reduction in on-target cutting, as compared with alternatives such as eCas9, Cas9-HF1, and HypaCas9 (Vakulskas et al. 2018). This enzyme was identified through an unbiased selection process in *E. coli*. Interestingly, using a very similar approach, a distinct and highly efficient high-fidelity *Sp*Cas9 variant called Sniper-Cas9 was developed (Lee et al. 2018). The authors recommend using both HiFi-Cas9 and Sniper-Cas9 base editor variants for increased discrimination of DNA off-targets. One unexplored route to reduce deaminase-dependent off-target editing would be to construct a split base editor, whereby two guide RNAs are required to bring together and complement the split deaminase domain before it is enzymatically active, akin to split-Cas9 approaches (Zetsche et al. 2015).

Coelho et al. have developed a tool to reduce persistent off-target editing by Cas9 and base editors called CRISPR-Cas9 guide RNA-assisted reduction of damage, or CRISPR GUARD (Coelho et al. 2020; Rose et al. 2020). This technology directs nuclease-dead Cas9 complexes formed with short guide RNAs to off-target sites, thus protecting them from editing by active Cas9. Importantly, CRISPR GUARD can be multiplexed to significantly reduce editing at multiple off-target sites simultaneously, without negatively impacting the on-target editing efficiency. The authors also provided a tool for automated design of GUARD RNAs for predicted gRNA off-target sites. An advantage of CRISPR GUARD is its simplicity and versatility as it works by co-delivery of a short gRNA, meaning it can be used with any Cas9 or base editor variant.

14.5 Prime Editing and Base Editing

Despite significant improvements in the targeting space for base editors within the genome, they still have the limitation of only being able to introduce the transition mutations, C → T (G → A) or A → G (T → C). A recently developed technology called prime editing massively increases the repertoire of precise edits (Anzalone et al. 2019). Prime Editor 3 (PE3) uses an optimized M-MLV reverse transcriptase fused to nickase Cas9(H840A), which installs edits encoded in the 3′ of an extended gRNA, or prime editing gRNA (pegRNA). Nicking of the unedited strand with an additional distal gRNA to initiate DNA repair leads to efficient incorporation of new genetic information around the pegRNA nick site, including deletions, insertions, transversions, and transitions. Prime editing is an useful alternative to base editing with CBE or ABE when bystander mutations are unacceptably high. However, if the region is accessible for base editing through proximity to a PAM and the target base is appropriately placed within the deaminase activity window, then base editing is recommended because it is more efficient and gives fewer indels than PE3.

14.6 Choosing the Right Editor

In Figure 14.2, we present a "Yes/No" decision flow chart to help readers to identify the most appropriate base editing technology for their genome editing task. Possible applications include, but are not limited to: functional genomics/screening, cell and disease model generation with SNVs, investigating SNVs in noncoding regions, investigating SNVs of unknown clinical significance, gene knock-outs, and therapeutic gene editing by correction of disease-causing mutations.

14.7 Therapeutic Uses of Base Editors

The first description of base editors by Komor et al. highlighted the therapeutic use of the platform, with a description of nearly 1000 clinically actionable mutations that could be reverted using BE3 (Komor et al. 2016). Since then, the breadth of coverage of the ClinVar database of annotated pathogenic human mutations (Landrum et al. 2018) that are treatable has steadily grown, as base editors have been applied to different PAM variants platforms (Hu et al. 2018; Huang et al. 2019; Liu et al. 2019) and with the advent of the adenosine base editing platform (Gaudelli et al. 2017); these platforms could now be employed to revert over 60% of pathogenic human SNPs. However, reversion is only one therapeutic use of the base editing platform. Elegant work by two groups have highlighted the potential for base editors for gene ablation, with the CRISPR-STOP (Kuscu et al. 2017) and iSTOP (Billon et al. 2017) platforms demonstrating the ability to introduce termination codons into the vast majority of human genes. The use of base editors for gene ablation becomes particularly attractive when one considers the need to target multiple genes. With the now extensive reports of large-scale genetic damage induced by induction of multiple DSBs when targeting CRISPR/Cas9 to single and multiple sites across the genome (Adikusuma et al. 2018; Cullot et al. 2019; Kosicki et al. 2018). The use of base editors to induce stop codons with their inherent lack of DSBs reduces these critical safety concerns (Carreras et al. 2019; Song et al. 2020b).

To date, there have been a number of studies demonstrating the therapeutic use of base editors across a number of murine models of human disease. David Liu's group have recently used base editors to correct the human mutation driving the Niemann–Pick neurological condition in murine models of the disease (Levy et al. 2020). These studies showed clear reversion of the mutation in cortex and cerebellum, with a marginal clinical benefit. Yeh and colleagues were able to treat a mouse model harboring a recessive loss of function allele in *TMC1* which leads to deafness in humans (Yeh et al. 2020). Using adenoviral delivery into the inner ear of these mice, CBEs were able to correct over 50% alleles in the cochlea and drive partial restoration of hearing (Yeh et al. 2020). Ryu et al. recently used *in-vivo* delivery in a Duchenne muscular dystrophy mouse model to demonstrate reversion of a premature stop codon in the *Dmd* gene (Ryu et al. 2018). While direct intramuscular injection of ABE resulted in less than 4% of allelic correction, this was enough to allow restored expression of dystrophin in 15% of myofibers, in line with earlier work using CRISPR/Cas9 nucleases for efficacious treatment of a murine Dmd model (Nelson et al. 2016). Song et al. used ABEs to treat mouse models of hereditary tyrosinemia type I (Song et al. 2020a). This model contains a splice site mutation in the fumarylacetoacetate hydrolase gene, leading to abrogated protein production. Fah is involved in tyrosine catabolism, and its reduced levels result in a lethal build-up in the blood. Tail vein injection of plasmids expressing ABE and sgRNA were able to correct low percentages of Fah alleles from liver samples, which were able to then generate sufficient protein expression to prevent terminal body weight loss (Song et al. 2020a). Finally, high cholesterol can be treated through inhibition of proprotein convertase subtilisin/kexin type 9 (PCSK9); using a mouse model expressing a human PSCK9 allele (Carreras et al. 2019), Carreras et al. demonstrated that adenoviral delivery of BE3 could drive 10% editing and inactivation of human PCSK9, while Chadwick et al. showed efficient and stable editing of mouse *Pcsk9* with a significant reduction in blood cholesterol (Chadwick et al. 2017).

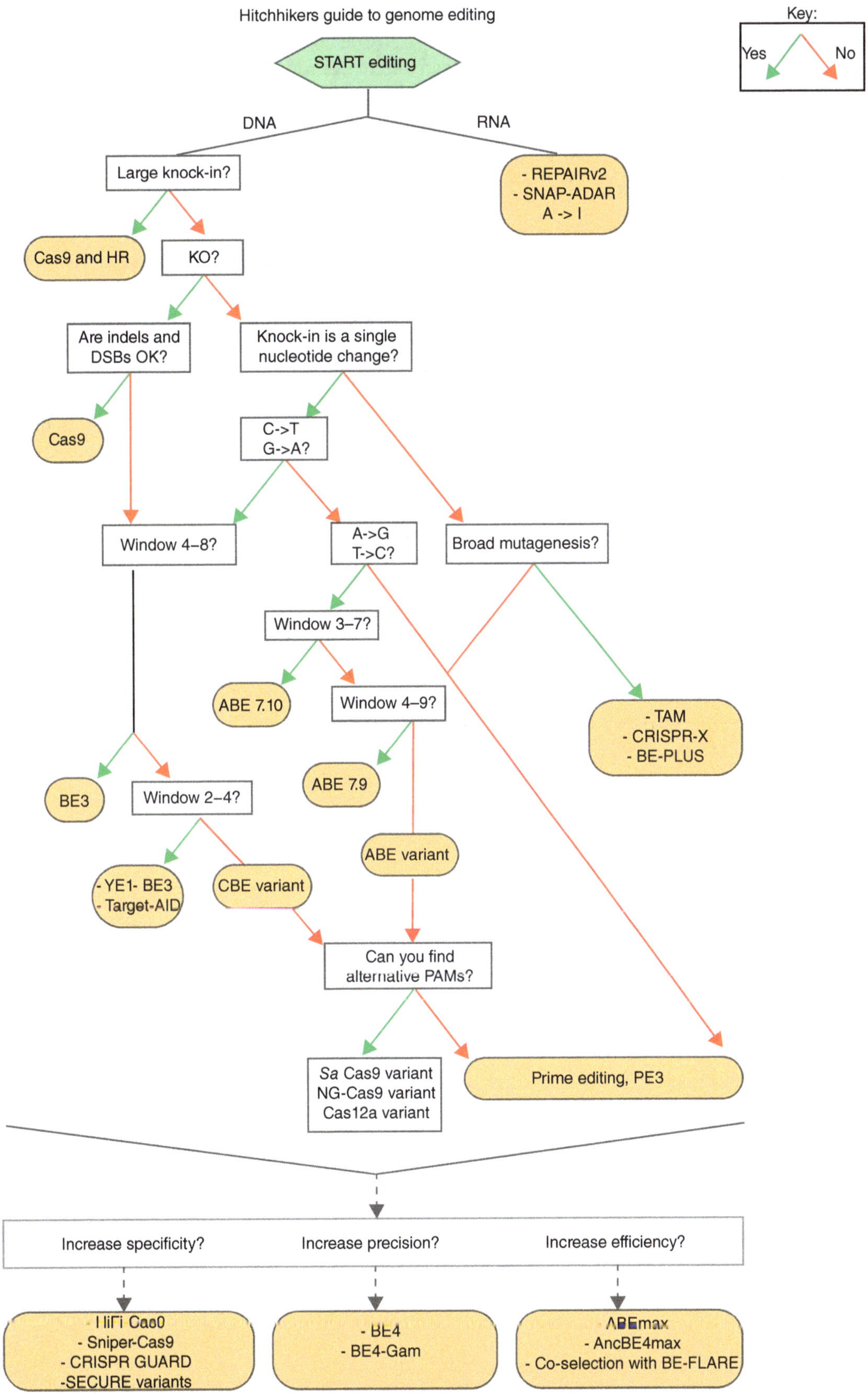

Figure 14.2 Decision flow chart for choosing base editing tools.

Currently, only Beam Therapeutics are progressing base editors to the clinic, with a broad disease pipeline with sickle cell anemia the most advanced programme. The United States has a large homogenous patient population carrying the disease driving E6V mutation in the HBB gene (Lin et al. 2019). This mutation can be recoded to a benign E6A variant using ABEs, with promising early studies showing 40% editing efficiencies on test case *ex-vivo* samples (Lin et al. 2019).

As will be discussed in later sections, delivery of genome engineering machinery remains a major obstacle for *in-vivo* therapeutics. The above examples rely on plasmid or viral delivery, and how this will translate to the clinic remains to be resolved. Adeno-associated viral delivery is a clinically approved route for *in-vivo* delivery; however, the large size of base editing complexes exceeds the cargo capacity of these systems. Using intein technology, Levy et al. have circumvented this issue by splitting the entire complex into two smaller parts, allowing efficient AAV packaging (Levy et al. 2020). Following co-delivery into a cell, protein splicing restores a fully functional base editor. Approaches such as these, or the use of smaller Cas proteins, such as the recently discovered Cas-phi (Pausch et al. 2020) will aid in solving viral-based *in-vivo* delivery solutions. *Ex-vivo* therapies offer simpler delivery opportunities, with T-cell-based immunotherapies an obvious setting to apply genome editing tools. The Moriarity lab has reported successful engineering of *ex-vivo* primary T and CAR-T cells using mRNA and ribonuclear delivered base editors (Webber et al. 2019). Importantly, the authors also showed that while multiplexed editing using Cas9 nucleases resulted in robust generation of translocations, virtually none were detected when using BE4 while maintaining high editing efficiencies.

Safety and immunogenicity of CRISPR, will be discussed in detail later in this book. With respect to base editing, the multicomponent complex adds extra layers of concern which will be briefly outlined here. In addition to bacterial-derived Cas9, adenosine base editors contain bacterial TadA (Gaudelli et al. 2017), while first-generation cytosine editors contain a bacteriophage uracil DNA glycosylase inhibitor and a rat Apobec1 deaminase (Komor et al. 2016), all potentially leading to novel immunogenic responses. Several groups have replaced some of these components with human equivalents, such as human APOBEC family counterparts (Coelho et al. 2018; Gehrke et al. 2018), which offers an alternative route to avoid immune responses. As with CRISPR/Cas9, unintended off-target editing effects of base editors have been solved using protein evolution techniques. The discovery of RNA editing by BE3, BE4, and ABEmax allowed the Joung lab to engineer new variants of rat Apobec1 which significantly reduced this burden for cytosine editors (Grunewald et al. 2019a). Yu et al. have similarly reported development of alternative variants with reduced RNA editing levels (Yu et al. 2020). This study, along with Doman et al. was also able to demonstrate variants with reduced genome-wide mutational burdens, an unwanted side effect of all DNA-modifying enzymes and particular for mutagenic base editors which have seen increased SNVs in human cell lines, mouse embryos, and plants (Doman et al. 2020; Jin et al. 2019; McGrath et al. 2019; Zuo et al. 2019). Finally, due to the mutagenic nature of base editors, on-target bystander mutations at nearby cytosines or adenosines can be problematic (Gehrke et al. 2018). Multiple strategies are now available to overcome this issue, with modifications of linker lengths and evolution of the deaminase domains leading to ever-increasing precision in the base editing window, with, for example, development of the YE2 BE3 variant able to edit only in a 2–3 nt window (Kim et al. 2017b). Alternative options are to use deaminase domains with nucleotide preferences which suppress bystander targeting (Gehrke et al. 2018), or the use of inactive CRISPR/Cas complexes to sterically block these sites from damage (Coelho et al. 2020; Rose et al. 2020).

14.8 Conclusions

In summary, both cytosine and adenosine base editors have progressed enormously in terms of efficiency and safety from their inception in 2016. The ability to directly revert of inactivate single or multiplexed targets in human systems offers enormous performance and safety advantages over nucleases. With a raft of data showing successful treatment of disease models and improved safety margins, the use of base editors to offer direct precise clinical therapies looks promising.

References

Adikusuma, F., Piltz, S., Corbett, M.A. et al. (2018). Large deletions induced by Cas9 cleavage. *Nature* 560: E8–E9.

Aguirre, A.J., Meyers, R.M., Weir, B.A. et al. (2016). Genomic copy number dictates a gene-independent cell response to CRISPR/Cas9 targeting. *Cancer Discov.* 6: 914–929.

Alexandrov, L.B., Nik-Zainal, S., Wedge, D.C. et al. (2013). Signatures of mutational processes in human cancer. *Nature* 500: 415–421.

Anzalone, A.V., Randolph, P.B., Davis, J.R. et al. (2019). Search-and-replace genome editing without double-strand breaks or donor DNA. *Nature* 576: 149–157.

Billon, P., Bryant, E.E., Joseph, S.A. et al. (2017). CRISPR-mediated base editing enables efficient disruption of eukaryotic genes through induction of STOP codons. *Mol. Cell* 67: 1068–1079. e4.

Carreras, A., Pane, L.S., Nitsch, R. et al. (2019). In vivo genome and base editing of a human PCSK9 knock-in hypercholesterolemic mouse model. *BMC Biol.* 17: 4.

Chadwick, A.C., Wang, X., and Musunuru, K. (2017). In vivo base editing of PCSK9 (proprotein convertase subtilisin/kexin type 9) as a therapeutic alternative to genome editing. *Arterioscler. Thromb. Vasc. Biol.* 37: 1741–1747.

Coelho, M.A., Li, S., Pane, L.S. et al. (2018). BE-FLARE: a fluorescent reporter of base editing activity reveals editing characteristics of APOBEC3A and APOBEC3B. *BMC Biol.* 16: 150.

Coelho, M.A., De Braekeleer, E., Firth, M. et al. (2020). CRISPR GUARD protects off-target sites from Cas9 nuclease activity using short guide RNAs. *Nat. Commun.* 11: 4132.

Cullot, G., Boutin, J., Toutain, J. et al. (2019). CRISPR-Cas9 genome editing induces megabase-scale chromosomal truncations. *Nat. Commun.* 10: 1136.

Doman, J.L., Raguram, A., Newby, G.A., and Liu, D.R. (2020). Evaluation and minimization of Cas9-independent off-target DNA editing by cytosine base editors. *Nat. Biotechnol.* 38: 620–628.

Gaudelli, N.M., Komor, A.C., Rees, H.A. et al. (2017). Programmable base editing of A*T to G*C in genomic DNA without DNA cleavage. *Nature* 551: 464–471.

Gehrke, J.M., Cervantes, O., Clement, M.K. et al. (2018). An APOBEC3A-Cas9 base editor with minimized bystander and off-target activities. *Nat. Biotechnol.* 36: 977–982.

Grunewald, J., Zhou, R., Garcia, S.P. et al. (2019a). Transcriptome-wide off-target RNA editing induced by CRISPR-guided DNA base editors. *Nature* 569: 433–437.

Grunewald, J., Zhou, R., Iyer, S. et al. (2019b). CRISPR DNA base editors with reduced RNA off-target and self-editing activities. *Nat. Biotechnol.* 37: 1041–1048.

Haapaniemi, E., Botla, S., Persson, J. et al. (2018). CRISPR-Cas9 genome editing induces a p53-mediated DNA damage response. *Nat. Med.* 24: 927–930.

Hess, G.T., Fresard, L., Han, K. et al. (2016). Directed evolution using dCas9-targeted somatic hypermutation in mammalian cells. *Nat. Methods* 13: 1036–1042.

Hu, J.H., Miller, S.M., Geurts, M.H. et al. (2018). Evolved Cas9 variants with broad PAM compatibility and high DNA specificity. *Nature* 556: 57–63.

Huang, T.P., Zhao, K.T., Miller, S.M. et al. (2019). Circularly permuted and PAM-modified Cas9 variants broaden the targeting scope of base editors. *Nat. Biotechnol.* 37: 626–631.

Ihry, R.J., Worringer, K.A., Salick, M.R. et al. (2018). p53 inhibits CRISPR-Cas9 engineering in human pluripotent stem cells. *Nat. Med.* 24: 939–946.

Jiang, W., Feng, S., Huang, S. et al. (2018). BE-PLUS: a new base editing tool with broadened editing window and enhanced fidelity. *Cell Res.* 28: 855–861.

Jin, S., Zong, Y., Gao, Q. et al. (2019). Cytosine, but not adenine, base editors induce genome-wide off-target mutations in rice. *Science* 364: 292–295.

Kim, S., Kim, D., Cho, S.W. et al. (2014). Highly efficient RNA-guided genome editing in human cells via delivery of purified Cas9 ribonucleoproteins. *Genome Res.* 24: 1012–1019.

Kim, D., Lim, K., Kim, S.T. et al. (2017a). Genome-wide target specificities of CRISPR RNA-guided programmable deaminases. *Nat. Biotechnol.* 35: 475–480.

Kim, Y.B., Komor, A.C., Levy, J.M. et al. (2017b). Increasing the genome-targeting scope and precision of base editing with engineered Cas9-cytidine deaminase fusions. *Nat. Biotechnol.* 35: 371–376.

Kim, D., Kim, D.E., Lee, G. et al. (2019). Genome-wide target specificity of CRISPR RNA-guided adenine base editors. *Nat. Biotechnol.* 37: 430–435.

Kleinstiver, B.P., Sousa, A.A., Walton, R.T. et al. (2019). Engineered CRISPR-Cas12a variants with increased activities and improved targeting ranges for gene, epigenetic and base editing. *Nat. Biotechnol.* 37: 276–282.

Koblan, L.W., Doman, J.L., Wilson, C. et al. (2018). Improving cytidine and adenine base editors by expression optimization and ancestral reconstruction. *Nat. Biotechnol.* 36: 843–846.

Komor, A.C., Kim, Y.B., Packer, M.S. et al. (2016). Programmable editing of a target base in genomic DNA without double-stranded DNA cleavage. *Nature* 533: 420–424.

Komor, A.C., Zhao, K.T., Packer, M.S. et al. (2017). Improved base excision repair inhibition and bacteriophage Mu Gam protein yields C:G-to-T: a base editors with higher efficiency and product purity. *Sci. Adv.* 3: eaao4774.

Kosicki, M., Tomberg, K., and Bradley, A. (2018). Repair of double-strand breaks induced by CRISPR-Cas9 leads to large deletions and complex rearrangements. *Nat. Biotechnol.* 36: 765–771.

Kuscu, C., Parlak, M., Tufan, T. et al. (2017). CRISPR-STOP: gene silencing through base-editing-induced nonsense mutations. *Nat. Methods* 14: 710–712.

Landrum, M.J., Lee, J.M., Riley, G.R. et al. (2014). ClinVar: public archive of relationships among sequence variation and human phenotype. *Nucleic Acids Res.* 42: D980–D985.

Landrum, M.J., Lee, J.M., Benson, M. et al. (2016). ClinVar: public archive of interpretations of clinically relevant variants. *Nucleic Acids Res.* 44: D862–D868.

Landrum, M.J., Lee, J.M., Benson, M. et al. (2018). ClinVar: improving access to variant interpretations and supporting evidence. *Nucleic Acids Res.* 46: D1062–D1067.

Lee, J.K., Jeong, E., Lee, J. et al. (2018). Directed evolution of CRISPR-Cas9 to increase its specificity. *Nat. Commun.* 9: 3048.

Levy, J.M., Yeh, W.H., Pendse, N. et al. (2020). Cytosine and adenine base editing of the brain, liver, retina, heart and skeletal muscle of mice via adeno-associated viruses. *Nat. Biomed. Eng.* 4: 97–110.

Lin, L., Rybak, A.P., Rinaldi, C. et al. (2019). Complementary base editing approaches for the treatment of sickle cell disease and beta thalassemia. *Blood* 134: 3352–3352.

Liu, Z., Shan, H., Chen, S. et al. (2019). Efficient base editing with expanded targeting scope using an engineered Spy-mac Cas9 variant. *Cell Discov.* 5: 58.

Ma, Y., Zhang, J., Yin, W. et al. (2016). Targeted AID-mediated mutagenesis (TAM) enables efficient genomic diversification in mammalian cells. *Nat. Methods* 13: 1029–1035.

Mcgrath, E., Shin, H., Zhang, L. et al. (2019). Targeting specificity of APOBEC-based cytosine base editor in human iPSCs determined by whole genome sequencing. *Nat. Commun.* 10: 5353.

Nelson, C.E., Hakim, C.H., Ousterout, D.G. et al. (2016). In vivo genome editing improves muscle function in a mouse model of Duchenne muscular dystrophy. *Science* 351: 403–407.

Nishida, K., Arazoe, T., Yachie, N. et al. (2016). Targeted nucleotide editing using hybrid prokaryotic and vertebrate adaptive immune systems. *Science* 353: aaf8729-1–aaf8729-8.

Nishimasu, H., Shi, X., Ishiguro, S. et al. (2018). Engineered CRISPR-Cas9 nuclease with expanded targeting space. *Science* 361: 1259–1262.

Pausch, P., Al-Shayeb, B., Bisom-Rapp, E. et al. (2020). CRISPR-CasPhi from huge phages is a hypercompact genome editor. *Science* 369: 333–337.

Rees, H.A., Wilson, C., Doman, J.L., and Liu, D.R. (2019). Analysis and minimization of cellular RNA editing by DNA adenine base editors. *Sci. Adv.* 5: eaax5717.

Rose, J.C., Popp, N.A., Richardson, C.D. et al. (2020). Suppression of unwanted CRISPR-Cas9 editing by co-administration of catalytically inactivating truncated guide RNAs. *Nat. Commun.* 11: 2697.

Ryu, S.M., Koo, T., Kim, K. et al. (2018). Adenine base editing in mouse embryos and an adult mouse model of Duchenne muscular dystrophy. *Nat. Biotechnol.* 36: 536–539.

Song, C.Q., Jiang, T., Richter, M. et al. (2020a). Adenine base editing in an adult mouse model of tyrosinaemia. *Nat. Biomed. Eng.* 4: 125–130.

Song, Y., Liu, Z., Zhang, Y. et al. (2020b). Large-fragment deletions induced by Cas9 cleavage while not in the BEs system. *Mol. Ther. Nucleic Acids* 21: 523–526.

Taylor, B.J., Nik-Zainal, S., Wu, Y.L. et al. (2013). DNA deaminases induce break-associated mutation showers with implication of APOBEC3B and 3A in breast cancer kataegis. *Elife* 2: e00534.

Vakulskas, C.A., Dever, D.P., Rettig, G.R. et al. (2018). A high-fidelity Cas9 mutant delivered as a ribonucleoprotein complex enables efficient gene editing in human hematopoietic stem and progenitor cells. *Nat. Med.* 24: 1216–1224.

Walton, R.T., Christie, K.A., Whittaker, M.N., and Kleinstiver, B.P. (2020). Unconstrained genome targeting with near-PAMless engineered CRISPR-Cas9 variants. *Science* 368 (6488): 290–296.

Webber, B.R., Lonetree, C.L., Kluesner, M.G. et al. (2019). Highly efficient multiplex human T cell engineering without double-strand breaks using Cas9 base editors. *Nat. Commun.* 10: 5222.

Yeh, W.H., Shubina-Oleinik, O., Levy, J.M. et al. (2020). In vivo base editing restores sensory transduction and transiently improves auditory function in a mouse model of recessive deafness. *Sci. Transl. Med.* 12.

Yu, Y., Leete, T.C., Born, D.A. et al. (2020). Cytosine base editors with minimized unguided DNA and RNA off-target events and high on-target activity. *Nat. Commun.* 11: 2052.

Zafra, M.P., Schatoff, E.M., Katti, A. et al. (2018). Optimized base editors enable efficient editing in cells, organoids and mice. *Nat. Biotechnol.* 36: 888–893.

Zetsche, B., Volz, S.E., and Zhang, F. (2015). A split-Cas9 architecture for inducible genome editing and transcription modulation. *Nat. Biotechnol.* 33: 139–142.

Zuo, E., Sun, Y., Wei, W. et al. (2019). Cytosine base editor generates substantial off-target single-nucleotide variants in mouse embryos. *Science* 364: 289–292.

15

RNA Base Editing Technologies for Gene Therapy

Sumit Deswal

Genome Engineering, Discovery Sciences, BioPharmaceuticals R&D, AstraZeneca, Gothenburg, Sweden

15.1 Introduction

A large number of human genetic diseases (about 58%) are caused by DNA point mutations (Rees and Liu 2018). The discovery of programmable nucleases such as meganucleases, ZFNs, TALENs, and CRISPR-Cas has provided tools for studying the function of genes and their role in human disease as well as gave hope to fix these mutations in the patients. However, the use of these nucleases for the correction of point mutations *in vivo* poses several challenges. These challenges include immunogenicity due to prokaryotic origin of the proteins (Chew et al. 2016), delivery issue due to their large size (Platt et al. 2014), and permanent nature of editing (including off-targets). To avoid these limitations of DNA nucleases, approaches that instead directly target RNA would be highly desirable, as these would enable tunability and reversibility and importantly no off-target mutations would be permanent. Thus, RNA base editing using adenosine deaminases acting on RNA (ADARs) of human origin could be an attractive approach for *in vivo* correction of disease-causing point mutations.

The concept of therapeutic RNA editing is not new. Already in 1995, researchers at Ribozyme Pharmaceuticals discovered that antisense oligonucleotides (ASOs) could recruit ADAR to make low efficiency edits on matching strands of RNA in the cells (Woolf et al. 1995). However, the initial progress in therapeutic RNA editing field has been very slow. Only in the last five to six years, the field has gained momentum with the development of several new strategies resulting in higher efficiency and lower off-target RNA editing (Figure 15.1).

ADARs are enzymes that catalyze Adenosine (A) to Inosine (I) conversion in RNA (Figure 15.2a). The converted Inosine is read as Guanosine by the translational machinery (Wong et al. 2001). Recently, an evolved version ADAR2 deaminase domain (ADAR2 DD) was described that can catalyze Cytosine (C) to Uracil (U) conversion in RNA (Figure 15.2b) (Abudayyeh et al. 2019). ADAR activity was first described as a dsRNA duplex unwinding activity in *Xenopus* oocytes (Bass and Weintraub 1987). In vertebrates, there are three different ADAR genes (ADAR1, ADAR2, and ADAR3) with conserved functional protein domains (Nishikura 2010) (Figure 15.2c). ADAR1 has two isoforms, ADAR1 p150 and ADAR1 p110 (with molecular weight of 150 and 110 kD, respectively) (Patterson and Samuel 1995). ADAR1 p110 is constitutively expressed, while ADAR1 p150 expression is induced by Interferon signaling (George and Samuel 1999; Patterson and Samuel 1995). There are two to three dsRNA-binding domains (dsRBD) present in each of the ADAR proteins. The dsRBD binds double-stranded RNA (dsRNA) directly and brings the C-terminal catalytic deaminase domain in contact with target adenosine. ADAR1 and ADAR2 are expressed in many tissues; however, expression of ADAR3 is restricted to certain regions in brain (Chen et al. 2000). ADAR3 is also shown to lack deaminase activity and rather has inhibitory role in RNA editing (Chen et al. 2000).

With the first engineering of a SNAP-tag fusion version of ADAR to recruit to the target RNA (Stafforst and Schneider 2012), there have been a number of efforts to engineer ADAR proteins and/or design strategies to harness endogenous ADAR protein to edit the desired RNA site. This chapter will discuss some of these approaches and future aspects of what is still needed to bring this promising technology to the clinics.

Genome Editing in Drug Discovery, First Edition. Edited by Marcello Maresca and Sumit Deswal.

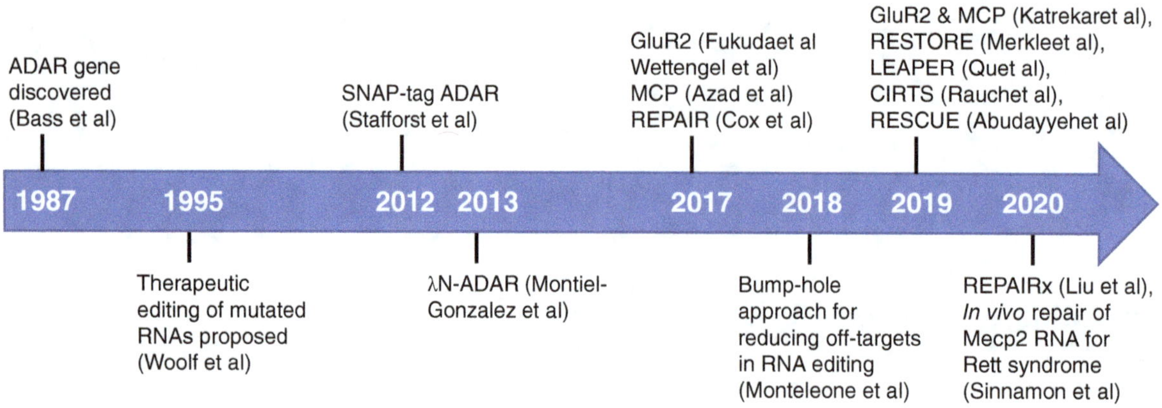

Figure 15.1 Brief history of developments in the RNA editing technologies.

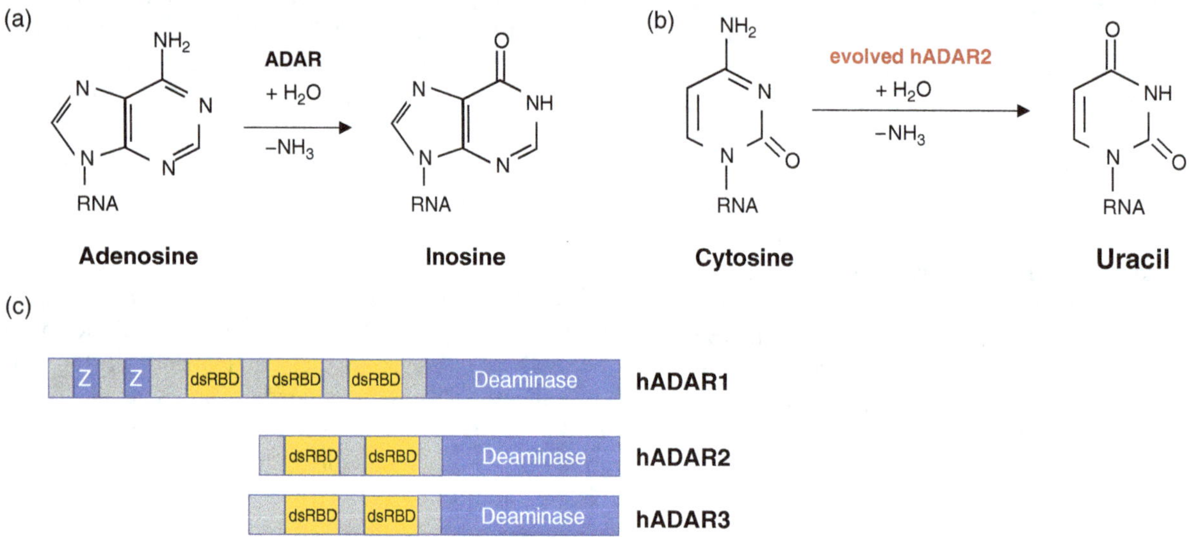

Figure 15.2 ADAR proteins structure and function. (a) Chemical reaction catalyzed by natural ADAR, resulting in conversion of an Adenosine into Inosine. (b) An evolved version of hADAR deaminase domain catalyzes Cytosine to Inosine conversion. (c) Three human ADAR proteins. Different domains are marked. Z-DNA binding domains (Z, only in ADAR1), the dsRNA binding (dsRBDs), and the deaminase domain.

15.2 RNA Editing Technologies

Broadly, the systems used for RNA editing can be grouped into three categories: (i) those using endogenous ADAR proteins for targeted RNA editing, (ii) engineered ADAR proteins employing fusion proteins for RNA targeting, and (iii) those employing Cas proteins for RNA targeting (technically they could be in group 2; however, a separate discussion may be needed due to special considerations, e.g. their large size).

15.2.1 Systems Employing Endogenous ADAR Proteins

There are three different published strategies for RNA editing that harness endogenous ADARs for targeted editing. These include gRNA design based on ADAR natural substrate (GluR2-ADAR), chemically stabilized ASOs (RESTORE), and use of long ADAR-recruiting RNAs (arRNAs; LEAPER).

15.2.1.1 GluR2-ADAR

The ionotropic AMPA glutamate receptor 2 (Glur2) belongs to a family of ligand-activated glutamate receptors that are sensitive to alpha-amino-3-hydroxy-5-methyl-4-isoxazole propionate (AMPA). Glutamate receptors function as the main excitatory neurotransmitter at many synapses in the central nervous system. In human brain, 100% of the GluR2 pre-mRNA undergoes adenosine-to-inosine editing by ADAR2 (Egebjerg et al. 1994; Stefl et al. 2010). Hence, GluR2 is a natural substrate for ADAR2, and was exploited to design ADAR-recruiting guide RNAs to target ADAR deaminase to specific sites on the RNA in a number of studies (Fukuda et al. 2017; Katrekar et al. 2019; Wettengel et al. 2017).

Fukuda et al. and Wettengel et al. designed Adar RNAs based on GluR2 pre-mRNA and tested the RNA editing efficiency using fluorescence and luciferase reporters (Fukuda et al. 2017; Wettengel et al. 2017). Additionally, system was shown to be effective in editing endogenous loci in the cell lines. Katrekar et al, designed and further optimized GluR2-ADAR systems bearing ADAR recruiting domains engineered from the GluR2 pre-messenger RNA (GluR2 adRNA) (Katrekar et al. 2019). The optimizations included variations in antisense domain length, mismatch position modifications, and the addition of editing inducer elements, etc. The system was tested in two independent mouse models of human disease. The *mdx* mouse model of Duchenne muscular dystrophy (DMD), which bears an ochre stop site (TAA) in exon 23 of the dystrophin gene and the male sparse fur ash (spf^{ash}) mouse model of ornithine transcarbamylase (OTC) deficiency which bears a guanine to adenine transition of the last nucleotide of the fourth exon of the OTC gene resulting in aberrant RNA splicing. Adeno-associated virus 8 (AAV8) was used as a delivery vehicle for the constructs in both mouse models. In the mdx mice, constructs were injected into the tibialis anterior or gastrocnemius muscle. Few weeks after the injection, tibialis anterior and gastrocnemius muscles were collected. Immunofluorescence staining and immunoblotting studies revealed clear restoration of dystrophin expression (Katrekar et al. 2019). In the spf^{ash}, NGS analysis was performed on the liver tissue and up to 0.8–4.7% edited mRNA was detected among the correctly spliced OTC mRNA when ADAR2 was expressed along with adRNA. In both models, adRNA alone also resulted in significant (less than 1%) RNA editing which was improved when ADAR2 or its hyperactive version was expressed along with adRNA.

15.2.1.2 RESTORE

Recruiting endogenous ADAR to specific transcripts for oligonucleotide-mediated RNA editing (RESTORE) uses chemically stabilized ASOs that recruit endogenous human ADARs to edit endogenous transcripts (Merkle et al. 2019). The ASOs comprised an ADAR-recruiting domain composed entirely of natural ribonucleotides and a specificity domain that was chemically modified (2′-O-methylations, phosphorothioate), containing a modification gap opposite the editing site. The approach was successfully tested on several endogenous transcripts in several human cell lines and human primary cells. Almost no off-target editing was observed (Merkle et al. 2019). As this technology harnesses endogenous ADAR for RNA editing, it circumvents the immunogenicity or delivery issues associated with big prokaryotic or fusion proteins; however, chemical modification of ASO can cause immunogenic response when administered repeatedly. Due to requirement of the chemical modifications on the ASO, gRNAs cannot be genetically encoded.

15.2.1.3 LEAPER

LEAPER short for "leveraging endogenous ADAR for programmable editing of RNA" uses relatively long (>70 nt) engineered arRNAs to recruit endogenous ADAR1 or ADAR2 enzymes to change a specific adenosine to inosine (Qu et al. 2019). arRNA can be delivered by a plasmid or viral vector or as a synthetic oligonucleotide. A high RNA editing efficiency was observed that correlated with the expression levels of ADAR1/2 protein in different cell lines. Several optimizations were performed that includes variations in the arRNAs' length and the placing of the A–C mismatch. The length of the arRNA correlates positively with the editing efficiency, peaking at 111–191 nt, with a minimal length of 71 nt for arRNA to work. It was found that placing the A–C mismatch in the middle region resulted in highest editing efficiency. Since no administration of exogenous protein or a synthetically modified RNA is needed, LEAPER represents an attractive approach for correcting pathogenic mutations at the transcript level. However, frequent local off-target editing at the bystander adenosines flanking the on-target adenosine is observed, which can be reduced to some extent following the design principles suggested by the authors. Another limitation of all endogenous ADAR-based approaches is variable editing efficiency in different cell types due to variable ADAR expression.

15.2.2 Systems Employing an Engineered Version of ADAR Protein

These systems employ fusion of ADAR deaminase domain (or a hyperactive mutant version of it) to one of the RNA recruiting domains, e.g. SNAP tag, λN peptide, and MS2 coat protein.

15.2.2.1 SNAP-ADAR

SNAP-tag is a 182 amino acid polypeptide (about 20 kDa) derived from the human DNA repair protein O^6-alkylguanine-DNA alkyltransferase (hAGT), which irreversibly transfers the alkyl group from its substrate, O^6-alkylguanine-DNA, to one of its cysteine residues (Keppler et al. 2003). SNAP-tag can be fused to a protein of interest and further specifically and covalently tagged with a suitable ligand, such as a fluorescent reporter (Keppler et al. 2003). Stafforst et al. for the first time developed the SNAP-ADAR system for targeted RNA editing by fusing deaminase domain (aa 798–1226) of hADAR1 to the C-terminus of a SNAP-tag domain (Stafforst and Schneider 2012). The ADAR1DD-SNAP-tag fusion then reacts with 5′-O-benzylguanine (BG)-modified gRNA to generate covalently bound gRNA–deaminase. This covalently ligated system showed successful repair of a Stop codon (UAG) to Tryptophan (UGG) nonsense mutation in the open reading frame (ORF) of the enhanced-fluorescent protein reporter (eCFP) in situ (Stafforst and Schneider 2012). In further improvement of this system, four different editase were developed, SNAP–ADAR1, SNAP–ADAR2, and their respective hyperactive E488Q variants (Vogel et al. 2018). Already with SNAP–ADAR1 and SNAP-ADAR2 editing efficiency up to 40–80% was achieved on endogenous mRNAs in the cell lines, which was further enhanced to 60–90% with the hyperactive versions. High efficiency and low off-targets are clearly a big advantage of SNAP-ADAR technology. Similar to RESTORE technology, due to requirement of chemical modification of Adar RNA, gRNAs cannot be genetically encoded.

15.2.2.2 λN-ADAR

The λ-phage N protein binds to boxB RNA with high affinity and specificity, to regulate antitermination during transcription of λ-phage mRNAs (Chattopadhyay et al. 1995). The λN peptide, which mediates the binding of the N protein, is only 22 amino acids long while the boxB RNA hairpin is only 17 nucleotides long (Austin et al. 2002). Montiel-Gonzalez et al. engineered a site-directed RNA editase by fusing the λN peptide to the deaminase domain of human ADAR2 (λN–ADAR) (Montiel-Gonzalez et al. 2013). The investigators corrected a premature termination codon in mRNAs encoding the cystic fibrosis transmembrane conductance regulator anion channel (CFTR) in Xenopus oocytes. Finally, a premature termination codon in enhanced green fluorescent protein (EGFP) reporter was corrected in N2A cells and primary neurons, using genetically encoded version of λN–ADAR and guide RNA (Montiel-Gonzalez et al. 2013).

In further studies, a nuclear localization signal was added to the editase to increase nuclear editing efficiency (Sinnamon et al. 2017; Vallecillo-Viejo et al. 2018). Nuclear localization resulted in lowering in the abundance of the off-target editing. In these studies, efficient editing was achieved for Methyl CpG Binding Protein 2 (Mecp2) RNA in primary neurons. Mutations in Mecp2 gene cause Rett syndrome, a neurological disease. In a 2020 study, Sinnamon et al. showed *in vivo* repair of the pathological Mecp2 RNA in a Rett syndrome mouse model using adeno-associated virus-mediated delivery of editase into the hippocampus of the mouse (Sinnamon et al. 2020). These encouraging results give hope for a potential therapy for Rett syndrome. Advantage of the λN-ADAR is its small size which can be packaged in AAV, one of the only viral-mediated delivery system that is approved so far for gene therapy.

15.2.2.3 MCP-ADAR

MS2 tagging is a technique based upon the natural interaction of the MS2 bacteriophage coat protein (MCP) with a stem-loop structure from the phage genome (Golding et al. 2005). This technique is commonly used for biochemical purification of RNA–protein complexes. Azad et al. for the first time fused the MCP to ADAR1 deaminase domain and designed gRNA based on the stem-loop (Azad et al. 2017). Using this system, the authors successfully showed editing of a premature stop codon in the EGFP in human cell lines. In another study by Azad et al. it was found that RNA editing is highest with MCP-ADAR1 DD compared with MCP-ADAR2 DD or MCP-ADAR2 (DD) hyperactive version (Azad et al. 2019). Katrekar et al. further optimized the MS2 adRNA design via systematic antisense domain length and mismatch position modifications, coupled with the use of hyperactive versions of the deaminase domains (Katrekar et al. 2019). Efficiency of RNA editing was checked by comparing editing of the endogenous RAB7A transcript by different versions of MCP-ADAR and an editing in the range of 20–40% was generally achieved in the human cell line. Further, *in vivo* application of the MCP-ADAR technology was shown by editing DMD gene in the mdx mouse model of DMD (Katrekar et al. 2019).

15.2.2.4 CIRTS

In the general scheme of CRISPR-Cas-inspired RNA targeting system (CIRTS), an effector protein is fused to a gRNA binding (via an adjacent hairpin structure) protein and an ssRNA binding protein (Rauch et al. 2019). In one of those systems, CIRTS8, E488Q mutated hADAR2 deaminase domain was fused to TBP (an evolved protein that binds to HIV-1 trans-activation response element hairpin RNA (Blakeley and McNaughton 2014; Crawford et al. 2016) for RNA hairpin binding and β-defensin 3 for ssRNA binding (Rauch et al. 2019). The system was tested in HEK293T cells and showed an efficiency of about 47% in editing a luciferase reporter. In further development of this system, it was engineered to edit endogenous targets in the cells in a small molecule (abscisic acid) inducible fashion (Rauch et al. 2020). The CIRTS system has the advantage of its small size and as only human protein parts were used for engineering, risk of immunogenicity is low.

15.2.3 CRISPR-Based RNA Editing Systems

RNA binding Cas proteins can be fused to ADAR deaminase domains and directed to a target RNA for editing. This was first achieved by Cox et al. when the authors fused E488Q-mutated hyperactive ADAR2 DD to catalytically inactive PspCas13b (Prevotella sp. P5-125 Cas13b) and demonstrated RNA editing of reporter and endogenous transcripts as well as disease-relevant mutations (Cox et al. 2017). This system termed RNA editing for programmable A to I replacement (REPAIR v1) was highly efficient; however, it showed many off-targeting edits as well. Further, Cox et al. used a rational mutagenesis to improve the specificity of dCas13b-ADAR2DD fusions in order to generate REPAIRv2 with about 1000-fold higher specificity (Cox et al. 2017).

In another effort, Zhang group used directed evolution to develop a cytosine deaminase domain from the ADAR2 adenosine deaminase domain (Abudayyeh et al. 2019). Again Cas13 was fused to the newly engineered ADAR2 cytosine deaminase. This system referred to as RNA Editing for Specific C-to-U Exchange (RESCUE), doubles the number of mutations targetable by RNA editing (Abudayyeh et al. 2019).

Using a bioinformatics approach, Konermann et al. screened prokaryotic genome sequences for CRISPR arrays and associated nucleases (Konermann et al. 2018). This screen discovered Cas13d that exhibits RNA-guided RNA-specific cleavage activity. Konermann et al. further engineered optimized versions of several Cas13d orthologs. Of those, CasRx, *Ruminococcus flavefaciens* Cas13d fused to a nuclear localization sequence, proved the most efficient at cleaving target sequences. An enzymatically inactive CasRx (dCasRx) retained binding to its target RNA, suggesting that it could be used to create a splice effector and RNA editor. Recently, Liu et al. fused ADAR2 DD to CasRx and achieved high efficiency in RNA editing without much off-targets (Liu et al. 2020). This system is named REPAIRx. Beside a good on-target vs off-target efficiency profile, another advantage of CasRx over known Cas13b is that CasRx is small enough (967 amino acids vs 1090 amino acids) to be packaged within an AAV delivery vehicle along with a guide RNA (Figure 15.3 and Table 15.1).

15.3 Potential Clinical Applications of RNA Editing

A large number of human genetic diseases are caused by point mutations in DNA and about 47% of them can be corrected by an A-T to G-C base pair change (Rees and Liu 2018). Among them, conditions for which a transient repair could provide clinical benefits are particularly suitable for RNA base editing. Some of these are discussed below.

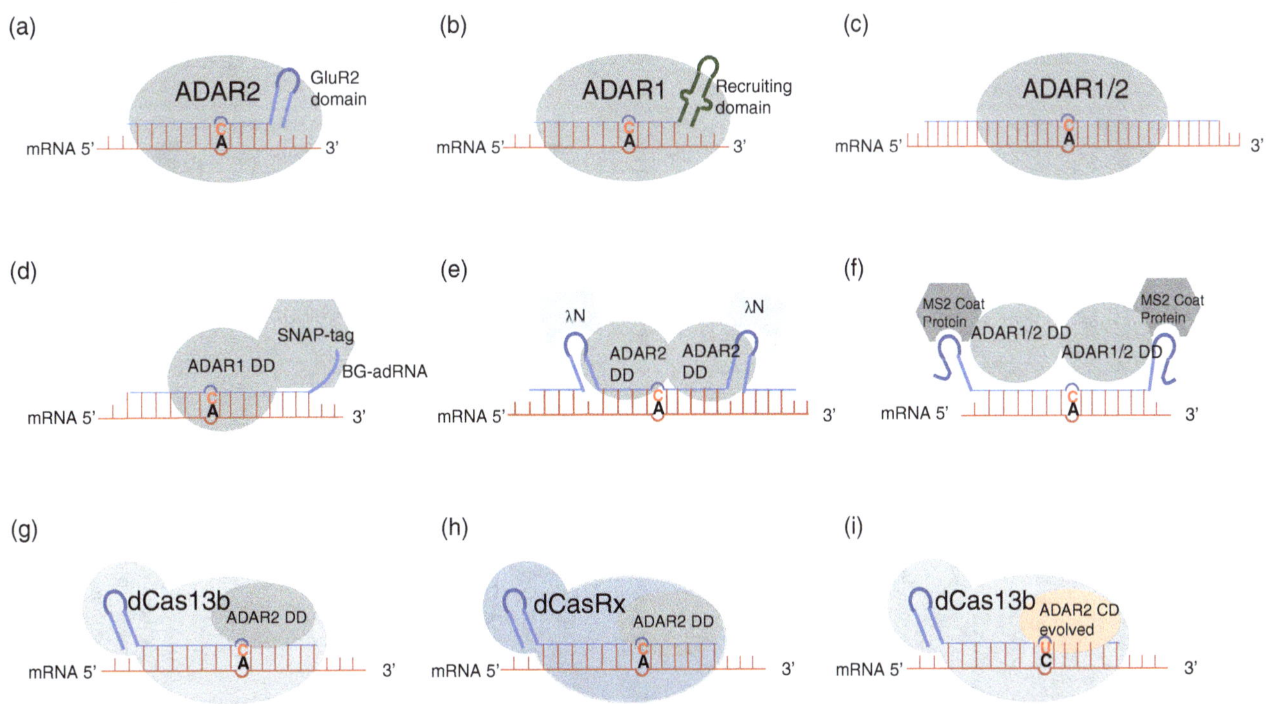

Figure 15.3 Different ADAR-based RNA editing systems. Top 3 (a–c) use endogenous ADAR proteins for editing. Middle row (d–f) use an engineered ADAR deaminase domain (wt or its mutant hyperactive version) fused to a recruiting domain to target to the site of interest. Bottom row (g–i) use CRISPR for RNA editing. (a) GluR2. (b) RESTORE. (c) LEAPER. (d) SNAP-tag. (e) λN. (f) MCP. (g) REPAIR. (h) REPAIRx. (i) RESCUE.

Table 15.1 A summary of the RNA editing technologies developed so far.

Method	Mechanism	ADAR RNA antisense region	Fusion protein size	Cellular system used	Pros	Cons	Reference
GluR2-ADAR	GluR2 hairpin used for recruiting endogenous ADAR2	20 nt	—	HELA, HEK293T cells, mouse	Harnesses endogenous ADAR, hence no immunogenicity or delivery issues	Very low (<1%) editing when exogenous ADAR is not expressed	(Fukuda et al. 2017; Katrekar et al. 2019; Wettengel et al. 2017)
RESTORE	Recruiting endogenous ADARs using chemosynthetic antisense oligonucleotides	40 nt	—	Several cell lines and primary cells	Harnesses endogenous ADAR, hence no immunogenicity or delivery issues	Chemical modification of ASO can cause immunogenic response when administered repeatedly	(Merkle et al. 2019)
LEAPAR	Long oligos recruiting endogenous ADAR to target editing	70 nt	—	Several cell lines and primary cells	Harnesses endogenous ADAR, hence no immunogenicity or delivery issues	Frequent local off-target editing at the bystander adenosines flanking the on-target adenosine	(Qu et al. 2019)
SNAP-ADAR	SNAP-tag fusion to ADAR DD	20 nt	182 aa	HEK293T cells	High efficiency, low off-target	Requires chemical modification of ADAR RNA	(Stafforst and Schneider 2012; Vogel et al. 2018)
λN-ADAR	λN peptide fusion to ADAR DD	50 nt	100 aa	N2A cells and primary neuron	High efficiency, small fusion protein, fits into AAV	Off-target issues	(Montiel-Gonzalez et al. 2013)
MCP-ADAR	MS2 Cap Protein fusion to ADAR DD	40 nt	130 aa	HEK293T cells, mouse	Small fusion proteins, fits into AAV	Needs protein engineering	(Azad et al. 2017, 2019; Katrekar et al. 2019)
CIRTS	hADAR2 DD fusion with TBP (for RNA hairpin binding) and β-defensin 3 (for ssRNA binding)	20–40 nt	~300 aa	HEK293T cells	Human protein parts used for engineering, relatively small protein parts which can fit in AAV, easily adaptable for other RNA modifications	Needs protein engineering	(Rauch et al. 2019, 2020)
REPAIR	dCas13b-ADAR DD fusion	50 nt	1100 aa	HEK293FT cells	Easy for target screening	Large protein size, bacterial origin, off-target issues	(Cox et al. 2017)
REPAIRx	dCasRx-ADAR DD fusion	22 nt	967 aa	N2A, HEK293T cells	Higher efficiency and specificity compared with REPAIR	Large protein size, bacterial origin, off-target issues	(Liu et al. 2020)
RESCUE	ADAR deaminase modified by directed evolution to convert into cytosine deaminase and fused to dCas13b	30 nt	1100 aa	HEK293FT, HUVEC cells	Only approach so far for direct C to U conversion of RNA	Large protein size, bacterial origin, off-target issues	(Abudayyeh et al. 2019)

15.3.1 Duchenne Muscular Dystrophy

DMD is a neuromuscular disease that primarily affects males. Muscle weakness begins by the age of four and most affected are unable to walk by the age of 12 (Bushby et al. 2010). DMD affects about one in 5000 males at birth (Stark 2015). It is caused by mutations in the DMD gene, the largest known human gene (Muntoni et al. 2003). The disorder is X-linked recessive as the gene is present on X-chromosome (Muntoni et al. 2003). The protein dystrophin encoded by DMD gene connects the cytoskeleton to the extracellular matrix in muscle cells. RNA base editors converting a premature stop codon to Tryptophan or causing an exon-skip could be an attractive therapy, and was recently demonstrated in a mouse model for DMD (Katrekar et al. 2019).

15.3.2 Rett Syndrome

Rett syndrome is a genetic disorder with symptoms including impairments in language and coordination and repetitive movements. The conditions affect about 1 in 10 000 females. Rett syndrome is commonly caused by a mutation in the Mecp2 gene on the X chromosome (Amir et al. 1999). Currently, there is no cure for Rett syndrome, treatment is directed toward improving function and addressing symptoms. Base editing to correct the Mecp2 mutation is an attractive approach for this disease and is currently under investigation in the animal models (Sinnamon et al. 2020).

15.3.3 Hearing Loss

RNA editing approaches are especially useful for conditions in which transient editing would be enough for therapeutic benefit. Damage to the sensory hair cells and cochlear supporting cells of the ear due to trauma such as exposure to loud noise and ear infections or aging leads to sensorineural hearing loss as these cells cannot regenerate once damaged. Stimulation of the Wnt signaling has been shown to facilitate reprogramming of supporting cells into sensory hair cells (McLean et al. 2017). Wnt signaling stimulation was recently shown using a cytidine-to-uridine RNA editing approach (Abudayyeh et al. 2019).

15.3.4 Inherited Retinal Degeneration

Inherited retinal degenerations (IRDs) are diverse group of progressively degenerating diseases that can eventually lead to blindness. Genetically, IRDs are heterogenous diseases with more than 260 disease genes identified so far (Duncan et al. 2018). These genes products are normally involved in retinal function. The best example of treatment for these diseases is gene therapy for the RPE65 gene (for cases in which RPE65 is mutated) that was recently approved by the United States Food and Drug Administration (US FDA) (Bainbridge et al. 2015; Bennett et al. 2016). Due to packaging limit of adeno-associated viruses (about 4.7 kb), which is currently the vector of choice for gene therapy due to its safety profile, a different strategy is needed for mutated genes which are difficult to package in an AAV. In IRDs, there are several large genes such as USH2A and ABCA4 that are often mutated. RNA base editing offers a nice alternate strategy for such cases and was recently shown to correct the pathogenic mutations in these genes in the human cell lines (Fry et al. 2020).

15.4 Challenges and Opportunities

In any therapeutic genome editing project, two criteria are of particular importance. First, genome editing tool must have a high on-target efficiency and minimal off-target activity. Secondly, it should be delivered to the organ of interest without a wide systemic distribution. To which degree these criteria are met may depend on the target gene and the affected organ. Some organs, for example liver, are relatively easier to target with adeno-associated virus (AAV) or lipid nano particle (LNP)-based delivery methods. Among the efforts to reduce the off-targets in RNA A to I editing, Monteleone et al. used a structure-guided strategy to develop combinations of mutant ADAR2 and guide RNAs (Monteleone et al. 2019). This strategy termed selective site-directed RNA editing (SDRE) was shown to reduce off-target activity in *in vitro* experiments as well as in human cells (Monteleone et al. 2019). Song et al. developed irCLASH to identify the dsRNA substrates of ADARs in human cells and provided important information for the development of optimized gRNAs to retarget endogenous ADARs (Song et al. 2020). It was found that ADAR proteins bind dsRNA in tandem, and each protein in the dimer has a 50 bp footprint.

A large number of studies in the RNA editing field are conducted using human cell lines due to their ease of use. Although the cell lines are great tool for research, it is important to select the right cell line for the target under investigation. Schaffer et al. studied RNA A to I editing in comprehensive details in the human cell lines and discovered that for almost every target of interest, a suitable cell line that mimics normal tissues exists (Schaffer et al. 2020). Surprisingly, in their study, ADAR genes expression alone is not a good predictor of RNA editing, suggesting additional factors are involved (Schaffer et al. 2020). Using the catalogue presented in this study (http://srv00.recas.ba.infn.it/atlas/claire.html), one can choose the right cell line for the experimental research.

15.5 Conclusions

RNA editing is an attractive therapeutic approach for pathogenic conditions where a transient correction of mutation will be beneficial to the patient. As RNA has a short half-life, no editing or any off-target would be permanent. Although most of the RNA editing technologies are in experimental development phase at the moment, due to small size of the cargo that need to be delivered (compared with large proteins like Cas9 or Cas13), several of these systems give hope for gene therapy. Of particular interest are the systems that exploit endogenous ADARs for target editing, as in these cases only gRNA need to be delivered. Only a handful of studies exist on the individual engineered RNA editing systems at the moment, as more research groups begin exploring these systems, especially in *in vivo* studies, more confidence will come for their clinical studies.

References

Abudayyeh, O.O., Gootenberg, J.S., Franklin, B. et al. (2019). A cytosine deaminase for programmable single-base RNA editing. *Science* 365: 382–386.

Amir, R.E., Van Den Veyver, I.B., Wan, M. et al. (1999). Rett syndrome is caused by mutations in X-linked MECP2, encoding methyl-CpG-binding protein 2. *Nat. Genet.* 23: 185–188.

Austin, R.J., Xia, T., Ren, J. et al. (2002). Designed arginine-rich RNA-binding peptides with picomolar affinity. *J. Am. Chem. Soc.* 124: 10966–10967.

Azad, M.T.A., Bhakta, S., and Tsukahara, T. (2017). Site-directed RNA editing by adenosine deaminase acting on RNA for correction of the genetic code in gene therapy. *Gene Ther.* 24: 779–786.

Azad, M.T.A., Qulsum, U., and Tsukahara, T. (2019). Comparative activity of adenosine deaminase acting on RNA (ADARs) isoforms for correction of genetic code in gene therapy. *Curr. Gene Ther.* 19: 31–39.

Bainbridge, J.W., Mehat, M.S., Sundaram, V. et al. (2015). Long-term effect of gene therapy on Leber's congenital amaurosis. *N. Engl. J. Med.* 372: 1887–1897.

Bass, B.L. and Weintraub, H. (1987). A developmentally regulated activity that unwinds RNA duplexes. *Cell* 48: 607–613.

Bennett, J., Wellman, J., Marshall, K.A. et al. (2016). Safety and durability of effect of contralateral-eye administration of AAV2 gene therapy in patients with childhood-onset blindness caused by RPE65 mutations: a follow-on phase 1 trial. *Lancet* 388: 661–672.

Blakeley, B.D. and Mcnaughton, B.R. (2014). Synthetic RNA recognition motifs that selectively recognize HIV-1 trans-activation response element hairpin RNA. *ACS Chem. Biol.* 9: 1320–1329.

Bushby, K., Finkel, R., Birnkrant, D.J. et al. (2010). Diagnosis and management of Duchenne muscular dystrophy, part 1: diagnosis, and pharmacological and psychosocial management. *Lancet Neurol.* 9: 77–93.

Chattopadhyay, S., Garcia-Mena, J., Devito, J. et al. (1995). Bipartite function of a small RNA hairpin in transcription antitermination in bacteriophage lambda. *Proc. Natl. Acad. Sci. U. S. A.* 92: 4061–4065.

Chen, C.X., Cho, D.S., Wang, Q. et al. (2000). A third member of the RNA-specific adenosine deaminase gene family, ADAR3, contains both single- and double-stranded RNA binding domains. *RNA* 6: 755–767.

Chew, W.L., Tabebordbar, M., Cheng, J.K. et al. (2016). A multifunctional AAV-CRISPR-Cas9 and its host response. *Nat. Methods* 13: 868–874.

Cox, D.B.T., Gootenberg, J.S., Abudayyeh, O.O. et al. (2017). RNA editing with CRISPR-Cas13. *Science* 358: 1019–1027.

Crawford, D.W., Blakeley, B.D., Chen, P.H. et al. (2016). An evolved RNA recognition motif that suppresses HIV-1 Tat/TAR-dependent transcription. *ACS Chem. Biol.* 11: 2206–2215.

Duncan, J. L., Pierce, E. A., Laster, A. M., Daiger, S. P., Birch, D. G., Ash, J. D., Iannaccone, A., Flannery, J. G., Sahel, J. A., Zack, D. J., Zarbin, M. A. & and The Foundation Fighting Blindness Scientific Advisory, B. 2018. Inherited retinal degenerations: current landscape and knowledge gaps. *Transl. Vis. Sci. Technol.*, 7, 6.

Egebjerg, J., Kukekov, V., and Heinemann, S.F. (1994). Intron sequence directs RNA editing of the glutamate receptor subunit GluR2 coding sequence. *Proc. Natl. Acad. Sci. U. S. A.* 91: 10270–10274.

Fry, L.E., Peddle, C.F., Barnard, A.R. et al. (2020). RNA editing as a therapeutic approach for retinal gene therapy requiring long coding sequences. *Int. J. Mol. Sci.* 21.

Fukuda, M., Umeno, H., Nose, K. et al. (2017). Construction of a guide-RNA for site-directed RNA mutagenesis utilising intracellular A-to-I RNA editing. *Sci. Rep.* 7: 41478.

George, C.X. and Samuel, C.E. (1999). Human RNA-specific adenosine deaminase ADAR1 transcripts possess alternative exon 1 structures that initiate from different promoters, one constitutively active and the other interferon inducible. *Proc. Natl. Acad. Sci. U. S. A.* 96: 4621–4626.

Golding, I., Paulsson, J., Zawilski, S.M., and Cox, E.C. (2005). Real-time kinetics of gene activity in individual bacteria. *Cell* 123: 1025–1036.

Katrekar, D., Chen, G., Meluzzi, D. et al. (2019). In vivo RNA editing of point mutations via RNA-guided adenosine deaminases. *Nat. Methods* 16: 239–242.

Keppler, A., Gendreizig, S., Gronemeyer, T. et al. (2003). A general method for the covalent labeling of fusion proteins with small molecules in vivo. *Nat. Biotechnol.* 21: 86–89.

Konermann, S., Lotfy, P., Brideau, N.J. et al. (2018). Transcriptome engineering with RNA-targeting type VI-D CRISPR effectors. *Cell* 173: 665–676. e14.

Liu, Y., Mao, S., Huang, S. et al. (2020). REPAIRx, a specific yet highly efficient programmable A > I RNA base editor. *EMBO J.* 39: e104748.

Mclean, W.J., Yin, X., Lu, L. et al. (2017). Clonal expansion of Lgr5-positive cells from mammalian cochlea and high-purity generation of sensory hair cells. *Cell Rep.* 18: 1917–1929.

Merkle, T., Merz, S., Reautschnig, P. et al. (2019). Precise RNA editing by recruiting endogenous ADARs with antisense oligonucleotides. *Nat. Biotechnol.* 37: 133–138.

Monteleone, L.R., Matthews, M.M., Palumbo, C.M. et al. (2019). A bump-hole approach for directed RNA editing. *Cell Chem. Biol.* 26: 269–277. e5.

Montiel-Gonzalez, M.F., Vallecillo-Viejo, I., Yudowski, G.A., and Rosenthal, J.J. (2013). Correction of mutations within the cystic fibrosis transmembrane conductance regulator by site-directed RNA editing. *Proc. Natl. Acad. Sci. U. S. A.* 110: 18285–18290.

Muntoni, F., Torelli, S., and Ferlini, A. (2003). Dystrophin and mutations: one gene, several proteins, multiple phenotypes. *Lancet Neurol.* 2: 731–740.

Nishikura, K. (2010). Functions and regulation of RNA editing by ADAR deaminases. *Annu. Rev. Biochem.* 79: 321–349.

Patterson, J.B. and Samuel, C.E. (1995). Expression and regulation by interferon of a double-stranded-RNA-specific adenosine deaminase from human cells: evidence for two forms of the deaminase. *Mol. Cell. Biol.* 15: 5376–5388.

Platt, R.J., Chen, S., Zhou, Y. et al. (2014). CRISPR-Cas9 knockin mice for genome editing and cancer modeling. *Cell* 159: 440–455.

Qu, L., Yi, Z., Zhu, S. et al. (2019). Programmable RNA editing by recruiting endogenous ADAR using engineered RNAs. *Nat. Biotechnol.* 37: 1059–1069.

Rauch, S., He, E., Srienc, M. et al. (2019). Programmable RNA-guided RNA effector proteins built from human parts. *Cell* 178: 122–134. e12.

Rauch, S., Jones, K.A., and Dickinson, B.C. (2020). Small molecule-inducible RNA-targeting systems for temporal control of RNA regulation. *ACS Cent. Sci.* 6: 1987–1996.

Rees, H.A. and Liu, D.R. (2018). Base editing: precision chemistry on the genome and transcriptome of living cells. *Nat. Rev. Genet.* 19: 770–788.

Schaffer, A.A., Kopel, E., Hendel, A. et al. (2020). The cell line A-to-I RNA editing catalogue. *Nucleic Acids Res.* 48: 5849–5858.

Sinnamon, J.R., Kim, S.Y., Corson, G.M. et al. (2017). Site-directed RNA repair of endogenous Mecp2 RNA in neurons. *Proc. Natl. Acad. Sci. U. S. A.* 114: E9395–E9402.

Sinnamon, J.R., Kim, S.Y., Fisk, J.R. et al. (2020). In vivo repair of a protein underlying a neurological disorder by programmable RNA editing. *Cell Rep.* 32: 107878.

Song, Y., Yang, W., Fu, Q. et al. (2020). irCLASH reveals RNA substrates recognized by human ADARs. *Nat. Struct. Mol. Biol.* 27: 351–362.

Stafforst, T. and Schneider, M.F. (2012). An RNA-deaminase conjugate selectively repairs point mutations. *Angew. Chem. Int. Ed. Eng.* 51: 11166–11169.

Stark, A.E. (2015). Determinants of the incidence of Duchenne muscular dystrophy. *Ann. Transl. Med.* 3: 287.

Stefl, R., Oberstrass, F.C., Hood, J.L. et al. (2010). The solution structure of the ADAR2 dsRBM-RNA complex reveals a sequence-specific readout of the minor groove. *Cell* 143: 225–237.

Vallecillo-Viejo, I.C., Liscovitch-Brauer, N., Montiel-Gonzalez, M.F. et al. (2018). Abundant off-target edits from site-directed RNA editing can be reduced by nuclear localization of the editing enzyme. *RNA Biol.* 15: 104–114.

Vogel, P., Moschref, M., Li, Q. et al. (2018). Efficient and precise editing of endogenous transcripts with SNAP-tagged ADARs. *Nat. Methods* 15: 535–538.

Wettengel, J., Reautschnig, P., Geisler, S. et al. (2017). Harnessing human ADAR2 for RNA repair - recoding a PINK1 mutation rescues mitophagy. *Nucleic Acids Res.* 45: 2797–2808.

Wong, S.K., Sato, S., and Lazinski, D.W. (2001). Substrate recognition by ADAR1 and ADAR2. *RNA* 7: 846–858.

Woolf, T.M., Chase, J.M., and Stinchcomb, D.T. (1995). Toward the therapeutic editing of mutated RNA sequences. *Proc. Natl. Acad. Sci. U. S. A.* 92: 8298–8302.

16

Genome Editing Applications in Cancer T Cell Therapy

William A. Nyberg[1,2] *and Justin Eyquem*[1–3]

[1] *Gladstone-UCSF Institute of Genomic Immunology, San Francisco, CA, USA*
[2] *Division of Hemato-Oncology, Department of Medicine, University of California San Francisco, San Francisco, CA, USA*
[3] *Department of Microbiology and Immunology & Parker Institute of Cancer Immunotherapy, University of California San Francisco, San Francisco, CA, USA*

16.1 Introduction

In the following chapter, we will cover basic principles of adoptive cell therapies (ACT) using T lymphocytes and the most recent breakthroughs in ACT of cancers that have been facilitated by the advancement of gene editing.

T lymphocytes are part of the adoptive immune system that develop in the thymus, where they are selected to recognize foreign antigen using their specific receptor, the T cell receptor (TCR). An important step of T cell development in the thymus is the clearing of any cells that develop TCRs binding self-antigens. This makes T cells especially good at clearing foreign infectious agents such as bacteria and viruses from the circulation but makes it harder for T cells to clear tumor-forming cells of self-origin. Nevertheless, T cells are capable of identifying and killing cells that express tumor-associated-antigens (TAA), originated from mutations or splicing events specific to tumor cells. The concept of immune surveillance of tumors was proposed in 1970 and has been widely accepted and included as a hallmark of cancer (Hanahan and Weinberg 2011). With the use of immunotherapies, researchers have been able to enhance these capabilities by either directing the immune system to target tumor cells or by restoring the anticancer effects using immune checkpoint blockade inhibitors. Adoptive T cell therapy, in which antigen-specific T cells are generated ex vivo and transferred back into patients, underwent successful clinical trials in the 1980s when tumor-infiltrating lymphocytes were harvested and used to treat patients with metastatic melanoma (Rosenberg et al. 1986). ACT using genetically engineered T cells represent a more recent class of T cell therapies. It includes the generation of T cells expressing TAA-specific recombinant TCR or chimeric antigen receptor (CAR). There are currently numerous clinical trials with CAR T cells targeting a range of different cancer types with remarkable results in patient with refractory hematological cancers. However, many aspects of this technology remain to be improved (Sadelain et al. 2017). First, clinical results have been moderate in solid tumors, notably due to the lack of ideal targets, inefficient homing at the tumor site, limited T cell persistence, and T cell dysfunction. Second, CAR therapies have been associated with toxicities, such as cytokine release syndrome (CRS) or cerebral edema, whose mechanism is poorly understood. Lastly, the wide implementation of T cell therapies is limited by the use of randomly integrating vectors in autologous patient cells, which involves high cost and unavoidably results in a high variability of the final product.

With recent technical advancements in gene editing, the ability to perform precise gene disruption (KO), gene targeting (KI), and large screens have transformed research in T cell immunology as well as adoptive T cell therapies. In this chapter, we will summarize the impact of gene editing on the field of CAR/TCR T cells and what future prospective of clinical applications might look like.

16.2 CAR T Cells

CARs are synthetic receptors using the antigen recognition moiety of an antibody with the intracellular domains of T cell receptors combined with co-stimulatory domains.

Genome Editing in Drug Discovery, First Edition. Edited by Marcello Maresca and Sumit Deswal.

These receptors are able to identify cell surface antigens on tumor cells and reprogram T lymphocytes to kill target cells, proliferate, and secrete effector cytokines in an MHC-independent matter.

16.2.1 Receptor Structure

The ability of a CAR to bind with high specificity to a predefined cell surface antigen is honed from plasma cell-derived antibodies. The antigen recognition domain of the CAR mainly consist of a single chain variable fragment (scFv), the results of variable light (V_L), and variable heavy (V_H) chains of an antibody linked together with a flexible spacer to create a highly specific antigen recognition domain (Sadelain 2017a). There have also been recent reports using single-domain antibody (VHH or nanobody), derived from camelid or selected with phage display technology (Xie et al. 2019). Other studies have reported the use of the antigen ligand as a binder, such as APRIL for BCMA (7) or IL13 muteins for IL13Ra1 (Kong et al. 2012). The binding moiety is joined by a hinge and a transmembrane domain to an intracellular signaling domain. While the extracellular domains of the CAR are derived from antibodies, the intracellular signaling domains are most commonly derived from TCR signaling domains. The intracellular domain of the first-generation CAR, published as early as the 1990s, simply contained a TCR-derived CD3ζ signaling domain (Gross et al. 1989). While this first proof of concept demonstrated the ability to redirect T cell cytotoxicity to a cell-surface antigen in an MHC-independent fashion, this receptor displayed poor T cell proliferation and persistence and was not able to induce long-term tumor control in a mouse model. The addition of a CD28 or a 4-1BB co-stimulatory domain to the intracellular part of the CAR brought the second generation of CARs that were the first CAR to reach FDA approval for treatment of hematological malignancies using anti-CD19 CARs (Finney et al. 2004; June and Sadelain 2018; Maher et al. 2002). A third generation of CARs was engineered by combining CD28 and 4-1BB co-stimulatory domains followed by a CD3ζ domain; these CARs have shown various results and it remains to be seen whether they can increase the therapeutic potential of CAR T cells (Tammana et al. 2010). As of today, the second generation of CARs using either CD28 or the 4-1BB remains the golden standard for clinical trials of CAR T cells. However, these CARs have been followed by a number of novel innovations in terms of CAR structure, design, and the introduction of armored CARs co-expressing cytokines (Figure 16.1).

16.2.2 Alternative Approaches

T cells require an input of a number of cytokines in order to proliferate, sustain, and function. These cytokines can be produced by T helper cells, other hematologic cells, or fibroblasts. In an attempt to boost CAR T cell function and persistence as well as engaging the innate immune system, CAR T cells have been engineered to co-express essential cytokines or ligands. These CARs are referred to as the Armored CARs or fourth-generation CARs and have been engineered to co-express IL-12, IL-7, IL-18, CD40L, CCL-19, and CCR2 ((Adachi et al. 2018; Avanzi et al. 2018; Kagoya et al. 2018; Koneru et al. 2015; Yeku et al. 2017). In the case of IL-12, it has been shown that IL-12 helps the CAR T cells sustain in the immunosuppressive tumor microenvironment of solid tumors by counteracting the suppressive effects of T regulatory cells and myeloid-derived suppressor cells (Koneru et al. 2015). Another

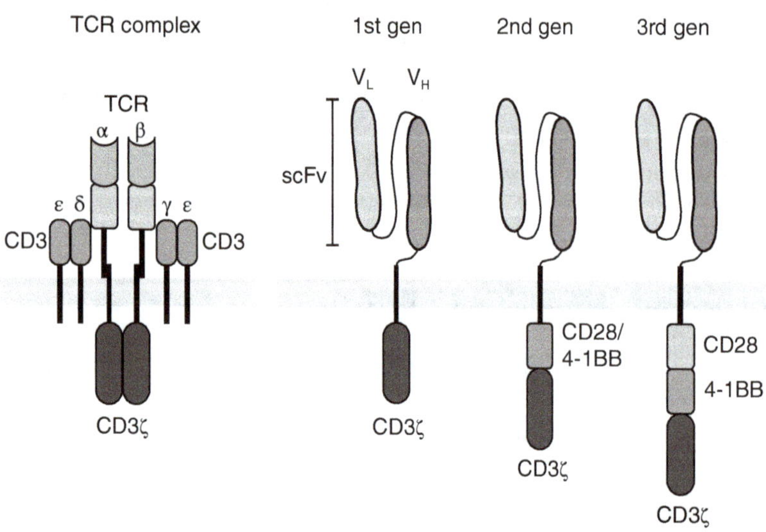

Figure 16.1 A schematic representation of the TCR complex and the three generation of established CARs.

example of armored CARs are the IL-18-expressing CAR T cells that like IL-12-expressing CARs persist longer in tumors, but are also able to engage an antitumor response from endogenous immune cells (Avanzi et al. 2018). As the signals provided by cytokines are essential for CAR T cell memory formation and long-term persistence, Kagoya et al. recently demonstrated that the addition of a truncated cytoplasmic domain from the IL-2 receptor β-chain (IL-2Rβ) and a STAT3-binding tyrosine-X-X-glutamine (YXXQ) motif to the CAR improved persistence and efficacy *in vivo* (Kagoya et al. 2018). The cells showed antigen-dependent activation of IL-2 downstream signaling pathways upon CAR activation leading to improved survival and proliferation.

While CD28-based CARs have been shown to display better effector functions at early time points (Zhao et al. 2015) and higher sensitivity to low antigen density than the 4-1BB-based CARs (Hamieh et al. 2019), they also induce a faster effector differentiation and less persistence. Different approaches have been studied to improve CD28-based CARs, either by co-expressing the 4-1BB ligand (Zhao et al. 2015) or mutating the second and third ITAMs from the CD3 zeta chains (1XX CAR) (Feucht et al. 2019). Both approaches resulted in an improved T cell persistence while maintaining the advantages offered by the CD28 signaling. A 1XX CAR targeting CD19 will soon be introduced to the clinic.

The success of CAR T cells have led to the development of alternative receptors for T cell therapies, one such example is the use of synthetic Notch (synNotch) receptors for T cell therapies. The synNotch receptor can contain a wide choice of extracellular domains as well as a variety of intracellular signaling domains, thus they function as extracellular sensors with a programmable response output. When coupled with an extracellular scFv domain, the synNotch receptors can be designed to bind tumor-associated antigen to induce the expression of either cytokines or specific transcription factors important for T cell effector activity against tumors (Roybal et al. 2016). The synNotch system can also be designed to drive expression of a CAR upon antigen recognition, allowing for local expression of CAR T cells to reduce of target effects or systemic targeting of the CAR.

16.2.3 Clinical Perspectives

In 2017, the FDA approved the first two CAR T cell therapies for treatment of acute-lymphoblastic leukemia (ALL) and non-Hodgkin's lymphoma (NHL). These two CAR T cell therapies are marketed under the names Kymriah and Yescarta and are both second-generation CARs targeting the B cell antigen CD19. While Kymriah is introduced to T cells using a lentivirus and contains the co-stimulatory domain 4-1BB, Yescarta is a retroviral drug with a CAR that contains a CD28 co-stimulatory domain (Sadelain 2017b, p. 19). Both treatments have shown remarkable results, respectively, and many more CAR T cell therapies are now undergoing clinical trials. The B cell surface antigen CD19 was selected as a target for the first CAR T cells developed as it is selectively expressed at high levels on B cells, making it suitable to treat B cell lymphomas and leukemias. CD19 has also previously been the target of FDA-approved treatments using antibodies in hematological cancers. While B cells express high levels of CD19, plasma B cells do not. However, plasma cells express the surface antigen B cell maturation antigen (BCMA), which has been the subject of many clinical trials using CAR T cells for treatment of patients with multiple myeloma (MM) (Raje et al. 2019). In contrast to clinical trials targeting CD19, these trials have shown very high rates of relapse and limited benefits on long-term survival (D'Agostino and Raje 2020).

While results have been impressive with >50% remission rates or disease-free survival treating ALL and NHL, as well as diffuse large B cell lymphoma (DLBL), clinical trials using CAR T cell therapies against solid tumors have shown poor results. Taken together with the results from the large number of clinical trials targeting MM with BCMA CARs, it becomes clear that many improvements are needed for CAR T cell therapies to increase their clinical range of use. Alternative approaches using T cell therapies with success against solid tumors have been displayed using T cells expressing recombinant TCRs targeting tumor-associated antigens. The most well-known recombinant TCR used in clinical trials today is the NY-ESO-1 targeting TCR. NY-ESO-1, also known as the tumor associated antigen CTAG1B, is expressed at high levels by many solid tumors. Initial studies using T cells expressing recombinant NY-ESO-1 targeting TCRs have shown positive results in patients with NY-ESO-1 positive tumors of synovial cell sarcoma and malignant melanoma (Robbins et al. 2015).

16.2.4 Challenges and Adverse Events

While adoptive CAR T cell therapies have shown great results against hematological cancers in clinical trials, many facets of the therapy may be improved to increase efficacy against a wider range of cancers and reduce associated toxicities. One such challenge are risk factors associated to random integration of viral vectors. The vast majority of CAR T cell therapies are generated with randomly integrating retroviral or lentiviral vectors, while integration efficiencies are high, the risk of insertional mutagenesis will always be present using these methods. A clinical case was recently reported in which a CAR cassette was integrated in the tumor suppressor *TET2* locus functionally generating *TET2*-KO CAR T cells (Fraietta et al. 2018). The one cell exerted dramatic clonal expansion and represented at some point close to 100% of the patient blood T cells before contracting. While the clonal population provided most of the antitumor effects, such an event represents a major risk factor for developing novel cancers. Similarly, another study reported a patient with a lentiviral CAR cassette integrated into the *CBL* locus and resulting in clonal expansion (Shah et al. 2019). Together, these reports raise serious concerns over random integration in clinical use of CAR T cells, especially in the context of next-generation T cell therapies in which additional genetic manipulations are performed (Stadtmauer et al. 2020).

Toxicities associated with CAR T cell use in clinical have been reported. The most common adverse effect of CAR T cells is CRS which in some rare cases have led to fatal outcomes for the patient. There have also been reports of neurological side effects, including cerebral edema with CD19 targeting CAR T cell. A recent report led by the Satpathy lab identified CD19 expression in brain mural cells, suggesting an on-target off-tumor toxicity (Parker et al. 2020). These adverse effects of CAR T cells are mostly associated to on-target effects and are correlated to the tumor burden, symptoms improve in patients as tumor cells are cleared (Park et al. 2018). However, while all cancer therapies are associated to adverse effects, these must be addressed to improve widespread use of CAR T cells. Strategies to reduce CRS include the use of anti-IL-6 receptor antibody (Tocilizumab) and the IL-1 receptor antagonist Anakinra. Both these drugs are commonly used for rheumatic diseases and appear to ameliorate symptoms of CRS. To generate CAR T cells with reduced CRS potential, T cells have been engineered to co-express an IL-1 receptor agonist with the CAR, mimicking the effects seen by addition of Anakinra (Giavridis et al. 2018; Norelli et al. 2018).

While anti-CD19 therapies have shown unprecedently high complete response rate, long-term efficacy relapse still remains an issue and for anti-BCMA therapies relapse occurs in 50–80% of patients (D'Agostino and Raje 2020). Relapse can occur through poor T cell persistence in patients, mutations, or low expression of antigens on the tumor cells (Ruella and Maus 2016). Improving CAR T cell persistence and efficacy against low antigen-expressing tumor cells remains a great challenge for researchers to solve. Other factors contributing to relapse in patients are immunosuppressive signaling that occurs in high frequencies in tumor microenvironment as well as T cell exhaustion and dysfunction, these issues will be discussed in the following chapters and how researchers have deployed gene editing of T cells to overcome these hurdles.

16.2.5 Manufacturing and Costs

Since the emergence of T cell therapies, reducing manufacturing costs has been an ever-present challenge. T cells are commonly using retroviral and lentiviral vectors or nucleases. Engineered T cells need to be isolated, activated, and expanded in facilities using approved good manufacturing practices (GMP). Each manufacturing step is complex and requires optimal reproducibility, safety, and efficacy. GMP facilities are limited by the use of GMP-approved reagents which contributes to difficulties in reproducing conditions used in academic research laboratories. The current FDA-approved CAR T cell therapies requires use of patient-specific T cells which further complicates manufacturing and increases costs, one infusion of CAR T cells for DLBL has been quoted at a cost between 373 000 and 475 000 USD (Hay and Cheung 2019), highlighting the need to reduce costs. Efforts to streamline and automate the manufacturing process are required to reduce costs of manufacturing T cell therapies. A major limiting factor to CAR T cell manufacturing is the need for patient-specific T cells and the prospect of generating allogenic CAR T cells for "off-the-shelf" clinical use is a very promising strategy to reduce overall costs for ACTs and to reduce patient-to-patient variability of treatments. The use of gene editing to generate allogenic T cell therapies will discussed in detail later in this chapter.

16.3 Gene Editing in T Cells

T cell engineering is a powerful tool to interrogate and control T cell specificity and function. It has also become the engine for effective immunotherapies based on the genetic instruction of targeted immune responses. Interestingly, T cell engineering may be more relied upon in therapeutic investigations than in experimental research, as immunologists have long preferred to modify T cells via the germline and seldom use this platform for mouse or human T cell investigation. Here, we review the vast potential of T cell engineering based on emerging gene editing technologies.

Together with hematopoietic stem cells (HSCs), T cells have been the most studied primary cell type for gene transfer and gene disruption, using tools that have shown steadily increasing efficacy and safety over the past two decades (Dunbar et al. 2018). Starting with the use of replication-defective retroviral vectors, T cell engineering has over time made use of various RNA vectors (γ-retroviral, lentiviral, spumaviral), DNA vectors (*Sleeping Beauty* transposon), mRNA electroporation, and more recently targeted endonucleases (Sadelain et al. 2017).

The safety profile of retroviral vectors in patients treated with genetically modified T cells has so far been mostly positive (Heslop et al. 2010; Marcucci et al. 2018; Newrzela et al. 2008), evading the concern of insertional oncogenesis that has plagued their use in HSCs (Dunbar et al. 2018). Albeit hitherto less efficient and inducing more cytotoxicity, the *Sleeping Beauty* transposon is emerging as another effective vector system for efficient transgene delivery to human T cells (Hudecek and Ivics 2018). Transient transfection of mRNA or siRNA bypass the need for viral vector production and the risk of insertional mutagenesis, but only afford transient effects typically lasting a few days.

Gene addition strategies initially made the introduction of receptors possible for antigens such as T cell receptors, CARs, chimeric costimulatory receptors, HIV decoys, suicide genes, cytokines, and chemokines in primary T cells. γ-Retroviral vectors and lentiviral vectors are the foundation of current CAR therapies (June and Sadelain 2018).

These gene addition tools have also been used to block or reduce functions through the use of dominant-negative receptors (Bollard et al. 2002), shRNAs, miRNAS, and miR sponges. However, the advent of effective targeted nucleases that enable specific gene disruption or repair now provides new means to achieve complete functional ablation or rewire T cell regulation and function. Following the Meganucleases and Zinc finger nucleases (ZFNs), Tale Effector Nucleases (TALENs) and CRISPR/Cas9 have now expanded the reach and ease of editing primary cells. Gene editing tools were initially used for T cell gene disruption and more recently to afford optimal gene relation upon targeted transgene insertion. This book chapter will focus on the advances that gene editing stands to offer to both therapeutic and experimental T cell engineering.

16.3.1 Cassette Design and Gene Targeting Strategies

Depending on the desired cellular phenotype, gene targeting can result in a knock-out (KO), a gene knock-in (KI) without target gene disruption, or simultaneous KO by KI. To accomplish any of these outcomes, all the required genetic elements are assembled in the targeting DNA construct. Several important aspects need to be considered when performing gene targeting experiments, such as genome location, maintenance of gene function, and use of regulatory sequences in expression cassettes.

A transgenic expression cassette can be integrated at genic regions, intergenic areas, or safe harbor regions without a genome function (Sadelain et al. 2011). When delivered in a genic region, the integration approach can be designed to keep the function of the targeted gene (KI only) or to disrupt it (KO by KI strategy). When choosing the nondisruptive KI approach, the transgenic gene cassette is typically placed under the control of the regulatory sequences of the targeted gene (e.g. insertion at the initiation codon (ATG) followed by a 2A sequence or after the last codon preceded by a 2A or an IRES). With the KO–KI approach, the possibilities are bigger in terms of targeting construct design. Endogenous regulatory sequences can be used to express the transgene, which can be placed at the ATG with its own stop codon a polyA sequence, thus resulting in disruption of the targeted transgene. In addition, exogenous promoter/enhancer can be inserted along with the transgene. This approach permits the use of promoters with different strength, thus allowing the fine-tuning of transgene expression.

We tested all the above approaches in human primary T cells to drive CAR gene expression from the *TRAC* and *B2M* loci using endogenous regulatory sequences or by expressing the CAR with a variety of exogenous promoter/enhancer elements (Eyquem et al. 2017). Other laboratories have used similar strategies, including the targeting of a CAR transgenic cassette at the *CCR5* locus (Hale et al. 2017; Sather et al. 2015).

16.3.2 Delivery of Gene Editing Tools

Multiple methods have been used to deliver endonucleases and donor templates into T cells. Depending on the editing platform, the size of the targeting construct and the final application, different combination of reagents will provide optimal editing efficiency with minimal toxicity.

16.3.3 Nuclease Delivery

Plasmids and viral vectors were the first delivery vehicles used for tailored nucleases. They can provide high and sustainable nuclease expression, allowing high on-target cleavage. However, these vehicles have several limitations. Plasmids can randomly integrate in the genome or can be toxic to certain cell types (e.g. hematopoietic cells). Viral vectors like integration-deficient lentiviruses (IDLVs) are more difficult to manufacture and they transduce diverse human cell types, though with different efficiencies, which results in variable nuclease expression. Stable expression of nucleases from nonhuman origins can also induce immunogenicity resulting in rejection of the engineered T cells. In vitro synthesized mRNA has become the most common platform to express all the above tailored nucleases. The two most important advantages of this form of delivery are the homogenous expression of the nuclease in the target population, and a shorter half-life which reduces the DNA cleavage at potential off-target sites. mRNA can be delivered by using chemical or physical methods. In T cells, for example, mRNA is typically delivered through electroporation and is the method of choice for ZFN and TALEN. Finally, some of the nucleases can be delivered as protein as well, with Cas9 being the most commonly used as protein. For optimal nuclease activity, Cas9 is incubated with the gRNA to form a ribonucleoprotein (RNP) that can be delivered into cells through electroporation, yielding high on-target cleavage with lower levels of off-target activity when compared with plasmids and mRNA delivery(Kim et al. 2014). This approach works very efficiently in human T cells (Schumann et al. 2015) and has now been widely adopted in the field.

16.3.4 Donor Template Delivery

Depending on the application and cell type, the DNA donor template can be delivered as viral vectors (IDLV (Lombardo et al. 2011) or AAV (Sather et al. 2015)), plasmid, dsDNA, ssDNA (Roth et al. 2018), or ssODN (Schumann et al. 2015). All these delivery platforms have already been tested in primary human T cells with different editing efficiencies and associated cytotoxicity, with AAV serotype 6 being the most commonly used vector to deliver targeting constructs (Eyquem et al. 2017). To date, AAV donor template delivery has proven to be the most efficient method in terms of KI efficiency and low cytotoxicity. Our laboratory is routinely achieving more than 70% targeted integration of large cassettes (~4 KB) with minimal toxicity associated. However, producing AAV can be time-consuming and requires an ultracentrifugation step which limits multiplexing possibilities. Alternatively, dsDNA donor templates are associated with a high dose-dependent toxicity when electroporated into primary T cell but are much easier and faster to produce allowing massive parallel production. Recent studies have shown that electroporating optimal doses of Cas9 RNP and dsDNA template can reduce toxicity, making nonviral delivery a flexible and viable method for gene editing of primary T cells (Roth et al. 2018). Further improvements have been made by stabilizing the RNP with the addition of polyglutamic acid which increases electroporation efficiency and reduces toxicity (Roth et al. 2018). Finally, the efficiency of dsDNA donor template delivery into the nucleus can be improved by adding truncated Cas9 target sequences to the ends of the HDR template. Thus, RNP efficiently binds the DNA donor template without nuclease activity and facilitates the transfer of the complex into the nucleuses. This method is referred to as the "shuttle" system which has been reported to increase nonviral KI efficiencies as well as reducing the amount of DNA template needed (Nguyen et al. 2020). While AAV delivery of donor templates remains the superior method of template delivery for precise gene editing in T cells at this time, nonviral delivery method allows for increased flexibility in terms of designing and generating templates at low cost for high-throughput applications (Table 16.1).

16.3.5 Genetic Screens in T Cells

One of the greatest contributions of gene editing to T cell biology and the advancement of T cell therapies have been the possibility to perform genetic screens. A wide range of CRISPR/Cas9 screens have been performed in T cells, both using targeted and genome-wide libraries. The introduction of the CRISPR/Cas9 system has helped researchers to design gRNA libraries with high on-target efficiencies for multiple applications. Pioneering work applying genetic screens in primary T cells to understand their function in immunosuppressive tumor microenvironments *in vivo* were first performed using large short hairpin RNA (shRNA) libraries in Kai Wucherpfennig's laboratory (Zhou et al. 2014). This study has since been validated and improved upon by performing gene disruption screens. Genetic screens in primary human T cells using CRISPR/Cas9 were initially problematic due to inefficient nuclease delivery methods. By combining viral delivery of the sgRNA with nuclease electroporation, the Marson laboratory were first to report genome-wide genetic screens in human primary T cells by performing sgRNA lentiviral infection with Cas9 protein electroporation (SLICE) to identify genes that enhance the proliferation and antitumor activity of CD8+ T cells (Shifrut et al. 2018). Finally, in a recent report, Roth et al. have used a pooled-knock-in (PoKI) targeting method to simultaneously study multiple synthetic receptors and transcription factors co-expressed with a recombinant TCR (Roth et al. 2020). Using a barcoding system, they developed pooled knockin sequencing (PoKI-seq), combining single-cell transcriptome analysis and pooled knockin screening to measure cell abundance ex vivo and *in vivo*.

16.4 Gene Editing in T Cell Therapy

16.4.1 Limiting GVHD and TCR Mispairing

The vast majority of current clinical protocols utilize autologous T cells that are collected by apheresis and engineered with retroviral vectors to stably express a CAR or a recombinant TCR. This approach therefore requires

Table 16.1 A decade of gene-targeting experiments in primary T cells.

Nuclease							
Type	Delivery	Donor template	Locus	GOI	Promoter	Efficiency (%)	Reference
ZFN	RNA	NILV	AAVS1	GFP	EF1a/PGK/SFFV	8	Lombardo et al. 2011
			CCR5	GFP	EF1a/PGK/SFFV	9	
MegaTal	RNA	AAV6	CCR5	GFP	MND	60	Sather et al. 2015
				CD19-CAR	MND	14	
ZFN	RNA	AAV6	CCR5	GFP	PGK	55	Wang et al. 2016
			AAVS1	GFP	PGK	41	
MegaTal	RNA	AAV6	CD40L	CD40L	Endogeneous	30	Hubbard et al. 2016
TALEN	RNA	AAV6	CCR5	HIV-CAR	MND	46	Hale et al. 2017
MegaTal	RNA	AAV6	CCR5	CD19-CAR	MND	37	Hale et al. 2017
			TRAC	CD19-CAR	MND	42	
CRISPR/TALEN	RNA	AAV6	TRAC	CD19-CAR	Endogeneous	40–60	Eyquem et al. 2017
					PGK		
					RV-LTR		
					EF1a		
			B2M		Endegenous		
					PGK		
					PGK100		
					EF1a		
Homing endonuclease	RNA	AAV6	TRAC	CD19-CAR	Jet	33	Macleod et al. 2017
CRISPR/Cas9	RNP	dsDNA /ssDNA	Multiple locus	GFP	Endogenous	15–50	Roth et al. 2018
			TRAC	NYESO TCR	Endogenous	20	
			IL2RA	Corrected IL2Ra	Endogenous	15	
TALEN	RNA	AAV6	TRAC + CD25	IL12	Endogenous	30	Sachdeva et al. 2019
			TRAC + PCDC1	IL12	Endogenous	11	
CRISPR/Cas9	RNP	dsDNA /ssDNA	Multiple locus	GFP	Endogenous	15–50	Nguyen et al. 2020
			TRAC	NYESO TCR	Endogenous	20	
			IL2RA	corrected IL2Ra	Endogenous	15	
CRISPR/Cas9	RNP	dsDNA	TRAC	NYESO TCR + Pool of 36 genes	Endogenous	15	Roth et al. 2020

patient-specific cell manufacturing, which unavoidably results in patient-to-patient variability in the final cell product and high costs. Widespread implementation of this approach will further require progress in automation and miniaturization of cell manufacturing to meet the demand for adoptive T cell therapies. For the past decade, many teams have put a lot of efforts in to generating so-called "universal T cells" that can be produced from a healthy donor and stored until needed. Such off-the-shelf approaches are very attractive as it would improve standardization of the manufacturing, lower cost through economy of scale and diminish the time between enrollment of patients into a protocol and engineered T cell injection.

A potential solution to generate off-the-shelf T cells is the editing of allogeneic T cells to offset graft versus host disease (GVHD) and/or eliminate their immunogenicity. To abolish the expression of the TCR/CD3 complex at the cell surface, ZFN (Torikai et al. 2012), TALEN (Berdien et al. 2014; Poirot et al. 2015), megaTAL (Boissel et al. 2014), and more recently CRISPR (Eyquem et al. 2017; Ren et al. 2017) have been used to disrupt the loci encoding the TCR alpha or beta chains, TRAC or TRBC. In 2015, Poirot et al. showed that TCR-negative CAR T cells exerted a reduced GVHD without affecting the antitumor function in a xenograft mouse model for leukemia (Poirot et al. 2015) and paved the way to the first gene-edited allogeneic CAR T cell clinical trial (Qasim et al. 2017).

As the highly polymorphic Human Leukocyte Antigen I (HLA-I) is an important mediator of the immune rejection, eliminating its expression from the T cell surface could limit their rejection from allogeneic host. Thus, the beta-2-microglobulin (B2M), the common chain of the HLA-I complex, has been efficiently disrupted in CAR T cells (Eyquem et al. 2017; Ren et al. 2017; Torikai et al. 2013). The effect of HLA-I disruption on CAR T cell function and persistence in an allogeneic host remains to be studied in relevant models. It is also important to note that the absence of HLA-I at the surface of CAR T cells makes them vulnerable to lysis by NK cells. It has been shown that expression of a class I molecule such as the minimal polymorphic HLA-E protein can bind to inhibitory CD94/NGK2A complex and prevent NK-mediated lysis (Braud et al. 1998; Lee et al. 1998). By expressing a single-chain HLA-E, Gornalusse et al. generated B2M-negative stem cells that escape NK rejection (Gornalusse et al. 2017) and such an approach can be adapted to allogeneic CAR T cells. More recently, the Mamonkin laboratory has developed an alloimmune defense receptor (ADR) which selectively recognizes 4-1BB, a cell surface receptor temporarily upregulated by activated lymphocytes (Mo et al. 2021). They demonstrated that allogeneic T cells co-expressing a CAR and an ADR maintained antitumor activity while resisting host T and NK cells rejection in tumor mouse models.

In order to extend lymphodepletion while maintaining infused allogeneic CAR-T cells, TALEN have been used to also disrupt CD52 (Poirot et al. 2015), a protein targetable with alemtuzumab, which is used as an antibody in the treatment of leukemia. Two infants suffering from relapsed B-ALL were successfully treated with such a CAR T cell product in combination with an anti-CD52 serotherapy (Qasim et al. 2017). Similarly, the deoxycytidine kinase (dCK) gene can be knocked out to make CAR-T cells resistant to alemtuzumab or purine nucleotide analogues (Valton et al. 2015).

In the context of T cell engineered to express a tumor-specific recombinant TCR, the disruption of the endogenous TCR genes has many advantages. The TCR complex is composed of the TCR alpha and beta chains associated with the CD3 delta, epsilon, gamma, and zeta chains. The assembly of all the chains in the endoplasmic reticulum is required for TCR complex to be addressed at the cell surface. Thus, the tumor-specific TCR competes with the endogenous TCR for the CD3 chains and dilutes of the TCRs resulting in a lower recombinant TCR expression and increased the risk of harmful TCR hybrids. In 2012, Chiara Bonini showed that sequentially disrupting the TCR alpha and beta chains with ZFN associated with a lentiviral delivery of a recombinant TCR results in a higher expression and avidity of a NY-ESO-1-specific TCR (Provasi et al. 2012). However, this protocol included four engineering steps and was not suited for clinical application. The same team eventually showed that eliminating the TCR alpha only was sufficient to reduce alloreactivity and improve NY-ESO-1 TCR expression as well as antitumor activity in a xenograft mouse model for myeloma (Mastaglio et al. 2017). More recently, a UPenn clinical trial lead by Carl June treated three patients (two with advanced refractory myeloma and one with metastatic sarcoma) with autologous T cells lentivirally transduced to express a NY-ESO-1-specific TCR and triple KO for the TCR alpha, TCR beta, and PD-1 using CRISPR/Cas9 (Stadtmauer et al. 2020). The KO efficiency of individual genes spanned between 10 and 50% with an average of 10% of the patient cells carrying the triple KO. While significant levels of translocations were detectable in all patients' T cells, no clonal expansion has been observed and their percentage decreased over time. Edited T cells were detectable in patients for up to nine months indicating minimal immunogenicity. While this clinical trial was not designed to demonstrate the superiority of specific gene edits, it paved the way to more next-generation T cell therapies by showing the feasibility of CRISPR-engineered T cells.

16.4.2 Limit Immunosuppressive Signaling

Although CD19-targeting CAR T cells have shown remarkable results against acute lymphoblastic leukemia (ALL) (Maude et al. 2018; Park et al. 2018), clinical results remain moderate in solid tumors due to a number of obstacles. First, it has been challenging to identify tumor-specific targets homogeneously expressed solid tumors. Second, there is an inefficient homing to the tumor site and poor penetration through the tumor extracellular matrix. Finally, engineered T cells are poorly persistent and often dysfunctional in the solid tumor microenvironment notably due to chronic TCR/CAR signaling and immunosuppressive signaling (Sadelain et al. 2017). These immunosuppressive signals are in part mediated by immune checkpoint including programmed cell death 1 (PD-1), cytotoxic T-lymphocyte antigen 4 (CTLA4), lymphocyte-activated gene-3 (LAG3), and T cell immunoglobulin and mucin domain 3 (TIM3). Antagonistic antibodies targeting these checkpoint receptors like anti-PD-1, anti-PD-L1, or anti-CTLA-4 allow the amplification of the endogenous antitumor T cell response (Sharma and Allison 2015) and dramatically improved the clinical outcome of patients with highly mutational tumors such as melanoma or non-small cell lung cancer (Lawrence et al. 2013). In tumor with lower mutational burden, disrupting these pathways can potentially improve T cells engineered to express a tumor-specific TCR or a CAR.

In 2015, Beane et al. used ZFN to disrupt PD-1 in tumor infiltrating lymphocytes (TILs) isolated from patient with melanoma (Beane et al. 2015). Using large-scale electroporation, they engineered up to 100 million T cells and showed that PD-1-engineered T cells displayed better tumor cell lysis and cytokine profile in vitro. The same year, Menger et al. used TALEN to disrupt PD-1 in melanoma-reactive CD8 mouse T cells and in fibrosarcoma-reactive polyclonal mouse T cells resulting in an improved T cell persistence and antitumor activity (Menger et al. 2016). In human CAR T cells, several teams have used CRISPR-Cas9 to efficiently knockout the PD-1 receptor (Liu et al. 2017; Ren et al. 2017; Rupp et al. 2017; Schumann et al. 2015; Su et al. 2016). Three of these studies showed that PD1 disruption enhanced antitumor activity of human CAR T cells in xenograft mouse models with a subcutaneous CD19-expressing tumor (Rupp et al. 2017), a systemic Burkitt lymphoma (Liu et al. 2017) or subcutaneous hepatocellular carcinoma (Guo et al. 2018). While these studies suggest that PD-1 disruption may enhance CAR T cell performance, better characterization of these engineered T cells remain to be performed. First, to ensure the relevance of the PD1 disruption, a fair comparison with an anti-PD1 or a PD1 dominant negative (Cherkassky et al. 2016; Liu et al. 2016) should be performed. Finally, the safety of such engineering has to be addressed as a study showed that haploid disruption of PD1 can promote T-cell non-Hodgkin's lymphoma (T-NHL) in a mouse model (Wartewig et al. 2017). Several phase I/II clinical trials exploring PD-1 knockout for adoptive T cell therapy have been initiated as a consequence.

In addition to PD-1, several studies have suggested that blocking PD-1 and CTLA-4, LAG-3 or TIM-3 has synergistically enhanced antitumor effects (Sakuishi et al. 2010; Woo et al. 2012; Zappasodi et al. 2018). Thus, disruption of these other pathways to enhance T cell function recently started to be studied and CRISPR-Cas9-mediated KO of CTLA-4 (Ren et al. 2017; Zhang et al. 2017, p. 3) and LAG-3 have been reported. While these studies report high gene editing efficiency in primary T cells, they do not perform conclusive functional studies on their therapeutic activity.

Finally, the tumor vasculature in ovarian cancer and other tumor types have been shown to express FasL, which induces apoptotic death of cells expressing the Fas receptor (Fas) and reduces T cell infiltration (Motz et al. 2014). Ren et al. reported a triple (TCR/B2M/Fas) KO using gRNAs co-delivered with a CAR-expressing cassette in a lentivirus, followed by Cas9 mRNA electroporation (Ren et al. 2017). Compared with conventional CD19 CAR T cells, the triple KO CD19 CAR T cells displayed a reduced cell death in an in vitro proliferation assay and delayed tumor relapses in NSG mice. The participation of each gene disruption in the therapeutic activity remains to be better studied as well as the occurrence of translocation in such multiplex approach.

16.4.3 Delaying Exhaustion of Intratumoral T Cells

T cell exhaustion and dysfunction lead to reduced effector function, poor long-term efficacy of ACTs, and contributes to tumor recurrence and relapse in patients. Efforts to understand the underlying intracellular mechanisms that regulate T cell exhaustion have led to multiple discoveries in recent years including TCF1 as a central transcription factor in balancing memory and exhaustion. Using a mouse model of chronic LCMV infection where T cells are unable to fully clear the infection, researchers have found that TCF1 is downregulated in exhausted and terminally differentiated CD8+ T cells compared with self-renewing memory populations of T cells (Wang et al. 2019, p. 1). These findings uncovered TCF1 as an important transcription factor that maintains T cells in a progenitor state and offers an insight into the possibilities of engineering exhaustion-resistant T cells.

Further insight to the progression of exhausted T cells can be found downstream of TCR signaling. Upon engagement with their specific antigen and co-stimulatory receptors, the TCR signal is relayed intracellularly mainly by three nuclear transcription factor complexes: NFAT, NF-κB, and AP1. Using mouse models of repeated NFAT activation either by chronic infection or tumor engagement, loss of TCF1 in CD8+ T cells has been accompanied by the increased expression of the transcriptional regulator TOX. Expression of TOX on its own has been reported to induce expression of gene signatures associated to T cell exhaustion (Alfei et al. 2019; Khan et al. 2019; Scott et al. 2019; Seo et al. 2019; Yao et al. 2019). Depletion of TOX in tumor models leads to an increase in non-exhausted T cells; however, these cells do not persist in the tumor and remains largely dysfunctional (Scott et al. 2019). While offering great insights to mechanisms behind T cell exhaustion, these findings pose questions over the suitability of TOX as a target for gene editing in T cells to generate exhaustion-resistant T cells.

In a report analyzing mouse CD8+ CAR TILs from mice-bearing human CD19+ tumors, Chen et al. found a striking transcriptional overlap between PD1- and TIM3-expressing exhausted CD8+ cells to NR4A target genes. They showed that KO three genes in the NR4A family (NR4A1, NR4A2, and NR4A3) in CD8+ CAR T cells improved antitumor activity and survival in mouse models (Chen et al. 2019), similar to the results seen in studies of TOX. Interestingly, it was later found that the NR4A transcription factors cooperate with TOX to regulate transcriptional programs (Seo et al. 2019), offering fascinating insights to the intricate transcriptional network that contributes to NFAT-induced T cell exhaustion.

Another mechanism of T cell exhaustion comes from studying the AP1 transcription factor complex, which like NFAT is also an effector complex downstream of the TCR. The AP1 protein complex consists of dimers of the JUN and FOS protein families. These dimers can be either homo- or heterodimers of multiple different combinations. A recent report studying tonic signaling in CAR T cells show upregulation of a large number of JUN and FOS genes in exhausted CAR T cells (Lynn et al. 2019). Interestingly, expression of these factors in T cells causes different outcomes for CAR T cells. Specifically, this study reported that overexpression of cJUN leads to the formation of cJUN-JUNB heterodimers in T cells, reducing the amount of JUNB-BATF heterodimers that contribute to T cell exhaustion. Thus, co-expressing cJUN and CAR generates a CAR T cell more resistant to exhaustion induced by tonic signaling.

With the development of gene editing in T cells, these exciting findings offer potential paths to control T cell differentiation and exhaustion with dramatic implications on the efficacy CAR T cell therapies.

16.4.4 Controlling T Cell Fate with Gene Editing

In previous sections, we have discussed ways to use gene editing to limit immunosuppressive signaling in T cells or delay T cell exhaustion. Another approach to engineer T cells is to control T cell fate by editing epigenetic factors or broad transcriptional regulators. One such example can be found in heterochromatin regulator SUV39H1. Histone methyltransferases, such as SUV39H1, are important to maintain chromatin organization and gene silencing that contributes to lineage specificity. Depletion of Suv39h1 in mice generated animals with reduced protection against repeated *Listeria monocytogenes* infections (Pace et al. 2018). Interestingly, it was found that transcriptional silencing by Suv39h1 is critical to maintain stemness and memory function of CD8+ T cells, thus leading to increased memory formation in Suv39h1-depleted mice coupled with a lack of CD8+ effector differentiation to mount a response to the infection. These findings have potential implications for immunotherapies where terminally differentiated CAR T cells often present an exhausted and nonreversible phenotype. Targeting epigenetic regulation and SUV39H1 by gene editing could potentially allow for induced stemness and plasticity in terminally differentiated CD8+ effector cells.

An alternative approach to target T cell fate is to generate more long-lived effector T cells. In a CRISPR/Cas9 screen performed *in vivo* using a gRNA library targeting metabolic genes and regulators, it was found that knocking out REGNASE-1 reprogrammed CD8+ T cells to long-lived TCF1-expressing CD8+ cells with better persistence and improved therapeutic efficacy in a solid tumor model (Wei et al. 2019). REGNASE-1 is an endoribonuclease that prevents aberrant T cell activity by degrading mRNAs associated to T cell function. By performing a secondary genome-wide CRISPR/Cas9 screen on REGNASE-1-deficient cells, the same study revealed that REGNASE-1 KO combined with disruption of PTPN2 or SOCS1 further improved therapeutic efficacy. Interestingly, PTPN2 is a phosphatase that in itself has been shown to play an important role in modulating terminally differentiated T cells (Brownlie et al. 2019; LaFleur et al. 2019). Editing the *PTPN2* gene has been shown to improve the function of CAR T cells by bolstering downstream signaling factors important for T cell function such as LCK and by increasing expression of CXCR3 important for tumor homing (Wiede et al. 2020). PTPN2 deletion was shown to improve the efficacy of HER2-specific CAR T cells against HER2-expressing solid tumors.

16.4.5 Broadening CAR Target Antigens in Blood Malignancies

Although CAR-T cell therapies have shown successful clinical outcomes, especially in B-cell malignancies and MM, it has been challenging to identify optimal target in other malignancies. Perna et al. reported that an ideal CAR target has to fulfill a range of criteria. First, it should be ubiquitously expressed on most if not all tumor cells (Perna et al. 2017). To prevent on-target/off-tumor toxicity, the optimal tumor target should not be expressed on any normal tissue/organ, or at least not in vital tissues nor in closely related normal cellular counterparts, such as hematopoietic stem/progenitor cells (HSPCs) in the case of blood malignancies. To limit fraternal killing, the target antigen should also not be expressed in CAR T cells.

In cases of T-cell malignancy, CD3, CD4, CD5, and CD7 have been targeted by CAR-based therapies because they are ubiquitously expressed in malignant T-cells but the fraternal killing significantly affected the manufacturing by reducing overall CAR-T cell yield (Chen et al. 2017; Cooper et al. 2018; Gogishvili et al. 2017; Pinz et al. 2016; Raikar et al. 2018; Rasaiyaah et al. 2018). Also, continuous interaction of CAR-T cells with target antigens during T-cell expansion significantly reduced their antitumor efficacy, presumably due to chronic signaling-mediated T-cell exhaustion (Cooper et al. 2018). To tackle this problem, several teams used CRISPR/Cas9 to disrupt CD3, CD5, or CD7 in CAR-T cells and thus preventing the self-killing effect of CAR-T cells (Cooper et al. 2018; Gomes-Silva et al. 2017; Pinz et al. 2016; Raikar et al. 2018). Collectively, disruption of CAR target antigens in T-cells not only prevents fratricide but also increases the antitumor efficacy of CAR T cells by circumventing T-cell exhaustion and CAR downregulation.

CD33 is an attractive target as it is highly expressed in AML. However, CD33 is also expressed in all normal myeloid cells downstream the common myeloid progenitor. In contrast to B-cells, the immunodeficiency that results from the depletion of normal myeloid cells cannot be overcome with pharmacologic supplementation and would leave patients vulnerable to life-threatening infections. Using CRISPR/Cas9 guide RNAs targeting CD33, Saar Gill edited HSCs to be devoid of CD33. These HSCs demonstrated normal myeloid compartment development and normal function of mature myeloid cells. T cells were then engineered to express CARs targeting CD33 and demonstrated activity against unedited myeloid cells but not CD33-deficient cells (Kim et al. 2018).

16.4.6 Targeted Integration

The vast majority of adoptive T cells therapies' protocols utilize autologous T cells that are collected by apheresis and engineered with retroviral vectors to stably express recombinant TCRs or CARs. These approaches utilize randomly integrating vectors, including gamma-retroviral, lentiviral, and transposons, which all result in semi-random integration and variable expression of the CAR owing to transgene variegation. Position effects may result in heterogeneous T cell function, transgene silencing, and, potentially, insertional oncogenesis. Although it has been thought for a long time that mature T lymphocytes were relatively "resistant" to insertional mutagenesis, two recent reports described patients in which the integration of the lentiviral vector coding for the CAR disrupted the *Tet2* and *CBL* genes and resulted in a dramatic clonal expansion (Fraietta et al. 2018; Shah et al. 2019). For these multiple reasons, the idea of targeting transgenes into specific loci has emerged.

While the initial reports using gene targeting in T cells were mainly technical advances (Table 16.2), Hubbard et al. introduced the first clinically relevant application for X-linked hyper-immunoglobulin (Ig)M syndrome in which CD40L is mutated. In this study, the authors targeted a functional CD40L cDNA at the CD40LG locus, upstream the translation start, thus restoring the endogenous gene function (Hubbard et al. 2016).

In 2017, we inserted a CAR transgene under the control of the TCR alpha (*TRAC*) promoter which simultaneously disrupted the expression of the endogenous gene, thus generating universal CAR+ / TCR− T cells in a single step (Eyquem et al. 2017). The *TRAC* promoter provides a highly homogeneous and predictable expression which greatly standardized the generation of CAR T cells. Additionally, we discovered that the endogenous *TRAC* promoter offers optimal CAR expression that prevents undesirable differentiation and exhaustion, which results in antitumor activity superior to any other expression level we studied in a pre-ALL xenograft mouse model. These results greatly improved the understanding of CAR biology by establishing the first link between CAR expression and antitumor function. Furthermore, this report added a criticality for the development of off-the-shelf CAR therapy as it combines the scalability of allogeneic T cell manufacturing with the uniformity and safety of targeted CAR gene integration.

A year later, Roth et al. demonstrated the possibility to target a recombinant TCR to the TRAC locus taking advantage of the endogenous transcriptional regulation while disrupting the endogenous TCR chains. In the same study, they used gene targeting to restore a pathogenic IL2RA mutation in cells from patients with monogenic

Table 16.2 Knock-out experiments performed on primary T cells.

Function	GOI	Nuclease	Reference
Confer HIV resistance	CCR5	ZNF	Urnov et al. 2005
Limiting GVHD	TRAC/TRBC	ZNF	Torikai et al. 2012
	TRAC/TRBC	MegaTal	Boissel et al. 2014
	TRAC/TRBC	TALEN	Berdien et al. 2014
	TRAC/TRBC	CRISPR/Cas9	Ren et al. 2017
	TRAC/TRBC	CRISPR/Cas9	Eyquem et al. 2017
	B2M	ZNF	Torikai et al. 2013
	B2M	CRISPR/Cas9	Eyquem et al. 2017
Improving engraftment	CD52	TALEN	Poirot et al. 2015
Reduce immunosuppressive signaling	PD-1	ZNF	Beane et al. 2015
	PD-1	TALEN	Menger et al. 2016
	PD-1	CRISPR/Cas9	Su et al. 2016
	CTLA4	CRISPR/Cas9	Ren et al. 2017
	LAG-3	CRISPR/Cas9	Zhang et al. 2017
	FAS	CRISPR/Cas9	Ren et al. 2017
Reduce T cell exhaustion	TOX	CRISPR/Cas9	Scott et al. 2019
	NR4A	CRISPR/Cas9	Chen et al. 2019
Transcriptional regulators	SUV39H1	CRISPR/Cas9	Pace et al. 2018
	REGNASE-1	CRISPR/Cas9	Wei et al. 2019
	PTPN2	CRISPR/Cas9	LaFleur et al. 2019
	SOCS1	CRISPR/Cas9	Wei et al. 2019

autoimmune disease, and demonstrate improved signaling function (Roth et al. 2020).

Finally, in a recent study, Sachdeva et al. targeted a transgene coding for IL-12 under the control of the endogenous promoters of the CD25 and PD-1 loci. Both endogenous genes are known to be activated upon T cell activation, thus generating T cells secreting IL-12 in response to CAR engagement (Sachdeva et al. 2019). They showed in vitro that the induction was maintained after multiple stimulation.

16.5 Conclusion

The recent FDA approval of the first engineered T cells is the result of more than two decades of development. The remarkable clinical efficacy observed in hematological malignancies brought a wave of newcomers to the field, combining basic immunology, synthetic biology, and gene editing. The first clinical trials with gene edited T cells mostly validated the ability to manufacture next-generation T cell therapies and appeared to be safe. The field is now ready for more complex T engineering with the hope of reduced production cost with universal off-the-shelf products and improve antitumor activity in solid tumors. While not as developed as primary T cells and not discussed in this chapter, other sources of cells are currently genetically engineered with exciting results, such as NK cells, Macrophages, and iPSC-derived immune cells.

References

Adachi, K., Kano, Y., Nagai, T. et al. (2018). IL-7 and CCL19 expression in CAR-T cells improves immune cell infiltration and CAR-T cell survival in the tumor. *Nat. Biotechnol.* 36: 346–351. https://doi.org/10.1038/nbt.4086.

Alfei, F., Kanev, K., Hofmann, M. et al. (2019). TOX reinforces the phenotype and longevity of exhausted T cells in chronic viral infection. *Nature* 571: 265–269. https://doi.org/10.1038/s41586-019-1326-9.

Avanzi, M.P., Yeku, O., Li, X. et al. (2018). Engineered tumor-targeted T cells mediate enhanced anti-tumor efficacy both directly and through activation of the endogenous immune system. *Cell Rep.* 23: 2130–2141. https://doi.org/10.1016/j.celrep.2018.04.051.

Beane, J.D., Lee, G., Zheng, Z. et al. (2015). Clinical scale zinc finger nuclease-mediated gene editing of PD-1 in tumor infiltrating lymphocytes for the treatment of metastatic melanoma. *Mol. Ther. J. Am. Soc. Gene Ther.* 23: 1380–1390. https://doi.org/10.1038/mt.2015.71.

Berdien, B., Mock, U., Atanackovic, D., and Fehse, B. (2014). TALEN-mediated editing of endogenous T-cell receptors facilitates efficient reprogramming of T lymphocytes by lentiviral gene transfer. *Gene Ther.* 21: 539–548. https://doi.org/10.1038/gt.2014.26.

Boissel, S., Jarjour, J., Astrakhan, A. et al. (2014). megaTALs: a rare-cleaving nuclease architecture for therapeutic genome engineering. *Nucleic Acids Res.* 42: 2591–2601. https://doi.org/10.1093/nar/gkt1224.

Bollard, C.M., Rössig, C., Calonge, M.J. et al. (2002). Adapting a transforming growth factor beta-related tumor protection strategy to enhance antitumor immunity. *Blood* 99: 3179–3187. https://doi.org/10.1182/blood.v99.9.3179.

Braud, V.M., Allan, D.S., O'Callaghan, C.A. et al. (1998). HLA-E binds to natural killer cell receptors CD94/NKG2A, B and C. *Nature* 391: 795–799. https://doi.org/10.1038/35869.

Brownlie, R.J., Wright, D., Zamoyska, R., and Salmond, R.J. (2019). Deletion of PTPN22 improves effector and memory CD8+ T cell responses to tumors. *JCI Insight* 5 https://doi.org/10.1172/jci.insight.127847.

Chen, K.H., Wada, M., Pinz, K.G. et al. (2017). Preclinical targeting of aggressive T-cell malignancies using anti-CD5 chimeric antigen receptor. *Leukemia* 31: 2151–2160. https://doi.org/10.1038/leu.2017.8.

Chen, J., López-Moyado, I.F., Seo, H. et al. (2019). NR4A transcription factors limit CAR T cell function in solid tumours. *Nature* 567: 530–534. https://doi.org/10.1038/s41586-019-0985-x.

Cherkassky, L., Morello, A., Villena-Vargas, J. et al. (2016). Human CAR T cells with cell-intrinsic PD-1 checkpoint blockade resist tumor-mediated inhibition. *J. Clin. Invest.* 126: 3130–3144. https://doi.org/10.1172/JCI83092.

Cooper, M.L., Choi, J., Staser, K. et al. (2018). An "off-the-shelf" fratricide-resistant CAR-T for the treatment of T cell hematologic malignancies. *Leukemia* 32: 1970–1983. https://doi.org/10.1038/s41375-018-0065-5.

D'Agostino, M. and Raje, N. (2020). Anti-BCMA CAR T-cell therapy in multiple myeloma: can we do better? *Leukemia* 34: 21–34. https://doi.org/10.1038/s41375-019-0669-4.

Dunbar, C.E., High, K.A., Joung, J.K. et al. (2018). Gene therapy comes of age. *Science* 359 https://doi.org/10.1126/science.aan4672.

Eyquem, J., Mansilla-Soto, J., Giavridis, T. et al. (2017). Targeting a CAR to the TRAC locus with CRISPR/Cas9 enhances tumour rejection. *Nature* 543: 113–117. https://doi.org/10.1038/nature21405.

Feucht, J., Sun, J., Eyquem, J. et al. (2019). Calibration of CAR activation potential directs alternative T cell fates and therapeutic potency. *Nat. Med.* 25: 82–88. https://doi.org/10.1038/s41591-018-0290-5.

Finney, H.M., Akbar, A.N., and Lawson, A.D.G. (2004). Activation of resting human primary T cells with chimeric receptors: costimulation from CD28, inducible costimulator, CD134, and CD137 in series with signals from the TCR zeta chain. *J. Immunol.* 1950 (172): 104–113. https://doi.org/10.4049/jimmunol.172.1.104.

Fraietta, J.A., Nobles, C.L., Sammons, M.A. et al. (2018). Disruption of TET2 promotes the therapeutic efficacy of CD19-targeted T cells. *Nature* 558: 307–312. https://doi.org/10.1038/s41586-018-0178-z.

Giavridis, T., van der Stegen, S.J.C., Eyquem, J. et al. (2018). CAR T cell-induced cytokine release syndrome is mediated by macrophages and abated by IL-1 blockade. *Nat. Med.* 24: 731–738. https://doi.org/10.1038/s41591-018-0041-7.

Gogishvili, T., Danhof, S., Prommersberger, S. et al. (2017). SLAMF7-CAR T cells eliminate myeloma and confer selective fratricide of SLAMF7+ normal lymphocytes. *Blood* 130: 2838–2847. https://doi.org/10.1182/blood-2017-04-778423.

Gomes-Silva, D., Srinivasan, M., Sharma, S. et al. (2017). CD7-edited T cells expressing a CD7-specific CAR for the therapy of T-cell malignancies. *Blood* 130: 285–296. https://doi.org/10.1182/blood-2017-01-761320.

Gornalusse, G.G., Hirata, R.K., Funk, S.E. et al. (2017). HLA-E-expressing pluripotent stem cells escape allogeneic responses and lysis by NK cells. *Nat. Biotechnol.* 35: 765–772. https://doi.org/10.1038/nbt.3860.

Gross, G., Waks, T., and Eshhar, Z. (1989). Expression of immunoglobulin-T-cell receptor chimeric molecules as functional receptors with antibody-type specificity. *Proc. Natl. Acad. Sci. U. S. A.* 86: 10024–10028. https://doi.org/10.1073/pnas.86.24.10024.

Guo, X., Jiang, H., Shi, B. et al. (2018). Disruption of PD-1 enhanced the anti-tumor activity of chimeric antigen receptor T cells against hepatocellular carcinoma. *Front. Pharmacol.* 9: 1118. https://doi.org/10.3389/fphar.2018.01118.

Hale, M., Mesojednik, T., Romano Ibarra, G.S. et al. (2017). Engineering HIV-resistant, anti-HIV chimeric antigen receptor T cells. *Mol. Ther. J. Am. Soc. Gene Ther.* 25: 570–579. https://doi.org/10.1016/j.ymthe.2016.12.023.

Hamieh, M., Dobrin, A., Cabriolu, A. et al. (2019). CAR T cell trogocytosis and cooperative killing regulate tumour antigen escape. *Nature* 568: 112–116. https://doi.org/10.1038/s41586-019-1054-1.

Hanahan, D. and Weinberg, R.A. (2011). Hallmarks of cancer: the next generation. *Cell* 144: 646–674. https://doi.org/10.1016/j.cell.2011.02.013.

Hay, A.E. and Cheung, M.C. (2019). CAR T-cells: costs, comparisons, and commentary. *J. Med. Econ.* 22: 613–615. https://doi.org/10.1080/13696998.2019.1582059.

Heslop, H.E., Slobod, K.S., Pule, M.A. et al. (2010). Long-term outcome of EBV-specific T-cell infusions to prevent or treat EBV-related lymphoproliferative disease in transplant recipients. *Blood* 115: 925–935. https://doi.org/10.1182/blood-2009-08-239186.

Hubbard, N., Hagin, D., Sommer, K. et al. (2016). Targeted gene editing restores regulated CD40L function in X-linked hyper-IgM syndrome. *Blood* 127: 2513–2522. https://doi.org/10.1182/blood-2015-11-683235.

Hudecek, M. and Ivics, Z. (2018). Non-viral therapeutic cell engineering with the sleeping beauty transposon system. *Curr. Opin. Genet. Dev.* 52: 100–108. https://doi.org/10.1016/j.gde.2018.06.003.

June, C.H. and Sadelain, M. (2018). Chimeric antigen receptor therapy. *N. Engl. J. Med.* 379: 64–73. https://doi.org/10.1056/NEJMra1706169.

Kagoya, Y., Tanaka, S., Guo, T. et al. (2018). A novel chimeric antigen receptor containing a JAK-STAT signaling domain mediates superior antitumor effects. *Nat. Med.* 24: 352–359. https://doi.org/10.1038/nm.4478.

Khan, O., Giles, J.R., McDonald, S. et al. (2019). TOX transcriptionally and epigenetically programs CD8+ T cell exhaustion. *Nature* 571: 211–218. https://doi.org/10.1038/s41586-019-1325-x.

Kim, S., Kim, D., Cho, S.W. et al. (2014). Highly efficient RNA-guided genome editing in human cells via delivery of purified Cas9 ribonucleoproteins. *Genome Res.* 24: 1012–1019. https://doi.org/10.1101/gr.171322.113.

Kim, M.Y., Yu, K.-R., Kenderian, S.S. et al. (2018). Genetic inactivation of CD33 in hematopoietic stem cells to enable CAR T cell immunotherapy for acute myeloid leukemia. *Cell* 173: 1439–1453.e19. https://doi.org/10.1016/j.cell.2018.05.013.

Koneru, M., Purdon, T.J., Spriggs, D. et al. (2015). IL-12 secreting tumor-targeted chimeric antigen receptor T cells eradicate ovarian tumors in vivo. *Oncoimmunology* 4: e994446. https://doi.org/10.4161/2162402X.2014.994446.

Kong, S., Sengupta, S., Tyler, B. et al. (2012). Suppression of human glioma xenografts with second-generation IL13R-specific chimeric antigen receptor-modified T cells. *Clin. Cancer Res. Off. J. Am. Assoc. Cancer Res.* 18: 5949–5960. https://doi.org/10.1158/1078-0432.CCR-12-0319.

LaFleur, M.W., Nguyen, T.H., Coxe, M.A. et al. (2019). PTPN2 regulates the generation of exhausted CD8+ T cell subpopulations and restrains tumor immunity. *Nat. Immunol.* 20: 1335–1347. https://doi.org/10.1038/s41590-019-0480-4.

Lawrence, M.S., Stojanov, P., Polak, P. et al. (2013). Mutational heterogeneity in cancer and the search for new cancer-associated genes. *Nature* 499: 214–218. https://doi.org/10.1038/nature12213.

Lee, N., Llano, M., Carretero, M. et al. (1998). HLA-E is a major ligand for the natural killer inhibitory receptor CD94/NKG2A. *Proc. Natl. Acad. Sci. U. S. A.* 95: 5199–5204. https://doi.org/10.1073/pnas.95.9.5199.

Liu, X., Ranganathan, R., Jiang, S. et al. (2016). A chimeric switch-receptor targeting PD1 augments the efficacy of second-generation CAR T cells in advanced solid tumors. *Cancer Res.* 76: 1578–1590. https://doi.org/10.1158/0008-5472.CAN-15-2524.

Liu, X., Zhang, Y., Cheng, C. et al. (2017). CRISPR-Cas9-mediated multiplex gene editing in CAR-T cells. *Cell Res.* 27: 154–157. https://doi.org/10.1038/cr.2016.142.

Lombardo, A., Cesana, D., Genovese, P. et al. (2011). Site-specific integration and tailoring of cassette design for sustainable gene transfer. *Nat. Methods* 8: 861–869. https://doi.org/10.1038/nmeth.1674.

Lynn, R.C., Weber, E.W., Sotillo, E. et al. (2019). c-Jun overexpression in CAR T cells induces exhaustion resistance. *Nature* 576: 293–300. https://doi.org/10.1038/s41586-019-1805-z.

MacLeod, D.T., Antony, J., Martin, A.J. et al. (2017). Integration of a CD19 CAR into the TCR alpha chain locus streamlines production of allogeneic gene-edited CAR T cells. *Mol. Ther.* 25 (4): 949–961.

Maher, J., Brentjens, R.J., Gunset, G. et al. (2002). Human T-lymphocyte cytotoxicity and proliferation directed by a single chimeric TCRzeta /CD28 receptor. *Nat. Biotechnol.* 20: 70–75. https://doi.org/10.1038/nbt0102-70.

Marcucci, K.T., Jadlowsky, J.K., Hwang, W.-T. et al. (2018). Retroviral and lentiviral safety analysis of gene-modified T cell products and infused HIV and oncology patients. *Mol. Ther. J. Am. Soc. Gene Ther.* 26: 269–279. https://doi.org/10.1016/j.ymthe.2017.10.012.

Mastaglio, S., Genovese, P., Magnani, Z. et al. (2017). NY-ESO-1 TCR single edited stem and central memory T cells to treat multiple myeloma without graft-versus-host disease. *Blood* 130: 606–618. https://doi.org/10.1182/blood-2016-08-732636.

Maude, S.L., Laetsch, T.W., Buechner, J. et al. (2018). Tisagenlecleucel in children and young adults with B-cell lymphoblastic leukemia. *N. Engl. J. Med.* 378: 439–448. https://doi.org/10.1056/NEJMoa1709866.

Menger, L., Sledzinska, A., Bergerhoff, K. et al. (2016). TALEN-mediated inactivation of PD-1 in tumor-reactive lymphocytes promotes intratumoral T-cell persistence and rejection of established tumors. *Cancer Res.* 76: 2087–2093. https://doi.org/10.1158/0008-5472.CAN-15-3352.

Mo, F., Watanabe, N., McKenna, M.K. et al. (2021). Engineered off-the-shelf therapeutic T cells resist host immune rejection. *Nat. Biotechnol.* 39: 56–63. https://doi.org/10.1038/s41587-020-0601-5.

Motz, G.T., Santoro, S.P., Wang, L.-P. et al. (2014). Tumor endothelium FasL establishes a selective immune barrier promoting tolerance in tumors. *Nat. Med.* 20: 607–615. https://doi.org/10.1038/nm.3541.

Newrzela, S., Cornils, K., Li, Z. et al. (2008). Resistance of mature T cells to oncogene transformation. *Blood* 112: 2278–2286. https://doi.org/10.1182/blood-2007-12-128751.

Nguyen, D.N., Roth, T.L., Li, P.J. et al. (2020). Polymer-stabilized Cas9 nanoparticles and modified repair templates increase genome editing efficiency. *Nat. Biotechnol.* 38: 44–49. https://doi.org/10.1038/s41587-019-0325-6.

Norelli, M., Camisa, B., Barbiera, G. et al. (2018). Monocyte-derived IL-1 and IL-6 are differentially required for cytokine-release syndrome and neurotoxicity due to CAR T cells. *Nat. Med.* 24: 739–748. https://doi.org/10.1038/s41591-018-0036-4.

Pace, L., Goudot, C., Zueva, E. et al. (2018). The epigenetic control of stemness in CD8+ T cell fate commitment. *Science* 359: 177–186. https://doi.org/10.1126/science.aah6499.

Park, J.H., Rivière, I., Gonen, M. et al. (2018). Long-term follow-up of CD19 CAR therapy in acute lymphoblastic leukemia. *N. Engl. J. Med.* 378: 449–459. https://doi.org/10.1056/NEJMoa1709919.

Parker, K.R., Migliorini, D., Perkey, E. et al. (2020). Single-cell analyses identify brain mural cells expressing CD19 as potential off-tumor targets for CAR-T immunotherapies. *Cell* https://doi.org/10.1016/j.cell.2020.08.022.

Perna, F., Berman, S.H., Soni, R.K. et al. (2017). Integrating proteomics and transcriptomics for systematic combinatorial chimeric antigen receptor therapy of AML. *Cancer Cell* 32: 506–519.e5. https://doi.org/10.1016/j.ccell.2017.09.004.

Pinz, K., Liu, H., Golightly, M. et al. (2016). Preclinical targeting of human T-cell malignancies using CD4-specific chimeric antigen receptor (CAR)-engineered T cells. *Leukemia* 30: 701–707. https://doi.org/10.1038/leu.2015.311.

Poirot, L., Philip, B., Schiffer-Mannioui, C. et al. (2015). Multiplex genome-edited T-cell manufacturing platform for "off-the-shelf" adoptive T-cell immunotherapies. *Cancer Res.* 75: 3853–3864. https://doi.org/10.1158/0008-5472.CAN-14-3321.

Provasi, E., Genovese, P., Lombardo, A. et al. (2012). Editing T cell specificity towards leukemia by zinc finger nucleases and lentiviral gene transfer. *Nat. Med.* 18: 807–815. https://doi.org/10.1038/nm.2700.

Qasim, W., Zhan, H., Samarasinghe, S. et al. (2017). Molecular remission of infant B-ALL after infusion of universal TALEN gene-edited CAR T cells. *Sci. Transl. Med.* 9 https://doi.org/10.1126/scitranslmed.aaj2013.

Raikar, S.S., Fleischer, L.C., Moot, R. et al. (2018). Development of chimeric antigen receptors targeting T-cell malignancies using two structurally different anti-CD5 antigen binding domains in NK and CRISPR-edited T cell lines. *Oncoimmunology* 7: e1407898. https://doi.org/10.1080/2162402X.2017.1407898.

Raje, N., Berdeja, J., Lin, Y. et al. (2019). Anti-BCMA CAR T-cell therapy bb2121 in relapsed or refractory multiple myeloma. *N. Engl. J. Med.* 380: 1726–1737. https://doi.org/10.1056/NEJMoa1817226.

Rasaiyaah, J., Georgiadis, C., Preece, R. et al. (2018). TCRαβ/CD3 disruption enables CD3-specific antileukemic T cell immunotherapy. *JCI Insight* 3 https://doi.org/10.1172/jci.insight.99442.

Ren, J., Zhang, X., Liu, X. et al. (2017). A versatile system for rapid multiplex genome-edited CAR T cell generation. *Oncotarget* 8: 17002–17011. https://doi.org/10.18632/oncotarget.15218.

Robbins, P.F., Kassim, S.H., Tran, T.L.N. et al. (2015). A pilot trial using lymphocytes genetically engineered with an NY-ESO-1-reactive T-cell receptor: long-term follow-up and correlates with response. *Clin. Cancer Res. Off. J. Am. Assoc. Cancer Res.* 21: 1019–1027. https://doi.org/10.1158/1078-0432.CCR-14-2708.

Rosenberg, S.A., Spiess, P., and Lafreniere, R. (1986). A new approach to the adoptive immunotherapy of cancer with tumor-infiltrating lymphocytes. *Science* 233: 1318–1321. https://doi.org/10.1126/science.3489291.

Roth, T.L., Puig-Saus, C., Yu, R. et al. (2018). Reprogramming human T cell function and specificity with non-viral genome targeting. *Nature* 559: 405–409. https://doi.org/10.1038/s41586-018-0326-5.

Roth, T.L., Li, P.J., Blaeschke, F. et al. (2020). Pooled knockin targeting for genome engineering of cellular immunotherapies. *Cell* 181: 728–744.e21. https://doi.org/10.1016/j.cell.2020.03.039.

Roybal, K.T., Williams, J.Z., Morsut, L. et al. (2016). Engineering T cells with customized therapeutic response programs using synthetic notch receptors. *Cell* 167: 419–432.e16. https://doi.org/10.1016/j.cell.2016.09.011.

Ruella, M. and Maus, M.V. (2016). Catch me if you can: leukemia escape after CD19-directed T cell immunotherapies. *Comput. Struct. Biotechnol. J.* 14: 357–362. https://doi.org/10.1016/j.csbj.2016.09.003.

Rupp, L.J., Schumann, K., Roybal, K.T. et al. (2017). CRISPR/Cas9-mediated PD-1 disruption enhances anti-tumor efficacy of human chimeric antigen receptor T cells. *Sci. Rep.* 7: 737. https://doi.org/10.1038/s41598-017-00462-8.

Sachdeva, M., Busser, B.W., Temburni, S. et al. (2019). Repurposing endogenous immune pathways to tailor and control chimeric antigen receptor T cell functionality. *Nat. Commun.* 10: 5100. https://doi.org/10.1038/s41467-019-13088-3.

Sadelain, M. (2017a). Chimeric antigen receptors: a paradigm shift in immunotherapy. *Annu. Rev. Cancer Biol.* 1: 447–466. https://doi.org/10.1146/annurev-cancerbio-050216-034351.

Sadelain, M. (2017b). CD19 CAR T cells. *Cell* 171: 1471. https://doi.org/10.1016/j.cell.2017.12.002.

Sadelain, M., Papapetrou, E.P., and Bushman, F.D. (2011). Safe harbours for the integration of new DNA in the human genome. *Nat. Rev. Cancer* 12: 51–58. https://doi.org/10.1038/nrc3179.

Sadelain, M., Rivière, I., and Riddell, S. (2017). Therapeutic T cell engineering. *Nature* 545: 423–431. https://doi.org/10.1038/nature22395.

Sakuishi, K., Apetoh, L., Sullivan, J.M. et al. (2010). Targeting Tim-3 and PD-1 pathways to reverse T cell exhaustion and restore anti-tumor immunity. *J. Exp. Med.* 207: 2187–2194. https://doi.org/10.1084/jem.20100643.

Sather, B.D., Romano Ibarra, G.S., Sommer, K. et al. (2015). Efficient modification of CCR5 in primary human hematopoietic cells using a megaTAL nuclease and AAV donor template. *Sci. Transl. Med.* 7: 307ra156. https://doi.org/10.1126/scitranslmed.aac5530.

Schumann, K., Lin, S., Boyer, E. et al. (2015). Generation of knock-in primary human T cells using Cas9 ribonucleoproteins. *Proc. Natl. Acad. Sci. U. S. A.* 112: 10437–10442. https://doi.org/10.1073/pnas.1512503112.

Scott, A.C., Dündar, F., Zumbo, P. et al. (2019). TOX is a critical regulator of tumour-specific T cell differentiation. *Nature* 571: 270–274. https://doi.org/10.1038/s41586-019-1324-y.

Seo, H., Chen, J., González-Avalos, E. et al. (2019). TOX and TOX2 transcription factors cooperate with NR4A transcription factors to impose CD8+ T cell exhaustion. *Proc. Natl. Acad. Sci. U. S. A.* 116: 12410–12415. https://doi.org/10.1073/pnas.1905675116.

Shah, N.N., Qin, H., Yates, B. et al. (2019). Clonal expansion of CAR T cells harboring lentivector integration in the CBL gene following anti-CD22 CAR T-cell therapy. *Blood Adv.* 3: 2317–2322. https://doi.org/10.1182/bloodadvances.2019000219.

Sharma, P. and Allison, J.P. (2015). The future of immune checkpoint therapy. *Science* 348: 56–61. https://doi.org/10.1126/science.aaa8172.

Shifrut, E., Carnevale, J., Tobin, V. et al. (2018). Genome-wide CRISPR screens in primary human T cells reveal key regulators of immune function. *Cell* 175: 1958–1971.e15. https://doi.org/10.1016/j.cell.2018.10.024.

Stadtmauer, E.A., Fraietta, J.A., Davis, M.M. et al. (2020). CRISPR-engineered T cells in patients with refractory cancer. *Science* 367 https://doi.org/10.1126/science.aba7365.

Su, S., Hu, B., Shao, J. et al. (2016). CRISPR-Cas9 mediated efficient PD-1 disruption on human primary T cells from cancer patients. *Sci. Rep.* 6: 20070. https://doi.org/10.1038/srep20070.

Takayama, K., et al. (2017). Highly efficient biallelic genome editing of human ES/iPS cells using a CRISPR/Cas9 or TALEN system. *Nucleic Acids Res.* 45: 5198–5207.

Tammana, S., Huang, X., Wong, M. et al. (2010). 4-1BB and CD28 signaling plays a synergistic role in redirecting umbilical cord blood T cells against B-cell malignancies. *Hum. Gene Ther.* 21: 75–86. https://doi.org/10.1089/hum.2009.122.

Torikai, H., Reik, A., Liu, P.-Q. et al. (2012). A foundation for universal T-cell based immunotherapy: T cells engineered to express a CD19-specific chimeric-antigen-receptor and eliminate expression of endogenous TCR. *Blood* 119: 5697–5705. https://doi.org/10.1182/blood-2012-01-405365.

Torikai, H., Reik, A., Soldner, F. et al. (2013). Toward eliminating HLA class I expression to generate universal cells from allogeneic donors. *Blood* 122: 1341–1349. https://doi.org/10.1182/blood-2013-03-478255.

Urnov, F.D., Miller, J.C., Lee, Y.-L. et al. (2005). Highly efficient endogenous human gene correction using designed zinc-finger nucleases. *Nature* 435 (7042): 646–651.

Valton, J., Guyot, V., Marechal, A. et al. (2015). A multidrug-resistant engineered CAR T cell for allogeneic combination immunotherapy. *Mol. Ther. J. Am. Soc. Gene Ther.* 23: 1507–1518. https://doi.org/10.1038/mt.2015.104.

Vispe, S., et al. (1998). Overexpression of Rad51 protein stimulates homologous recombination and increases resistance of mammalian cells to ionizing radiation. *Nucleic Acids Res.* 26: 2859–2864.

Wang, J., DeClercq, J.J., Hayward, S.B. et al. (2016). Highly efficient homology-driven genome editing in human T cells by combining zinc-finger nuclease mRNA and AAV6 donor delivery. *Nucleic Acids Res.* 44 (3): e30.

Wang, Y., Hu, J., Li, Y. et al. (2019). The transcription factor TCF1 preserves the effector function of exhausted CD8 T cells during chronic viral infection. *Front. Immunol.* 10: 169. https://doi.org/10.3389/fimmu.2019.00169.

Wartewig, T., Kurgyis, Z., Keppler, S. et al. (2017). PD-1 is a haploinsufficient suppressor of T cell lymphomagenesis. *Nature* 552: 121–125. https://doi.org/10.1038/nature24649.

Wei, J., Long, L., Zheng, W. et al. (2019). Targeting REGNASE-1 programs long-lived effector T cells for cancer therapy. *Nature* 576: 471–476. https://doi.org/10.1038/s41586-019-1821-z.

Wiede, F., Lu, K.-H., Du, X. et al. (2020). PTPN2 phosphatase deletion in T cells promotes anti-tumour immunity and CAR T-cell efficacy in solid tumours. *EMBO J.* 39: e103637. https://doi.org/10.15252/embj.2019103637.

Woo, S.-R., Turnis, M.E., Goldberg, M.V. et al. (2012). Immune inhibitory molecules LAG-3 and PD-1 synergistically regulate T-cell function to promote tumoral immune escape. *Cancer Res.* 72: 917–927. https://doi.org/10.1158/0008-5472.CAN-11-1620.

Xie, Y.J., Dougan, M., Jailkhani, N. et al. (2019). Nanobody-based CAR T cells that target the tumor microenvironment inhibit the growth of solid tumors in immunocompetent mice. *Proc. Natl. Acad. Sci. U. S. A.* 116: 7624–7631. https://doi.org/10.1073/pnas.1817147116.

Yao, C., Sun, H.-W., Lacey, N.E. et al. (2019). Single-cell RNA-seq reveals TOX as a key regulator of CD8+ T cell persistence in chronic infection. *Nat. Immunol.* 20: 890–901. https://doi.org/10.1038/s41590-019-0403-4.

Yeku, O.O., Purdon, T.J., Koneru, M. et al. (2017). Armored CAR T cells enhance antitumor efficacy and overcome the tumor microenvironment. *Sci. Rep.* 7: 10541. https://doi.org/10.1038/s41598-017-10940-8.

Zappasodi, R., Merghoub, T., and Wolchok, J.D. (2018). Emerging concepts for immune checkpoint blockade-based combination therapies. *Cancer Cell* 33: 581–598. https://doi.org/10.1016/j.ccell.2018.03.005.

Zhang, Y., Zhang, X., Cheng, C. et al. (2017). CRISPR-Cas9 mediated LAG-3 disruption in CAR-T cells. *Front. Med.* 11 (4): 554–562.

Zhao, Z., Condomines, M., van der Stegen, S.J.C. et al. (2015). Structural design of engineered costimulation determines tumor rejection kinetics and persistence of CAR T cells. *Cancer Cell* 28: 415–428. https://doi.org/10.1016/j.ccell.2015.09.004.

Zhou, P., Shaffer, D.R., Alvarez Arias, D.A. et al. (2014). In vivo discovery of immunotherapy targets in the tumour microenvironment. *Nature* 506: 52–57. https://doi.org/10.1038/nature12988.

17

Genome-Editing Applications in Stem Cell Engineering and Regenerative Medicine

Pragya Gupta, Shashank Jaitly, Priya Thakur, Sangam Giri Goswami, Nupur Bhargava, and Sivaprakash Ramalingam

Genomics and Molecular Medicine, CSIR-Institute of Genomics and Integrative Biology, New Delhi, India

17.1 Introduction

In the early 1960s, Till and McCulloch reported the discovery of self-renewal property of a living cell, which became one of the major breakthroughs in stem cell biology. It was discovered after subjecting the mice with irradiation followed by intravenous injection of bone marrow cells from healthy donor animals. In these experiments, the growth of blood progenitor cells could salvage the mice from death. Their discovery led to the basic foundation for bone marrow transplantation (BMT) to treat blood disorders (Till and McCulloch 1961). Subsequent studies described the differentiation potential of these bone marrow cells into various other cell types along with the ability to self-renew without senescence and designated them as stem cells.

Stem cells have the ability of self-renewal and are capable of proliferation and differentiation into different cell types. Based on the differentiation potential, stem cells can be classified as totipotent (cells have the ability to form an entire organism), pluripotent (cells can give rise to all cell types of the body), multipotent (most adult stem cells are multipotent; it has the ability to divide and develop into multiple specialized cell types), and unipotent (capable of developing into only one type of cell) (Fortier 2005; Mason and Dunnill 2008). Autologous stem cells play a crucial role in personalized regenerative medicine by providing an important cell source. Moreover, these cells provide a favorable microenvironment and stimulate tissue regeneration and help the body to form new functional tissue to substitute faulty ones. Various types of stem cells can be isolated from tissues, such as adipose tissue (adipose-derived stem cells, ASCs), bone marrow (mesenchymal stem cells, MSCs), blood (hematopoietic stem progenitor cells, HSPCs), embryo (embryonic stem cells, ESCs), etc. ASCs and BMSCs can further differentiate into adipocytes, osteocytes, chondrocytes and are frequently used in the regeneration of bone and cartilages (Fortier 2005).

Four decades later, after the pioneering work of Till and McCollach's, JJ Thomson and his colleagues isolated ESCs from blastocysts (Thomson 1998). ESCs have an enormous potential in regenerative therapy but their clinical use is still restricted due to ethical issues. Subsequently, a revolutionary study published by Yamanaka and his colleagues in 2006 on generation of induced pluripotent stem cells (iPSCs) started a movement revolving around the potential that lies in this technology (Takahashi and Yamanaka 2006). The self-renewal and differentiation ability of iPSCs into any cell type from all three germ layers (ectoderm, mesoderm, and endoderm), is at the center of why iPSCs hold the key to the future of disease modeling and therapeutics. In addition, it bypasses a lot of ethical issues associated with the use of ESCs in research and clinical use. Furthermore, the development of iPSC technology can allow for a near-unlimited amount of either healthy or disease-specific cell lineages; however, it is extremely difficult to obtain such cells (e.g. cardiomyocytes and neural cells) in routine clinical settings (Elitt et al. 2018). Due to these above-mentioned properties, iPSCs hold great potential for applications in biomedical research and medicine.

Recently, a number of clinical trials have been started for iPSC-based cell therapy for various diseases such as macular degeneration (Mandai et al. 2017), ischemic heart disease (Menasché et al. 2018), and Parkinson's disease (PD; Takahashi and Sawamoto 2018). However, especially for blood disorders, the recent enthusiasm of generating transplantable CD34+ HSPCs *in vitro* from patient-specific iPSCs for autologous cell therapy has been hampered due to lack of efficient differentiation protocols. Over the past couple of decades, mostly conventional gene therapy approach has been used to manipulate these autologous

Genome Editing in Drug Discovery, First Edition. Edited by Marcello Maresca and Sumit Deswal.

stem cells so as to achieve permanent cure for various human diseases.

Two of the conventional therapeutic technologies that are in use for gene therapy, one where an exogenous "therapeutic" gene is used to replace the defective gene in the genome of those who suffer from hereditary diseases using viral vectors. Another approach is RNA interference (RNAi), which mediates repression of defective genes by knockdown of the target mRNA (Kay 2011; Vaishnaw et al. 2010). Gene therapy is achieving progressively more success as the technology advances, allowing treatment of more diseases. The new generation of vectors in current clinical trials may show improved safety and efficacy. However, there remain significant challenges to advancing gene therapy for more disorders. It has major limitations such as lack of precision, acute immune response, and potentially tumorigenic. The random introduction of a genetic modification by gene therapy could lead to unintended side effects such as disrupting host metabolism, producing toxic or allergenic compounds, and activation of oncogenes due to insertional mutagenesis or position effects. Thus, continued improvements in methods for effective and safe genome modification are needed. Despite promise and recent success, RNAi has limitations that restrict their utility for a large number of diseases. Meanwhile, the use of RNAi is limited to targets for which gene knockdown is beneficial. Also, RNAi often cannot fully repress gene expression and is therefore unlikely to provide a benefit for diseases in which complete ablation of gene function is necessary for gene therapy. RNAi may also have poor specificity, posing potential safety concerns and sometimes decreasing the effectiveness of treatment (Tiemann and Rossi 2009).

Recent strides in the development of genome-editing technologies based on programmable nucleases have substantially improved our ability to make precise changes in the genomes of eukaryotic cells (Figure 17.1). The two main advantages of genome editing-based correction of the faulty gene are as follows: (i) the modified gene prevails under control of its endogenous promoter, allowing lifelong expression and natural adjustment in the cell; (ii) gene repair has the ability to bypass the engagement of the external DNA, hence reducing the chance of insertional mutagenesis and providing a broader scope of therapeutic applications (Bogdanove and Voytas 2011; Chandrasegaran and Carroll 2016; Hsu et al. 2014).

Genome editing coupled with iPSC technology is having a great impact on how we are addressing biomedical research and the future of medicine. Mutant and isogenic control iPSC-lines have paved the way to model multiple human diseases and further our understanding for its therapeutic applications. Patient-derived iPSC banks are expected to further enhance our understanding of heterogeneity between patients, increase accessibility to such model systems, and further the research for better regenerative therapies. Genome editing and stem cell-based technologies have already become an invaluable tool for personalized medicine, disease modeling, drug screening,

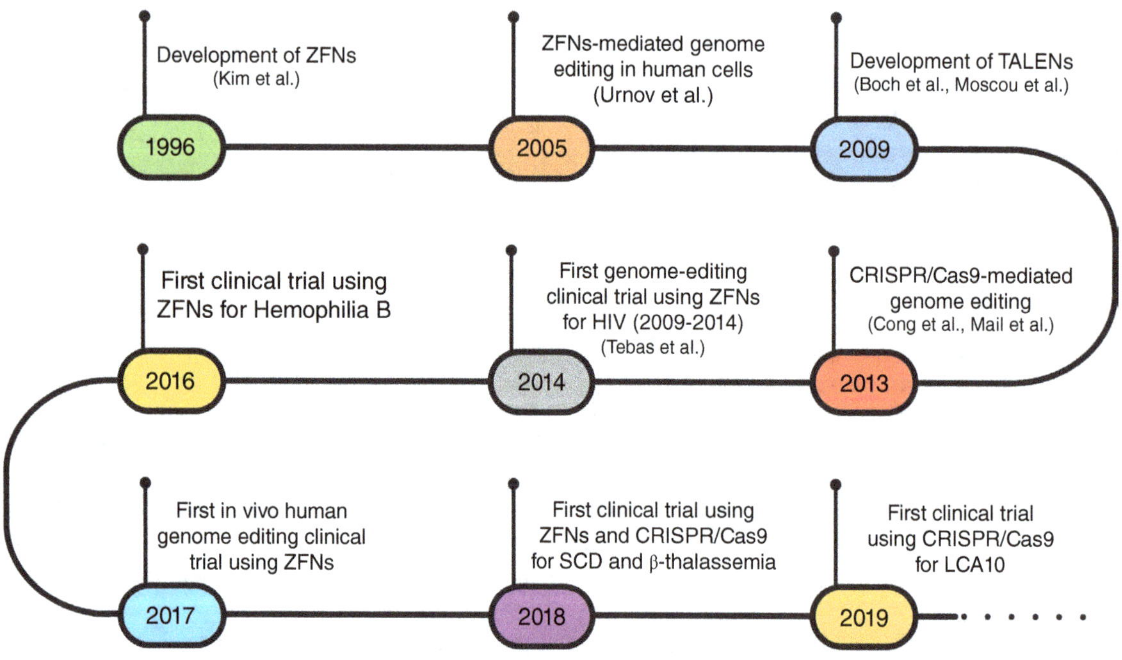

Figure 17.1 **Timeline of genome-editing nucleases and genome-editing research.** Key developments in the field of genome-editing-mediated clinical translations are shown.

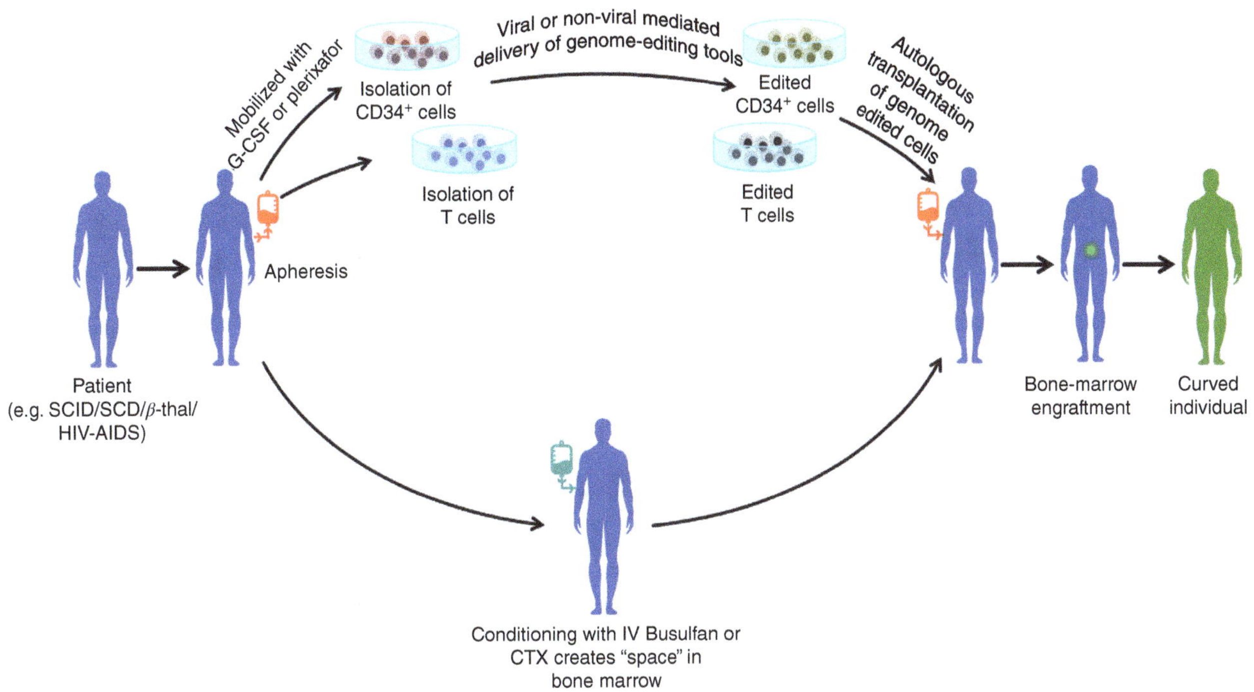

Figure 17.2 Ex vivo genome editing-based autologous cell replacement therapy. Target cells are mobilized in the patient, using G-CSF or plerixafor. The mobilized target cells are isolated and defective gene is repaired using gene-editing nucleases through viral or nonviral approaches. The patient is conditioned with immunosuppressive drugs such as busulfan or Cytoxan (CTX) intravenously prior to bone marrow engraftment. The edited cells are infused back into the bone marrow of the patient and engraftment and recovery is monitored.

development of novel therapeutics, and toxicity testing (Figure 17.1). Moreover, these technologies have already been tested in clinics for their therapeutic benefit for certain diseases as *ex vivo* or *in vivo* gene therapy. In *ex vivo* genome editing-based cell therapy, defective gene is repaired *in vitro* in patient-derived stem cells and then transplanted back into the patient (Figure 17.2). In the case of *in vivo* genome-editing therapy, gene-editing tools packaged in suitable delivery vehicles (viral vectors/nanoparticles) are delivered directly into the patient to achieve *in vivo* genome editing at the target site in the patient genome within the desired tissues (Figure 17.3). In this book chapter, we discuss potential *ex vivo* and *in vivo* clinical applications of genome editing for various human diseases and look at the future prospects in the context of cell and gene therapy and regenerative medicine. We will end by highlighting some of the challenges to translate genome editing to regenerative medicine.

17.2 Hematological Disorders

Hematological disorders encompass a range of blood-related diseases, many of them occurring due to genetic mutations, thus contributing to the global health burden significantly. Currently, allogenic hematopoietic stem cell transplantation (HSCT) is the only curative therapy available for a range of blood disorders. Over the last several years, the number of treated patients using HSCT has significantly increased. Though it can cure numerous blood disorders but has significant complications. An emerging alternative to allogenic HSCT is the precise modification of HSCs using a powerful targeted genome-editing approach. In this section, we discuss the genome-editing therapeutic strategies for some of the most common blood disorders such as sickle cell anemia, β-thalassemia (BT), and Hemophilia.

17.2.1 β-hemoglobinopathies: Sickle Cell Disease and β-thalassemia

β-hemoglobinopathies are a group of autosomal recessive blood disorders caused by genetic mutations in the β-globin (HBB) gene on chromosome 11. HBB gene encodes for the β-globin protein which is a vital component in the functional hemoglobin present inside red blood cells (RBCs) and is essential for oxygen delivery to the tissues. Mutations in the gene result in deficiency of functional RBCs (anemia), multi-organ dysfunction, and death. Sickle Cell disease (SCD) and BT are the most prevalent disorders in

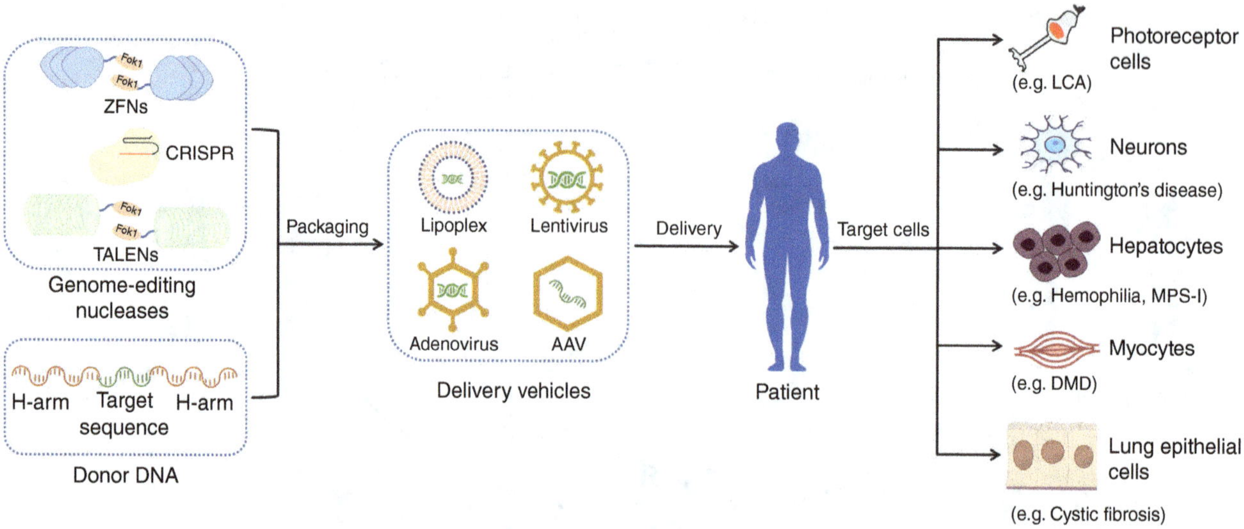

Figure 17.3 **In vivo genome-editing approach for various human diseases.** In this approach, gene-editing nucleases (ZFNs/TALENs/CRISPR-Cas) and donor DNA (target sequence) with appropriate homology arms (H-arm) are packaged in suitable delivery vehicles (viral vectors/nanoparticles). The components are delivered to the patient to achieve *in vivo* genome editing at the target site in the patient genome within the desired tissues.

β-hemoglobinopathies. SCD is the result of production of dysfunctional hemoglobin (HbS) protein, while BT is the result of insufficient production of functional β-globin (HbA) protein. More than 300 000 children are born with SCD every year worldwide, mostly in Africa and India (Thein and Thein 2016). Around 60 000 children are born with BT every year worldwide (Galanello and Origa 2010).

Current treatments for β-hemoglobinopathies are blood transfusion and hydroxyurea (HU), Volexator, crizanlizumab, and L-glutamine. Based on severity, patients get a combination of treatments, regular healthy blood transfusions along with iron chelating drugs to cope with the anemic phenotype and iron-related complications due to regular blood transfusions. HU, a Food and Drug Administration (FDA)-approved drug (for SCD), is primarily being used to induce fetal hemoglobin expression in SCD patients to alleviate the anemic phenotype. Volexator is an anti-sickling drug while crizanlizumab is a monoclonal antibody that blocks P-selectin on the wall of blood vessels leading to no adhesion of sickle RBCs on the wall of blood vessel preventing Vaso-ocular crisis (VOC). L-glutamine helps in reducing oxidative stress in sickle RBCs leading to relief in the pain caused by sickle RBCs. However, there is no single drug available for the BT. Until now, the only curative therapy for both the above diseases is BMT/allogenic HSCT with transient but prolonged immunodeficient state of the patient post transplantation. Success of this approach depends on highly compatible human leukocyte antigen (HLA)-matched donor and their availability is still a substantial challenge. Hence, there is still a huge unmet medical need to develop alternate therapeutic strategies for patients with β-hemoglobinopathies. In this context, gene therapy is another promising and potential option where patient's own HSCs are corrected/edited using genome-editing tools to produce healthy hemoglobin and transplanted back to the same patient; autologous transplantation. This approach of autologous transplantation of corrected HSCs eliminates the need of allogenic transplantation.

There are two ways to develop genome editing-based therapeutic strategies for β-hemoglobinopathies, which are as follows: (i) Genetic correction of the *HBB* gene and (ii) Increasing γ-globin expression in patient-derived hematopoietic stem cells.

17.2.1.1 Genetic Correction of the *HBB* Gene

As the HBB gene possesses several disease-causing mutations, the genetic correction of this gene is the first and obvious approach for genome editing-based therapy for β-hemoglobinopathies. In this approach, Sickle Cell mutation in patient-derived HSCs is corrected by donor DNA-based homology-directed repair (HDR) *ex vivo*, with the help of targeted endonucleases. The corrected HSCs are expanded and transplanted back to the same patients (Figure 17.2). This autologous transplantation of genetically corrected HSCs eliminates the need of allogenic transplantation. ZFNs were the first nucleases to be used for gene editing in clinical trials. Some successful attempts have been reported by various groups of scientists to develop an effective strategy to correct SCD mutation in HSPCs. First successful correction of SCD mutation in patient-derived HSPCs using ZFNs was demonstrated by Kohn and colleagues (Hoban et al. 2015). In this study, the corrected HSPCs maintained their ability to engraft non-obese diabetic (NOD)/SCID/gamma mice (NSG mice) and showed multi-lineage differentiation potential. Subsequently, Porteus and colleagues showed efficient correction of SCD mutation in patient-derived HSPCs using CRISPR/Cas9, sgRNA ribonucleoprotein, and rAAV6 Homologous recombination (HR) donor delivery (Dever et al. 2016). This was the first report of correcting SCD mutation using the CRISPR/Cas9 system. The corrected HSPCs successfully differentiated into erythrocytes expressing normal β-globin (HbA) levels. The above gene editing studies in HSCs have performed on the use of viral vector donors for HDR. However, DeWitt et al have demonstrated the selection-free HBB mutation correction in SCD patient-derived HSPCs using optimized design and delivery of ribonucleoprotein (RNP) complex consisting of SpCas9 and sgRNA, along with single-stranded DNA oligonucleotide (ssODN). Recently, high efficiency HBB correction (24.5 ± 7.6%) was achieved in patient-derived HSPCs using high-fidelity SpyCas9 with reduced off-target activity. The corrected HSPCs were transplanted into NSG mice and reduced sickle cells were observed 16 weeks post transplantation (DeWitt et al. 2016).

BT is caused by the number of mutations in the HBB gene, which is the reason it is cumbersome to develop a mutation correction approach for each BT mutation. The situation gets worse in case of patients having two compound heterozygous mutations. However, very recently, Xu et al. have successfully targeted IVS-110 G > A and IVS2-654C > T using CRISPR/Cas9 and CRISPR Cas9/Cpf1, respectively, in BT patient's HSPCs (Xu et al. 2019b). Multiple studies have demonstrated the genetic correction of HBB gene mutations in SCD and BT patient-derived iPSCs. Nevertheless, the major bottleneck in the iPSC field is generating mature to definitive and adult CD34+ HSPCs *in vitro* from patient-specific iPSCs for cell and gene therapy. However, till date no attempt has been made to correct the BT mutation in patient-derived HSPCS using HDR-based genetic correction approach.

17.2.1.2 Increasing γ-globin Expression

The production of γ-globin subunit leads to the synthesis of fetal hemoglobin (HbF) in early life (fetal stage), but is silenced and replaced by the production of β-globin resulting in the synthesis of adult hemoglobin (HbA). Usually, production of γ-globin is very nominal in adults but there are some healthy individuals with higher HbF levels and this condition is called hereditary persistence of fetal hemoglobin (HPFH). Interestingly, observations suggest that the patients of SCD and BT show alleviation in clinical manifestation if there is coinheritance with heterozygous for HPFH. HPFH occurs due to certain mutations that lead to an increased expression of the γ-globin genes: *HBG1* and *HBG2*. These mutations disrupt either expression of γ-globin repressor genes or their binding sites and thereby help in reactivation of HbF. Most of the reported HPFH mutations have been basically classified into two subtypes; (i) Non-deletional HPFH mutations are single nucleotide polymorphisms (SNPs) found upstream of *HBG1/2* genes. (ii) Deletional HPFH mutations are the deletions of a wide range (13–106 kb), mostly found in the β-globin gene cluster at chromosome 11 encompassing γ-globin repressor binding sites and competitive genes (*HBD* and *HBB*). Hence, generating HPFH or HPFH-like genotype to reactivate HbF in patient-derived HSCs using genome-editing tools has enormous potential to be a promising curative treatment for β-hemoglobinopathies.

Genome-wide association studies (GWAS) identified a major loci strongly associated with HbF expression, which is present in a region of chromosome 2 within the BCL11A gene (Basak and Sankaran 2016). In the past several years, multiple orthogonal lines of evidence have corroborated BCL11A, a C2H2-type zinc finger containing transcription factor, as a crucial regulator of switching from fetal-to-adult-hemoglobin and as a promising therapeutic target for the induction of HbF in HSPCs. Furthermore, studies have shown that disruption of erythroid-specific enhancer of the BCL11A gene locus by gene-editing nucleases led to enhanced expression of HbF. Bauer and colleagues targeted +58 *BCL11A* erythroid enhancer to disrupt GATA1 binding site by creating NHEJ using selection-free Cas9 RNP in SCD/BT patient-derived HSPCs. The edited HSPCs-derived erythroblast showed a reduction in BCL11A expression and showed 32.2% HbF reactivation. Reduced sickling in edited SCD-HSC-derived erythroblasts and restored globin chain balance with efficient erythropoiesis were observed in edited BT-HSC-derived erythroblasts (Wu et al. 2019). The same group also targeted +58 BCL11A erythroid enhancer using another genome-editing tool, base editor A3A (N57Q) BE3 in plerixafor-mobilized SCD-HSC and BT-HSC. They found similar promising results as in the case of Cas9-mediated editing. Subsequently, Tans and colleagues have shown targeted disruption of exon 2 and GATAA motif of *BCL11A* using ZFNs in human bone marrow (BM)-derived HSPCs. Exon 2-edited HSPCs demonstrated significantly reduced engraftment potential in NSG mice while GATAA motif-edited HSPCs achieved robust long-term engraftment and the derived erythroid cells showed reactivation of HbF when chimeric BM was cultured *ex vivo* (Chang et al. 2017). Furthermore, targeted disrupted *BCL11A* gene in mobilized β-thalassemia major HSPCs using ZFNs was shown. The study showed that the HSPCs can be successfully edited post expansion for engraftment. *BCL11A* disruption in mobilized BT major

HSPCs resulted in restored globin chain balance, improved phenotype, survival advantage, and mature erythroid cells (Psatha et al. 2018). The above reports convincingly prove that reducing BCL11A expression is a viable approach for the treatment of β-hemoglobinopathies.

BCL11A protein binding sites in the β-globin gene cluster seem to be another potential target for *HBG* reactivation. Weiss and colleagues targeted the BCL11A binding sites at the upstream of *HBG1/2* genes in SCD patient-derived HSPCs using Cas9 RNP. The edited HSPCs were engrafted in NSG mice with multi-lineage potency and the derived erythroblasts were analyzed 16 weeks post engraftment and observed reactivation of HbF (Métais et al. 2019). The same group tried recapitulating a HPFH-associated 13-nucleotide deletion mutation, which is one of the most effective binding sites for HBG repressors. These 13-nucleotides were deleted using Cas9 in SCD patient-derived HSPCs. The edited HSPCs-derived erythroblasts were found to have sufficient levels of HbF to inhibit pathological hypoxia-induced RBC morphology (Traxler et al. 2016). Miccio and colleagues targeted the disruption of BCL11A and LRF binding site at the upstream of *HBG1/2* genes in SCD patient-derived HSPCs using Cas9 and the edited cells were engrafted into NSG mice and reactivated HbF with corrected SCD phenotype was observed 16 weeks post engraftment (Weber et al. 2020).

Along with these repressor-centered targets, naturally occurring large deletional HPFH mutations too are drawing scientists' attention to be used for β-hemoglobinopathies treatment. Annarita Miccio's group successfully recapitulated 13.6 kb HPFH deletion encompassing *HBD* and some parts of *HBB* gene using Cas9. Deletion/inversion of this mutation in SCD patient-derived HSPCs resulted in significant reactivation of HbF synthesis and amelioration of the SCD phenotype in edited SCD-HSPCs-derived erythroblasts (Antoniani et al. 2018).

Based on the above promising studies, recently various pharma/biotech companies have started multiple phase I clinical trials aiming to enhance the production of fetal hemoglobin by editing erythroid lineage-specific enhancer of the BCL11A in patient-derived HSPCs using either ZFNs or CRISPR/Cas9 (Table 17.1). The phase 1/2 clinical trial by Sangamo Therapeutics and Bioverative recruited 6 transfusion-dependent BT (TDT) patients, targeting BCL11A in HSPCs using ZFNs to enhance fetal hemoglobin (NCT03432364). Similarly, Bioverativ, a Sanofi Company, initiated a clinical trial, in which they recruited 8 adult SCD patients wherein BCL11A was targeted in HSPCs using ZFNs to enhance fetal hemoglobin (NCT03653247). Subsequently, Vertex Pharmaceuticals and CRISPR Therapeutics initiated phase 1/2 clinical trials for SCD (NCT03745287) and BT (NCT03655678) where 45 severe SCD and 45 transfusion-dependent BT patients are being recruited. The patient subjects will receive the autologous CD34+ HSPCs modified with CRISPR/Ca9 at the BCL11A gene. Currently, all the above studies are evaluating the safety and efficacy of the genome editing-based cell replacement therapy for the treatment of SCD and BT. Recently, Vertex and CRISPR therapeutics have released the initial outcome of their SCD study (NCT03745287) and shown that the subject receiving the treatment has not reported any VOC in the last nine months and 2 TDT patients have been free of peripheral RBC transfusion for more than 14 and 3 months, respectively (Corbacioglu et al. 2020).

For the last few years, scientists were trying to develop strategies with the potential of clinical implementation. Considering nuclease-mediated genome editing, introducing small or large deletions in the human genome by NHEJ is relatively more efficient than HDR-mediated DNA repair. In the case of BT, there are more than 200 mutations in the *HBB* gene and developing therapeutic strategies for each of them would be a cumbersome process. Therefore, reactivation of γ-globin strategy holds an enormous potential in genome editing-mediated therapies for β-hemoglobinopathies.

17.2.2 Hemophilia

Hemophilia is an X-linked monogenic disorder which occurs due to deficiency in coagulation factors in the intrinsic coagulation cascade (Furie and Furie 1988; Giannelli and Green 1996). Hemophilia is broadly categorized into two types namely Hemophilia A and Hemophilia B based on the site of genetic mutation. Hemophilia A is a more prevalent form of Hemophilia and occurs in 1 in 5000 live male births due to a mutation in the gene coding for factor VIII (FVIII), resulting in the loss of functional FVIII protein. FVIII is an integral cofactor for the serine protease factor IX (FIX), which is found deficient in patients with Hemophilia B. Clinically, Hemophilia A and hemophilia B patients are usually categorized into three groups based on residual coagulation factor activity: severe (<1%), moderate (1–5%), and mild (5–40%). Patients who suffer from Hemophilia who go untreated are at a high risk for either morbidity or mortality from spontaneous or trauma-induced bleeds.

The existing therapy for Hemophilia includes prophylactic administration of exogenous coagulation factors derived from pooled plasma or recombinant protein (Mariani et al. 2003; Di Michele 2011). The short biological half-lives of FVIII and FIX proteins ascertain regular infusions (two to three times per week) for maintenance of threshold levels (above 1%), which minimally alleviates the clinical

Table 17.1 List of clinical trials for various human diseases using genome-editing tools.

Disease		Study title	Editing tool	Sponsor	Status	NCT number
Blood disorders	Sickle cell disease	A study to assess the safety, tolerability, and efficacy of BIVV003 for autologous hematopoietic stem cell transplantation in patients with severe sickle cell disease (PRECIZN-1)	ZFNs	Bioverativ Therapeutics Inc.	Recruiting; I/II	NCT0365 3247
		A safety and efficacy study evaluating CTX001 in subjects with severe sickle cell disease	CRISPR-SpCas9	Vertex Pharmaceuticals Incorporated	Recruiting; I/II	NCT0374 5287
	β-thalassemia	A study to assess the safety, tolerability, and efficacy of ST-400 for treatment of transfusion-dependent beta-thalassemia (TDT)	ZFNs	Sangamo Therapeutics and Bioverativ Therapeutics Inc.	Active; I/II	NCT0343 2364
		A safety and efficacy study evaluating CTX001 in subjects with transfusion-dependent β-thalassemia	CRISPR-SpCas9	Vertex Pharmaceuticals Incorporated	Recruiting; I/II	NCT0365 5678
	β-thalassemia and severe sickle cell disease	A long-term follow-up study in subjects who received CTX001	CRISPR-SpCas9	Vertex Pharmaceuticals Incorporated	Enrolling by invitation; Follow-up	NCT04208 529
	Hemophilia B	Ascending dose study of genome editing by zinc finger nuclease therapeutic SB-FIX in subjects with severe hemophilia	ZFNs	Sangamo Therapeutics	Active; I	NCT0269 5160
Immunodeficiency disorder	HIV/AIDS	Autologous T-cells genetically modified at the CCR5 gene by zinc finger nucleases SB-728 for HIV (zinc-finger)	ZFNs	University of Pennsylvania and Sangamo Therapeutics	Completed; I	NCT0084 2634
		Study of autologous T-cells genetically modified at the CCR5 gene by zinc finger nucleases in HIV-infected subjects	ZFNs	Sangamo Therapeutics	Completed; I/II	NCT0125 2641
		Phase 1 dose escalation study of autologous T-cells genetically modified at the CCR5 gene by zinc finger nucleases in HIV-infected patients	ZFNs	Sangamo Therapeutics	Completed; I	NCT0104 4654
		Dose escalation study of cyclophosphamide in HIV-infected subjects on HAART receiving SB-728-T	ZFNs	Sangamo Therapeutics	Completed; I/II	NCT0154 3152
		Repeat doses of SB-728mR-T after cyclophosphamide conditioning in HIV-infected subjects on HAART	ZFNs	Sangamo Therapeutics	Completed; I/II	NCT0222 5665
		A phase I study of T-cells genetically modified at the CCR5 gene by zinc finger nucleases SB-728mR in HIV-infected patients	ZFNs	University of Pennsylvania	Completed; I	NCT0238 8594
		Safety study of zinc finger nuclease CCR5-modified hematopoietic stem/progenitor cells in HIV-1-infected patients	ZFNs	City of Hope Medical Centre, Sangamo Therapeutics, California Institute for Regenerative Medicine (CIRM)	Active; I	NCT0250 0849

(*Continued*)

Table 17.1 (Continued)

Disease		Study title	Editing tool	Sponsor	Status	NCT number
Immunodeficiency disorder	HIV/AIDS	CCR5-modified CD4+ T cells for HIV infection (TRAILBLAZER)—T-cell reinfusion after interfering with lymphocyte binding location of AIDS virus through zinc-finger-nuclease elimination of CCR5 receptors	ZFNs	Case Western Reserve University	Recruiting; I/II	NCT0366 6871
		A pilot study of T cells genetically modified by zinc finger nucleases SB-728mR and CD4 chimeric antigen receptor in HIV-infected subjects	ZFNs	University of Pennsylvania	Recruiting; I	NCT0361 7198
		Long-term follow-up of HIV subjects exposed to SB-728-T or SB-728mR-T	ZFNs	Sangamo Therapeutics	Enrolling by invitation; Follow-up	NCT0420 1782
		Safety of transplantation of CRISPR CCR5-modified CD34+ cells in HIV-infected subjects with hematological malignancies	CRISPR-SpCas9	Affiliated hospital to Academy of Military Medical Sciences	Recruiting; I/II	NCT0316 4135
Inborn errors	MPS-I	Ascending dose study of genome editing by the ZFN therapeutic SB-318 in subjects with MPSI	ZFNs	Sangamo Therapeutics	Active not recruiting; I/II	NCT0270 2115
	MPS-II	Ascending dose study of genome editing by the Zinc Finger Nuclease (ZFN) therapeutic SB-913 in subjects with MPSII (CHAMPIONS)	ZFNs	Sangamo Therapeutics	Active; I/II	NCT0304 1324
Ocular disease	LCA10	Single ascending dose study in participants with LCA10 (BRILLIANCE)	CRISPR-SaCas9	Allergan and editas medicine	Recruiting; I/II	NCT0387 2479

Source: The above details were taken from Clinicaltrials.gov, available at: <https://clinicaltrials.gov> [Accessed 17 July 2020].

symptoms. Due to the high treatment costs, frequencies of infusions, and potential health burden, there is a need for alternative, more cost-effective therapies for Hemophilia. Such alternative strategies are especially pertinent for patients in developing countries with comorbidities and mortalities resulting from inadequate factor supply.

Hemophilia is ideal for gene therapy approaches as the clinical symptoms like bleeding phenotype is responsive to a wide range of factors, and precise regulation is not necessary (Perrin et al. 2019) allowing for development of multiple strategies. Majorly, adeno-associated viral (AAV) vectors have made significant contribution in disease treatment with hepatic *in vivo* gene transfer that have progressed to some success in preclinical and clinical studies, with some clinical studies being in phase 3 for both Hemophilia A and B enrolled patients. Furthermore, there are other potential gene targets such as the clotting factor proteins which are secreted into the circulation, it is possible to correct the bleeding diathesis with gene delivery to a fraction of total hepatocytes. These regulators include FVIII and FIX that can be synthesized in non-native cells and tissues. For example, even though FVIII is naturally secreted by specialized endothelial cells such as LSEC and extrahepatic endothelial cells, expression in hepatocytes generates a functional protein that has restored hemostasis in animal models and human patients.

More studies are employing gene-editing nucleases for the correction of Hemophilia mutations. One of the first studies that showed the possibility of genome editing as a viable therapeutic strategy to treat Hemophilia was by Li et al. 2011. They showed that ZFNs are able to induce DSBs efficiently when delivered directly to mouse liver and that, when co-delivered with an appropriately designed gene-targeting vector, they can stimulate gene replacement through both homology-directed and homology-independent targeted gene insertion at the ZFN-specified locus. They have demonstrated that the level of gene targeting achieved was sufficient to correct the prolonged clotting times in a mouse model of hemophilia B, and remained persistent after induced liver regeneration.

Subsequently, Park et al. used a TALEN pair to invert a 140-kbp chromosomal segment that spans the portion of the F8 gene in a human iPSC line to create a Hemophilia A model cell line. They then reverted the inverted segment back to its normal orientation in the Hemophilia model iPSCs using the same TALEN pair (Park et al. 2014). Similar to the previous study, they showed the potential of gene editing nucleases in gene therapy applications by showing that F8 mRNA is expressed in cells derived from the reverted iPSCs lines, but not in those derived from the clones with the inverted segment. Another study used the CRISPR/Cas9 approach to revert these chromosomal segments back to the WT phenotype. Upon differentiation, endothelial cells from corrected iPSCs expressed the F8 gene and functionally rescued factor VIII deficiency in an otherwise lethal mouse model of Hemophilia was shown (Zhang et al. 2019).

The most advanced approach, however, to genome editing in Hemophilia is usage of ZFNs. Sangamo Therapeutics currently has two clinical trials running for Hemophilia B, one is an ascending dose study of genome editing by SB-FIX in Subjects With severe Hemophilia B and a Dose-Ranging Study of Recombinant AAV2/6 Human Factor 8 Gene Therapy SB-525 in Subjects with severe Hemophilia A (Figure 17.3). In the case of Hemophilia B study (NCT02695160), pair of ZFN and the correction gene (FIX) was intravenously delivered using AAV vectors, which inserts a correct copy of the Factor 9 gene into the albumin locus in hepatocytes. This genome editing strategy aims for a lifelong therapeutic production of the Factor IX clotting factor. In 2018, Sangamo announced treatment of the first patient in the Phase 1/2 clinical trial evaluating SB-FIX therapeutic strategy against Hemophilia B (Table 17.1).

The SB-525 clinical study (NCT03061201) developed as part of collaboration agreement between Sangamo Therapeutics and Pfizer for Hemophilia A, on the other hand uses a recombinant AAV2/6 vector encoding the cDNA for the B-domain deleted human F8 (hF8). The SB-525 vector encodes a liver-specific promoter module and AAV2/6 exhibits liver tropism, thus providing the potential for long-term hepatic production of FVIII in Hemophilia A subjects. The constant production of FVIII after a single SB-525 administration may provide potential benefit in durable protection against bleeding and the complications thereof without lifelong repetitive IV factor replacement administration.

Recently, updated results of the study were released by Sangamo and Pfizer; the patients showed a dose-dependent increase in FVIII levels and a dose-dependent reduction in the use of FVIII replacement therapy across the dose cohorts. Both firms are now planning to advance SB-525 to a registrational study, and Pfizer will take responsibility for SB-525 late-stage development and manufacturing. A process has been initiated to transfer SB-525 manufacturing process from Sangamo to Pfizer. Although gene therapy has paved a way for therapeutics for diseases such as Hemophilia, genome-editing strategies are really driving the field forward into a new era of therapy solutions.

Keeping into account the serious adverse effects in clinical trials which utilized genome-integrating viral vectors, a study did systemic evaluation of potential genotoxicity of ZFN-mediated transgenesis using different techniques such as deep sequencing of predicted off-target sites, copy number analysis, whole-genome sequencing, and

RNA-seq in primary human umbilical cord-lining epithelial cells (CLECs) with ZFN-mediated FVIII transgene integration at AAVS1 safe-harbor locus. They showed a low frequency of ZFN-associated indels, and no detectable off-target transgene integrations or chromosomal rearrangements in the above cells, thus reaffirming the safety and efficacy of ZFN-mediated F8 integration. Further research has also happened around uses of iPSCs as a disease model system and in one such study *in vitro* data use a transcription activator-like effector nuclease strategy, demonstrating correction of the intron-22 inversion in Hemophilia A iPSCs (Sivalingam et al. 2016).

17.2.3 Human Immunodeficiency Virus

Human immunodeficiency virus (HIV) belongs to genus lentivirus that targets the immune system of the host. There are two types of HIV namely HIV-1 and HIV-2 categorized based on the genetic characteristics and viral load differences. Among them, HIV-1 is the most common and pathogenic and spreads across the globe. On the contrary, HIV-2 is less pathogenic and is restricted to West Africa (Nyamweya et al. 2013). According to WHO, nearly 37.9 million people were living with HIV/AIDS worldwide by the end of 2018.

The entry of HIV in the host cells is mediated by CD4 receptors present on T-lymphocytes. Further, C-X-C chemokine receptor type 4 (CXCR4) or C-C chemokine receptor 5 (CCR5) are co-receptors facilitating HIV-1 entry through membrane fusion. Both the receptors, CCR5 and CXCR4, belong to the G-protein coupled receptors (GPCRs) superfamily and govern HIV tropism. The viral strain majorly utilizing CCR5 as co-receptors are R5 tropic strains and those utilizing CXCR4 are X4 tropic strains. There are some strains that utilize both the receptors and are termed as dual tropic viruses. It has been observed that the R5 strains are responsible for initial infection (Lopalco 2010) and subsequently they can develop dual tropism with both CCR5 and CXCR4 receptors. At a later stage, the virus predominantly interacts with CXCR4 which are majorly expressed on T-lymphocytes. This interaction leads to progressive loss of CD4+ cells and thus deteriorating the immune system of the host.

The initial infection with HIV might result in influenza-like symptoms after three to four weeks of contraction. However, at this stage, people might not be aware about the infection as they do not feel sick right away. Subsequently, the virus can either actively replicate and give symptoms or enter into a latent state. When in latent state, the virus is present in the body, not dividing but can be transmitted to others. This latency period can differ from individual to individual with no defined timeframe. However, once the virus is in the active phase, it deteriorates the immune system to such an extent that even an opportunistic pathogen can infect that otherwise could be defended by the body's immune system. Further, people have been shown to develop cancer, tuberculosis, and other secondary infections due to the deteriorated immune system. This leads to the last stage of HIV infection called Acquired immunodeficiency syndrome (AIDS) (Naif 2013).

Currently, HIV-infected patients are treated with Antiretroviral therapy (ART). The main aim of ART is to target the major viral characteristics such as high replication rate, low reverse transcriptase fidelity, and the recombination capability that leads to diverse HIV infection. There are various antiretroviral agents approved by the FDA such as those targeting reverse transcriptase, protease, CCR5 antagonists, integrase inhibitors, and others (Simon et al. 2006). Combination of these antiretroviral agents can reduce the viral replication and thus can prevent HIV infection or if not prevent, it can at least reduce the progression. This combinatorial approach is known as highly active antiretroviral therapy (HAART). Nonetheless, HAART has its own disadvantages, for example, the most common is the emergence of drug resistance (Deeks 2003). The other major concern associated with HAART is that the antiretroviral agents to date do not target integrated provirus from target cells and affected patients should take continuous and lifelong drug treatment due to the latency of HIV.

17.2.3.1 Genome Editing Using CCR5 as Target

Gene therapy and genome-editing approaches for HIV came into existence largely due to finding a small subset of population across the world that naturally harbors 32 bp deletion in CCR5 gene (CCR5Δ32). It creates premature stop codon and thereby resulting in the truncated, non-functional protein. Individuals with CCR5Δ32 seem to be permanently resistant to HIV infection (Liu et al. 1996; Samson et al. 1996). Based on the above observation, an HIV-1-infected patient, Timothy Brown, received HSCT from the donor having homozygous CCR5Δ32 deletion. Timothy Brown or so-called "Berlin patient" became the first person to be cured from HIV-1 infection. It has raised excitement, enormous hope, and curiosity in the field. This led to the emergence of gene editing technologies for HIV-1 infection treatment (Hütter et al. 2009) and prompted various preliminary studies in this direction (Li et al. 2013; DiGiusto et al. 2016). Based on the above findings, a targeted genome-editing approach has been applied to phenotypically mimic CCR5 mutation in human cells to cure HIV infection. The first proof-of-concept on CCR5 genome editing was established by Perez et al. in 2008 via delivering adenoviral vectors carrying both ZFNs into primary human CD4+ T cells (Perez et al. 2008) and subsequently in human CD34+ HSPCs by Holt et al. in 2010 (Holt et al. 2010). Holt et al. have shown that CCR5-depleted mice develop resistance against R5 tropic strains of HIV infection. In this study, CCR5 was disrupted using ZFNs with the mean modification efficiency of 17% in CD34+ HSCs. Mouse studies have suggested that modified HSCs when transplanted demonstrate a selective advantage from HIV infection (Holt et al. 2010).

Eventually, the joint effort of Sangamo Therapeutics and University of Pennsylvania led to the first gene editing nuclease-based clinical trial in humans (NCT00842634) that started in 2009. In this study, CD 4+ T-cells from 12 patients were edited *ex-vivo* using adenoviral vectors carrying ZFNs (SB-728) targeting CCR5 gene and 5–10 billion ZFNs-edited CD4+ T cells were reinfused back to the participated patient (Figure 17.2) (Tebas et al. 2014). This study demonstrated the efficacy and safety of ZFNs-based approach and thus giving rise to possibilities for future trials (16). Subsequent clinical trials with CCR5 disruption aimed at increasing efficiency by varying input dose (NCT01044654), multiple infusions of edited cells (NCT02225665). Another clinical trial was done using an increasing concentration of cyclophosphamide to augment engraftment along with variable infusions of ZFNs-modified CD4+ T cells in order to improve edited T-cell engraftment (NCT01543152). Further, keeping safety measures in mind, delivery methods shifted from viral-based systems to electroporation (NCT02388594) as the latter offer various advantages (Maggio and Gonçalves 2015).

TALENs have been used for CCR5 ablation against HIV-1 infection. There are various *in vitro* studies that suggest that efficiency and the cytotoxicity of TALENs is less as compared with that of ZFNs (Mussolino et al. 2011, 2014; Shi et al. 2017). However, there are till now no clinical trials started using TALENs-modified cells for HIV. With the discovery of CRISPR, researchers have gained interest in exploring CRISPR-modified HSPCs for HIV resistance as various studies have suggested the effectiveness in terms of editing efficiency and the delivery of CRISPR is better as compared with ZFNs and TALENs. Various *in vitro* studies have demonstrated efficient ablation of CCR5 using CRISPR/Cas9 as compared with its predecessor gene editing technologies. Currently, there is one ongoing clinical trial wherein CRISPR-edited CD34+ cells are transplanted in patients with hematological malignancy (Xu et al. 2017, 2019a) (NCT03164135). All the HIV clinical trials, completed and ongoing, are mentioned in Table 17.1.

In 2019, Adam Castillejo ("London patient") became the second person to be cured from HIV. Castillejo underwent transplantation for his leukemia by receiving CD34+ HSPC from CCR5Δ32 mutant donor. This recent announcement of London patient cure provided a reassuring proof of concept that genome editing of CCR5 has valuable potential for HIV cure (Mandavilli 2019).

Recently, a Chinese scientist Jiankui He used the CRISPR approach to edit CCR5 in human embryos to produce HIV-resistant babies, Lulu and Nana (Li et al. 2019). He presented his research in the second International Summit on human genome editing held at Hong Kong in 2018. His research was criticized by most of the scientific community around the world violating various standard regulation and ethical procedures. The research outcome of He's work was not peer reviewed and is not yet published. Genome editing is a powerful technique where one needs to be careful about what is at stake and what are the consequences for the same. Apart from that, Jiankui and his team did not explore genome-wide off-target cleavage effects and the health consequences associated with it.

17.2.3.2 Genome Editing Using CXCR4 as Target

CXCR4 signaling plays various roles including thymic differentiation, bone marrow retention of HSCs, and progenitor cell movement. Therefore, the use of CXCR4 as a target for gene editing in HSCs is debatable. Homozygous mutation in CXCR4 has proved to be embryonic lethal in mice model (Ma et al. 1998). On the contrary, one report suggests that T-cell-specific CXCR4 depletion in mice is immunologically tolerant (Chung et al. 2010). There are several studies suggesting editing of CXCR4 confers resistance against X4 tropic strains *in vitro* (Hou et al. 2015; Wang et al. 2017). For instance, Yuan et al. has shown ZFNs-based disruption of CXCR4 is relatively effective against HIV-1 infection. In this study, a comparison between shRNA suppression of CXCR4 and ZFNs disruption of CXCR4 was studied. Both *in vitro* and *in vivo* data suggest ZFNs-based strategy conferred better resistance against HIV infection (Yuan et al. 2012).

To date, there is no ongoing clinical trial with edited CXCR4. It is necessary to evaluate all the possible outcomes of CXCR4 disruption before going ahead with clinical trials. In view of the critical role of CXCR4, the majority of studies related to HIV-1 resistance targeting CXCR4 has been restricted to *in vitro* level and could not advance to the clinical level. However, a study by Liu S et al. might have potential at the clinical level. In the study, CRISPR/Cas9 approach has been used in combination with PiggyBac transposon system to replace wild-type CXCR4 with CXCR4 P191A (Liu et al. 2018). Previously, one report has suggested that mutation P191A, present on the second extracellular loop, hinders binding of HIV. Further, normal functions of CXCR4 are not disrupted *in vivo* (Tian et al. 2005).

As mentioned previously, the study conducted by Holt et al. demonstrated CCR5 disruption provides selective advantage against HIV infection. However, their group has also demonstrated that mice transplanted with ZFNs-modified HSCs are not resistant from X4 strains using CXCR4 as a co-receptor for infection (Holt et al. 2010). Since CCR5 and CXCR4 both play an important role in HIV-1 entry, Didigu et al. attempted to disrupt both the co-receptors using ZFNs. Both *in vitro* and mice study suggest that dual disruption of co-receptors might possess potential against HIV infection (Didigu et al. 2014). Another study by Yu et al. was conducted by using CRISPR/Cas9 to disrupt CXCR4 and CCR5 in CD4+ T cells. Their results suggested that CRISPR-modified cells were resistant from both R5 and X4 tropic strains (Yu et al. 2018).

Overall, the current preclinical studies and ongoing clinical trial suggests immense potential of genome editing for HIV therapy. Currently, ZFNs are the most used genome editing nuclease that is being used in the ongoing Phase I/II clinical trial for HIV infection. However, one of the major limitations with genome editing is off-target cleavage that needs to be addressed and therefore optimization of the nuclease efficiency and safety is required in order to precisely edit the genome. Hence, it is very early to realize the potential of genome editing in the clinical setting.

17.3 Neurodegenerative Diseases

Neurodegenerative diseases (NDs) are a group of heterogeneous, age-related disorders characterized by progressive loss of functions of neurons in the central and the peripheral nervous system. PD and Alzheimer's disease (AD) particularly have high rates of cognitive impairment, morbidity, and mortality among the elderly group of the population. Some of the other NDs are frontotemporal dementia, multiple system atrophy (MSA), repeat expansion disorders, Huntington's disease (HD), Fragile X Syndrome, and Spinocerebellar ataxia (SCAs). Each of these disorders has a varied epidemiology, overlapping clinical symptoms and mutated functional genes. Despite being a heterogenous group of disorders, they have one common hallmark of pathogenesis, accumulation of disease-specific protein leading to insoluble aggregates possibly causing neuronal dysfunction accompanied by death. Developing therapeutic approaches for neurodegenerative disease is quite challenging with traditional pharmacological drugs because of the hurdles such as the complexity of the nervous system, physical barriers such as the blood–brain barrier that limit the distribution of the drug to the brain when peripherally administered. Therefore, researchers are now looking at alternative strategies such as gene therapy (copy of wild-type gene or copy of gene encoding for nerve growth factor to compensate for nerve loss) and genetic correction of the mutated gene to potentially treat these disorders.

17.3.1 Parkinson's Disease

PD is a heterogeneous, age-related, progressive neurodegenerative and the most common movement disorder affecting the central nervous system of the body. The knowledge of etiology in the majority of the cases remains elusive. Thus, leading to a postulation of multifactorial theories of both genetic and environmental involvement (Maiti et al. 2017). It affects 1–2 per 1000 of the population at any given time (Tysnes and Storstein 2017). Extensive research in finding the causative gene has shown mutations in alpha-synuclein (*SCNA)* and Leucine-rich repeat kinase 2 (*LRRK*) responsible for autosomal-dominant PD whereas mutations in other genes like parkin, *GBA*, ATPase132A2, phosphatase, and tensin-induced homolog kinase 1 are associated with autosomal recessive PD (Hernandez et al. 2016). It is pathologically characterized by the presence of alpha-synuclein containing Lewy bodies in the substantia nigra of the brain. This leads to the loss of dopamine-producing neurons in the substantia nigra of the brain which results in reduced facilitatory and voluntary movements. As the disease progresses, the alpha synuclein accumulation expands from that area to the other parts of the brain which then results in manifestation of advanced clinical symptoms like frequent uncontrolled tremors, bradykinesia, rigidity, and postural instability (Triarhou 2020). Confirmatory diagnosis of PD can be done by autopsy of midbrain showing Lewy bodies. Lack of molecular understanding of the disease mechanism has been a hurdle in finding an effective cure for the disease.

The progression of the disease can usually be delayed by compensating the deficiency of dopamine in the patients with anti-Parkinson drugs; Levodopa (dopamine precursor), COMT inhibitors and MAO-B inhibitors (enzyme degrading dopamine), Dopamine agonists. However, prolonged usage of these drugs has shown to have direct consequences on the development of involuntary movements and behavioral changes. However, there is a need for more promising therapeutics which has encouraged researchers to develop robust therapeutics using genetic factors of the disease.

Gene therapy-based therapeutic approaches have shown positive outcomes at preclinical stages in *in-vitro* PD models and this has taken forward to clinical trials. All of these clinical trials have primarily focused on single-dose administration as one-stop solution. To enlist few, Voyager therapeutics (RESTORE-1), Neurocrine Biosciences (VY-AADC01 NCT01973543), a phase-1 clinical trial in which aromatic-l-amino acid decarboxylase (AADC) gene is administered locally in PD patients. Similarly, Oxford BioMedica is running a clinical trial at phase 1–2 with a drug named Prosavin, Glial cell line-derived neurotrophic factor (*GDNF*) (NCT04167540), Glutamate decarboxylase (GAD) (NCT00643890), and Neutrin gene (NCT00252850).

17.3.1.1 iPSCs Cell Therapy

17.3.1.1.1 Healthy Donor Cell Therapy Recently, an iPSCs cell therapy-based approach was initiated in 2018 to achieve permanent cure for PD patients. A team of clinical researchers led by Takahashi from Kyoto University in Japan have started a first human clinical trial (UMIN000033564) of transplantation of iPSCs-generated dopaminergic neurons into PD patients (Takahashi and Sawamoto 2018). The donor iPSCs are derived from the donor with a matching HLA and was additionally tested in rat models (Takahashi 2019) before transplanting them into seven volunteered PD patients. These patients will be monitored for two years. First patient in this clinical trial had not shown any sign of tumorigenesis indicating the safety and efficacy of the treatment (Fan et al. 2020) and the group is yet to publish the results of the study.

17.3.1.1.2 Autologous Genetically Corrected Cell Therapy Limitations in conventional gene transfer approaches such as inability of target-specific expression of transgene in certain CNS cell type, the efficacy of vector delivery in order to obtain adequate expression of transgene to reach therapeutic effect has led the research community to focus on gene-editing approach. Several proof-of-concept preclinical studies have shown using gene-editing nucleases and pluripotent stem cells for PD. Genetic correction of *LRRK2* mutation G2019S harboring in PD patient-derived iPSC using ZFNs has successfully rescued the disease phenotype in differentiated dopaminergic neurons (Reinhardt et al. 2013). Similarly, another research group led by Harushia Inoue from Kyoto University has shown *in vitro* correction of gene variants [6-pyruvoyltetrahydropterin synthase *(PTPS)* and Dihydropteridine reductase (*DHPR*)] involved in BH4 metabolism using CRISPR/Cas9 in patient-derived iPSC. BH4 is a cofactor of tyrosine hydroxylase, a rate-limiting enzyme in dopamine biosynthesis (Ishikawa et al. 2016). These preclinical studies using patient-derived iPSCs and transplantation of genetically corrected autologous differentiated neurons holds a great promise as a better therapeutic approach.

17.3.2 Alzheimer's Disease

AD is the most common cause of dementia affecting approximately 24 million people worldwide (Reitz et al. 2011). Typically, AD is characterized by insidious onset and gradual progression of memory loss with loss of other cognitive functions due to death of cholinergic neurons in the brain. Mutations in presenilin genes *PSEN1* and *PSEN2* and amyloid precursor protein (APP) are known for early onset of autosomal dominant familial AD and presence of specific allele variants of Apolipoprotein E4 (*ApoE*) gene, with epsilon-4 being at significantly higher risk, known for late onset of sporadic AD (Bekris et al. 2010). Familial AD is characterized by the presence of amyloid plaques caused due to the formation of insoluble aggregates of amyloid beta protein and sporadic AD is characterized by the formation of neurofibrillary tangles of hyperphosphorylated Tau protein (Murphy and LeVine 2010). Thus far there is no cure and neither the etiology of the disease is completely understood.

Currently, drugs like acetylcholinesterase inhibitors and *N*-methyl-D-aspartate receptor agonists targeting both amyloid and tau protein aggregate formation are used in managing the cognitive problems. The clinical benefits from their use are small and they have not clearly shown a delay or halt in the progression. Unavailability of cure and inefficiency of the above drugs in managing the disease progression has led to finding out an alternative therapeutic approach.

Recently, gene therapy-based clinical trials have been initiated at different academic institutions and pharma/biotech companies to achieve permanent cure for AD. Weil medical college of Cornell University (NCT03634007), University of California (NCT00017940), Sangamo Therapeutics (NCT00087789), and Libella gene therapeutics (NCT04133454) are currently running clinical trials using AAV-based gene therapy vectors as a therapeutic approach in AD patients.

17.3.2.1 CRISPR/Cas9 Correction in Patient-Derived iPSCs

Keeping in mind the hurdles of gene therapy, scientists have also explored genome-editing approaches. Preclinical studies have successfully shown correction of A79V mutation in *PSEN1* gene in AD patient-derived iPSC using CRISPR/Cas9 (Pires et al. 2016). Similarly, correction of N141I mutation in *PSEN2* (Sproul A et al.) and L150P in *PSEN1* gene using CRISPR/Cas9 in patient-derived iPSCs have shown reduction in APP aggregates in differentiated neurons (Ortiz-Virumbrales et al. 2017). The Swedish mutation in APP allele was successfully knockout and has shown reduction in protein aggregate formation (György et al. 2018). Zhang lab has made use of CRISPR-directed base editors to correct the APOE4 mutation, the most common mutation responsible for sporadic and late onset of AD in the population. RNA Editing for Specific U to C Exchange (RESCUE), a cytidine-base editor, was used in converting CC to UU in APOE4 mRNA which will then translated for the APOE2 variant known to be non-pathological for AD (Li et al. 2020).

17.3.3 Huntington's Disease

HD is a progressive fatal, autosomal dominant neurodegenerative disorder caused due to the expansion of CAG repeats in exon 1 of the huntingtin gene (*HTT*) (Warby et al. 2009). These repeats in the mutant HTT (mHTT) protein exhibit a toxic gain of function mutation and are responsible for the formation of insoluble protein aggregates of huntingtin protein resulting in neuronal death in the striatum regions of the brain. The onset of the disease is directly correlated to the number CAG/polyQ repeats. Repeats of more than 36 are considered as a threshold for pathological hallmarks (Reiner et al. 2011). HD patients suffer from progressive motor dysfunction, psychological problems, and cognitive decline.

There is no cure for the disease but severity of the symptoms can be minimized and the disease progression can be delayed with the help of physical therapy, occupational therapy, and speech therapy. Tetrabenazine is an approved drug for HD to alleviate chorea symptoms. Psychiatric symptoms can be managed with selective serotonin reuptake inhibitors. As all of these approaches have shown modest results in managing the disease, alternative therapeutics approaches have been explored. Preclinical studies in animal models done by Martin *et.al* have shown that RNAi (miHTT) can mitigate the protein aggregates of mHTT and alleviates the motor and neuropathological symptoms in the HD mouse model. The success in the preclinical experiments have led to the viral vector-mediated clinical trial (NCT04120493) by the name of AMT-130, consisting of rAAV-miHTT at Phase 1-2 stage which has started in 2019.

17.3.3.1 Haplotype-Specific Approaches of Inactivating Dominant Mutant Allele

A study done in mice showed depletion of HTT in the adult does not affect normal growth, survival, and neuronal viability (Yang et al. 2017). Hence, different research groups have used genome-editing approach to inactivate the endogenous HTT. CRISPR/Cas9 was used to selectively and permanently inactivate only mutant allele of HTT gene in HD patient-derived fibroblast cells (Shin et al. 2016). They have demonstrated an *in vitro* strategy of inactivating the mutant allele by using haplotype-specific dual sgRNA at target sites in HD. By using CRISPR/Cas9-guided dual sgRNA that depends on PAM sites generated by SNP alleles on mutant chromosome showed selective excision of ~44 kb of DNA including promoter, transcription start site, and CAG repeats resulting in complete inactivation of the mutant allele. Subsequently, a study in MSCs derived from HD mice model has also shown a similar *in vitro* approach using CRISPR/Cas9 disruption of mHTT gene (Dunbar et al. 2019). Using two plasmids targeting two different sites of the HTT gene, one at the upstream of the open reading frame and another at the exon 1 intron boundary, thus disrupting the mHTT gene. Several preclinical studies have demonstrated the feasibility of genome editing in neurodegenerative disease models both *ex vivo* and *in vivo* mouse models (Figure 17.3) (Monteys et al. 2017). However, still a major bottleneck is the delivery of genome-editing reagents or edited cells into the specific tissue.

17.3.4 Cystic Fibrosis

Cystic fibrosis (CF) is a monogenic autosomal recessive disorder caused by mutations in the cystic fibrosis transmembrane conductance regulator (*CFTR*) gene which encodes a chloride/bicarbonate channel expressed on the apical surface of the secretory cells. Lack of CFTR activity results in multi-organ dysfunction and ultimately mortality from respiratory sequelae. Almost 2000 mutations in the CFTR gene have been found to cause CF; they decrease the flow of Cl- and HCO3- through the epithelia of multiple organs, including the lung, pancreas, sweat glands, vas deferens, liver, and intestine. As a consequence, they interfere with their normal functioning. In the respiratory tracts, the lack of a CFTR function drives the accumulation of abnormally thick and sticky mucus that underlies chronic lung inflammation and recurrent bacterial infections, leading to progressive lung degradation. CF-associated lung disease is responsible for >95% of CF deaths (Davis et al. 1996; Stoltz et al. 2015). CF is the most prevalent and destructive inherited disorder among Caucasian population with around 80 000–100 000 affected worldwide; prompting efforts with research in the CF domain and better therapy development being targeted at minimizing lung disease.

For more than seven decades, therapies for CF focused on treatment of the symptoms rather than the origin of the disease which is the CFTR protein dysfunction. However, since 2011, a number of small-molecule therapies known as potentiators and correctors have been developed and licensed that can functionally correct at least 33 different CF-causing mutations, accounting for a significant percentage of the individuals with CF. However, this still means 90% of disease-causing variants cannot be corrected by any available small-molecule therapies.

Gene therapy is particularly attractive for monogenic diseases such as CF. Despite the fact that CF is a multi-organ disease, improving respiratory manifestations will lead to a significant improvement in the patient's quality of life and may be associated with a decrease in mortality. The inhaled route is the easiest way to access the targeted abnormal zones. One therapeutic approach that has been extensively studied over the last 25 years is the use of the CFTR cDNA, delivered directly to the lung, to complement the genetic defect in *CFTR*. This approach could potentially be curative, at least for the CF lung disease, in a manner that is independent of the disease-causing variants. However, after more than 20 clinical trials, the only clinical benefit reported so far has been a stabilization of lung function in patients who received at least nine doses of CFTR cDNA delivered by cationic liposomes over a 12-month period.

Although gene therapy carries promise, it has several limitations. First, finding the appropriate plasmid DNA molecule model is important in terms of clinical potency (Pringle et al. 2009; Dhand 2017). Second, natural barriers such as mucus, versatile immune responses, and intracellular limitations considerably impair gene transfer into the lungs (Osman et al. 2018). Finally, because the airway epithelium is constantly renewing, genetic therapies necessitate repeated administration. Nevertheless, a number of studies have shown that precise editing of the genome can be more efficient than cDNA addition approach at least in cellular and animal models, which are discussed below.

ZFNs were first used for genetic correction of the most common CFTR mutation ΔF508 in patient-derived epithelial cells. HDR-mediated targeted insertion of a 4.5 kb genomic donor fragment harboring CFTR exons 10-24 was achieved with the efficiency of <1%. Subsequently, CRISPR/Cas9 was used to develop a potential therapy for CF that was published in 2013. Schwank et al. (2013) tested the recovery of a functional CFTR protein in intestinal organoids obtained from CF pediatric patients carrying the ΔF508 mutation. After lipofectamine-mediated transfection of the intestinal stem cells, CRISPR/Cas9 gene editing repaired the mutation at the CFTR locus through the CFTR gene substitution by HDR-mediated approach, improving the forskolin-induced response and prompting organoid swelling. The same group suggested colonic transplantation of genetically corrected organoids as a probable perspective (Yui et al. 2012). Another successful delivery of the CRISPR/Cas9 system was reported by Bellec et al. (2015). They knocked-down the CFTR gene in HAE cells and Calu-3 cells using the CRISPR/Cas9 approach delivered with HIV-1 lentivirus. CRISPR/Cas9-directed gene modification was associated with a decline of transepithelial Cl- secretion and a decrease in response to a CFTR inhibitor, as measured in polarized cells using chambers (Figure 17.3).

The biggest obstacle to overcome in regenerative medicine for CF is to determine the relevant cells that will be capable of repairing a defect. The desired cell should be non-immunogenic, patient specific, easy to proliferate, and easy to modify (Guo et al. 2017). Because endogenous progenitor cells are difficult to recognize and insufficient in their quantity and activity, focus has been directed on exogenous cell sources (McQualter et al. 2010; Rock and Königshoff 2012). Indeed, repairing a CFTR mutation has been demonstrated in human iPSC cells by the means of CRISPR/Cas9. Firth et al. (2015) have used CRISPR technology and a piggyBac-based donor to obtain footprint-free gene correction at the CFTR locus in patient-derived iPSCs. The recovery of expression and function of CFTR, as measured by patch-clamp technique, in lung epithelial cells that were differentiated from edited iPSCs was demonstrated (Firth et al. 2015). Though iPSC-based cell therapy approach is promising, there are some safety concerns such as cell grafts, including the transmission of possible genetic abnormalities and tumor risk, hampering potential clinical therapeutic purposes, that need to be addressed.

17.4 Duchenne Muscular Dystrophy

Duchenne muscular dystrophy (DMD) is the most severe form of muscular dystrophies, among more than 800 monogenic disorders that lead to skeletal muscle dysfunction. It is caused by mutation in the X-linked dystrophin gene which is approximately 2–3 mega bases long. Dystrophin functions in muscle as a mechanical link to tether the actin cytoskeleton to the inner surface of the sarcolemma. Lack of dystrophin in skeletal muscle results in fragility of the sarcolemma, impaired intracellular signaling, myocyte necrosis, inflammatory infiltration, and, eventually, replacement of muscle with fibrotic and fatty tissue. Skeletal muscle degeneration leads to skeletal deformities, respiratory insufficiency, cardiomyopathies and in some cases cognitive impairment. DMD is a X-linked disorder, hence it is more prominent in males which is around 1 : 5000 men worldwide (Long et al. 2015). DMD is diagnosed in early years of life due to muscle weakness which can begin as early as two to three years, first affecting the proximal limbs then the distal limbs.

During DMD progression, satellite cells, an injury-responsive stem cell population beneath the muscle basal lamina, becomes activated and fuses with damaged myofibers. Ultimately, this regenerative cell population becomes depleted and muscle degeneration ensues. In contrast, the adult heart lacks a robust stem cell population, so the absence of dystrophin in the heart ultimately results in myocyte dysfunction, consequent loss of pump function, and fatal cardiomyopathy (Gao and McNally 2015; Chang et al. 2016).

There is no permanent cure to treat DMD presently. But certain drugs are administered to reduce its symptoms. Corticosteroids are small molecules resembling human hormones called cortisol's and help in reducing inflammation. Corticosteroids have a greater and longer lasting effect on maintaining muscle strength in DMD patients. They also suppress the immune system from targeting the muscle cells. Emflaza and Prednisone are the drugs usually prescribed for muscular dystrophies. Sarepta Therapeutics' Eteplirsen, a morpholino antisense oligo, was the first therapy to be approved for DMD treatment in September 2016. The drug helps in splicing out the mutation hence leading to enhanced DMD production. Ataluren is another drug focusing on treating DMD by aiding the protein machinery to skip premature stop codons in the patients. The drug has been approved by European regulators but is yet to be approved by the FDA (Flavell 2020).

The bottleneck in genetically correcting DMD is its diversity of mutations, spanning around 3000 variants. Roughly two-third of the mutations involve exon deletions and disrupt the dystrophin open reading frame. Also, it is difficult to correct the complete dystrophin gene due to its sheer size. Therefore, people used naked DNA constructs as intramuscular injections. But the effect was short lived and expression was region specific. Researchers later shifted to using AAV and HIV constructs for administration (Figure 17.3). The disadvantage of HIV is its integrative nature whereas AAV has a 4.5 kb intake capacity (Mollanoori et al. 2021). Hence, the challenge is to develop a therapy that would be applicable to a larger cohort of patients than single individuals.

17.4.1 Approaches to Enhance the Dystrophin Levels Using Genome Editing

The lack of a viable long-term therapy for DMD pushes the focus of research toward gene editing. There are certain advantages pertaining to the usage of gene editing in DMD. First, it is possible to delete the exons and restore ORF in the rod domain of dystrophin. Second, as the dystrophin gene is present on the X chromosome, males possess only a single copy of the affected gene. Hence, a monoallelic correction is enough to restore the function in males. Third, it is observed that a minor increase in dystrophin levels is enough to achieve therapeutic benefits. Gene-editing nuclease-mediated genome editing has shown promising results and restoration of dystrophin in DMD animal models and patient-derived iPSCs. Different strategies are used for the correction of dystrophin gene (Figure 17.4) as follows.

17.4.1.1 Exon Deletion

Exon deletion works by simultaneously cutting by two sgRNAs, leading to the loss of the intervening sequence. Exon deletion leads to a truncated protein but partially functional; hence, being to alleviate the symptoms. More than 50% of DMD mutations lie in the hotspot region between exon 45 and 55 of the dystrophin gene (Bladen et al. 2015). The advantage of targeting hotspot regions is that a single set of sgRNA can be used on a variety of patients. Exon deletion strategy can also be applied to correct duplication of a region, in this scenario only a single sgRNA is required to delete one copy of the duplicated region. Around 5% of the DMD mutations are duplicative in nature. Moretti et al. enhanced shortened dystrophin (DMDΔ51–52) level via deletion of exon 51 in a DMDΔ52 Pig. Excision of exon 51 restored dystrophin expression and restored skeletal myotube formation and abnormal Ca^{2+} handling in cardiomyocytes (Moretti et al. 2020).

17.4.1.2 Exon Skipping and NHEJ-Derived Reframing

This method allows to skip the mutated part and create a BMD-like phenotype which is less severe in nature (Wilson et al. 2017). The disadvantage of presently available antisense oligonucleotide therapy is that it targets mRNA and hence is short lived. Gene editing can provide a one-time solution as it leads to a permanent change at the genetic level. sgRNAs are designed to target the splice acceptor site or the splice donor site, this allows the skipping of the targeted exon. Studies have been done to skip exon 50 by disrupting its splice acceptor site, leading to functional fusion of codon 49 and 51. As only single sgRNA is used, the efficiency of this system is more than exon deletion (Amoasii et al. 2018).

NHEJ-derived reframing works on the concept of addition or deletion of nucleotides to correct the reading frame. In this method also a single sgRNA is required. As indels generated are small, hence this approach tends to preserve most of the targeted dystrophin region. Deletion of exon 44 generates a premature stop codon in exon 45. Hence, Zhang et al. designed a sgRNA targeting the splice acceptor site in exon 45 to correct out of frame mutation in exon 44 in a ΔEx44 DMD mice (Zhang et al. 2020).

17.4.1.3 HDR-Mediated Repairing

HDR allows precise correction of the gene. Unlike gene therapy which focuses primarily on the usage of exogenous

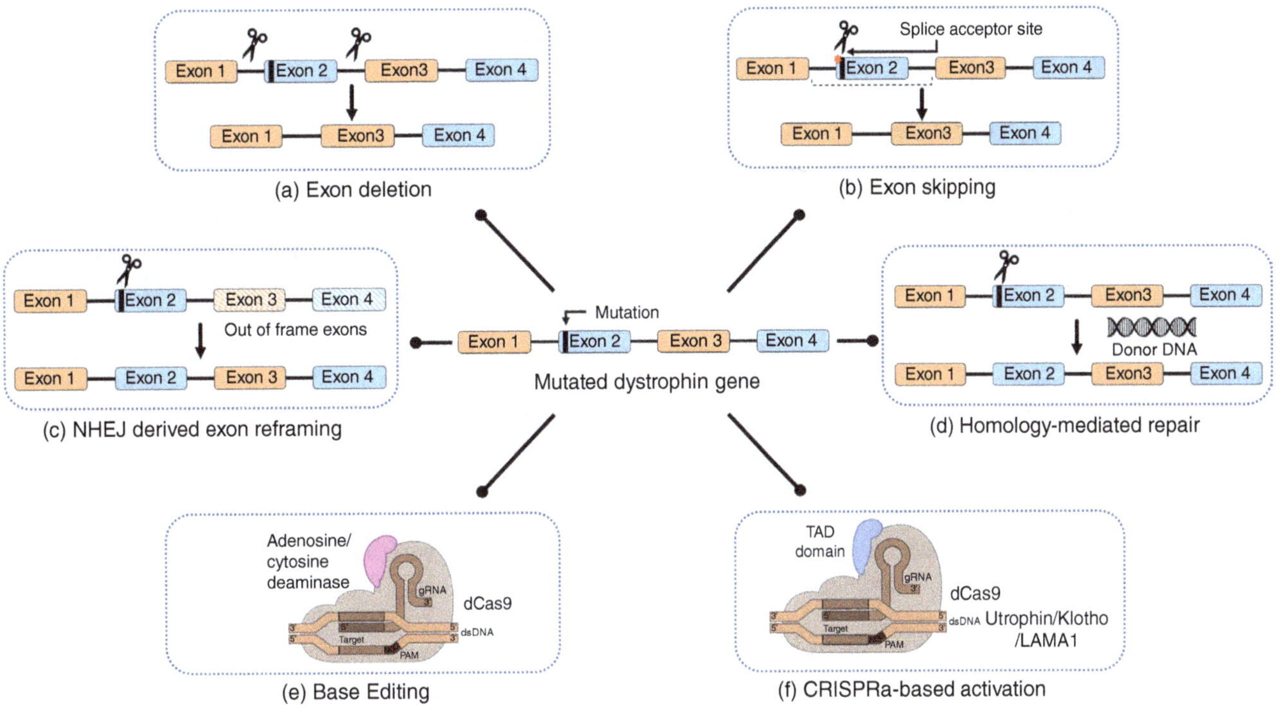

Figure 17.4 **Genome-editing approaches to enhance dystrophin levels.** DMD mutations are spread widely across the dystrophin gene, hence hypothetical exon numbers are used to explain the following approaches: (a) Mutated exon is deleted using a set of two sgRNAs, (b) Skipping of mutated exon via deletion of splice acceptor/donor site, (c) Repositioning of out of frame exon via NHEJ-mediated repair, (d) HDR-mediated repair of the mutation via homologous recombination using a donor DNA, (e) CRISPR-Base editor system is used to fix appropriate single nucleotide changes, (f) CRISPRa system is employed to enhance compensatory genes via the topologically associated domain (TAD).

genes; hence, leaving the nonfunctional gene behind. The disadvantage of using HDR is that it occurs at a lower frequency than NHEJ, it is a post mitotic event; hence, more prominent in dividing cells and it is difficult to ensure delivery of all the constituents namely Cas9, sgRNA, and donor DNA to the same cell. Also, there is a limitation to the length of the DNA that can be used for HDR. Bengtsson et al. used mdx^{4cv} mice for establishing a HDR-mediated repair to correct small mutations in exon 53 harboring essential domains of dystrophin which cannot be corrected using INDELs or by excising out the exon (Bengtsson et al. 2017).

17.4.1.4 Base Editing

Base editors work by conversion of a C-G to T-A or A-T to G-C. This is achieved by using a cytidine deaminase or adenosine deaminase (ADA) fused to a dead Cas9. An estimated 25–30% mutations in DMD are point mutations (Aartsma-Rus et al. 2006; Min et al. 2019). The advantage of base editors lies in the fact that they do not make any cut in the DNA. Ryu et al. used an adenine base editor (ABE) to correct a stop codon mutation in exon 20 of a DMD mice. As ABE and its sgRNA could not be packaged in a single AAV vector due to size constraints, they were packaged in a trans splicing AAV (tsAAV) vector. tsAAV vector allows heterodimerization of the packaged protein post transfection in the cell via ITR sequences. Dystrophin expression was enhanced up to 17% in myofibers. Restoration of 4% of the normal dystrophin expression is enough to improve muscle function (Ryu et al. 2018).

17.4.1.5 CRISPRa

CRISPRa is a technique to use Cas9 for the activation of a gene. This is achieved by the fusion of Cas9 with transcription activation domains such as VP16, VP64, or VP160 (Mali et al. 2013). CRISPRa has been used to reduce the progression of DMD by activating compensatory genes such as Utrophin, Klotho, and Laminin subunit alpha 1 (LAMA1). CRISPRa tends to have very less off-target effects and has the potential to be a single treatment for various types of DMD mutations. Liao et al. used CRISPR/Cas9 transgene activator system to enhance the expression of Klotho and Utrophin to ameliorate DMD phenotype. Klotho is found to be epigenetically silenced in muscle cells of mdx mice, whereas utrophin produces a protein having high similarity to dystrophin and has been previously reported to alleviate DMD symptoms in mice (Liao et al. 2017).

17.4.2 Using iPSCs Toward DMD Treatment

Since the discovery of iPSCs, one of the most important applications of iPSC has been in disease modeling, with low availability of samples and usage of invasive techniques; iPSCs surpass these issues and help in drug screening, studying the disease and testing cell-based therapies.

One of the first studies of generating iPSC and transplantation in DMD mice was carried out by Darabi et al. which led to the successful engraftment and improved contraction of the targeted muscle (Darabi et al. 2011). Maffioletti et al. generated an artificial 3D model system from iPSCs of human skeletal muscle tissues which was engrafted in immunodeficient mice. This model system will help in better understanding the DMD pathology (Maffioletti et al. 2018).

The strategies used to correct mutations *in vivo* have also been used to correct mutations in iPSCs. Ousterout et al. used single and multiple sgRNAs to correct DMD mutations in the exon 45–55 by generating shifts in the exon or by deleting multiple exons. They also transplanted these iPSC-derived edited myoblasts in immunodeficient mice (Ousterout et al. 2015). Li et al. used patient-derived iPSCs to correct DMD mutations via three strategies; exon skipping, frameshifting, and exon knock in using CRISPR and TALEN gene editing system. They also differentiated these iPSCs to skeletal muscle cells and found enhanced expression, exon knock in approach was found to be the most effective (Li et al. 2015).

With the growing research toward DMD treatment using gene editing tools and iPSCs reaching preclinical stages, there is a good hope that gene editing for the treatment of DMD may soon reach the clinical stages as well.

17.5 Ocular Diseases

Eye being an immune privileged site presents as a perfect organ for gene manipulations. Also, the cells which are usually targeted, i.e. retinal cells and photoreceptors, are terminally differentiated cells. Hence, the effect of gene editing tends to stay lifelong. The other advantage being that the effect of the treatment can be easily measured by using noninvasive ophthalmic tools.

Leber's congenital amaurosis (LCA) is an autosomal recessive disorder and the primary cause of blindness at childhood and people generally suffer with severe visual impairment since infancy. The disease has an occurrence of around 2–3 cases per 100 000 (Leber congenital amaurosis 2020). It can be caused due to a mutation in at least 14 different genes, giving rise to different subtypes of LCA. The genes involved are primarily essential for ocular development and functioning of the retina. Some of the most common mutations are found in the CEP290, CRB1, GUCY2D, and RPE65 genes. LCA10 is caused by mutations in CEP290 accounting for about 30% of all the LCA cases (den Hollander et al. 2006). Like dystrophin, CEP290 is a very large gene with 54 exons and 7440 bp in ORF.

To overcome this hurdle and to find a lifelong cure, AGN-151587 (EDIT-101) was developed. AGN-151587 (EDIT-101) is an experimental CRISPR-based medicine developed by US-based Allergan, a pharmaceutical company and Editas medicine, a leading genome editing company in the United States (Maeder et al. 2019). The medicine is the world's first *in vivo* genome-editing study for eye disease using the CRISPR tool and is presently in phase 1/2 of the clinical trial which is also termed as the BRILLIANCE clinical trial (NCT03872479) (Table 17.1). CEP290 with an IVS26 mutation is being treated using a set of two sgRNAs, targeting the flanking regions of the mutation. The CRISPR system is being administered via a subretinal injection and targeted using an AAV5 vector (Figure 17.3) (CRISPR treatment inserted directly into the body for first time 2020). The trial is expected to be completed by 2024.

Many preclinical studies are being done pertaining to the usage of iPSC-derived cells for regenerative medicine. But the first-ever iPSC-related clinical trial was carried out in an attempt to treat an age-related macular disorder (AMD) in 2013. Sheets of retinal pigment epithelial (RPE) cells differentiated from iPSCs were transplanted into the eyes of an AMD patient. The patient's vision improved with no safety-related concerns. However, two mutations were found in the RPE cells of the recipient patient which were not present in the donor fibroblast cells used for iPSC reprogramming (Garber 2015).

Unedited cells can be effective in treating age-related degeneration but for genetic diseases, gene editing-mediated correction is essential for the engrafted cells. Bassuk et al. were the first to precisely correct c.3070G > T mutation in the Retinitis Pigmentosa GTPase Regulator (RPGR) which leads to X-linked retinitis pigmentosa (XLRP) using a CRISPR/Cas9 construct in iPSC. The repair was achieved using a donor homology template (Bassuk et al. 2016).

Eye is a complex organ and a single cell-type lineage cannot be relied on to study disease pathology. Hence, Deng et al. generated retinal organoids from three Retinitis pigmentosa patients with different frameshift mutations in the RPGR gene. It was then observed that there is a significant defect in the photoreceptors in terms of morphology, localization, transcriptional profiling, and electrophysiological activity. These defects were restored to normal phenotype on CRISPR/Cas9-mediated gene editing of the defective gene (Deng et al. 2018).

17.6 Inborn Errors of Metabolism

Inborn errors of metabolism (IEMs) are a diverse group of congenital, mostly inherited disorders which are caused due to mutation in genes encoding for a metabolic enzyme or a transport protein playing a role in any biochemical or metabolic pathway. IEMs are commonly caused due to mutations in a single gene involved in carbohydrates or proteins or lipid metabolism. IEMs are individually rare but their collective occurrences are common. Number of people affected in a population varies across the globe. In one study done by British Columbia, Canada reported that IEMs collectively occur at a frequency of 1 in 2500 births (Applegarth et al. 2000). Another study done in Australian population reported the incidence rate of 15.7 in 100 000 births (Wilcken et al. 2003) whereas in the United Kingdom, 1 in 784 births was observed (Sanderson et al. 2006). They are mostly inherited as an autosomal recessive trait.

They can be broadly categorized on the basis of metabolites accumulated, disrupted canonical metabolic pathways, or organelle that is affected. Disorders involving metabolic pathways to organelle, cytoplasm, or to intracellular transport can be common in different types of cell and different organ systems. These disorders can be broadly classified into three subgroups: (i) accumulation of toxic metabolites due to inhibition in intermediary steps of anabolic or catabolic metabolic pathway of biomolecules (amino acid metabolism defect, urea cycle defect, lysosomal storage disorders, peroxisomal diseases), (ii) mito chondrial enzymes involved in energy homeostasis (mitochondrial disorders, fatty acid oxidation disorders), (iii) biochemical reactions involved in absorption, transport, and pathways (methionine or lysine malabsorption syndrome, disorders of amino acid or carbohydrate transport, disorders of autophagy) (Ferreira et al. 2018; Guerrero et al. 2018).

IEMs involve varying symptoms from appearing physically normal to having a distinctive dysmorphic physical feature. Depending upon the particular disease, they could vary from affecting a single organ or multi system such as visceral organs, immunological system, endocrine system to varying range of complicated clinical symptoms like involving both visceral and neurological symptoms, septicemia, and failure to thrive. Age of onset varies from infancy to adolescence with the most severe form appearing in early childhood accompanied by significant morbidity and mortality (Agana et al. 2018).

Traditional therapies of IEMs aim at diet restrictions, supplementing deficient enzymes, preventing the formation of toxic products with help of recombinant enzyme replacement therapy (ERT), and prevention of toxic products formation with the help of substrate reduction therapy (Camp et al. 2012; Cagas 1976, Yue et al. 2019). Both involve regular intravenous transfusions of FDA-approved drugs which requires skilled medical staff. In addition, these treatment options could be of least help in alleviating neurological symptoms, as these recombinant enzymes are unable to cross the blood–brain barrier (Fukao and Nakamura 2019). Other options in which the pharmaceutical sector is focusing on is the screening of small molecule chaperones which can increase enzyme activity and can also easily cross the blood–brain barrier (DeWeerdt 2016).

In very severe cases, organ transplantation has also been used (Corrigendum for Hudgins et al. 2017). Research focus is also toward HSCT, in which patient HSCs are isolated from the patient (Burlina et al. 2014; Tan et al. 2019). With the help of gene editing, the wild-type gene is introduced or the mutation is corrected, generating corrected patient HSCs which can then be transplanted back to the patient. But, the correction of mutation in the underlying gene has always been the ultimate therapeutic goal (Schneller et al. 2017). With the advent of site-specific endonucleases, the idea of permanent cure for IEMs as a potential therapeutic approach could be possible. Preclinical and clinical studies of gene editing as a therapeutic approach for various IEMs have been done. Here, we will enlist and briefly talk about each of them.

17.7 Lysosomal Storage Disorders (LSDs)

Lysosomes are well known as the recycling machinery of the cell, carrying out a sequence of biochemical reactions involved in the breaking of biomolecules into simpler metabolites which can be used to synthesize new biomolecules, thereby, maintaining the equilibrium inside the cell. Functional deficiency of enzymes at any such step would block the subsequent reactions resulting in the disruption of metabolic flux and accumulation of toxic substrate which ultimately results in lysosomal defects (Platt et al. 2018; Lysosomal Storage Disorders - NORD (National Organization for Rare Disorders) 2020).

17.7.1 Hurler Syndrome Mucopolysaccharidosis Type I (MPS-I)

It is a rare, autosomal recessive LSD which affects multiple organs. It is caused due to mutation in the alpha-L-iduronidase (*IDUA)* gene, responsible for the production of the enzyme involved in the breakdown of complex sugar molecules called glycosaminoglycans (GAGs), also known as mucopolysaccharides. Defect in the enzyme leads to the accumulation of GAGs in the lysosomes causing an obstacle in normal functioning of the lysosomes. Depending upon the severity of the clinical symptoms, it is divided into severe and attenuated types (Coutinho et al. 2012).

Currently, the disease can be managed by ERT of Aldurazyme which can manage the systemic symptoms. In order to achieve, long-term viable therapy, currently research is focused toward genome editing. Preclinical studies have successfully shown an *ex vivo* genome-editing approach of the *IDUA* gene using CRISPR/Cas9 into the safe harbor locus (CCR5) of patient-derived $CD34^+$ stem cells (Mathew Porteus et al.). Thus, demonstrating the possibility of genome editing with stem cell transplantation to treat MPS-I. Currently, Sangamo Therapeutics is running a ZFN-based genome editing phase 1/2 clinical trial (NCT02702115) (Table 17.1) with 3 participants of attenuated MPS-I. Evaluation of the safety and dose tolerability of a study drug SB-318 consisting of two ZFNs targeting the albumin locus and human *IDUA* gene packaged in AAV2/6 targeting liver would lead to the secretion of functional protein into blood (Figure 17.3).

17.7.2 Hunter Syndrome or Mucopolysaccharidosis Type II (MPS-II)

It is a rare, X-linked recessively inherited, lysosomal storage disorder which is caused due to mutation in iduronate 2-sulphatase (*IDS)* gene responsible for the production of the I2S enzyme involved in the breakdown of GAGs. Because of the enzyme deficiency, there is an accumulation of GAGs in the lysosome, leading to the lysosomal enlargement and ultimately lysosomal defects. Approved therapy for MPS-II is ERT (Elaprase), but it is unable to alleviate neurological symptoms of the disease. First *in vivo* genome editing was initiated in 2017 for Hunter's syndrome using ZFNs. The trail's first patient, Brian Medax, an American citizen took a bold decision to take that risk. Sangamo Therapeutics' two ZFNs targeting the albumin locus and the correction gene (IDS) was delivered to him at UCSF children's hospital in a single intravenous infusion using AAV vectors (AAV2/6) that target the liver cells. Subsequently, Phase 1/2 clinical trial (NCT03041324) (Table 17.1) was started in 2017 by Sangamo Therapeutics using the drug SB-913, involving 9 participants. SB-913 (CHAMPIONS) is an intravenously delivered ZFNs therapeutic for gene editing. It inserts the correct copy of the *IDS* gene into albumin locus in hepatocytes with the goal of a lifelong therapeutic production of the IDS enzyme (Figure 17.3). Early results of the first *in vivo* genome-editing trial are encouraging (Coutinho et al. 2012).

17.8 Hereditary Tyrosinemia Type I (HT-1)

It is a fatal, congenital disorder in which the catabolic pathway of amino acid tyrosine is affected. HT-1 is caused due to the mutations in fumarylacetoacetate hydrolase (*FAH*) gene. Total or partial deficiency of FAH causes accumulation of fumarylacetoacetic acid in the liver cells resulting in severe liver damage (Chinsky et al. 2017; GARD 2020). Individuals affected with HT-1 show symptoms like failure to thrive and enlarged liver, kidney, and nerve damage. Currently, there is no cure but the disease can be managed by diet restrictions such as a low protein diet with the absence of tyrosine and administration of drug nitisinone which acts as a substrate reduction therapy. In severe cases where the above approaches are of no help, liver transplantation is the primary treatment.

For alternative better therapeutic approaches there are preclinical studies done using CRISPR/Cas9-based *in vivo* HDR-mediated genetic correction of FAH mutation in mouse model via tail vein injection (Yin et al. 2014). Another group (Yin et al. 2016) has similarly shown HDR-based genetic correction of FAH gene in which the guide RNA and the donor template were packaged in AAV8 and mRNA of SpCas9 was packaged in lipid nanoparticles to avoid off-target effects. This could alleviate the pathological symptoms and lead to complete restoration of the liver in one month. Alternative approach to treat HT-1 is via deleting an enzyme upstream of FAH, 4-hydroxypehnylpyruvate dioxygenase (*HPD*). Deletion of HPD along with FAH alleviates the HT-1 symptoms (Endo et al. 1990). This was demonstrated by Pankowicz *et al.* using CRISPR/Cas9 double knockout of FAH and HPD gene *in vivo* HT-1 mice model and was successful in restoring the liver after eight weeks of treatment (Pankowicz et al. 2018).

17.9 Ornithine Transcarbamylase Deficiency (OTCD)

OTCD is an inherited, X-linked recessive disorder, and the most common urea cycle disorder in humans. It is caused due to mutations in ornithine transcarbamylase enzyme

(*OTC*) which results in elevated levels of ammonia in the blood. OTCD patients have very little or no production of OTC enzyme, restoration of as little as 3% could alleviate the clinical symptoms. Severe OTCD could cause lethargy, ataxia, and death without rapid intervention. Current therapeutic approaches include a low protein diet and sodium benzoate as nitrogen scavenging agent. Gene therapy has been efficient at the preclinical stage and there are clinical trials going on but not yet approved. Gene editing as an alternative therapeutic approach has also been investigated and is at a preclinical trial stage in which Yang et al. has shown genetic correction using SaCas9 under liver specific promoter packaged in AAV8 and sgRNA and donor template in other AAV8 *in vivo* OTC mice model. Study showed that 10% of allele correction restores enzyme activity to 20 and 16% at 3rd and 8th week after treatment (Yang et al. 2016).

17.10 Primary Immune Deficiencies

Primary Immune deficiencies (PIDs) are a heterogenous group of rare genetic disorders with more than 300 mutations, resulting in patients having immune system dysfunction and underlying severe immune deficiencies resulting in high morbidity and mortality. PIDs can be treated by allogeneic HSC transplantation, but when histocompatibility leukocyte antigen-matched donors are lacking, this can be a high-risk procedure. Hence, correcting a patient's own HSCs with gene therapy and genome editing offers an attractive alternative.

X-linked severe combined immunodeficiency (X-SCID) is the most prevalent form of SCID and is caused by mutations in the *IL2RG* gene, which leads to a reduced production or absence of circulating NK and T cells, and functional impairment of the B cells. In the first trials of gene therapy for PIDs, retroviral vectors were used in which expression of the normal transgene was driven by the viral promoter LTR. With this approach, successful and durable T cell reconstitution was achieved in patients with X-linked SCID [X-SCID (Gaspar et al. 2011; Hacein-Bey-Abina et al. 2010), ADA deficiency (Aiuti et al. 2009; Blaese et al. 1995; Kohn et al. 1995), and Wiskott–Aldrich syndrome [WAS (Moratto et al. 2011). Unfortunately, several patients developed leukemia. These serious adverse events were caused by preferential integration of retroviral vectors in proximity of transcription initiation sites of the genes (including oncogenes) and by the strong enhancer activity of the viral LTR, leading to increased and deregulated expression of the targeted oncogenes (Boztug et al. 2010; Hacein-Bey-Abina et al. 2008).

To counter these adverse effects, much effort has gone in the development of safer viral vectors. A gene therapy trial to correct X-SCID using a self-inactivating retroviral (SIN-RV) vector, in which the U3 enhancer was deleted from the LTR and expression was driven by the weaker eukaryotic human elongation factor 1α (EF1α) short promoter, is currently underway in Europe and in the United States. Preliminary results from this trial show a similar kinetics of T cell recovery compared with that of the previous trial, but with significantly less integration within proto-oncogenes. Long-term safety effects remain to be studied (Hacein-Bey-Abina 2003). In addition to the safer SIN-RV vectors, lentiviral vectors are promising. *In vitro* (Braun et al. 2014) and *in vivo* studies (Montini et al. 2006) have demonstrated that lentiviral vectors integrate randomly in actively transcribed genes, without any preference for the transcription initiation sites and regulatory elements; thus, they are potentially safer (Montini et al. 2006). Therefore, several trials using SIN lentiviral (SIN-LV) vectors have been initiated (Montini et al. 2009; Biffi et al. 2011), including a treatment trial for WAS that shows promising results (Schiroli et al. 2017).

Recently, gene correction strategies have been successfully demonstrated in a study in which a mutated *IL2RG* gene was targeted for correction with ZFNs in HSCs obtained from a patient suffering from SCID-X1 (Genovese et al. 2014). First, HSCs were transduced using an integration-deficient lentivirus containing a HDR template encoding a therapeutic cDNA for IL2RG. Following transduction, cells were electroporated with mRNA encoding ZFNs targeting a mutational hotspot in *IL2RG* to stimulate HDR-mediated gene correction (Figure 17.2). To increase HDR frequency, culture conditions were optimized with small molecules to encourage HSC division. This strategy resulted in gene-corrected HSCs from the SCID-X1 patient being obtained in culture at therapeutically relevant rates. HSCs from unaffected individuals that underwent the same gene correction procedure could sustain long-term hematopoiesis in mice (Weissman and Shizuru 2008).

Successful differentiation of hiPSCs into T cells *in vitro* has been recently reported (Kennedy et al. 2012; Chang et al. 2014; Themeli et al. 2013), making it possible to test the ability of genome editing to restore T cell differentiation capacity of iPSCs from patients with SCID. However, *in vivo* use of gene-edited hiPSCs for correction of human PIDs is not yet ready for the clinic because of safety concerns related to the tumorigenic potential of hiPSCs and because of the difficulty in generating definitive HSCs from iPSCs. Nonetheless, these recent achievements with *in vitro* targeted differentiation of hiPSCs have a great value; they enable preclinical efficacy and safety studies of genome-editing approaches that may eventually be applied to hiPSCs.

17.11 Current Challenges in Therapeutic Genome Editing

For the last two decades, remarkable progress has been made in addressing the confrontations of conventional viral vector-based gene therapy by developing new genome-editing technologies for site-specific modification of the human genome. This has helped to conquer some of the limitations inherent to gene therapy methods. However, several challenges still need to be addressed to completely appreciate the potential of therapeutic genome editing for regenerative medicine. A number of universal challenges need to be addressed, which are tenacious issues largely relating to safety concerns of gene-editing nuclease associated off-target cleavage effects, editing efficiency at the desired site, human immune response, and delivery of genome-editing tools.

Off-target cleavage by endonucleases likely occurs at sites whose sequences differ slightly from the target site or at partial target sites within the genome. Such cleavage effects can cause undesired genetic modifications that may lead to functional impairment or enhancement of particular genes or pathways, cellular toxicity, and tumorigenesis, which could have adverse effects and obstruct the potential of therapeutic benefits. It was observed that CRISPR/Cas9-mediated DSBs induce large deletions and complex genomic rearrangements in mouse and human cells (Kosicki et al. 2018). Though several approaches have been developed to identify the whole genome off-target effects, methods to assess the accurate off-target effects are still lacking. Furthermore, several key questions endure about how the human immune system will respond to *in vivo* delivery of gene-editing nucleases. Recent studies have shown that a considerable percentage of humans may harbor a pre-existing adaptive immune response to Cas9 nucleases (Charlesworth et al. 2019). These data raise critical considerations that should be addressed as these therapies progress to clinical trials.

The efficiency of genome editing is mainly dependent on the cell cycle stage. Precise genetic correction of mutated DNA sequence uses HDR-based approach, which is appropriate for a large subset of disease-causing mutations. HDR is naturally a less efficient mode of DNA repair than NHEJ, especially in primary cells like HSPCs, neurons, hepatocytes, cardiomyocytes, and epithelial cells because HDR is restricted to the S and G2 phase of the cell cycle (Chandrasegaran and Carroll 2016). HDR requires a DNA donor template while NHEJ occurs throughout the cell cycle and does not require any exogenous DNA donor. Moreover, HDR-based genetic correction approaches may also expand and enrich the corrected cell population, usually limited to *ex vivo* strategies. Although *ex vivo* genome editing may be possible in diseases as those affecting hematopoietic systems, it would restrict applications in other diseases where transplantation is not an option.

The final obstacle pertains to the delivery of gene-editing nucleases and exogenous donor DNA template in the case of HDR-mediated genetic correction (Annaluru et al. 2015). A range of delivery vehicles such as nucleic acids, proteins, viral vectors, and nanoparticles are explored for *ex vivo* genome editing. However, viral vectors are the most promising delivery system for *in vivo* genome editing. Currently, both non-integrating and integrating viral vector systems have been explored for *in vivo* editing but the latter has safety concerns such as random integration and insertional mutagenesis. Non-integrating AAV vectors, with extensive range of serotypes, are promising candidates for both *ex-vivo* and *in-vivo* genome editing. However, small packaging capacity is a major limitation for large proteins such as various Cas9 and TALENs.

17.12 Conclusions

Therapeutic genome-editing was a dream of the nineties, but development of programmable nucleases has altered the definition of cell and gene therapy. Remarkable progress has been made toward therapeutic genome editing in stem cells and applications in regenerative medicine. Genome editing and stem cell engineering is accelerating expeditiously, with sustained focus on pioneering and improving strategies. There are strong reasons to believe that in the next decade, the lives of patients will be transformed by the use of either *ex vivo* or *in vivo* genome editing.

Acknowledgements

SR acknowledges Department of Biotechnology (DBT), Government of India for the Ramalingaswami Fellowship grant (BT/RLF/Re-entry/45/2013) and Science and Engineering and Research Board (SERB) for the Young Investigator grant award (ECR/2017/002212) and Council for Scientific and Industrial Research (CSIR) (HCP008). PG is a recipient of Junior Research Fellowship (JRF) from CSIR, Government of India. SG is a recipient of JRF from University Grants Commission, Government of India. PT is a recipient of JRF from DBT, Government of India. We would like to apologize to all the authors whose studies in this field we are unable to cite due to space limitations.

References

Aartsma-Rus, A., Van Deutekom, J., Fokkema, I. et al. (2006). Entries in the Leiden Duchenne muscular dystrophy mutation database: an overview of mutation types and paradoxical cases that confirm the reading-frame rule. *Muscle & Nerve* 34 (2): 135–144.

Agana, M., Frueh, J., Kamboj, M. et al. (2018). Common metabolic disorder (inborn errors of metabolism) concerns in primary care practice. *Annals of Translational Medicine* 6 (24): 469–469.

Aiuti, A., Cattaneo, F., Galimberti, S. et al. (2009). Gene therapy for immunodeficiency due to adenosine deaminase deficiency. *New England Journal of Medicine* 360 (5): 447–458.

Amoasii, L., Hildyard, J., Li, H. et al. (2018). Gene editing restores dystrophin expression in a canine model of Duchenne muscular dystrophy. *Science* 362 (6410): 86–91.

Annaluru, N., Ramalingam, S., and Chandrasegaran, S. (2015). Rewriting the blueprint of life by synthetic genomics and genome engineering. *Genome Biology* 16: 125. (Page number not provided).

Antoniani, C., Meneghini, V., Lattanzi, A. et al. (2018). Induction of fetal hemoglobin synthesis by CRISPR/Cas9-mediated editing of the human β-globin locus. *Blood* 131 (17): 1960–1973.

Applegarth, D., Toone, J., and Lowry, R. (2000). Incidence of inborn errors of metabolism in British Columbia, 1969–1996. *Pediatrics* 105 (1): e10–e10.

Basak, A. and Sankaran, V. (2016). Regulation of the fetal hemoglobin silencing factor BCL11A. *Annals of the New York Academy of Sciences* 1368 (1): 25–30.

Bassuk, A., Zheng, A., Li, Y. et al. (2016). Precision medicine: genetic repair of retinitis pigmentosa in patient-derived stem cells. *Scientific Reports* 6 (1).

Bekris, L., Yu, C., Bird, T., and Tsuang, D. (2010). Review article: genetics of Alzheimer disease. *Journal of Geriatric Psychiatry and Neurology* 23 (4): 213–227.

Bellec, J., Bacchetta, M., Losa, D. et al. (2015). CFTR inactivation by lentiviral vector-mediated RNA interference and CRISPR-Cas9 genome editing in human airway epithelial cells. *Current Gene Therapy* 15 (5): 447–459.

Bengtsson, N., Hall, J., Odom, G. et al. (2017). Muscle-specific CRISPR/Cas9 dystrophin gene editing ameliorates pathophysiology in a mouse model for Duchenne muscular dystrophy. *Nature Communications* 8 (1): 14454.

Biffi, A., Bartolomae, C., Cesana, D. et al. (2011). Lentiviral vector common integration sites in preclinical models and a clinical trial reflect a benign integration bias and not oncogenic selection. *Blood* 117 (20): 5332–5339.

Bladen, C., Salgado, D., Monges, S. et al. (2015). The TREAT-NMD DMD global database: analysis of more than 7,000 Duchenne muscular dystrophy mutations. *Human Mutation* 36 (4): 395–402.

Blaese, R., Culver, K., Miller, A. et al. (1995). T lymphocyte-directed gene therapy for ADA- SCID: initial trial results after 4 years. *Science* 270 (5235): 475–480.

Bogdanove, A. and Voytas, D. (2011). TAL effectors: customizable proteins for DNA targeting. *Science* 333 (6051): 1843–1846.

Boztug, K., Schmidt, M., Schwarzer, A. et al. (2010). Stem-cell gene therapy for the Wiskott–Aldrich syndrome. *New England Journal of Medicine* 363 (20): 1918–1927.

Braun, C., Boztug, K., Paruzynski, A. et al. (2014). Gene therapy for Wiskott-Aldrich syndrome--long-term efficacy and genotoxicity. *Science Translational Medicine* 6 (227): 227ra33.

Burlina, A., Bordugo, A., Hoffmann, G., and Meyburg, J. (2014). Cell and organ transplantation for inborn errors of metabolism. In: *Inborn Errors of Metabolism* (eds. B. Lee and F. Scaglia), 271–279. Oxford University Press.

Cagas, C. (1976). Treatment of inborn errors of metabolism. Current treatment and future prospects. *Southern Medical Journal* 69 (12): 1610.

Camp, K., Lloyd-Puryear, M., and Huntington, K. (2012). Nutritional treatment for inborn errors of metabolism: indications, regulations, and availability of medical foods and dietary supplements using phenylketonuria as an example. *Molecular Genetics and Metabolism* 107 (1–2): 3–9.

Chandrasegaran, S. and Carroll, D. (2016). Origins of programmable nucleases for genome engineering. *Journal of Molecular Biology* 428 (5): 963–989.

Chang, C., Lai, Y., Lamb, L., and Townes, T. (2014). Broad T-cell receptor repertoire in T-lymphocytes derived from human induced pluripotent stem cells. *PLoS One* 9 (5): e97335.

Chang, N., Chevalier, F., and Rudnicki, M. (2016). Satellite cells in muscular dystrophy – lost in polarity. *Trends in Molecular Medicine* 22 (6): 479–496.

Chang, K., Smith, S., Sullivan, T. et al. (2017). Long-term engraftment and fetal globin induction upon BCL11A gene editing in bone-marrow-derived CD34 + hematopoietic stem and progenitor cells. *Molecular Therapy - Methods & Clinical Development* 4: 137–148.

Charlesworth, C., Deshpande, P., Dever, D. et al. (2019). Identification of preexisting adaptive immunity to Cas9 proteins in humans. *Nature Medicine* 25 (2): 249–254.

Chinsky, J., Singh, R., Ficicioglu, C. et al. (2017). Diagnosis and treatment of tyrosinemia type I: a US and Canadian consensus group review and recommendations. *Genetics in Medicine* 19 (12): 1380–1380.

Chung, S., Seki, K., Choi, B. et al. (2010). CXC chemokine receptor 4 expressed in T cells plays an important role in the development of collagen-induced arthritis. *Arthritis Research & Therapy* 12 (5): R188.

Corbacioglu, S., Cappellini, M., Chapin, J., et al. 2020. Initial safety and efficacy results with a single dose of autologous CRISPR-Cas9-modified CD34+ hematopoietic stem and progenitor cells in transfusion-dependent B-thalassemia and sickle cell disease. Crisprtx.gcs-web.com. https://crisprtx.gcs-web.com/static-files/84c48306-670f-4f98-b489-fa0dda90d9f4 (accessed 17 July 2020).

Coutinho, M., Lacerda, L., and Alves, S. (2012). Glycosaminoglycan storage disorders: a review. *Biochemistry Research International* 2012: 1–16.

Darabi, R., Pan, W., Bosnakovski, D. et al. (2011). Functional myogenic engraftment from mouse iPS cells. *Stem Cell Reviews and Reports* 7 (4): 948–957.

Davis, P., Drumm, M., and Konstan, M. (1996). Cystic fibrosis. *American Journal of Respiratory and Critical Care Medicine* 154 (5): 1229–1256.

Deeks, S. (2003). Treatment of antiretroviral-drug-resistant HIV-1 infection. *The Lancet* 362 (9400): 2002–2011.

Deng, W., Gao, M., Lei, X. et al. (2018). Gene correction reverses ciliopathy and photoreceptor loss in iPSC-derived retinal organoids from retinitis pigmentosa patients. *Stem Cell Reports* 10 (4): 1267–1281.

Dever, D., Bak, R., Reinisch, A. et al. (2016). CRISPR/Cas9 β-globin gene targeting in human haematopoietic stem cells. *Nature* 539 (7629): 384–389.

DeWeerdt, S. (2016). Drug development: through the barrier. *Nature* 537 (7621): S154–S157.

DeWitt, M., Magis, W., Bray, N. et al. (2016). Selection-free genome editing of the sickle mutation in human adult hematopoietic stem/progenitor cells. *Science Translational Medicine* 8 (360): 360ra134.

Dhand, R. (2017). Inhaled drug therapy 2016: the year in review. *Respiratory Care* 62 (7): 978–996.

Di Michele, D. (2011). Immune tolerance induction in haemophilia: evidence and the way forward. *Journal of Thrombosis and Haemostasis* 9: 216–225.

Didigu, C., Wilen, C., Wang, J. et al. (2014). Simultaneous zinc-finger nuclease editing of the HIV coreceptors ccr5 and cxcr4 protects CD4+ T cells from HIV-1 infection. *Blood* 123 (1): 61–69.

DiGiusto, D., Cannon, P., Holmes, M. et al. (2016). Preclinical development and qualification of ZFN-mediated CCR5 disruption in human hematopoietic stem/progenitor cells. *Molecular Therapy - Methods & Clinical Development* 3: 16067.

Dunbar, G., Koneru, S., Kolli, N. et al. (2019). Silencing of the mutant huntingtin gene through CRISPR-Cas9 improves the mitochondrial biomarkers in an in vitro model of Huntington'S disease. *Cell Transplantation* 28 (4): 460–463. https://doi.org/10.1177/0963689719840662. Epub 2019 Apr 4. PMID: 30947515; PMCID: PMC6628560.

Elitt, M., Barbar, L., and Tesar, P. (2018). Drug screening for human genetic diseases using iPSC models. *Human Molecular Genetics* 27 (R2): R89–R98.

Endo, F., Katoh, H., and Matsuda, I. (1990). Putative genetic deficiency of 4-hydroxyphenylpyruvic acid dioxygenase in mice: a murine model for hereditary tyrosinaemia type III. *Journal of Inherited Metabolic Disease* 13 (5): 780–782.

Fan, Y., Winanto, and Ng, S. (2020). Replacing what's lost: a new era of stem cell therapy for Parkinson's disease. *Translational Neurodegeneration* 9 (1): 2. https://doi.org/10.1186/s40035-019-0180-x. PMID: 31911835; PMCID: PMC6945567.

Ferreira, C., van Karnebeek, C., Vockley, J., and Blau, N. (2018). A proposed nosology of inborn errors of metabolism. *Genetics in Medicine* 21 (1): 102–106.

Firth, A., Menon, T., Parker, G. et al. (2015). Functional gene correction for cystic fibrosis in lung epithelial cells generated from patient iPSCs. *Cell Reports* 12 (9): 1385–1390.

Flavell, L. (2020). Treatment options for muscular dystrophy. Muscular Dystrophy News. https://musculardystrophynews.com/treatment-options-for-muscular-dystrophy/ (accessed 20 May 2020).

Fortier, L. (2005). Stem cells: classifications, controversies, and clinical applications. *Veterinary Surgery* 34 (5): 415–423.

Fukao, T. and Nakamura, K. (2019). Advances in inborn errors of metabolism. *Journal of Human Genetics* 64 (2): 65–65.

Furie, B. and Furie, B. (1988). The molecular basis of blood coagulation. *Cell* 53 (4): 505–518.

Galanello, R. and Origa, R. (2010). Beta-thalassemia. *Orphanet Journal of Rare Diseases* 5 (1): 11. https://doi.org/10.1186/1750-1172-5-11.

Gao, Q. and McNally, E. (2015). The dystrophin complex: structure, function, and implications for therapy. *Comprehensive Physiology* 5 (3): 1223–1239.

Garber, K. (2015). RIKEN suspends first clinical trial involving induced pluripotent stem cells. *Nature Biotechnology* 33 (9): 890–891.

Gaspar, H., Cooray, S., Gilmour, K. et al. (2011). Long-term persistence of a polyclonal T cell repertoire after gene therapy for X-linked severe combined immunodeficiency. *Science Translational Medicine* 3 (97): 97ra79.

Genovese, P., Schiroli, G., Escobar, G. et al. (2014). Targeted genome editing in human repopulating haematopoietic stem cells. *Nature* 510 (7504): 235–240.

Giannelli, F. and Green, P. (1996). The molecular basis of haemophilia A and B. *Baillière's Clinical Haematology* 9 (2): 211–228.

Guerrero, R., Salazar, D., and Tanpaiboon, P. (2018). Laboratory diagnostic approaches in metabolic disorders. *Annals of Translational Medicine* 6 (24): 470.

Guo, L., Karoubi, G., Duchesneau, P. et al. (2017). Generation of induced progenitor-like cells from mature epithelial cells using interrupted reprogramming. *Stem Cell Reports* 9 (6): 1780–1795.

György, B., Lööv, C., Zaborowski, M. et al. (2018). CRISPR/Cas9 mediated disruption of the swedish APP allele as a therapeutic approach for early-onset Alzheimer's disease. *Molecular Therapy - Nucleic Acids* 11: 429–440.

Hacein-Bey-Abina, S. (2003). LMO2-associated clonal T cell proliferation in two patients after gene therapy for SCID-X1. *Science* 302 (5644): 415–419.

Hacein-Bey-Abina, S., Garrigue, A., Wang, G. et al. (2008). Insertional oncogenesis in 4 patients after retrovirus-mediated gene therapy of SCID-X1. *Journal of Clinical Investigation* 118 (9): 3132–3142.

Hacein-Bey-Abina, S., Hauer, J., Lim, A. et al. (2010). Efficacy of gene therapy for X-linked severe combined immunodeficiency. *New England Journal of Medicine* 363 (4): 355–364.

Hernandez, D., Reed, X., and Singleton, A. (2016). Genetics in Parkinson disease: mendelian versus non-mendelian inheritance. *Journal of Neurochemistry* 139: 59–74.

Hoban, M., Cost, G., Mendel, M. et al. (2015). Correction of the sickle cell disease mutation in human hematopoietic stem/progenitor cells. *Blood* 125 (17): 2597–2604.

den Hollander, A., Koenekoop, R., Yzer, S. et al. (2006). Mutations in the CEP290 (NPHP6) gene are a frequent cause of leber congenital amaurosis. *The American Journal of Human Genetics* 79 (3): 556–561.

Holt, N., Wang, J., Kim, K. et al. (2010). Human hematopoietic stem/progenitor cells modified by zinc-finger nucleases targeted to CCR5 control HIV-1 in vivo. *Nature Biotechnology* 28 (8): 839–847.

Hou, P., Chen, S., Wang, S. et al. (2015). Genome editing of CXCR4 by CRISPR/cas9 confers cells resistant to HIV-1 infection. *Scientific Reports* 5 (1): 15577. https://doi.org/10.1038/srep15577.

Hsu, P., Lander, E., and Zhang, F. (2014). Development and applications of CRISPR-Cas9 for genome engineering. *Cell* 157 (6): 1262–1278.

Corrigendum forHudgins, E.J., Liebhold, A.M., and Leung, B. (2017). *Ecology Letters*, 2018 21 (11): 1752–1754. https://doi.org/10.1111/ele.12741.

Hütter, G., Nowak, D., Mossner, M. et al. (2009). Long-term control of HIV byCCR5Delta32/Delta32 stem-cell transplantation. *New England Journal of Medicine* 360 (7): 692–698.

Ishikawa, T., Imamura, K., Kondo, T. et al. (2016). Genetic and pharmacological correction of aberrant dopamine synthesis using patient iPSCs with BH4 metabolism disorders. *Human Molecular Genetics* 25 (23): 5188–5197. https://doi.org/10.1093/hmg/ddw339. PMID: 27798097; PMCID: PMC5886044.p.ddw339.

Kay, M. (2011). State-of-the-art gene-based therapies: the road ahead. *Nature Reviews Genetics* 12 (5): 316–328.

Kennedy, M., Awong, G., Sturgeon, C. et al. (2012). T lymphocyte potential marks the emergence of definitive hematopoietic progenitors in human pluripotent stem cell differentiation cultures. *Cell Reports* 2 (6): 1722–1735.

Kohn, D., Weinberg, K., Nolta, J. et al. (1995). Engraftment of gene–modified umbilical cord blood cells in neonates with adenosine deaminase deficiency. *Nature Medicine* 1 (10): 1017–1023.

Kosicki, M., Tomberg, K., and Bradley, A. (2018). Erratum: repair of double-strand breaks induced by CRISPR–Cas9 leads to large deletions and complex rearrangements. *Nature Biotechnology* 36 (9): 899–899.

Li, H., Haurigot, V., Doyon, Y. et al. (2011). In vivo genome editing restores haemostasis in a mouse model of haemophilia. *Nature* 475 (7355): 217–221.

Li, L., Krymskaya, L., Wang, J. et al. (2013). Genomic editing of the HIV-1 coreceptor CCR5 in adult hematopoietic stem and progenitor cells using zinc finger nucleases. *Molecular Therapy* 21 (6): 1259–1269.

Li, H., Fujimoto, N., Sasakawa, N. et al. (2015). Precise correction of the dystrophin gene in duchenne muscular dystrophy patient induced pluripotent stem cells by TALEN and CRISPR-Cas9. *Stem Cell Reports* 4 (1): 143 154.

Li, J., Walker, S., Nie, J., and Zhang, X. (2019). Experiments that led to the first gene-edited babies: the ethical failings and the urgent need for better governance. *Journal of Zhejiang University-SCIENCE B* 20 (1): 32–38.

Li, H., Yang, Y., Hong, W. et al. (2020). Applications of genome editing technology in the targeted therapy of human diseases: mechanisms, advances and prospects. *Signal Transduction and Targeted Therapy* 5 (1): 1. https://doi.org/10.1038/s41392-019-0089-y. PMID: 32296011; PMCID: PMC6946647.

Liao, H., Hatanaka, F., Araoka, T. et al. (2017). In vivo target gene activation via CRISPR/Cas9-mediated trans-epigenetic modulation. *Cell* 171 (7): 1495–1507.e15.

Liu, R., Paxton, W., Choe, S. et al. (1996). Homozygous defect in HIV-1 coreceptor accounts for resistance of some multiply-exposed individuals to HIV-1 infection. *Cell* 86 (3): 367–377.

Liu, S., Wang, Q., Yu, X. et al. (2018). HIV-1 inhibition in cells with CXCR4 mutant genome created by CRISPR-Cas9 and piggyBac recombinant technologies. *Scientific Reports* 8 (1): 8573. https://doi.org/10.1038/s41598-018-26894-4.

Long, C., Amoasii, L., Mireault, A. et al. (2015). Postnatal genome editing partially restores dystrophin expression in a mouse model of muscular dystrophy. *Science* 351 (6271): 400–403.

Lopalco, L. (2010). CCR5: from natural resistance to a new anti-HIV strategy. *Viruses* 2 (2): 574–600.

Ma, Q., Jones, D., Borghesani, P. et al. (1998). Impaired B-lymphopoiesis, myelopoiesis, and derailed cerebellar neuron migration in CXCR4- and SDF-1-deficient mice. *Proceedings of the National Academy of Sciences* 95 (16): 9448–9453.

Maeder, M., Stefanidakis, M., Wilson, C. et al. (2019). Development of a gene-editing approach to restore vision loss in leber congenital amaurosis type 10. *Nature Medicine* 25 (2): 229–233.

Maffioletti, S., Sarcar, S., Henderson, A. et al. (2018). Three-dimensional human iPSC-derived artificial skeletal muscles model muscular dystrophies and enable multilineage tissue engineering. *Cell Reports* 23 (3): 899–908.

Maggio, I. and Gonçalves, M. (2015). Genome editing at the crossroads of delivery, specificity, and fidelity. *Trends in Biotechnology* 33 (5): 280–291.

Maiti, P., Manna, J., and Dunbar, G. (2017). Current understanding of the molecular mechanisms in Parkinson's disease: targets for potential treatments. *Translational Neurodegeneration* 6: 28. (Page number not provided).

Mali, P., Aach, J., Stranges, P. et al. (2013). CAS9 transcriptional activators for target specificity screening and paired nickases for cooperative genome engineering. *Nature Biotechnology* 31 (9): 833–838.

Mandai, M., Watanabe, A., Kurimoto, Y. et al. (2017). Autologous induced stem-cell–derived retinal cells for macular degeneration. *New England Journal of Medicine* 376 (11): 1038–1046.

Mandavilli, A. (2019). H.I.V. is reported cured in a second patient, a milestone in the global AIDS epidemic. Nytimes.com. https://www.nytimes.com/2019/03/04/health/aids-cure-london-patient.html (accessed 14 July 2020).

Mariani, G., Siragusa, S., and Kroner, B. (2003). Immune tolerance induction in hemophilia A: a review. *Seminars in Thrombosis and Hemostasis* 29 (1): 069–076.

Mason, C. and Dunnill, P. (2008). A brief definition of regenerative medicine. *Regenerative Medicine* 3 (1): 1–5.

McQualter, J., Yuen, K., Williams, B., and Bertoncello, I. (2010). Evidence of an epithelial stem/progenitor cell hierarchy in the adult mouse lung. *Proceedings of the National Academy of Sciences* 107 (4): 1414–1419.

Menasché, P., Vanneaux, V., Hagège, A. et al. (2018). Transplantation of human embryonic stem cell–derived cardiovascular progenitors for severe ischemic left ventricular dysfunction. *Journal of the American College of Cardiology* 71 (4): 429–438.

Métais, J., Doerfler, P., Mayuranathan, T. et al. (2019). Genome editing of HBG1 and HBG2 to induce fetal hemoglobin. *Blood Advances* 3 (21): 3379–3392.

Min, Y., Bassel-Duby, R., and Olson, E. (2019). CRISPR correction of duchenne muscular dystrophy. *Annual Review of Medicine* 70 (1): 239–255.

Mollanoori, H., Rahmati, Y., Hassani, B. et al. (2021). Promising therapeutic approaches using CRISPR/Cas9 genome editing technology in the treatment of Duchenne muscular dystrophy. *Genes & Diseases* 8 (2): 146–156.

Monteys, A., Ebanks, S., Keiser, M., and Davidson, B. (2017). CRISPR/Cas9 editing of the mutant huntingtin allele in vitro and in vivo. *Molecular Therapy* 25 (1): 12–23.

Montini, E., Cesana, D., Schmidt, M. et al. (2006). Hematopoietic stem cell gene transfer in a tumor-prone mouse model uncovers low genotoxicity of lentiviral vector integration. *Nature Biotechnology* 24 (6): 687–696.

Montini, E., Cesana, D., Schmidt, M. et al. (2009). The genotoxic potential of retroviral vectors is strongly modulated by vector design and integration site selection in a mouse model of HSC gene therapy. *Journal of Clinical Investigation* 119 (4): 964–975.

Moratto, D., Giliani, S., Bonfim, C. et al. (2011). Long-term outcome and lineage-specific chimerism in 194 patients with Wiskott-Aldrich syndrome treated by hematopoietic cell transplantation in the period 1980-2009: an international collaborative study. *Blood* 118 (6): 1675–1684.

Moretti, A., Fonteyne, L., Giesert, F. et al. (2020). Somatic gene editing ameliorates skeletal and cardiac muscle failure in pig and human models of Duchenne muscular dystrophy. *Nature Medicine* 26 (2): 207–214.

Murphy, M. and LeVine, H. (2010). Alzheimer's disease and the amyloid-β peptide. *Journal of Alzheimer's Disease* 19 (1): 311–323.

Mussolino, C., Morbitzer, R., Lütge, F. et al. (2011). A novel TALE nuclease scaffold enables high genome editing activity in combination with low toxicity. *Nucleic Acids Research* 39 (21): 9283–9293.

Mussolino, C., Alzubi, J., Fine, E. et al. (2014). TALENs facilitate targeted genome editing in human cells with high specificity and low cytotoxicity. *Nucleic Acids Research* 42 (10): 6762–6773.

Naif, H. (2013). Pathogenesis of HIV infection. *Infectious Disease Reports* 5 (1S): 6.

Nature.com (2020). CRISPR treatment inserted directly into the body for first time. https://www.nature.com/articles/d41586-020-00655-8 (accessed 24 May 2020).

NORD (National Organization for Rare Disorders). (2020). Lysosomal storage disorders - NORD (National Organization For Rare Disorders). https://rarediseases.org/rare-diseases/lysosomal-storage-disorders/ (accessed 15 July 2020).

Nyamweya, S., Hegedus, A., Jaye, A. et al. (2013). Comparing HIV-1 and HIV-2 infection: lessons for viral immunopathogenesis. *Reviews in Medical Virology* 23 (4): 221–240.

Ortiz-Virumbrales, M., Moreno, C., Kruglikov, I. et al. (2017). CRISPR/Cas9-correctable mutation-related molecular and physiological phenotypes in Ipsc-derived Alzheimer'S PSEN2 N141I neurons. *Acta Neuropathologica Communications* 5 (1): 77. https://doi.org/10.1186/s40478-017-0475-z. PMID: 29078805; PMCID: PMC5660456.

Osman, G., Rodriguez, J., Chan, S. et al. (2018). PEGylated enhanced cell penetrating peptide nanoparticles for lung gene therapy. *Journal of Controlled Release* 285: 35–45.

Ousterout, D., Kabadi, A., Thakore, P. et al. (2015). Multiplex CRISPR/Cas9-based genome editing for correction of dystrophin mutations that cause Duchenne muscular dystrophy. *Nature Communications* 6 (1): 6244.

Pankowicz, F., Barzi, M., Kim, K. et al. (2018). Rapid disruption of genes specifically in livers of mice using multiplex CRISPR/Cas9 editing. *Gastroenterology* 155 (6): 1967–1970.e6.

Park, C., Kim, J., Kweon, J. et al. (2014). Targeted inversion and reversion of the blood coagulation factor 8 gene in human iPS cells using TALENs. *Proceedings of the National Academy of Sciences* 111 (25): 9253–9258.

Perez, E., Wang, J., Miller, J. et al. (2008). Establishment of HIV-1 resistance in CD4+ T cells by genome editing using zinc-finger nucleases. *Nature Biotechnology* 26 (7): 808–816.

Perrin, G., Herzog, R., and Markusic, D. (2019). Update on clinical gene therapy for hemophilia. *Blood* 133 (5): 407–414.

Pires, C., Schmid, B., Petræus, C. et al. (2016). Generation of a gene-corrected isogenic control cell line from an Alzheimer's disease patient iPSC line carrying a A79V mutation in PSEN1. *Stem Cell Research* 17 (2): 285–288.

Platt, F., d'Azzo, A., Davidson, B. et al. (2018). Lysosomal storage diseases. *Nature Reviews Disease Primers* 4 (1): 27. https://doi.org/10.1038/s41572-018-0025-4. Erratum in: Nat Rev Dis Primers. 2018 Oct 18;4(1):36. Erratum in: Nat Rev Dis Primers. 2019 May 17;5(1):34. PMID: 30275469.

Pringle, I., Hyde, S., and Gill, D. (2009). Non-viral vectors in cystic fibrosis gene therapy: recent developments and future prospects. *Expert Opinion on Biological Therapy* 9 (8): 991–1003.

Psatha, N., Reik, A., Phelps, S. et al. (2018). Disruption of the BCL11A erythroid enhancer reactivates fetal hemoglobin in erythroid cells of patients with β-thalassemia major. *Molecular Therapy - Methods & Clinical Development* 10: 313–326.

Rarediseases.info.nih.gov. (2020). Tyrosinemia type 1 | Genetic And Rare Diseases Information Center (GARD) – an NCATS program. https://rarediseases.info.nih.gov/diseases/2658/tyrosinemia-type-1 (accessed 15 July 2020).

Reference, G. (2020). Leber congenital amaurosis. Genetics Home Reference. https://ghr.nlm.nih.gov/condition/leber-congenital-amaurosis#statistics (accessed 24 May 2020).

Reiner, A., Dragatsis, I., and Dietrich, P. (2011). Genetics and neuropathology of Huntington's disease. *International Review of Neurobiology* 98: 325–372. https://doi.org/10.1016/B978-0-12-381328-2.00014-6.

Reinhardt, P., Schmid, B., Burbulla, L. et al. (2013). Genetic correction of a LRRK2 mutation in human iPSCs links parkinsonian neurodegeneration to ERK-dependent changes in gene expression. *Cell Stem Cell* 12 (3): 354–367.

Reitz, C., Brayne, C., and Mayeux, R. (2011). Epidemiology of Alzheimer disease. *Nature Reviews Neurology* 7 (3): 137–152.

Rock, J. and Königshoff, M. (2012). Endogenous lung regeneration. *American Journal of Respiratory and Critical Care Medicine* 186 (12): 1213–1219.

Ryu, S., Koo, T., Kim, K. et al. (2018). Adenine base editing in mouse embryos and an adult mouse model of Duchenne muscular dystrophy. *Nature Biotechnology* 36 (6): 536–539.

Samson, M., Libert, F., Doranz, B. et al. (1996). Resistance to HIV-1 infection in caucasian individuals bearing mutant alleles of the CCR-5 chemokine receptor gene. *Nature* 382 (6593): 722–725.

Sanderson, S., Green, A., Preece, M., and Burton, H. (2006). The incidence of inherited metabolic disorders in the West Midlands, UK. *Archives of Disease in Childhood* 91 (11): 896–899.

Schiroli, G., Ferrari, S., Conway, A. et al. (2017). Preclinical modeling highlights the therapeutic potential of hematopoietic stem cell gene editing for correction of SCID-X1. *Science Translational Medicine* 9 (411): eaan0820.

Schneller, J., Lee, C., Bao, G., and Venditti, C. (2017). Genome editing for inborn errors of metabolism: advancing towards the clinic. *BMC Medicine* 15: 43. (Page number not provided).

Schwank, G., Koo, B., Sasselli, V. et al. (2013). Functional repair of CFTR by CRISPR/Cas9 in intestinal stem cell organoids of cystic fibrosis patients. *Cell Stem Cell* 13 (6): 653–658.

Shi, B., Li, J., Shi, X. et al. (2017). TALEN-mediated knockout of CCR5 confers protection against infection of human immunodeficiency virus. *JAIDS Journal of Acquired Immune Deficiency Syndromes* 74 (2): 229–241.

Shin, J., Kim, K., Chao, M. et al. (2016). Permanent inactivation of Huntington's disease mutation by personalized allele-specific CRISPR/Cas9. *Human Molecular Genetics* 25 (20): 4566–4576. https://doi.org/10.1093/hmg/ddw286. PMID: 28172889; PMCID: PMC6078600., p.ddw286.

Simon, V., Ho, D., and Abdool Karim, Q. (2006). HIV/AIDS epidemiology, pathogenesis, prevention, and treatment. *The Lancet* 368 (9534): 489–504.

Sivalingam, J., Kenanov, D., Han, H. et al. (2016). Multidimensional genome-wide analyses show accurate FVIII integration by ZFN in primary human cells. *Molecular Therapy* 24 (3): 607–619.

Stoltz, D., Meyerholz, D., and Welsh, M. (2015). Origins of cystic fibrosis lung disease. *New England Journal of Medicine* 372 (4): 351–362.

Takahashi, J. (2019). Preparing for first human trial of induced pluripotent stem cell-derived cells for Parkinson's disease: an interview with Jun Takahashi. *Regenerative Medicine* 14 (2): 93–95.

Takahashi, N. and Sawamoto, N., 2018. Kyoto trial to evaluate the safety and efficacy if Ipsc-derived dopaminergic progenitors in the treatment of Parkinson's disease HpscregHpscreg.eu. https://hpscreg.eu/browse/trial/54 (accessed 14 July 2020).

Takahashi, K. and Yamanaka, S. (2006). Induction of pluripotent stem cells from mouse embryonic and adult fibroblast cultures by defined factors. *Cell* 126 (4): 663–676.

Tan, E., Boelens, J., Jones, S., and Wynn, R. (2019). Hematopoietic stem cell transplantation in inborn errors of metabolism. *Frontiers in Pediatrics* 7: 433. https://doi.org/10.3389/fped.2019.00433.

Tebas, P., Stein, D., Tang, W. et al. (2014). Gene Editing ofCCR5in Autologous CD4 T Cells of Persons Infected with HIV. *New England Journal of Medicine* 370 (10): 901–910.

Thein, S. and Thein, M. (2016). World sickle cell day 2016: a time for appraisal. *Indian Journal of Medical Research* 143 (6): 678.

Themeli, M., Kloss, C., Ciriello, G. et al. (2013). Generation of tumor-targeted human T lymphocytes from induced pluripotent stem cells for cancer therapy. *Nature Biotechnology* 31 (10): 928–933.

Thomson, J. (1998). Embryonic stem cell lines derived from human blastocysts. *Science* 282 (5391): 1145–1147.

Tian, S., Choi, W., Liu, D. et al. (2005). Distinct functional sites for human Immunodeficiency virus type 1 and stromal cell-derived factor 1α on CXCR4 transmembrane helical domains. *Journal of Virology* 79 (20): 12667–12673.

Tiemann, K. and Rossi, J. (2009). RNAi-based therapeutics–current status, challenges and prospects. *EMBO Molecular Medicine* 1 (3): 142–151.

Till, J. and McCulloch, E. (1961). A direct measurement of the radiation sensitivity of normal mouse bone marrow cells. *Radiation Research* 14 (2): 213.

Traxler, E., Yao, Y., Wang, Y. et al. (2016). A genome-editing strategy to treat β-hemoglobinopathies that recapitulates a mutation associated with a benign genetic condition. *Nature Medicine* 22 (9): 987–990.

Triarhou, L. (2020). Dopamine and Parkinson's disease. Ncbi.nlm.nih.gov. https://www.ncbi.nlm.nih.gov/books/NBK6271/ (accessed 14 July 2020).

Tysnes, O. and Storstein, A. (2017). Epidemiology of Parkinson's disease. *Journal of Neural Transmission* 124 (8): 901–905.

Vaishnaw, A., Gollob, J., Gamba-Vitalo, C. et al. (2010). A status report on RNAi therapeutics. *Silence* 1 (1): 14.

Wang, Q., Chen, S., Xiao, Q. et al. (2017). Genome modification of CXCR4 by Staphylococcus aureus Cas9 renders cells resistance to HIV-1 infection. *Retrovirology* 14 (1): 51. https://doi.org/10.1186/s12977-017-0375-0.

Warby, S., Montpetit, A., Hayden, A. et al. (2009). CAG expansion in the huntington disease gene is associated with a specific and targetable predisposing haplogroup. *The American Journal of Human Genetics* 84 (3): 351–366.

Weber, L., Frati, G., Felix, T. et al. (2020). Editing a γ-globin repressor binding site restores fetal hemoglobin synthesis and corrects the sickle cell disease phenotype. *Science Advances* 6 (7): eaay9392.

Weissman, I. and Shizuru, J. (2008). The origins of the identification and isolation of hematopoietic stem cells, and their capability to induce donor-specific transplantation tolerance and treat autoimmune diseases. *Blood* 112 (9): 3543–3553.

Wilcken, B., Wiley, V., Hammond, J., and Carpenter, K. (2003). Screening newborns for inborn errors of metabolism by tandem mass spectrometry. *New England Journal of Medicine* 348 (23): 2304–2312.

Wilson, K., Faelan, C., Patterson-Kane, J. et al. (2017). Duchenne and becker muscular dystrophies: a review of animal models, clinical end points, and biomarker quantification. *Toxicologic Pathology* 45 (7): 961–976.

Wu, Y., Zeng, J., Roscoe, B. et al. (2019). Highly efficient therapeutic gene editing of human hematopoietic stem cells. *Nature Medicine* 25 (5): 776–783.

Xu, L., Yang, H., Gao, Y. et al. (2017). CRISPR/Cas9-mediated CCR5 ablation in human hematopoietic stem/progenitor cells confers HIV-1 resistance in vivo. *Molecular Therapy* 25 (8): 1782–1789.

Xu, L., Wang, J., Liu, Y. et al. (2019a). CRISPR-edited stem cells in a patient with HIV and acute lymphocytic leukemia. *New England Journal of Medicine* 381 (13): 1240–1247.

Xu, S., Luk, K., Yao, Q. et al. (2019b). Editing aberrant splice sites efficiently restores β-globin expression in β-thalassemia. *Blood* 133 (21): 2255–2262.

Yang, Y., Wang, L., Bell, P. et al. (2016). A dual AAV system enables the Cas9-mediated correction of a metabolic liver disease in newborn mice. *Nature Biotechnology* 34 (3): 334–338.

Yang, S., Chang, R., Yang, H. et al. (2017). CRISPR/Cas9-mediated gene editing ameliorates neurotoxicity in mouse model of huntington's disease. *The Journal of Clinical Investigation* 127 (7): 2719–2724. https://doi.org/10.1172/JCI92087. Epub 2017 Jun 19. PMID: 28628038; PMCID: PMC5490741.

Yin, H., Xue, W., Chen, S. et al. (2014). Genome editing with Cas9 in adult mice corrects a disease mutation and phenotype. *Nature Biotechnology* 32 (6): 551–553.

Yin, H., Song, C., Dorkin, J. et al. (2016). Therapeutic genome editing by combined viral and non-viral delivery of CRISPR system components in vivo. *Nature Biotechnology* 34 (3): 328–333.

Yu, S., Yao, Y., Xiao, H. et al. (2018). Simultaneous knockout of CXCR4 and CCR5 genes in CD4+ T cells via CRISPR/Cas9 confers resistance to both X4- and R5-tropic human immunodeficiency virus type 1 infection. *Human Gene Therapy* 29 (1): 51–67.

Yuan, J., Wang, J., Crain, K. et al. (2012). Zinc-finger nuclease editing of human cxcr4 promotes HIV-1 CD4+ T cell resistance and enrichment. *Molecular Therapy* 20 (4): 849–859.

Yue, W., Mackinnon, S., and Bezerra, G. (2019). Substrate reduction therapy for inborn errors of metabolism. *Emerging Topics in Life Sciences* 3 (1): 63–73.

Yui, S., Nakamura, T., Sato, T. et al. (2012). Functional engraftment of colon epithelium expanded in vitro from a single adult Lgr5+ stem cell. *Nature Medicine* 18 (4): 618–623.

Zhang, J., Cheng, X., Zhao, M. et al. (2019). Curing hemophilia A by NHEJ-mediated ectopic F8 insertion in the mouse. *Genome Biology* 20 (1): 276. https://doi.org/10.1186/s13059-019-1907-9.

Zhang, Y., Li, H., Min, Y. et al. (2020). Enhanced CRISPR-Cas9 correction of Duchenne muscular dystrophy in mice by a self-complementary AAV delivery system. *Science Advances* 6 (8): eaay6812.

18

Delivery and Formulation Methods for Therapeutic Genome Editing

Erik Oude Blenke[1] *and Venkata R. Krishnamurthy*[2]

[1] *Advanced Drug Delivery, Pharmaceutical Sciences, AstraZeneca R&D, Molndal, Sweden*
[2] *Advanced Drug Delivery, Pharmaceutical Sciences, AstraZeneca R&D, Boston, MA, USA*

18.1 Introduction

The very first mention of gene therapy for human genetic disease in scientific literature goes back to 1972 (by Friedman and Roblin in *Science*). It was in fact a warning, stating that we did not know enough about human genetics and the short- and long-term side effects of genetic manipulation and that it was unethical to pursue attempts in human patients until we have increased our understanding (Friedmann and Roblin 1972).

In the past decade, our understanding of genetics has *greatly* increased and the methodologies to manipulate the genome have been refined. Thanks to the CRISPR/Cas technology (but also Zinc Finger Nucleases and TALENs) we are no longer simply adding a piece of DNA, not knowing where it will integrate or how long it persists at all, but we are capable of precisely editing or even replacing genes. But despite the major progress in this area, it remains a significant challenge to apply any of these editing techniques inside living organisms. Mammalian cells have evolved to compartmentalize their cellular machinery and created membranes to keep foreign material out of the cell and its own genetic material inside an additional compartment (the nucleus). These membranes literally form a barrier that prevents us from delivering the payloads that will manipulate the genome, to their intracellular site of action. But in mammals, this compartmentalization recurs on a larger scale as well, as each organ is essentially a new barrier and the tissue types that are most exposed are not necessarily the types that are the most suitable for genetic therapy. On top of that, the human body has evolved numerous immune response mechanisms against foreign genetic and biological material and the compounds that can be used to engineer genes are extremely susceptible to enzymatic degradation. To overcome these challenges, most delivery strategies comprise of packaging these compounds in some form of nanosized delivery vehicle that protects them from degradation and helps them cross the biological barriers.

However, compared with the advances in genetic engineering, the progress to overcome these delivery challenges is somewhat lagging behind. The discovery of CRISPR/Cas has revolutionized the field of genome editing. But to unlock its great potential in medicine, a delivery system that is able to safely and effectively deliver the CRISPR/Cas components into target cells is essential. Encouraged by the success of the first-ever FDA-approved lipid nanoparticle (LNP) encapsulating a nucleic acid (siRNA) drug, both academics and pharmaceutical companies have renewed their efforts to develop novel formulations for the *in vivo* delivery of gene editing constructs. (ONPATTRO® or patisiran is an siRNA formulated in a LNP, developed by Alnylam Pharmaceuticals (Akinc *et al.* 2019)). Delivery technologies such as lipid nanoparticles (LNPs) have accelerated the development of nucleic acid-based drugs and enabled the clinical translation of nucleic acid-based therapeutics such as siRNA, mRNA, DNA, and genome editing agents. In the context of genome editing, a quick look at the pipeline of the front-running genome editing companies (CRISPR Therapeutics, Intellia Therapeutics, Editas Medicine, Sangamo Therapeutics, Beam Therapeutics, Verve Therapeutics) warrants the conclusion that we are no longer limited by the ability to safely and effectively edit the genome, but by the cell types we can apply this to. The most advanced program that is currently in clinic for gene editing is via *ex-vivo* delivery of the reagents. That is, in cells isolated from the patient that are manipulated outside the body and then reinfused into the patient as a treatment. In this case, delivery of the gene editing constructs can be done using "conventional" molecular biology techniques, such as electroporation, microinjection, and lipofection.

Genome Editing in Drug Discovery, First Edition. Edited by Marcello Maresca and Sumit Deswal.

In-vivo applications require more advanced delivery technology and these programs are typically a bit further behind in clinical development.

In general, delivery technologies can be distinguished into three main categories (see Figure 18.1): (i) *ex-vivo* technique (e.g. electroporation), (ii) viral delivery, and (iii) non-viral delivery.

Of all *ex-vivo* manipulation techniques, electroporation is the most widely used technique. It is shown to be scalable, robust, and reliable method that can be easily applied in a GMP-setting but limited mostly to *ex-vivo* applications.

Viral vectors are based on naturally occurring viruses. Although their effect is mostly pathogenic, they are very efficient in delivering their genome to other cells. Due to this pathogenic nature, many patients have already acquired a pre-existing immunity against certain types of viruses and even when this is not the case, they will elicit an immune response that limits their use to a single administration. In addition, some viruses are shown to have the risk of genome integration. However, thanks to recent advancements that make these viral vectors safer and cheaper to produce and due to the inability of nonviral vectors to access certain tissues, for certain applications viral vectors are still the preferred delivery mode.

Non-viral vectors are the answer to many of the problems associated with viral vectors but are not completely free of immunogenic reactions either and have long been plagued by a lack of potency. However, recent advances in the development of cationic lipid and lipid-like materials have made LNPs the state-of-the-art delivery technology for non-viral delivery and *in-vivo* genome editing. The aforementioned patisiran/ONPATTRO® siRNA LNP formulation is testimony to the increased potency of non-viral delivery systems. Since patisiran was taken into clinical development, many even more potent delivery materials, mostly tailored for mRNA-delivery, have been developed. This demonstrates the versatility of lipid-based materials and together with the relative ease of scale-up compared with viral vectors, this has made LNPs one of the most sophisticated non-viral delivery technologies to enable gene-based therapies.

This chapter is focused on designing optimized carriers to deliver various forms of CRISPR/Cas components *in vivo*. First, the different types of payloads; plasmid, RNA, and ribonucleoprotein (RNP) are discussed and then the delivery technologies used for delivering these different payloads will be summarized. Since *in vivo* application is the main challenge here, the focus will be on viral and non-viral delivery vehicles. Efficiency and safety aspects when designing these delivery systems for CRISPR/Cas will be highlighted and finally we will provide our perspective on future research.

18.2 Payloads and Modalities

For CRISPR-based gene knock-down or base editing, at least two components are required: the Cas enzyme/base editor and the guide RNA. For gene knock-in, a DNA template needs to be delivered as a third component. See Figure 18.2. Protein, RNA, and DNA, all can be delivered to the target cell either via mechanical methods or using a delivery vehicle. In the past decades, many different types of delivery systems have been investigated that are designed for a specific payload. For certain applications, the chances of success may be higher when, for example, the nuclease is expressed from an mRNA or pDNA instead of being delivered as the protein. At the same time, duration of

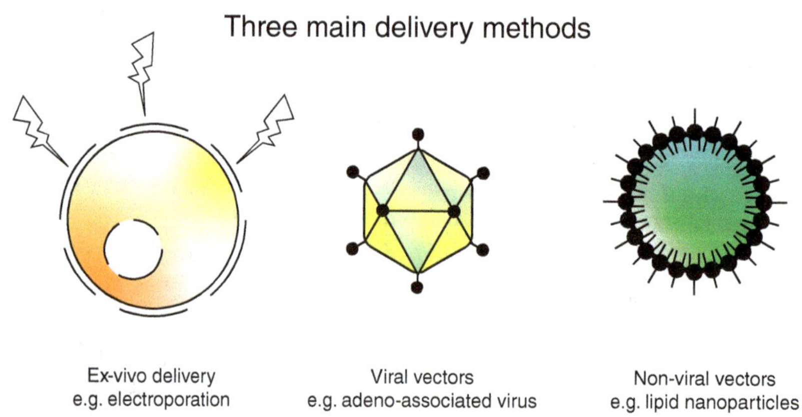

Figure 18.1 Three main pillars of delivery methods. Ex-vivo manipulation of cells is relatively straight-forward, electroporation is by far the most utilized technique. For *in vivo* and clinical application, delivery vehicles are used, that can be classified as viral and non-viral. Adeno-associated viral vectors are the most used viral vector and of non-viral vectors, lipid-based nanoparticles are the most used.

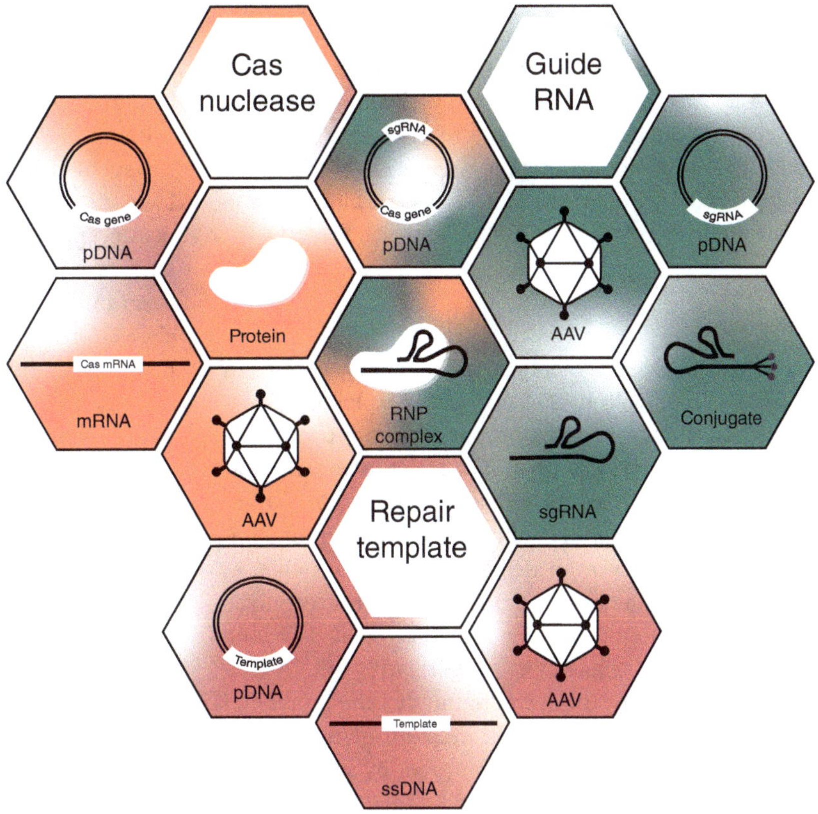

Figure 18.2 The different components required for therapeutic gene editing. Each component can be delivered in different ways and for different applications, a different delivery strategy will be required. Depending on the application, the desirable strategy can be to combine components, for example all components encoded in a single plasmid (although not desirable for *in vivo* application) or the complexation of the nuclease and guide RNA, to deliver it as a ribonucleoprotein complex (RNP).

expression may be an important factor determining the safety of the gene editing treatment, making one type of payload more suitable than the other. Therefore, the delivery strategy needs to be tailored to each specific application.

In the following section, each type of modality or payload will be described, along with their potential use and most feasible application. The focus of this chapter will be on CRISPR-based gene editing as this is currently the most widely investigated method. But as TALENs and ZFNs have been around much longer, some of the delivery approaches have been developed specifically for these methods. A few of these will be highlighted as well, when it is likely they can also be applied to CRISPR-based applications in the future.

18.2.1 Plasmid DNA (pDNA)

Initial efforts in gene delivery employed plasmid DNA (pDNA) as the payload. Plasmids can be delivered using mechanical methods or with synthetic delivery vehicles comprising of cationic lipids or polymers with abundant amino groups. Plasmids are relatively cheap and easy to produce and they can contain up to 200 kb, so they are typically not limited by the amount of genetic information they can encode. They can also be employed for CRISPR/Cas gene editing as both the nuclease and guide RNA can be expressed from a plasmid and the repair template can be encoded on a plasmid as well. Theoretically, encoding all components on a single plasmid would be the easiest option, but this is not a strategy that is often used. The main reason for this is that the long-lasting expression of a plasmid poses a major safety risk, as extended presence of both the nuclease and guide RNA in the same cell leads to off-target cuts. Furthermore, if successful, gene editing is a one-off event in each cell, so it may be counterproductive to prolong the expression of the editing components. Another reason is that DNA needs to reach the nucleus of the cell to be expressed, but plasmids are too large to cross the nuclear membrane. Therefore, their use is limited to dividing cells as the nuclear envelope is temporarily dissolved during mitosis.

Delivering the DNA repair template to the nucleus is of utmost importance for gene replacement approaches but this remains a significant challenge with synthetic carrier systems, especially *in vivo*. So altogether, pDNA is not often used because of the safety concerns associated with extended expression of the Cas nuclease. Delivery of the repair template could also be done with a smaller single-stranded DNA (ssDNA), but because of the limited efficacy of DNA delivery to the nucleus with synthetic systems, this is simply not the best choice. For this, it is often turned to viral delivery systems. In addition, expression of the nuclease from mRNA provides a more transient alternative to pDNA and is therefore more desirable.

18.2.2 RNA (mRNA and sgRNA)

The field of mRNAs as therapeutics has rapidly evolved in the recent years, with mRNA now being explored for protein replacement, vaccination, and immune-oncology applications. Contrary to DNA, it does not require nuclear localization to give rise to expression and thanks to its transient expression, Cas mRNA and sgRNA are safer for clinical therapy. However, mRNA is much less stable in biological environments than DNA and easily recognized as a foreign material by immune cells, so it was long thought to be too challenging to work with mRNA as a therapeutic modality. But it was found that the expression of the encoded protein and the stability of the mRNA inside the cell could be greatly improved by selecting the right structural elements such as 5′- and 3′-untranslated regions, 5′-cap and length of poly(A)-tail (Vallazza *et al.* 2015). Furthermore, incorporation of modified nucleosides such as 1-pseudo-uridine (Ψ) or 5-methoxy-uridine (5m-Ψ) could suppress the immune responses, thereby further enhancing the expression (Karikó *et al.* 2008). This mRNA can be produced using *in vitro* translation (IVT) from a production plasmid in a process that is fast, flexible, and scalable. Such mRNAs are often referred to as modified- or "mod-RNAs" to emphasize that they are structurally quite different molecules than naturally occurring mRNAs. See Figure 18.3.

Nevertheless, these mRNAs are still extremely susceptible to degradation in biological fluids and therefore therapeutic applications are almost exclusively developed in tandem with a synthetic delivery system to encapsulate them in. LNPs are now considered the delivery system of choice for mRNA delivery, pretty much regardless of the therapeutic application. LNPs can also be tailored for the delivery of Cas9 mRNA, either alone or in combination with the guide RNA.

The most commonly used guide RNA is the single-guide RNA, or sgRNA. Optimized designs are usually 100 nucleotides long and could therefore be produced either by IVT or by solid-phase synthesis. Solid-phase synthesis allows incorporation of modified bases and other backbone modifications, an approach that increases the intracellular stability and half-life of the guide (Yin *et al.* 2017). It is speculated in literature and patent applications that this could be used to develop sgRNA-conjugates, for example conjugated to GalNAc-sugars to enhance uptake in hepatocytes. However, to date this has not been demonstrated in the public domain and it is unclear if this approach has any advantage, as a sgRNA in the target cell alone does not have any effect until the Cas nuclease is also present.

Co-encapsulation of the Cas9 mRNA and sgRNA has the advantage that they can be delivered to the same cell. This

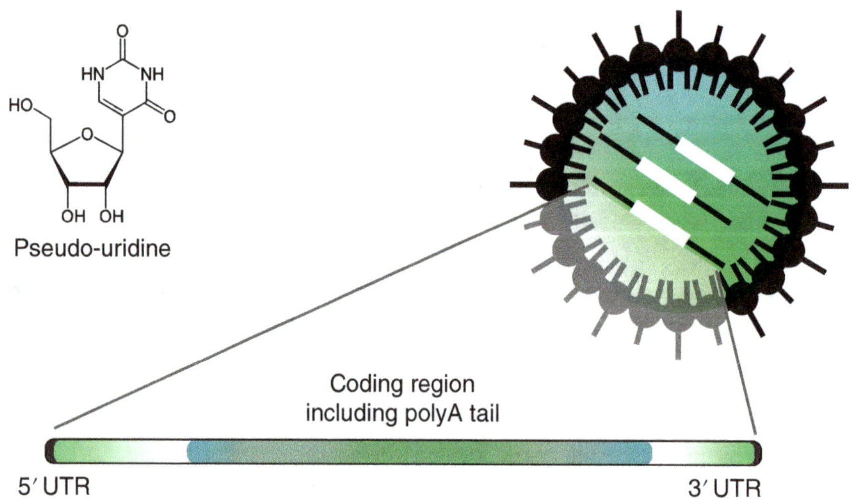

Figure 18.3 **Modified mRNA in a lipid nanoparticle.** Modified mRNA or "mod-RNA" is often encapsulated in a lipid (nano)particle for delivery to cells *in vitro* or *in vivo*. The term "modified" refers to the engineered and optimized sequences in the coding region and the untranslated regions or UTR's at each end of the mRNA, as well as to the modified uridine bases, for example pseudo-uridine (Ψ) shown in the box.

is of course crucial for the editing and especially for *in vivo as* it can be challenging to deliver multiple vectors to the same cell. Therefore, for these applications it is almost always chosen to co-encapsulate the cargos in the same LNP. *In vivo* proof of concept for this approach was shown by Intellia Therapeutics, where they knocked-down transthyretin (TTR) expression in the mouse liver. After a single administration of LNPs with the Cas9 mRNA and a TTR sgRNA, a 97% reduction of TTR serum levels was measured, that persisted during the 12-months of follow-up, despite the transient nature of all CRISPR components used here (Finn *et al.* 2018).

mRNA delivery technology has only really matured *after* the discovery of CRISPR/Cas and many companies now focus completely on CRISPR-based technologies. However, Sangamo Therapeutics is one of the few companies pioneering on ZFNs and they have also demonstrated preclinical mRNA-based delivery in mice. ZFNs and TALENs have their specificity built into their sequence and therefore do not require a guide RNA, which means they can be delivered via mRNAs alone. Both ZFNs and TALENs work in pairs, so two mRNA sequences need to be delivered to the same cell for them to be effective. Sangamo showed that encapsulation of the two ZFN mRNAs in 1 : 1 ratio in the same LNP enabled >90% knock-out of TTR and PCSK9 genes in mice. They also showed that repeated administration of the LNP was tolerated and increased levels of genome editing with each additional administration (Conway *et al.* 2019).

The immense interest in gene editing technologies has had a huge impact on the development of modified mRNA technology and LNPs for delivery. Combined, they are already being used in several clinical programs for various applications. LNP-based mRNA delivery is currently seen as the key technology that will also enable therapeutic gene editing in clinic.

18.2.3 Protein/Ribonucleoprotein Complex

Delivering the RNP complex (ex., Cas protein and sgRNA) is the most straightforward method for genome editing. Among the three different payload modalities, RNP-based CRISPR/Cas systems cause the least off-target effects because the number of RNPs delivered into target cell is limited and can be degraded quickly after entering the cell. Mechanical methods can be used to deliver proteins *in vitro/ex vivo,* but *in vivo* it is currently an even larger challenge than delivery of DNA or mRNA due to the lack of delivery systems that are tailored towards intracellular protein delivery. This challenge has also slowed down the clinical translation of protein-based gene editors such as ZFNs and TALENs.

For *ex vivo* applications, it was found that due to the intrinsic positive charge of the zinc-finger domain, ZFNs can be delivered directly into the cytosol of *in vitro* cultured cells without any delivery systems or transfection reagent (Gaj *et al.* 2012). TALENs (Liu *et al.* 2014) and Cas9 nucleases have been conjugated to cell penetrating peptides (CPP) to deliver them to *in vitro* cultures, whereas for the Cas9 application, the guide RNA was delivered by complexation to a similar, positively charged CPP (Ramakrishna *et al.* 2014).

But Cas9 protein itself is positively charged so the guide RNA can be complexed to it without any modifications and the RNP complex can be delivered to the cell via electroporation (Kim *et al.* 2014). Delivery as the RNP complex guarantees both components end up in the same cell and this approach could also be used with different direct delivery, or mechanical methods. For example, direct delivery of the RNP complex was demonstrated using a method called iTop, that employs a combination of hypertonicity-induced macropinocytosis and the transduction compound propanebetaine (induced transduction by osmocytosis and propanebetaine (D'Astolfo *et al.* 2015)). Alternatively, a mechanical method called TRIAMF, short for TRansmembrane Internalization Assisted by Membrane Filtration, was developed for the delivery of Cas9 RNP complex to human hematopoietic stem cells that temporarily permeabilize the cells by passing them through a filter membrane (Yen *et al.* 2018). All these methods are applicable for the manipulation of stem cells or T-cells with the Cas9 RNP outside the body, after which those cells can then be (re-)infused into a patient as therapy.

However, none of these methods are directly applicable *in vivo*. One of the features that makes gene editing nucleases so difficult to deliver is that proteins that bind nucleic acids are natively cationic. Upon complexation with the sgRNA, the Cas9 protein becomes a bit more anionic and it was shown that the complex could be delivered by the same type of cationic transfection reagents used for nucleic acid delivery (Zuris *et al.* 2014; Wang *et al.* 2016; Wei *et al.* 2020). This is a relatively new approach that is not widely adopted yet but it was demonstrated recently that the RNP complex can be delivered to several tissues in mice. This enables *in vivo* application with the desired profile of short persistence of the editing enzyme in the cell (Wei *et al.* 2020).

Of note, all described approaches with RNP complex delivery aim to knock-out a gene, not to edit it. It seems unlikely that a protein/sgRNA complex and the repair template can be delivered with a single particle, so if gene editing is the goal, either a combination strategy or nonprotein-based approach should be chosen.

18.3 Delivery Technologies

In a laboratory setting, all previously described modalities can be used and it is mostly practical considerations such as ease and cost of production or compatibility with high-throughput screens that influence the selected approach. For applications in test animals or later in patients, options are much more limited by how effectively the editing components can be delivered to the desired tissues. With an eye toward clinical application, safety and manufacturability are additional factors that need to be considered. Delivery technologies for *in vivo* application can be classified as viral-based and nonviral. Looking at the clinical landscape, it can be concluded that there is no clear preference yet for either of these technologies and time will tell which approach will prove to be the most favorable. Within each class, however, some clear trends can be identified. For non-viral delivery, LNPs are clearly the most advanced technology. For viral delivery, most activity nowadays seems to revolve around adeno-associated viruses (AAVs). Both classes of delivery technology will be described in the following sections, highlighting the most important issues for clinical development and application in gene editing.

18.3.1 Viral Vectors

The original concept of Gene Therapy was to deliver a functional copy of a gene to cells that did not have one. It seemed logical to use viruses for this as they are naturally capable of doing that, but there are risks involved with using pathogens for therapy. The commonly mentioned disadvantages of viral delivery are that they are pathogenic, replicate in their host, and (uncontrollably) integrate into the host's genome. A comparison of the main properties of viral vectors used for gene therapy is shown in Table 18.1.

More recently, almost all clinically approved viral vectors are of the AAV-type (AAVs). Two gene therapy products based on AAVs are currently on the market. One is Spark Therapeutics' LUXTURNA that delivers a functional copy of the *RPE65* gene aiming to treat Leber's congenital amaurosis, after intravitreal injection directly into the eye (Russell *et al.* 2017). The second one is Novartis' ZOLGENSMA, designed to deliver a functional copy of the human survival motor neuron (SMN) gene to the motor neuron cells of patients with spinal muscular atrophy (SMA), currently registered for intravenous administration (Hoy 2019). AAVs are reasonably safe, less immunogenic, replication deficient, and minimally integrating, thereby mitigating many of the previous risks. Problems that remain are difficulty to produce these vectors and their limited packing size. AAVs can only package a maximum of ~4.7 kb which is not enough for the typical ZFN or Cas9 nucleases, let alone the much larger base editors (Rees and Liu 2018) and "prime editors" (Anzalone *et al.* 2019) that are emerging. There are some workarounds to this, like using an intein-mediated split-Cas9 system that delivers the gene split between two viral vectors and recombines it inside the cell (Truong *et al.* 2015), or delivering the ZFN pair with two separate vectors (Weber *et al.* 2014).

But there is a more important reason why viruses are not used for expression of the nuclease. Compared with DNA, mRNA, and protein delivery, delivery via a viral vector has by far the longest lasting expression. As mentioned, viruses have traditionally been used to "add" genes. A recent report showed that Factor VIII gene replacement with an AAV in Hemophilia A patients still had a clinically relevant beneficial effect three years after administration (Pasi *et al.* 2020). Clearly, there is no point in having such long expression of the nuclease and the guide and it would only impose a huge safety risk.

Therefore, solutions need to be found to limit the duration of expression when viral vectors are used to deliver the nuclease. And to address the limited packaging capacity of AAVs, strategies need to be devised to divide the editing components over different vectors or to split a single construct over multiple AAVs. In the next section, the

Table 18.1 Overview of viral vectors used for gene therapy.

Vector type	Packaging capacity (kb)	Diameter (nm)	Genome type	Advantages	Disadvantages
Lentivirus	<8.5	80–120	ssRNA	Large packaging capacity	Potential insertional mutagenesis
Adenovirus	>8	80–100	dsDNA	Large packaging capacity, Transient Cas expression (non-integrating)	Pre-existing antibodies, high immunogenicity
Adeno-associated virus	<4.4	20–22	ssDNA	Large variety of target tissues, low immunogenicity on first injection	Low packaging capacity

production process of AAVs will be described first, which will help explain the differences between AAVs and other viral vectors and why they are challenging to scale-up. This will serve as an introduction to all efforts that are related to AAV engineering to make them suitable for gene editing.

18.3.1.1 Production

AAVs can infect dividing and nondividing human cells but are not known to cause any disease. Unlike retroviruses that randomly integrate into the genome which could in some cases lead to cancer, AAV genes will primarily remain as episomal DNA in the nucleus. Therefore, AAVs are regarded as the safest viral vectors available, albeit never completely free of risk. AAVs are part of the family of Parvoviruses and the genus *dependoparvovirus*, named like this because they depend on other viruses' genes to replicate. "Helper" viruses could be adenoviruses, herpesviruses, or vaccinia viruses so in the rare event that a host's cell is also infected with one of these viruses, the AAV could still replicate, making it a small but not impossible risk.

The dependence on additional genes for replication also makes the production process challenging. Typically a "helper-free" method is used, where the helper genes are coded by an additional plasmid rather than a live virus. The AAV genome contains ~4.7 kb of ssDNA and encodes two genes: *rep* and *cap*. The *rep* gene encodes proteins related to replication and integration and the *cap* gene contains the three proteins that form the viral capsid (VP1, VP2, VP3). The genome is flanked by inverted terminal repeats (ITRs) on each side. These ITRs contain palindromic sequences that normally make it fold upon themselves, to form a T-shaped hairpin. This structural element is required to load the gene into the capsid proteins during packaging (Wang *et al.* 2019). Therefore, the complete length of the *rep* and *cap* genes can be replaced with the recombinant gene of interest up to 4.7 kb, flanked by the ITRs to be packaged into a recombinant AAV. By removing the full *rep* and *cap* genes, the recombinant vector also becomes replication-deficient. But this 4.7 kb is relatively small and does pose challenges to encode the full gene- or base editing constructs, see the following sections.

Recombinant production nowadays is mostly done in HEK293 host cells because they already constitutively express certain adenovirus genes. Remaining helper genes are expressed from a helper plasmid. Because the wild-type *rep* and *cap* genes have been replaced by the transgene, a second plasmid is used to express these. Finally, a third plasmid will encode the gene of interest, flanked by the ITR sequences. Transfection with three different plasmids in adherent HEK293 cells is an established process in academic labs and in Drug Discovery, but it is challenging to scale up to several hundreds or thousands of liters of culture medium. A solution that is looked at for production at the industrial scale is to use producer cell lines that stably express all required genes under an inducible promoter and preferably grow in suspension for optimal compatibility with all sizes of bioreactors. Another challenge is the down-stream processing, as recombinant AAVs could be difficult to separate from process-related impurities such as host protein and DNA or product-related impurities, such as empty AAVs or AAV capsids containing foreign DNA (Wang *et al.* 2019). Addressing these challenges is beyond the scope of this chapter. Instead, several approaches that are used to engineer the vectors for specific gene editing applications will be described.

18.3.1.2 Smaller Cas9 Variants

One way to fit the transgene into the small packing size of 4.7 kb is to simply make the gene shorter. This concept works for the human dystrophin gene that is normally 11.5 kb long, but several truncated versions called micro-dystrophins were found that are smaller than 4.0 kb and could be fit into AAV vectors. These truncated versions are – naturally - less functional than the full-length protein, but have been shown to be beneficial in a large number of animal studies for Duchenne's Muscular Dystrophy (DMD; Duan 2018) and have been investigated in clinic (Bowles *et al.* 2012). The commonly used *Streptococcus Pyogenes* Cas9 (SpCas9) is ~4.1 kb which would leave just enough space to include one sgRNA in an AAV, but no space for regulatory elements such as tissue-specific promoters. To this end, Cas9 orthologs were screened and several "naturally truncated" variants of Cas9 were identified like *Staphylococcus Aureus* Cas9 (SaCas9) of ~3.2 kb (Ran *et al.* 2015) and *Campylobacter jejuni* Cas9 (CjCas9) of ~3.0 kb (Kim *et al.* 2017). Each of these variants was shown to fit into an AAV, with additional space for one or more sgRNAs and both were shown to edit their targets *in vivo*.

18.3.1.3 Split-Constructs

An alternative strategy to encode longer genes or include more regulatory elements is to split the gene over separate vectors. There are several ways to do this and depending on the strategy, the two vectors can be recombined inside the target cell on DNA level, mRNA level, and protein level. On the DNA level, the ITRs of two different vectors can be designed so that they overlap with each other and recombine inside the target cell and when flanked by appropriate splicing signals, the two pieces are joined into a single strand of DNA. This strategy was used to deliver an adenine base editor (ABE) split over dual AAVs in a DMD mouse model (Ryu *et al.* 2018). On the RNA level, two pre-mRNAs can also be recombined into one. This concept was demonstrated with two AAVs each encoding a half part of the CFTR gene, that were translated in two different pre-mRNAs that were then spliced into a single full-length CFTR mRNA in *in vitro* cell culture of human lung epithelial cells (Song *et al.* 2009). Recombination on the protein level is a more recent approach which requires the protein of interest to be able to be split into two parts that are smaller than the packaging limit of an AAV. With this system, two protein halves are expressed first, each fused with a short intein-sequence that is spliced out when the two halves are joined, resulting in a single full-length protein. This was first shown for Cas9 nuclease (Truong *et al.* 2015) and later for adenine (ABE) and cytosine base editors (CBE) with optimized AAV vectors for different tissues (Levy *et al.* 2020). It should be noted that the editing with split-intein Cas9 was a lot less efficient than with normal Cas9 and it took some effort to find where the base editors could be optimally split into two. However, especially for the large constructs that are required for base editing, this seems to be a necessary approach if delivery with AAVs is desired (Koblan *et al.* 2021). In combination with AAV vectors of different serotypes, this concept was demonstrated in the brain, liver, retina, heart, and skeletal muscles of mice (Levy *et al.* 2020).

18.3.1.4 Tissue Tropism

To enter the target cell, the capsid proteins first bind specific proteins or receptors on the outside of the target cells and different capsid proteins can bind different target tissues. Hundreds of different naturally occurring capsid sequences have been identified that are currently clustered in 12 different serotypes (AAV1 to AAV12). Originally, a new serotype was defined if an antibody that was known to bind to the capsid of one serotype would not bind to the capsid protein of the newly identified AAV. This is of relevance because parts of the population may already have pre-existing immunity against certain serotypes, making those less suitable than others. In addition, patients that will get treated with one serotype will most likely develop antibodies against it, so if repeated administration is ever considered, it should be done with different serotypes. However, with the identification of more and more naturally occurring variants and the engineering of new/mutant capsids, differences are becoming increasingly smaller and the true distinction has become a bit blurred (Choi *et al.* 2005).

The capsid proteins are encoded by the *cap* gene and the three proteins VP1, VP2, and VP3 together form the capsid in a 1 : 1 : 10 ratio with 60 proteins in total forming a single viral capsid. Different capsids bind different proteins on the target cells and the binding and internalization of AAV2 is probably the best characterized of all serotypes. The primary binding receptor for AAV2 is heparan sulfate proteoglycan (HSPG) and it was shown to use several other receptors as co-receptors for cellular entry such as $\alpha_V\beta5$ integrin, fibroblast growth factor receptor 1 (FGFR-1), and hepatocyte growth factor receptor (HGFR). Presence of these surface receptors correlates roughly with infectivity of AAV2 for these tissues, so AAV2 infects only cells with HSPG on the surface and further preference/tropism is dictated by the presence and density of one of the co-receptors. Indeed, this precludes some tissues but certainly does not limit infectivity to a single tissue or organ; AAV2 was shown to infect hepatocytes as well as retinal cells, different types of muscle cells, and neurons (Choi *et al.* 2005). Of note, the LUXTURNA gene therapy for retinal dystrophy is an AAV2 vector (Russell *et al.* 2017). Other serotypes may be slightly more specific for one tissue, like AAV6 for lung epithelial cells (Limberis *et al.* 2009), AAV7 for skeletal muscle, and AAV8 for liver (Gao *et al.* 2002).

In an attempt to improve the potency of AAV vectors in general, or in certain tissues specifically, capsids have been engineered, either by combining genes or by engineering of the capsid proteins.

18.3.1.5 Chimeric/Hybrid Capsids and Designer Capsids

Several approaches to engineer viral capsids to combine sequences or features from different serotypes have been described. One of the first employed methods was to mix the helper plasmids encoding the capsids of different AAV serotypes. The well-characterized AAV2 was combined with AAV1 to exploit the improved muscle transduction capacity of AAV1 and retain the heparin binding capacity of AAV2, which proved useful for purification using heparin columns. It was found that the ratio of incorporation of capsids from each subtype more or less followed the ratio in which the plasmids were added. But that meant that higher amounts of AAV1 capsid and less AAV2 led to vectors better capable of transducing muscle and less heparin binding and vice versa, so it was impossible to optimally exploit both traits. Furthermore, this approach yielded heterogenic products and different results per batch, making it less suitable for clinical manufacturing (Hauck *et al.* 2003). A more elegant approach used genetic engineering to introduce 5 amino acids from AAV1 into the capsid of AAV2 producing a new vector that was called AAV2.5 that retained both the muscle transduction capacity and heparin binding. At the same time, these mutations reduced the cross-reactivity with serum containing antibodies against either of the parent serotypes. This AAV2.5 vector was evaluated in a trial for DMD, delivering a version of the above-mentioned minidystrophin (Bowles *et al.* 2012).

Modern recombinant DNA techniques such as DNA shuffling and "error-prone PCR" can be used to create capsid libraries based on existing serotypes or to generate new sequences and using a process called directed evolution, one can select for features such as lower cross-reactivity with known serotypes or improved tropism for certain tissues or cell types. An alternative approach to directed evolution is the rational design of capsids, by introducing sequences or functionalities at a specific position on the capsid. For example, an Arg-Gly-Asp (RGD) motif was inserted into the capsid of AAV2. The RGD sequence is known to bind $\alpha_V\beta5$ integrin on endothelial cells and this way the AAV2 could be "retargeted" to infect integrin-expressing cells independent of the HSPG receptor (Shi and Bartlett 2003). Examples of chemical modification of the capsid are conjugation with a PEG-polymer to protect it from neutralization by antibodies. One such approach used the already existing lysine residues on the capsid to conjugate to (Lee *et al.* 2005). Another approach introduced a non-natural amino acid containing an azide into the capsid and used click-chemistry to attach a cyclic RGD peptide for retargeting to integrin receptors (Kelemen *et al.* 2016). From a production perspective, encoding the targeting moiety in the sequence is favored over the conjugation method, but the click-chemistry approach could also be applied to larger or synthetic targeting moieties (such as antibody fragments or a tri-antennary GalNAc). For the time being, however, all these capsid engineering techniques are just part of a large toolbox and many efforts are still ongoing to produce vectors that more specifically interact with certain tissues or evade the immune system better. AAVs may have the ability to transfect tissues that LNPs are not very likely to transfect, but for both types of vectors, it is still mostly their biodistribution that dictates which tissues they can reach.

18.3.1.6 Site-Specific Promoters and Inducible Expression

Although achieving high-enough expression in target tissues remains a challenge, several solutions have been proposed to limit expression in nontarget tissues. A commonly used strategy is to use tissue-specific promoters and the native regulatory elements with the transgene. Again lending an example from the DMD trials with AAVs, several muscle specific promoters have been developed (where "developed" means in this case, reducing their size so they can be used in AAVs) (Duan 2018). This way, the impact of the inevitable spillover to the liver can be minimized because the transgene expression is under the control of a muscle specific promoter. This would limit the expression of the Cas nuclease in nontarget tissues.

Another option to limit undesired expression is to add a chemically inducible control element. Several approaches have been described, for example a tetracycline-inducible promoter that only expresses the Cas9 enzyme in the presence of doxycycline (Dow *et al.* 2015) or an inactivated form of Cas9 that only adopts its active conformation in the presence of the cell-permeable small molecule 4-hydroxytamoxifen (Davis *et al.* 2015). Every additional regulatory element increases the size of the gene, so if it will turn out to be necessary to split the gene over two separate AAV vectors, an inducible split-Cas9 system can be used that assembles only in the presence of Rapamycin (Zetsche *et al.* 2015). Alternatively, a self-restricting feedback loop could be built into the constructs. For example, an additional guide sequence could be added that cuts the Cas9 gene after the protein is expressed. Or more elegantly, a sequence that is recognized by the on-target sgRNA can be built into the Cas9 promoter sequence without disrupting it, so no additional guide sequences need to be added, reducing the risk of off-target cuts and the total size of the construct (Wang *et al.* 2020).

18.3.1.7 Summary

To summarize, there is a large number of activities underway to improve the safety and specificity of AAV vectors and to adapt them for gene editing. At the moment, however, all different aspects are addressed separately and everything has yet to fall into place in one feasible delivery strategy. Advancements in capsid engineering will help to improve interaction with target tissues and minimize interaction with nontarget tissues and could help to evade the immune system, i.e. the adverse reactions against the vector and potential pre-existing immunity. This will be of great use for all clinical AAV applications.

But the biggest obstacle to bring this all together for gene editing is the limited packaging capacity of AAVs. For CRISPR-based applications, several strategies have been proposed to make the expression more tissue specific and more short-lived but all of these additional regulatory elements also take up space in the vector. Smaller Cas9 variants like SaCas9 could be pivotal here, but with regulatory elements included, it is still possible that the construct has to be split over two vectors. A split-construct strategy would also put a burden on the production capacity, as twice the number of virus particles is required (not even taking into account the lower editing efficiency, which may require higher doses). Needless to say, intravitreal injection would also require a very different amount of total vectors than application in a larger organ like liver, lung, or muscle, so production capacity (and costs) is another factor that will determine which delivery strategy will be feasible in clinic.

At this moment, there are no ongoing clinical activities with AAVs to deliver the editing components. One strategy that is pursued by multiple companies, that does employ viral vectors, uses AAVs to deliver the editing template, while the editing components are delivered with LNPs. This combination strategy will be described in more detail in the final section.

18.3.2 Lipid Nanoparticles (LNPs)

Lipid-based nanoparticles such as liposomes have been used since the late 1970s and lipoplexes have been used extensively for nucleic acid delivery. However, their low potency and poor tolerability has limited their application in clinic for a long time. But the field has evolved and after decades of fundamental research in the area, LNPs are now clinically validated delivery systems for nucleic acids as evidenced by the approval of ONPATTRO® in 2018, the first LNP-formulated siRNA for the treatment of hereditary amyloidosis. Since then, LNPs have gained much attention for the delivery of other nucleic acid-based modalities including mRNA therapeutics and CRISPR-Cas components. Then, in the wake of the SARS-CoV2 (COVID-19) pandemic, mRNA-based vaccines have leapfrogged all other LNP applications, demonstrating a flexibility and speed of development that few people had imagined to be possible.

An LNP consists of a mostly spherical particle, which has packed the nucleic acids in the core, complexed with a cationic lipid. This way, the nucleic acids are protected from degradation by nucleases and can be transported through the bloodstream. Other components in an LNP are cholesterol and "helper- or fusogenic lipids" such as DSPC or DOPE and PEG-lipids (See Figure 18.4). The function of phospholipids and cholesterol is described in the later section. But the key component in an LNP is the cationic lipid. When nanoparticles are taken up inside a cell, they are collected in endo-lysosomal compartments and their release into the cytosol where they can exert their function is seen as the bottleneck for efficient delivery and potency. It is thought that the cationic component mediates the "escape" of the LNP or its cargo, from the endosomal compartments to the cytosol and better cationic lipids will increase the potency of the LNP.

18.3.2.1 Ionizable Cationic Lipid for LNPs

The ionizable cationic lipid is the critical component in an LNP that determines the delivery potency and safety. Initially, permanently charged cationic lipids have been employed in LNP formulations because they aid complexation of the nucleic acids and could achieve high encapsulation. However, these formulations suffered from poor pharmacokinetics and tolerability upon intravenous administration due to high cationic charge and were cleared rapidly from the blood by the reticuloendothelial system (RES). Permanently charged cationic lipids introduce a certain degree of cellular toxicity, stimulate the immune system, and could aggregate in the blood. The landmark work in this area came from Heyes *et al.* that identified a pH-sensitive cationic lipid, DLin-DMA, and developed the first LNP system that was able to potently deliver siRNA *in vivo* in non-human primates after intravenous administration (Heyes *et al.* 2005; Zimmermann *et al.* 2006). These pH-sensitive lipids are neutral under physiological conditions but attain positive charge at endosomal pH. This feature enables the lipid to efficiently complex and induce endosomal fusion at acidic pH without the drawback of rapid elimination and poor tolerability as the lipid particles are net neutral at physiological pH. This lack of substantial positive charge in ionizable lipid at physiological pH enhances the circulation time (~30 minutes), thereby allowing better accumulation in target tissues.

Another key advancement was the discovery that increasing the degree of unsaturation in the hydrophobic domain of the ionizable lipid dramatically increased the potency. Notably, lipids with hydrophobic tails derived from linoleic acid (C18 chain with two double bonds) showed better intracellular delivery compared with oleic acid (C18 chain with one double bond). It was postulated that increasing the number of double bonds in close proximity induces a kink in the chain that gives the lipid a more conical shape. Increasing conical shapes correlated with

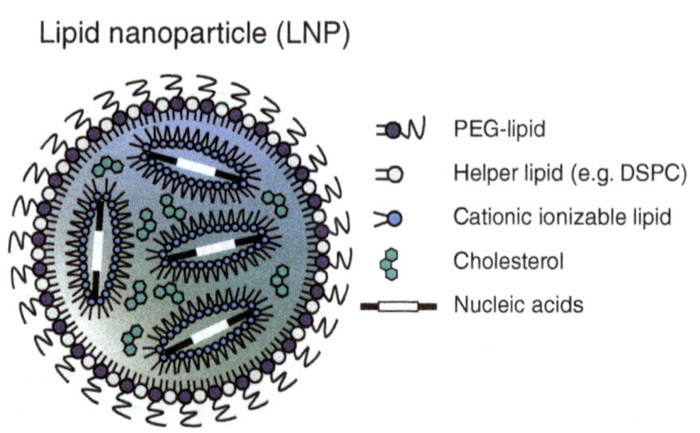

Figure 18.4 Internal structure and components of a lipid nanoparticle. Nucleic acids are encapsulated in the core of the LNP, complexed with the cationic ionizable lipid that forms a reversed hexagonal phase (H_{II}). The phospholipid helper lipid (DSPC) is localized on the surface of the LNP, together with the PEG-conjugated lipid that provides colloidal stability. Cholesterol is packed mostly in the core of the LNP.

the increased formation of reversed hexagonal phases (H_{II}) in the particle which indicates increased fusogenicity. It was hypothesized that unsaturated hydrocarbon chains in the lipid tail are able to induce membrane fusion and help the siRNA payload to escape from the endosome and allow effective cytosolic delivery (Heyes *et al.* 2005).

From a structure–activity relationship (SAR) perspective, there are several important factors in ionizable lipid design associated with potency and tolerability. First, the apparent pKa value of the LNPs is affected by the ionizability of the headgroup, known to be a key determinant of potency. It must be clarified here that pKa values for ionizable lipids estimated by, e.g. Chemdraw software are not a predictor of apparent pKa in LNPs as contributions by the supramolecular structures in aqueous environment must also be taken into consideration. Apparent pKa values of LNPs are typically measured by TNS assay. In most cases, the apparent pKa values of LNPs determined by TNS assay are found to be lower than pKa values of the ionizable lipid estimated by a Chemdraw software, suggesting a substantial but only partial contribution coming from the ionizable lipid itself in the apparent pKa of LNPs. Further findings led to the suggestion that the apparent pKa should be between 6.2 and 6.5 for optimum delivery to hepatocytes in the liver (Jayaraman *et al.* 2012). These features were optimized in ionizable lipids that resulted in the development of Dlin-MC3-DMA or "MC3," the ionizable lipid used in the ONPATTRO® LNP-siRNA product.

The biodegradability of lipids upon intracellular delivery is also an important factor in order to improve the tolerability. For example, the serum half-life of an MC3 LNP is approximately eight hours but the half-life of the MC3 molecule in the liver is more than 48 hours and accumulates after repeated dosing (Sabnis *et al.* 2018). In mice, the MC3 LNP is not tolerated at doses above 0.3 mg/kg which is especially problematic for genome editing applications where the anticipated dose levels are much higher, at least 2.0 mg/kg. Therefore, ester bonds were introduced in the hydrophobic tails to make the pH-sensitive cationic lipids biodegradable. The L319 lipid, for example, is the biodegradable version of MC3 (Maier *et al.* 2013). Another biodegradable lipid, Lipid 5, was designed for mRNA delivery and was shown in literature to be more potent than MC3 with a significant improvement in safety (Sabnis *et al.* 2018).

Finally, a subset of ionizable lipids termed as "lipidoids" have also been designed and synthesized by Anderson and colleagues in a combinatorial fashion that were shown to deliver nucleic acids to hepatocytes and endothelial cells. They reported that hydrocarbon chains with 12–14 carbon atoms were preferred for lipidoids and in addition, the DOPE helper lipid is better suited in their formulation as opposed to DSPC used in other ionizable cationic lipid formulations. Two published examples of lipidoids are C12-200 (Love *et al.* 2010) and cKK-E12 (Dong *et al.* 2014) that showed improved potency and performance compared with MC3. See Figure 18.5.

18.3.2.2 PEG-Lipids

Polyethylene glycol conjugated lipids, or PEG-lipids, play an important role in particle formation, stability, and transfection efficiency. PEG-lipids help prevent particle aggregation by creating a steric barrier around the particle surface. The steric barrier of the PEG-lipid prevents particle fusion during the LNP manufacturing process and helps to produce a homogenous particle population with narrow polydispersity and size usually between 50 and 100 nm. Typically, an LNP formulation contains 0.5–3.0 molar percentage of PEG-lipids and because the PEG-lipid can only reside on the surface of the particle, higher amounts of PEG-lipids force the particles to become smaller (Belliveau *et al.* 2012). In addition, PEG-lipids can be further exploited to balance circulation time and cellular uptake. When administered to the bloodstream, the PEG-lipids dissociate from the LNP and the length of the acyl chains can be varied to shorten the residence time of the PEG-lipid on the LNP and thereby the circulation time of the LNP in the blood. For example, PEG-lipids containing two C14 chains are found to be optimal for delivery to hepatocytes due to shorter residence time and quick uptake in the liver. This is in contrast for LNP-mediated delivery to the tumor where PEG-lipids with C18 or C20 chains are used in order to increase the circulatory time in blood to allow accumulation of the particles in solid tumor sites.

18.3.2.3 Phospholipids and Cholesterol

Phospholipids and cholesterol provide structural integrity and fusogenicity to the LNPs. PEG-lipids and DSPC reside mostly on the surface of the LNP, where the PEG-lipid provides colloidal stability and the amount of DSPC seems to influence interaction with target membranes. Cholesterol was reported to be mostly in the core of the particle (Arteta *et al.* 2018). The function of cholesterol is not completely clear, but it could be versatile. It was initially seen as "filler lipid" but it is also thought to help ordering the lipid membranes, providing structural integrity and make them more fluid, thereby adding to the fusogenicity. Recent insights have shown that cholesterol and analogs may also be important signaling molecules that determine how an LNP is processed inside the cell. A recent study showed that incorporation of C-24 sterols in LNPs enhanced the transfection efficiency considerably *in vitro* (Patel *et al.* 2020).

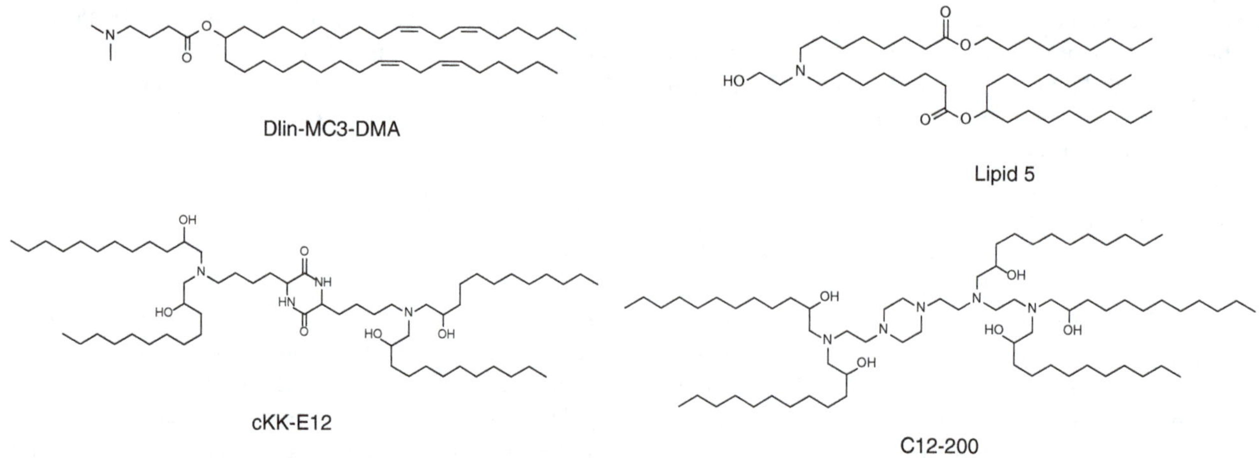

Figure 18.5 **Chemical structures of cationic ionizable lipids.** Dlin-MC3-DMA or "MC3" is the ionizable lipid in the clinically approved siRNA LNP formulation. L319 is the biodegradable variant of MC3, with hydrolysable esters in the acyl tail. Lipid 5 is a biodegradable lipid designed for mRNA delivery. C12-200 and cKK-E12 are "lipidoids" that are characterized by more than one ionizable amine group.

18.3.2.4 Formulations with Microfluidic Mixing

An important improvement that has helped transition LNP technology into the clinic is the formulation of these nanoparticles using microfluidic mixing (Evers *et al.* 2018). Traditionally, lipid particles were always formed by mixing the lipid components in organic solvent, with the drug compound in aqueous solution. In the case of nucleic acids and ionizable lipids, the aqueous solution would be an acidic buffer, to ensure the ionizable lipid is fully charged. But it was found that the conical shape of the cationic lipid does not make it mix well with bilayer forming lipids like DSPC and PEG-lipids, which leads to nonhomogeneous mixtures of various sized particles with unequal distribution of the excipients over all particles. The solution to this problem was to use microfluidic mixing devices that use a controlled flow rate and a fixed ratio (typically 3 : 1 of acidic aqueous buffer to ethanol) to gradually dilute the lipid mixture in the aqueous phase. The mixture is then dialyzed to remove ethanol and restore the pH to physiologically relevant levels. During this stage, the particles reorganize themselves to display a very monodisperse size distribution, for which the size is mostly dictated by the amount of PEG-lipids in the composition (Kulkarni *et al.* 2018). Preparation of LNPs with microfluidic mixing allowed the reproducible preparation of prototype formulations and enabled the thorough characterization and understanding that is required for such complex systems to progress beyond preclinical development. Another advantage of microfluidic mixing is that this technique is a continuous process that can be scaled, unlike other methods that have traditionally been used to prepare LNPs (such as sonication or high-pressure extrusion). This guarantees the same quality and control over the formulation throughout all steps of scale-up and development (Evers *et al.* 2018). Again, the 2020 SARS-CoV2 pandemic provided an unanticipated opportunity to demonstrate that clinical-grade material for mRNA-based vaccines could be produced in record time (Corbett *et al.* 2020) and that production could be scaled to millions of doses.

18.3.2.5 ApoE Adsorption and Tissue Interaction

Nanoparticles administered to the bloodstream predominantly end up in the liver. This is mainly due to the liver's high perfusion and high permeability and its physiological role to filter out nanosized particles and foreign substances from the blood. Initially, it was found that intravenously administered nanomedicines are taken up mostly by Kupffer cells in the liver. But, more serendipitously than by design, it was found that the LNPs with a near-neutral surface charge (pKa of the ionizable cationic lipid in the range of 6.4) and a size-range of 60–100 nm are predominantly taken up by hepatocytes in liver with high selectivity (Jayaraman *et al.* 2012).

Part of the (pre-)clinical development was the more thorough investigation of how this type of LNPs targets the hepatocytes so efficiently. Alnylam reported that apolipoprotein E (ApoE) that normally mediates the metabolism and clearance of physiological lipoproteins also plays an important role in the hepatocellular uptake of intravenously administered ionizable LNPs. Ionizable LNPs, but not permanently cationic LNPs, adsorb ApoE similar to low-density lipoproteins (LDL) and use the LDL receptor for uptake in hepatocytes. The dependence on both was demonstrated employing *LDLR-/-* mice and

ApoE-/- mice in which silencing by siRNA LNPs in the liver was abolished. *In vivo* delivery to the hepatocytes of *ApoE-/-* mice could be rescued by pre-adsorbing the LNPs with recombinant ApoE. As an alternative to this "endogenous targeting," an exogenous targeting approach was developed, which employs the addition of a multivalent N-acetylgalactosamine (GalNAc) ligand to the LNPs. This also rescued the delivery to *ApoE-/-* mice and performed similar in WT and *LDLR-/-* mice, demonstrating this mechanism is also not dependent on LDLR. Instead, using *ASGR2-/-* mice, it was demonstrated that GalNAc-targeted LNPs are dependent on the asialoglycoprotein-binding receptor, providing two different mechanisms for effective uptake into the liver (Akinc *et al.* 2010). Of note, using the same knock-out mouse models, the lipidoid cKK-E12 was shown to be also dependent on ApoE but not on LDLR for uptake (Dong *et al.* 2014).

This strong dependence on ApoE and in most cases also LDLR raises the question how applicable LNPs are in other tissues or administration routes. An emerging trend in the field is the manipulation of LNP's surface chemistry and the protein corona it adsorbs when administered to the bloodstream. It was demonstrated that changing the surface composition and charge had profound impact on *in vivo* biodistribution after intravenous injection and enabled targeting to disease-relevant tissues like lung and spleen (Cheng *et al.* 2020).

In addition to tissue distribution, the formulation strategy also needs to be tuned for each specific application and delivery route. Moderna Therapeutics has described the optimization of lipid formulations for intramuscular (IM) administration of mRNA-based vaccines. They selected a different ionizable lipid than the one they used for IV administrations and found that a slightly higher pKa of 6.9–7.1 is ideal for IM administration (Hassett *et al.* 2019). This could indicate that some positive charge is required to enable cellular uptake in the absence of ApoE and LDLR. The slightly charged surface of the LNP can interact with the negatively charged proteoglycans that are present on the outside of all cell types, to enable uptake. Obviously, this makes the cell interaction less specific, but that is less of an issue when injected locally. Of note, with the IM route of administration, protein expressed from mRNA is likely produced only very locally, so it is not clear whether this would be a suitable administration route for gene editing therapies in, e.g. DMD, where all muscles would have to be treated.

Another feasible administration route would be inhalation of LNPs to the lung. For therapeutic gene editing that would be important to treat cystic fibrosis (CF) but for this application the epithelial cells need to be targeted. These make up only a small part of the large surface area of the lung, so here some strategy should be devised to selectively target the epithelial cells (for example, with an "exogenous" targeting ligand) or to avoid uptake by phagocytic cells. In addition, making an LNP formulation inhalable is another technological challenge, although beyond the scope of this chapter. Translate Bio is a company that is focusing on mRNA therapeutics and they are using an inhaled LNP formulation in their CF clinical program, but not much is known about that formulation.

To summarize, LNPs are at present the preferred non-viral delivery vehicle for nucleic acids. Their track-record for siRNA and mRNA therapeutics has made them the most studied delivery vehicle for gene editing approaches as well, but application beyond the liver remains challenging. The key factors to the clinical success of LNPs are ionizable lipids and controlled manufacturing using microfluidics. Most clinically used LNPs adsorb ApoE in the bloodstream which facilitates uptake into hepatocytes, but if this is not the target tissue, manipulation of protein coronas can be exploited to alter the biodistribution of LNPs. Alternative administration routes such as local injection or inhalation also grant access to different tissues, but likely the LNP formulation needs to be tweaked to achieve equally effective uptake.

18.3.3 Non-LNP-Based Delivery Systems

Besides LNPs, other non-viral delivery systems such as polymers and polypeptide-based synthetic materials have also been explored for genome editing applications. Such polymers typically consist of multiple building blocks, combining different functionalities in a single polymer chain. These polymers should complex and encapsulate the nucleic acid cargo and assemble into nanoparticles suitable for intravenous injection. Many different polymers have been designed and synthesized, exploring different types of chemistry, branching, and charge density (see Figure 18.6). All these polymeric materials have common physiochemical characteristics such as amphiphilicity, protonatable amines, and hydrolysable or bio-reducible bonds to make the materials biodegradable. The advantage of such materials compared with ionizable lipids is that they combine all features in a single component, (usually) not needing multiple components to form a nanosized particle like LNPs do. The downside to this is that it makes the polymer chains heterogeneous in length and composition. Better control over these features is necessary to create well-defined structures that allow for the clinical development of polymer-based carriers for nucleic acid delivery.

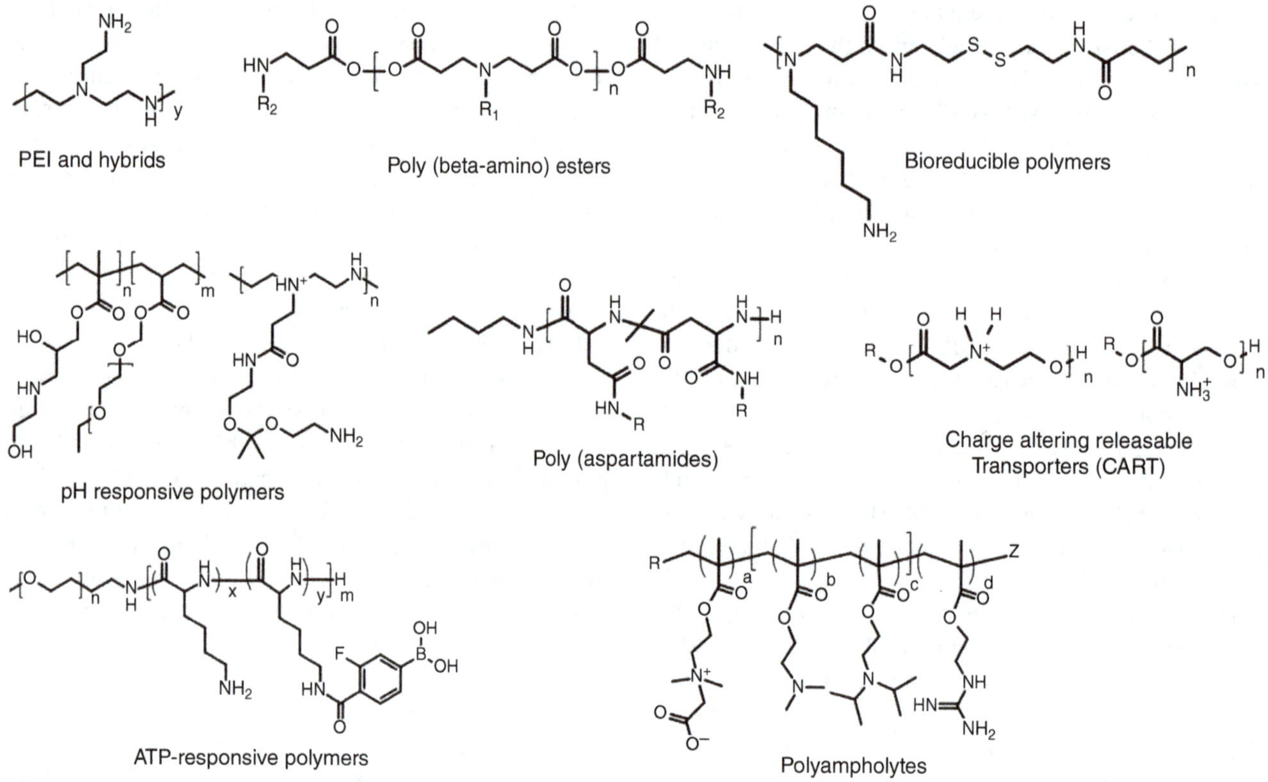

Figure 18.6 **Chemical structures of polymers for nucleic acid delivery.** Different functionalities can be combined in a single polymer chain, typically via amide, ester, or disulfide bonds. Polymers can be linear or branched and contain various types of amines to introduce charge, for complexation of the nucleic acids and to facilitate endosomal release.

18.4 *In vivo* Delivery Strategies

Different editing strategies require different delivery strategies, because different strategies require different numbers of components that need to be delivered. In the case of CRISPR-mediated gene knock-down or base-editing, the engineered enzyme and at least one guide RNA need to be delivered. In the case of gene correction by insertion, the DNA donor template needs to be delivered as well. ZFN and TALEN knock-down strategies do not require RNA guides, but employ a *pair* of nucleases that need to be delivered together. Most components could also be delivered as different modalities, most notably the editing enzyme that could be delivered as the protein directly, or expressed from a plasmid or an mRNA.

In theory, all options shown in Figure 18.2 are applicable. In practice, however, some modalities are used a lot more than others, due to characteristics such as duration of expression and ease of delivery. For example, due to the risk of off-target editing, the duration of expression of the editing enzyme should be as short as possible, so for that reason a plasmid is less favorable than expression from an mRNA. From this perspective, the delivery of the nuclease or RNP complex may be optimal, but there are no efficient methods to deliver proteins to cells *in vivo*. *In vitro* and *ex-vivo* applications are mostly enabled by "simple" electroporation which allows a lot more flexibility to optimize the approach for efficiency of editing/knock-down and minimal off-target effects, or even for process-related factors such as scalability and costs. *In vivo* applications on the other hand are almost always limited by efficiency of delivery and dictated by which tissues can be reached or modality can be delivered.

18.4.1 Clinically Feasible Delivery Strategies

In many cases, more than one delivery vehicle is required to successfully deliver all components to the right cell. It should be noted that there is currently not a single approach that has proven itself in clinic, but in Figure 18.7, several strategies will be outlined that could feasibly be applied.

ZFN and TALEN editing technologies are regarded as being a lot less flexible than CRISPR-based (RNA guided) strategies, but from a delivery perspective, these are certainly advantageous as the nuclease pair can be delivered with a single vector, either viral (Sharma *et al.* 2015) or non-viral encoded as an mRNA (Conway *et al.* 2019) (both examples of ZFNs).

As mentioned, CRISPR-based editing is much more flexible and versatile, but it does require the additional delivery of the guide RNA(s) as well. For CRISPR gene knock-down or base-editing, it is possible to deliver the enzyme and a limited number of guides using the same vehicle. This has been described using a single viral vector (Ran *et al.* 2015) as well as with a single LNP (Finn *et al.* 2018). But with LNPs being more tailorable than viruses, when a "synthetic strategy" (with LNP) is chosen, a mixture of LNPs could be used that are optimized for delivery of either the long mRNA or the short sgRNAs. Reasons to deliver the mRNA with a different LNP than the guide RNA(s) could be related to timing, for example to ensure the enzyme is actively expressed at the time of guide RNA delivery. Also, in preclinical research, separate formulations could be used to quickly combine one mRNA encoding the enzyme with several different sgRNAs in a different LNP or to test different ratios of mRNA to sgRNA by simply varying the administered amounts. The disadvantage of this approach is that two different LNPs need to reach the same target cell, which could be more challenging *in vivo* and a head-to-head comparison showed that the approach where the Cas9 mRNA and guide RNAs were delivered separately, indeed achieved a lower knock-down rate than co-encapsulated mRNA and guide RNAs (Finn *et al.* 2018).

When another layer of complexity is added, as in the case of gene replacement, it is almost certainly necessary to use more than one vector, because the editing components and the DNA template simply cannot be fitted in a single vector that is currently available. The components could be split in various ways, but there is one strategy emerging that is being pursued by several gene editing companies independently. It employs a combination of AAV and LNP, to exploit the favorable characteristics of each of the vectors. For delivery of the DNA donor template, AAVs offer the advantage that they can deliver the template directly to the nucleus. But as mentioned previously, the long-lasting expression from AAVs would pose a significant safety risk when used to deliver the nuclease. At the same time, because AAV vectors are designed to be non-integrating, transgene expression delivered by this type of viruses is not everlasting and will *still* be gradually lost in mitotic cells over time. Therefore, the commonly used strategy is to deliver the transgene with a viral vector and integrate it in the genome with the editing components that are delivered as more transient modalities. This strategy takes advantage of the virus' capability to deliver the transgene to the nucleus much more efficiently than when it would be delivered as DNA with a synthetic vector. The editing components can then be transiently expressed from mRNA delivered with an LNP and they should ensure integration and give control over the exact location. For example, to excise the defective gene and integrate the transgene downstream of its natural promoter, identical to wild-type cells. This AAV-LNP hybrid approach was first demonstrated in a mouse model of hereditary tyrosinemia type I (HTI) where the repair template and the sgRNAs were delivered via AAV and the Cas9 mRNA via an LNP (Yin *et al.* 2016). (This approach gives prolonged expression of the sgRNA as well, but because the Cas9 was expressed from mRNA, the period that both components are present at the same time is still short.) Sangamo Therapeutics used this approach to deliver a promoter-less hIDS gene with an AAV and use ZFNs expressed from mRNA in an LNP, to integrate the transgene under the *ALB* promoter in mice. They showed that with a single administration of the AAV and repeated administration of the LNP, increased percentages of integration and protein expression levels could be achieved, with each additional LNP dosing, at intervals of 14 days (Conway *et al.* 2019).

It could be argued that repeated dosing for gene editing will be too expensive for clinical use, but the published work by Sangamo Therapeutics demonstrates several important concepts. First, it illustrates the dramatic difference in duration of expression from a viral vector and an mRNA/LNP. Second, it demonstrates that LNPs can be used for repeated administration if necessary, which is perceived not to be the case for viral vectors of the same subtype. Finally, it proves the concept of *in vivo* gene insertion using the AAV-LNP hybrid approach, a strategy that exploits the characteristics of both types of delivery vehicles and is being pursued in clinic by several gene editing companies now.

Importantly, it also shows that viral and nonviral delivery systems are not competing technologies but could even work synergistically and that each of them can be developed for specific applications.

18.5 Conclusion

In conclusion, clinical efforts in therapeutic gene editing are currently limited to applications where the delivery is less of a challenge. Most applications manipulate cells *ex-vivo* to study whether the editing itself is safe and effective before attempting such revolutionary treatments *in vivo*.

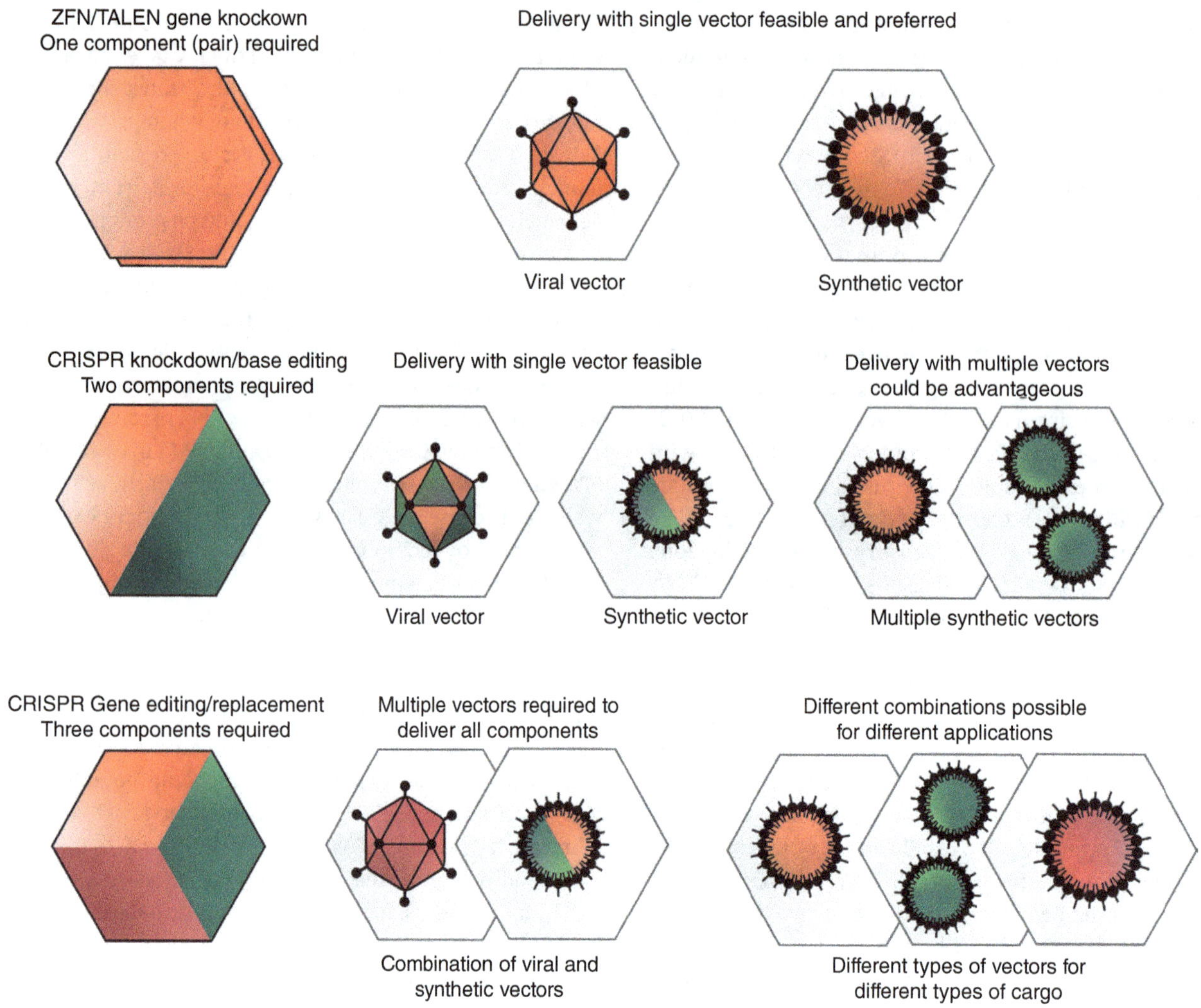

Figure 18.7 Different applications require different delivery strategies. Zinc-finger nuclease and TALE-based nuclease pairs could be fitted in a single vector, either viral or non-viral. For CRISPR-based knock-down, besides the nuclease at least one guide RNA needs to be delivered, which could also be fitted in a single vector of either viral or non-viral origin. In certain cases, there could be an advantage of delivering the compounds separately. For gene editing with addition of a new gene, it will almost certainly not be possible to deliver all components with one vector. Different combinations of vectors could be feasible, but one approach that is used more often is an AAV-LNP hybrid approach where the donor template is delivered with a viral vector and the editing components with an LNP.

In the meantime, clinical application of LNPs has gotten a tremendous boost thanks to the COVID-19 mRNA vaccines and important progress is also made to further improve viral vectors for gene editing purposes. This has led to several high-profile publications that leave no doubt that it will eventually be possible to apply this technology *in vivo*.

Based on the insights provided in this chapter, it is most likely that the first application of gene editing in the human body will be with CRISPR/Cas-based knock-down of a gene in the liver, where the editing components are delivered as mRNA/sgRNA with an LNP. Viral vectors could have the potential to edit genes in different tissues outside the liver and perhaps one day it will be possible to excise a whole gene and replace it with a new one. With the pace at which the fields of gene editing *and* drug delivery are advancing, the path to the clinic could look completely different in a couple of years. But one thing is certain, recent progress in drug delivery & formulation techniques has brought therapeutic gene editing a lot closer to clinical application and it will be exciting to see what the future holds for the field.

References

Akinc, A., Querbes, W., De, S. et al. (2010). Targeted delivery of RNAi therapeutics with endogenous and exogenous ligand-based mechanisms. *Molecular Therapy: The Journal of the American Society of Gene Therapy*. Nature Publishing Group 18 (7): 1357–1364. https://doi.org/10.1038/mt.2010.85.

Akinc, A., Maier, M.A., Manoharan, M. et al. (2019). The Onpattro story and the clinical translation of nanomedicines containing nucleic acid-based drugs. *Nature Nanotechnology*. Springer US 14 (12): 1084–1087. https://doi.org/10.1038/s41565-019-0591-y.

Anzalone, A.V., Randolph, P.B., Davis, J.R. et al. (2019). Search-and-replace genome editing without double-strand breaks or donor DNA. *Nature*. Springer US 576 (7785): 149–157. https://doi.org/10.1038/s41586-019-1711-4.

Arteta, M.Y., Kjellman, T., Bartesaghi, S. et al. (2018). Successful reprogramming of cellular protein production through mRNA delivered by functionalized lipid nanoparticles. *Proceedings of the National Academy of Sciences of the United States of America* 115 (15): E3351–E3360. https://doi.org/10.1073/pnas.1720542115.

Belliveau, N.M., Huft, J., Lin, P.J. et al. (2012). Microfluidic synthesis of highly potent limit-size lipid nanoparticles for in vivo delivery of siRNA. *Molecular Therapy--Nucleic Acids* 1: e37. https://doi.org/10.1038/mtna.2012.28.

Bowles, D.E., McPhee, S.W., Li, C. et al. (2012). Phase 1 gene therapy for Duchenne muscular dystrophy using a translational optimized AAV vector. *Molecular Therapy* 20 (2): 443–455. https://doi.org/10.1038/mt.2011.237.

Cheng, Q., Wei, T., Farbiak, L. et al. (2020). Selective organ targeting (SORT) nanoparticles for tissue-specific mRNA delivery and CRISPR–Cas gene editing. *Nature Nanotechnology* 15 (4): 313–320. https://doi.org/10.1038/s41565-020-0669-6.

Choi, V., McCarty, D., and Samulski, R. (2005). AAV hybrid serotypes: improved vectors for gene delivery. *Current Gene Therapy* 5 (3): 299–310. https://doi.org/10.2174/1566523054064968.

Conway, A., Mendel, M., Kim, K. et al. (2019). Non-viral delivery of zinc finger nuclease mRNA enables highly efficient in vivo genome editing of multiple therapeutic gene targets. *Molecular Therapy*. The American Society of Gene and Cell Therapy 27 (4): 866–877. https://doi.org/10.1016/j.ymthe.2019.03.003.

Corbett, K.S., Edwards, D.K., Leist, S.R. et al. (2020). SARS-CoV-2 mRNA vaccine design enabled by prototype pathogen preparedness. *Nature* 586 (7830): 567–571. https://doi.org/10.1038/s41586-020-2622-0.

D'Astolfo, D.S., Pagliero, R.J., Pras, A. et al. (2015). Efficient intracellular delivery of native proteins. *Cell*. Elsevier Inc. 161 (3): 674–690. https://doi.org/10.1016/j.cell.2015.03.028.

Davis, K.M., Pattanayak, V., Thompson, D.B. et al. (2015). Small molecule–triggered Cas9 protein with improved genome-editing specificity. *Nature Chemical Biology* 11 (5): 316–318. https://doi.org/10.1038/nchembio.1793.

Dong, Y., Love, K.T., Dorkin, J.R. et al. (2014). Lipopeptide nanoparticles for potent and selective siRNA delivery in rodents and nonhuman primates. *Proceedings of the National Academy of Sciences of the United States of America* 111 (11): 3955–3960. https://doi.org/10.1073/pnas.1322937111.

Dow, L.E., Fisher, J., O'Rourke, K.P. et al. (2015). Inducible in vivo genome editing with CRISPR-Cas9. *Nature Biotechnology* 33 (4): 390–394. https://doi.org/10.1038/nbt.3155.

Duan, D. (2018). Systemic AAV micro-dystrophin gene therapy for Duchenne muscular dystrophy. *Molecular Therapy* 26 (10): 2337–2356. https://doi.org/10.1016/j.ymthe.2018.07.011.

Evers, M.J.W., Kulkarni, J.A., van der Meel, R. et al. (2018). State-of-the-art design and rapid-mixing production techniques of lipid nanoparticles for nucleic acid delivery. *Small Methods* 2 (9): 1700375. https://doi.org/10.1002/smtd.201700375.

Finn, J.D., Smith, A.R., Patel, M.C. et al. (2018). A single administration of CRISPR/Cas9 lipid nanoparticles achieves robust and persistent in vivo genome editing. *Cell Reports*. ElsevierCompany 22 (9): 2227–2235. https://doi.org/10.1016/j.celrep.2018.02.014.

Friedmann, T. and Roblin, R. (1972). Gene therapy for human genetic disease? *Science* 175 (4025): 949–955. https://doi.org/10.1126/science.175.4025.949.

Gaj, T., Guo, J., Kato, Y. et al. (2012). Targeted gene knockout by direct delivery of zinc-finger nuclease proteins. *Nature Methods* 9 (8): 805–807. https://doi.org/10.1038/nmeth.2030.

Gao, G.-P., Alvira, M.R., Wang, L. et al. (2002). Novel adeno-associated viruses from rhesus monkeys as vectors for human gene therapy. *Proceedings of the National Academy of Sciences* 99 (18): 11854–11859. https://doi.org/10.1073/pnas.182412299.

Hassett, K.J., Benenato, K.E., Jacquinet, E. et al. (2019). Optimization of lipid nanoparticles for intramuscular administration of mRNA vaccines. *Molecular Therapy--Nucleic Acids* 15: 1–11. https://doi.org/10.1016/j.omtn.2019.01.013.

Hauck, B., Chen, L., and Xiao, W. (2003). Generation and characterization of chimeric recombinant AAV vectors. *Molecular Therapy* 7 (3): 419–425. https://doi.org/10.1016/S1525-0016(03)00012-1.

Heyes, J., Palmer, L., Bremner, K., and MacLachlan, I. (2005). Cationic lipid saturation influences intracellular delivery of encapsulated nucleic acids. *Journal of Controlled Release* 107 (2): 276–287. https://doi.org/10.1016/j.jconrel.2005.06.014.

Hoy, S.M. (2019). Onasemnogene abeparvovec: first global approval. *Drugs* 79 (11): 1255–1262. https://doi.org/10.1007/s40265-019-01162-5.

Jayaraman, M., Ansell, S.M., Mui, B.L. et al. (2012). Maximizing the potency of siRNA lipid nanoparticles for hepatic gene silencing in vivo. *Angewandte Chemie (International Ed. in English)* 51 (34): 8529–8533. https://doi.org/10.1002/anie.201203263.

Karikó, K., Muramatsu, H., Welsh, F.A. et al. (2008). Incorporation of pseudouridine into mRNA yields superior nonimmunogenic vector with increased translational capacity and biological stability. *Molecular Therapy*. The American Society of Gene Therapy 16 (11): 1833–1840. https://doi.org/10.1038/mt.2008.200.

Kelemen, R.E., Mukherjee, R., Cao, X. et al. (2016). A precise chemical strategy to alter the receptor specificity of the adeno-associated virus. *Angewandte Chemie International Edition* 55 (36): 10645–10649. https://doi.org/10.1002/anie.201604067.

Kim, S., Kim, D., Cho, S.W. et al. (2014). Highly efficient RNA-guided genome editing in human cells via delivery of purified Cas9 ribonucleoproteins. *Genome Research* 24 (6): 1012–1019. https://doi.org/10.1101/gr.171322.113.

Kim, E., Koo, T., Park, S.W. et al. (2017). In vivo genome editing with a small Cas9 orthologue derived from Campylobacter jejuni. *Nature Communications*. Nature Publishing Group 8 https://doi.org/10.1038/ncomms14500.

Koblan, L.W., Erdos, M.R., Wilson, C. et al. (2021). In vivo base editing rescues Hutchinson–Gilford progeria syndrome in mice. *Nature* 589 (7843): 608–614. https://doi.org/10.1038/s41586-020-03086-7.

Kulkarni, J.A., Darjuan, M.M., Mercer, J.E. et al. (2018). On the formation and morphology of lipid nanoparticles containing ionizable cationic lipids and siRNA. *ACS Nano* 12 (5): 4787–4795. https://doi.org/10.1021/acsnano.8b01516.

Lee, G.K., Maheshri, N., Kaspar, B., and Schaffer, D.V. (2005). PEG conjugation moderately protects adeno-associated viral vectors against antibody neutralization. *Biotechnology and Bioengineering* 92 (1): 24–34. https://doi.org/10.1002/bit.20562.

Levy, J.M., Yeh, W.H., Pendse, N. et al. (2020). Cytosine and adenine base editing of the brain, liver, retina, heart and skeletal muscle of mice via adeno-associated viruses. *Nature Biomedical Engineering*. Springer US 4 (1): 97–110. https://doi.org/10.1038/s41551-019-0501-5.

Limberis, M.P., Vandenberghe, L.H., Zhang, L. et al. (2009). Transduction efficiencies of novel AAV vectors in mouse airway epithelium in vivo and human ciliated airway epithelium in vitro. *Molecular Therapy* 17 (2): 294–301. https://doi.org/10.1038/mt.2008.261.

Liu, J., Gaj, T., Patterson, J.T. et al. (2014). Cell-penetrating peptide-mediated delivery of TALEN proteins via bioconjugation for genome engineering. *PLoS One* 9 (1): 1–7. https://doi.org/10.1371/journal.pone.0085755.

Love, K.T., Mahon, K.P., Levins, C.G. et al. (2010). Lipid-like materials for low-dose, in vivo gene silencing. *Proceedings of the National Academy of Sciences of the United States of America* 107 (5): 1864–1869. https://doi.org/10.1073/pnas.0910603106.

Maier, M.A., Jayaraman, M., Matsuda, S. et al. (2013). Biodegradable lipids enabling rapidly eliminated lipid nanoparticles for systemic delivery of RNAi therapeutics. *Molecular Therapy*. The American Society of Gene & Cell Therapy 21 (8): 1570–1578. https://doi.org/10.1038/mt.2013.124.

Pasi, K.J., Rangarajan, S., Mitchell, N. et al. (2020). Multiyear follow-up of AAV5-hFVIII-SQ gene tTherapy for hemophilia A. *The New England Journal of Medicine* 382 (1): 29–40. https://doi.org/10.1056/NEJMoa1908490.

Patel, S., Ashwanikumar, N., Robinson, E. et al. (2020). Naturally-occurring cholesterol analogues in lipid nanoparticles induce polymorphic shape and enhance intracellular delivery of mRNA. *Nature Communications* 11 (1): 1–13. https://doi.org/10.1038/s41467-020-14527-2.

Ramakrishna, S., Kwaku Dad, A.-B., Beloor, J. et al. (2014). Gene disruption by cell-penetrating peptide-mediated delivery of Cas9 protein and guide RNA. *Genome Research* 24 (6): 1020–1027. https://doi.org/10.1101/gr.171264.113.

Ran, F.A., Cong, L., Yan, W.X. et al. (2015). In vivo genome editing using Staphylococcus aureus Cas9. *Nature* 520 (7546): 186–191. https://doi.org/10.1038/nature14299.

Rees, H.A. and Liu, D.R. (2018). Base editing: precision chemistry on the genome and transcriptome of living cells. *Nature Reviews Genetics* 19 (12): 770–788. https://doi.org/10.1038/s41576-018-0059-1.

Russell, S., Bennett, J., Wellman, J.A. et al. (2017). Efficacy and safety of voretigene neparvovec (AAV2-hRPE65v2) in patients with RPE65 -mediated inherited retinal dystrophy: a randomised, controlled, open-label, phase 3 trial. *The Lancet* 390 (10097): 849–860. https://doi.org/10.1016/S0140-6736(17)31868-8.

Ryu, S.-M., Koo, T., Kim, K. et al. (2018). Adenine base editing in mouse embryos and an adult mouse model of Duchenne muscular dystrophy. *Nature Biotechnology* 36 (6): 536–539. https://doi.org/10.1038/nbt.4148.

Sabnis, S., Kumarasinghe, E.S., Salerno, T. et al. (2018). A novel amino lipid series for mRNA delivery: improved endosomal escape and sustained pharmacology and safety in non-human primates. *Molecular Therapy* 26 (6): 1509–1519. https://doi.org/10.1016/j.ymthe.2018.03.010.

Sharma, R., Anguela, X.M., Doyon, Y. et al. (2015). In vivo genome editing of the albumin locus as a platform for protein replacement therapy. *Blood* 126 (15): 1777–1784. https://doi.org/10.1182/blood-2014-12-615492.

Shi, W. and Bartlett, J.S. (2003). RGD inclusion in VP3 provides adeno-associated virus type 2 (AAV2)-based vectors with a heparan sulfate-independent cell entry mechanism. *Molecular Therapy* 7 (4): 515–525. https://doi.org/10.1016/S1525-0016(03)00042-X.

Song, Y., Lou, H.H., Boyer, J.L. et al. (2009). Functional cystic fibrosis transmembrane conductance regulator expression in cystic fibrosis airway epithelial cells by AAV6.2-mediated segmental trans -splicing. *Human Gene Therapy* 20 (3): 267–281. https://doi.org/10.1089/hum.2008.173.

Truong, D.-J.J., Kühner, K., Kühn, R. et al. (2015). Development of an intein-mediated split–Cas9 system for gene therapy. *Nucleic Acids Research* 43 (13): 6450–6458. https://doi.org/10.1093/nar/gkv601.

Vallazza, B., Petri, S., Poleganov, M.A. et al. (2015). Recombinant messenger RNA technology and its application in cancer immunotherapy, transcript replacement therapies, pluripotent stem cell induction, and beyond. *Wiley Interdisciplinary Reviews: RNA* 6 (5): 471–499. https://doi.org/10.1002/wrna.1288.

Wang, M., Zuris, J.A., and Meng, F. (2016). Efficient delivery of genome-editing proteins using bioreducible lipid nanoparticles. *Proceedings of the National Academy of Sciences of the United States of America* 113: 10–15. https://doi.org/10.1073/pnas.1520244113.

Wang, D., Tai, P.W.L., and Gao, G. (2019). Adeno-associated virus vector as a platform for gene therapy delivery. *Nature Reviews Drug Discovery* 18 (5): 358–378. https://doi.org/10.1038/s41573-019-0012-9.

Wang, H., Lu, H., Lei, Y.-S. et al. (2020). Development of a self-restricting CRISPR-Cas9 system to reduce off-target effects. *Molecular Therapy. Methods & Clinical Development* 18: 390–401. https://doi.org/10.1016/j.omtm.2020.06.012.

Weber, N.D., Stone, D., Sedlak, R.H. et al. (2014). AAV-mediated delivery of zinc finger nucleases targeting hepatitis B virus inhibits active replication. *PLoS One* 9 (5) https://doi.org/10.1371/journal.pone.0097579.

Wei, T., Cheng, Q., Min, Y.-L. et al. (2020). Systemic nanoparticle delivery of CRISPR-Cas9 ribonucleoproteins for effective tissue specific genome editing. *Nature Communications* 11 (1): 3232. https://doi.org/10.1038/s41467-020-17029-3.

Yen, J., Fiorino, M., Liu, Y. et al. (2018). TRIAMF: a new method for delivery of Cas9 ribonucleoprotein complex to human hematopoietic stem cells. *Scientific Reports* 8 (1): 16304. https://doi.org/10.1038/s41598-018-34601-6.

Yin, H., Song, C.-Q., Dorkin, J.R. et al. (2016). Therapeutic genome editing by combined viral and non-viral delivery of CRISPR system components in vivo. *Nature Biotechnology*. Nature Publishing Group 34 (3): 328–333. https://doi.org/10.1038/nbt.3471.

Yin, H., Song, C.-Q., Suresh, S. et al. (2017). Structure-guided chemical modification of guide RNA enables potent non-viral in vivo genome editing. *Nature Biotechnology* 35 (12): 1179–1187. https://doi.org/10.1038/nbt.4005.

Zetsche, B., Volz, S.E., and Zhang, F. (2015). A split-Cas9 architecture for inducible genome editing and transcription modulation. *Nature Biotechnology* 33 (2): 139–142. https://doi.org/10.1038/nbt.3149.

Zimmermann, T.S., Lee, A.C.H., Akinc, A. et al. (2006). RNAi-mediated gene silencing in non-human primates. *Nature* 441 (7089): 111–114. https://doi.org/10.1038/nature04688.

Zuris, J.A., Thompson, D.B., Shu, Y. et al. (2014). Cationic lipid-mediated delivery of proteins enables efficient protein-based genome editing in vitro and in vivo. *Nature Biotechnology* 33 (1): 73–80. https://doi.org/10.1038/nbt.3081.

19

Safety Aspects of Genome Editing

Immunogenicity

Rakesh Kantilal Chandode and Roberto Nitsch

Gene Therapy, Functional and Mechanistic Safety, Clinical Pharmacology and Safety Sciences, BioPharmaceuticals R&D, AstraZeneca, Gothenburg, Sweden

19.1 Introduction

Our knowledge of the genetic basis of disease has improved our understanding of disease mechanisms which lead to the emergence of therapeutic strategies that can modify nucleic acids within disease-affected cells. Although there are various nucleases used for genome editing, in this chapter, we will address the immunogenicity challenges of the CRISPR-Cas9 system.

The CRISPR/Cas9 nucleases allow to modify the DNA in cells and tissues in a site-specific manner but the host immunity against the endonuclease Cas9 puts its utility in question. "Immunogenicity" is the capacity of an antigenic substance to provoke immunity; this process provides protection to mammals from infectious diseases but can be detrimental and limiting against therapeutics derived from microbial origin. The most commonly used Cas9 orthologs are derived from *Staphylococcus aureus* (SaCas9) or *Streptococcus pyogenes* (SpCas9), both of which are prevalent human commensals. These pathogens are reported to induce both humoral, antibody-mediated, and cellular, T cell-mediated immunity. In fact, antibodies (Abs) have been detected against Cas9 from *S. aureus* and *S. pyogenes* in 80% of healthy individuals (Charlesworth et al. 2019; Wagner et al. 2019).

Today's biologics, and specifically CRISPR/Cas9, could be highly immunogenic and capable to trigger significant and sometimes sustained cellular perturbation often culminating in the activation of the major histocompatibility complex (MHC) genes. To minimize the chance of an adverse immune response after CRISPR/Cas9 therapeutic intervention, several strategies are under considerations such as using Cas9 orthologs from nonpathogenic bacteria, structural modification masking immunogenic epitopes of Cas9, altering antigen presentation routes of Cas9 epitopes. It is recently been demonstrated that Cas9 protein can be modified to eliminate immunodominant epitopes through targeted mutation while preserving its function and specificity (Wagner et al. 2021).

In this chapter, we will discuss various aspects of Cas9 and/or its delivery-mediated host immunogenicity, the assessment strategies, and we will further summarize mitigation strategies to reduce the Cas9 immunogenicity.

19.2 Immunogenicity

Immunogenicity is characterized by the ability of a substance such as antigen or an epitope to induce a humoral and/or cell-mediated immune response in the host. By definition, all immunogens are antigens, but not all antigens are necessarily immunogens. The majority of antigens are proteins, but some are lipids, carbohydrates, and nucleic acids. Bacteria and archaea have evolved various defense systems to protect against phages and other parasites. Many of these defense systems, such as restriction-modification and abortive-infection systems, provide innate immunity but an adaptive immunity also exists, and it is encoded by the CRISPR/Cas system. The origin and evolution of the CRISPR/Cas system is well described (Chew 2018).

The human body's natural immunological barriers are evolutionarily different from those of prokaryotes and they represent the main hurdles to an efficient use of all the molecular tools adopted from, or delivered via, prototypical human pathogens like the CRISPR/Cas9 system. Although there are several preclinical cell and gene therapy pipelines with CRISPR/Cas9 genome editing strategies, either through *ex vivo* editing of cells followed by transplantation to the patient, or *in vivo* editing of patient's cells concerns regarding their safety and efficacy remain partly unanswered.

Genome Editing in Drug Discovery, First Edition. Edited by Marcello Maresca and Sumit Deswal.

The host innate and adaptive arms of the immune system may respond to components of the delivery vehicle or the transgene products and eliminate cells producing the therapeutic protein, thus causing loss of efficiency and/or immunogenic response. In 1999, a subject with ornithine transcarbamylase deficiency developed lethal systemic inflammation following intra-arterial administration of an adenovirus (Ad) vector (Baker and Herzog 2020). The pre-existing anti-HAdV5 immune response was responsible for a heightened host dendritic cells (DCs) activation, which was identified as one of the driving factor in the lethal inflammatory response (Somanathan et al. 2020). This triggered the request from Health Authorities for more detailed assessments of immune memory toward the delivery vector and the transgene to warrant immunologically safer future gene and cell therapies. After 20 years, the genome editing field has made tremendous advancements in the understanding of immunogenicity but there are still many questions unanswered.

19.2.1 Pre-Existing Immunogenicity in Human Adults

19.2.1.1 CRISPR/Cas9

The most commonly used Cas9 orthologs for research and therapy are derived from *S. aureus* (SaCas9) or *S. pyogenes* (SpCas9), both of which are human commensals (Simhadri et al. 2018; Charlesworth et al. 2019; Wagner et al. 2019). Pre-existing Abs to Cas9 may not be a significant obstacle in some specific clinical interventions. Nevertheless, the presence of Abs at high titers suggests that individuals likely have memory T cells and B cells that are capable of mounting an adaptive immune response to Cas9 or to cells presenting Cas9 antigenic epitopes. This could represent a potential safety, more than efficacy, concern.

The cellular and humoral immune responses could be dependent on how Cas9 is delivered, nonetheless, even if DNA- or RNA-encoded, Cas9 protein and relative peptides could potentially be processed by intracellular MHC class I molecules to elicit a cytotoxic CD8$^+$ T cell response. On the other hand, directly delivered exogenous Cas9 protein could potentially be processed by MHC class II molecules and engage with CD4$^+$ T cells to elicit an antibody response (Figure 19.1).

In addition to Cas9, the gRNA delivery can elicit immune responses, although chemical modifications of the gRNA increase their stability and specificity reducing TLR activation (Kim et al. 2018).

19.2.1.2 Adeno-Associated Virus (AAV)

The human population has been widely exposed to the wild-type (WT) AAV. After primary infection, AAV genomes can persist for years in host cells, either

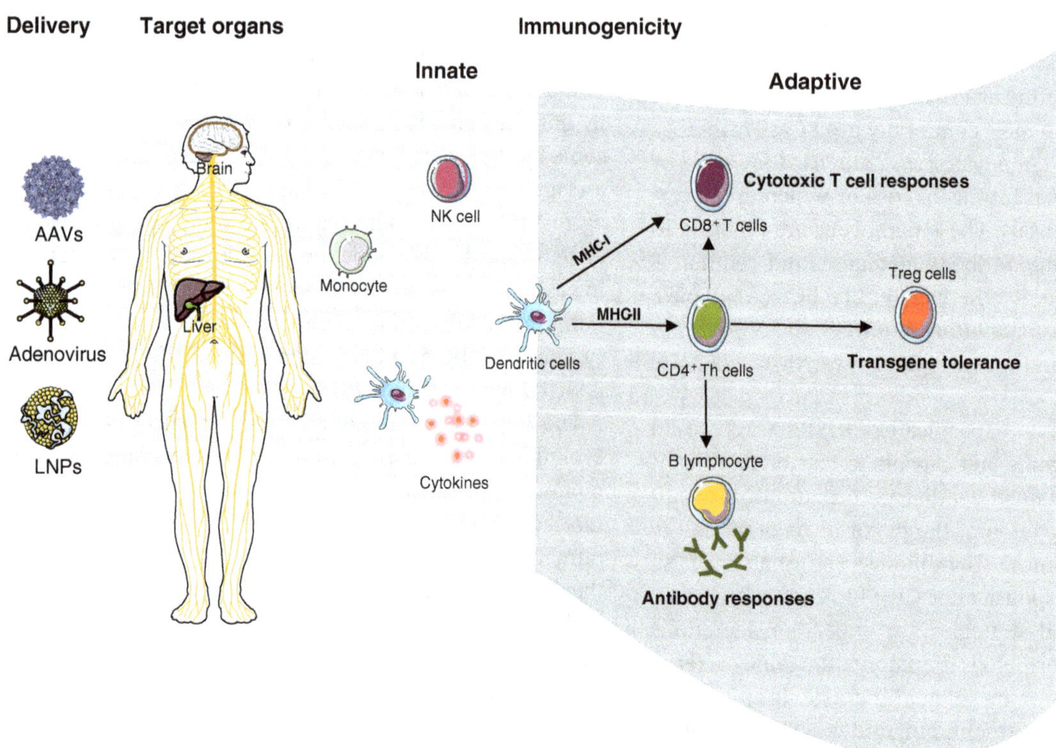

Figure 19.1 Innate and adaptive immune responses to various CRISPR delivery.

episomally or at incredibly low rates, integrated within the host DNA, and be reactivated by a helper virus, such as adenovirus, or a genotoxic reagent. Recent studies show that there may be link between the integration of the AAV genome and the development of hepatocellular carcinoma in gene therapy-treated subjects (Shirley et al. 2020; Verdera et al. 2020). The seroprevalence of neutralizing Abs directed against all WT AAVs in humans has been thoroughly investigated. Although seroprevalence varies geographically, with anti-AAV2 neutralizing Abs displaying the highest prevalence, ranging from 30% to 60% of the population. Due to the broad cross-reactivity between AAV serotypes, neutralizing Abs recognizing virtually all serotypes can be found in almost all subjects. The prevalence of total anti-AAV Abs is close to 70% of the population for AAV1 and AAV2, 45% for AAV6 and AAV9, and 38% for AAV8.The AAV-specific memory T cells have been shown to produce IFNγ, interleukin (IL)-2, and TNF-α (Verdera et al. 2020). Pre-existing immunity is a severe setback to the use of AAVs as a delivery vehicle for classical and CRISPR/Cas9 gene therapies. Nonetheless, the relatively acceptable safety profile makes AAV the preferred delivery choice to date.

19.2.2 Role of Antigen Presenting Cells (APCs)

After the discovery of DCs, Ralph M. Steinman described them as "nature's adjuvants *in vivo*". As professional antigen presenting cells, only DCs can initiate all antigen-specific T cell responses *in vivo*. Peptides generated by proteasomal degradation of endogenous cytosolic and nuclear antigens are loaded onto MHC class I molecules and presented to $CD8^+$ T cells through distinct intracellular pathways. In most APC, peptides derived from exogenous antigens that are taken up from the extracellular milieu are preferentially presented on MHC class II molecules to $CD4^+$ T cells. In DCs, however, internalized antigens are also presented to $CD8^+$ T cells through a process referred to as cross-presentation. MHC class I molecules are expressed by all nucleated cells and present protein fragments of cytosolic and nuclear origin at the cell surface (Alloatti et al. 2016).

The sensing of microbial stimuli causes DCs to enter a developmental program called "maturation", which modifies DC morphology and functions. Maturation enhances transiently the capacity of DCs to take up antigens, the expression of co-stimulatory molecules (CD40, CD80, and CD86), levels of MHC class I and class II, and it causes the secretion of a wide variety of inflammatory cytokines and chemokines (Reis e Sousa 2006). Finally, maturation leads to migration of DCs from tissues to draining lymph nodes.

Once the processed peptides loaded onto the peptide-binding groove of MHC class I molecules are then transported to the cell surface, where they can be recognized by T cells, via the T-cell receptor (TCR), called "signal 1" for T-cell activation. The interaction between peptide-loaded MHC class I molecules and the TCR, however, is insufficient for the initiation of T-cell responses. The induction of $CD8^+$ T-cell responses *in vivo* by antigens that are internalized by DCs. Additional signals, "2" and "3", consisting of co-stimulatory molecules and cytokines, respectively, are indispensable to the development of effective T-cell responses. Delivery of signals 2 and 3 by DCs requires their activation by innate signaling through pattern recognition receptors (PRR), such as toll-like receptors (TLR). PRRs recognize conserved motifs in microbes or danger signals, such as bacterial lipopolysaccharide (LPS), methylated oligonucleotides (CpG), or viral double-stranded RNA (Alloatti et al. 2016).

Given the microbial nature of Cas9, Adenovirus and AAV, the innate sensing and adaptive immune activation by DCs play central and limiting role in the immunogenicity of genome editing.

19.2.3 Role of T Cells

Neutralizing Abs and cytotoxic T cells can have a dramatic impact on the efficacy and safety of human gene therapies. On the other hand, regulatory T cells (Treg) can promote tolerance toward transgenes thereby enabling long-term benefits of *in vivo* gene therapy after a single administration (Wagner et al. 2021). Delivery of viral vectors to the liver induces Treg cells and can be exploited to drive tolerance during co-delivery to other organs (Mingozzi et al. 2003; Loduca et al. 2009). As commensal bacteria are known to promote Treg cells at mucosal surfaces, colonization with Cas9-expressing bacterial strains could have induced Cas9-specific Treg cells in humans. Given the limitations of *in vitro* human PBMCs (peripheral blood mononuclear cells) assessments of Cas9 pre-existing immunity, there is need for animal models that allows addressing the questions of mixed pre-existing immunity to Cas9.

The immune monitoring of both Cas9-directed effector and Treg responses in CRISPR/Cas9 clinical trials could provide information on whether a gene therapy-directed immune response is tolerogenic/suppressive or immunogenic based on Teff (effector)/Treg (regulatory) ratio in peripheral blood. A misbalanced Teff/Treg ratio skewed toward high frequency of inflammatory cells or pronounced $CD8^+$ T cell responses may allow defining safety parameters for future trials. Similarly, Teff/Treg ratios have been used in other clinical indications to predict clinical outcomes, e.g. after allogeneic stem cell transplantation (Wagner et al. 2021).

19.3 Delivery-Dependent Immunogenicity

The understanding of Cas9 immunogenicity is pivotal but the delivery method is central to the success of gene therapy (Doudna and Charpentier 2014). The major challenge for clinical translation is the delivery of editing systems to target cell types without provoking host immune system. A variety of nucleic acid or protein delivery methods are under considerations to introduce genome editing nucleases into target cells *ex vivo* or *in vivo*.

19.3.1 Viral Delivery

Immune-mediated rejection represents one of the most significant hurdles to human gene therapy. Whereas adeno-associated viral (AAV) vectors are known to be less immunogenic than other viral vectors such as adenoviruses, preclinical and clinical studies have demonstrated dose-dependent inflammation, which can reduce efficacy and lead to dose-limiting toxicity (Mingozzi et al. 2009; Chan et al. 2021).

For *in vivo* applications, the most promising delivery systems are viral vectors, particularly AAV vectors which have been approved for clinical use (Wirth et al. 2013). The AAV offers the ability to transduce the cells more efficiently than nonviral vectors, to exert long-term transgene expression, and to benefit from a minimal immune response to the vector with little integration of therapeutic genes into the patient genome. Both B cell and T cell host responses specific to either the transgene or the viral components of adenoviral (AdV) and AAV vectors have been observed, despite relatively lower immunogenicity was observed for AAV vectors. Specific AAV neutralizing Abs and T cells are detected in healthy donors and specific $CD8^+$ T cells expansion have been shown following gene delivery. The potential consequences of immune responses to expressed proteins from viral vectors or transgenes include (i) neutralization of the gene product, destruction of the edited cell, leading to loss of therapeutic activity and tissue destruction; (ii) induction of immune memory that prevents readministration of the therapeutic and consequent rapid innate inflammatory responses. The immune responses to gene therapies in humans and nonhuman primate models were more potent compared with mice. Therefore, immune-privileged organs became an attractive target to avoid strong immunity against the therapy and delivery vector or method.

19.3.1.1 New Antigen Formation Risk

Ideally, the most effective biologics for therapy should be delivered, and should persist in the host cell in a "stealth mode," thus avoiding any activation of surveillance systems. The MHC class I antigen presentation alerts the immune system to infection or oncogenic transformation and constitutively occurs in all nucleated cells of the body. This pathway has highly skewed immunopeptidome from the cellular degradome, indicating high selectivity of the access of protease-generated peptides to class I molecules. Peptide-loaded class I molecules are exported from the ER through the Golgi complex to the plasma membrane, thus enabling $CD8^+$ T-cell immunosurveillance. The mechanism of antigen presentation is extremely complex, and the same peptide degraded by 26S proteasomes in a degron-controlled manner could follow different kinetics of presentation based on whether the source protein is synthesized from plasmid or viral-encoded mRNA. Therefore, to the intrinsic immunogenic property of each biologics amenable for therapy, the delivery vehicle of the biologics itself can synergize and enhance the net immunogenicity.

19.3.2 NonViral Delivery

The direct delivery of CRISPR/Cas9-encoding DNA into cells by intramuscular electroporation drives infiltration of myeloid cells, elevates the expression of innate immunity genes, and can subsequently induce extensive T-cell cytotoxicity. In human embryonic stem cells, delivery of Cas9-gRNA RNPs induced efficient genome editing, less genomic off-targeting, and less cytotoxicity than DNA transfection, thus making RNPs an attractive alternative (Kim et al. 2014). Electroporation involves pulsing cells with high-voltage currents that create transient nanometer-sized pores in the cell membrane. This process allows negatively charged DNA or mRNA molecules or CRISPR/Cas9 RNPs to enter the cells. This method is a primary method of Cas9–sgRNA delivery to cells *ex vivo* (Doudna 2020).

The nonviral delivery of the Cas9 ribonucleoproteins (RNPs, or recombinant Cas9 pre-complexed with a gRNA) via mRNA encoding Cas9 requires delivering an RNA strand, which is challenging because of RNA's instability in serum and immunogenicity. The nonmicrobial nature of the lipid nano particles (LNPs) as a delivery vehicle is currently under considerations for gene therapy. The LNP biodistribution to off target organs leads to innate and adaptive immune activation, marked toxicity of the lipid-mediated nanoparticles, and potentially undesired selectivity of cell-type-specific uptake of the particles (Doudna 2020).

19.4 Methods to Study Cas9 Immunogenicity

19.4.1 *In vitro* Assays: ELISA, ELISpot, FluoroSpot, Immunopeptidome

ELISA-based assays are the foremost used methods to study pre-existing humoral immunity and for detecting Abs to Cas9 from *S. aureus* (SaCas9) and *S. pyogenes* (SpCas9) in human sera. The usefulness of assays to measure anti-Cas9 Abs (pre- and post-treatment) will probably depend on the specific clinical application and influencing parameters like route of administration (*ex vivo* vs. *in vivo*), target organ, and delivery vector. The use of INF-γ ELISpot to study Cas9-specific T cell responses is demonstrated to potently predict cellular responses toward Cas9 in healthy individuals (Charlesworth et al. 2019; Wagner et al. 2019). The multiplex FluoroSpot could identify different parameters from the same samples. The Flow cytometry-based PBMC analysis of B and T cells is used to characterize Cas9-specific responses in healthy individuals and also to study and identify less immunogenic Cas9 variants (Charlesworth et al. 2019; Wagner et al. 2019).

The identification of T cell epitopes and characterization of T cell responses, in conjunction with the identification of anti-Cas9 Abs, would be a more comprehensive immunogenicity assessment. T cell responses are also useful in making histocompatibility leukocyte antigen (HLA) mediated. The HLA repertoire is one of the most diverse in the human genome, and it exhibits considerable variations among racial, ethnic, and geographic groups. Identification of T cell epitopes presented by Cas9 as well as their HLA restriction could, therefore, provide useful information about immunogenicity risk in sub populations as well as in single individuals (Simhadri et al. 2018). The MHC immunopeptidome analysis might help identify CD4, CD8 T cell specific and most immunogenic Cas9 peptides.

19.4.2 *in vivo* Strategies

As classical animal models for gene therapy including non-human primates do not host human commensal bacteria like *S. pyogenes*, they do not reflect human immune sensitization. Thus, they are suboptimal for evaluating the impact of immunity on CRISPR/Cas *in vivo* treatments (Wagner et al. 2021). Recently, it was generated a mouse model with pre-existing Teff cells by subcutaneous preimmunization with small amounts of SaCas9 protein. One week after immunization, a liver targeting AAV-CRISPR was systemically delivered to the animal. While genome editing and AAV-mediated GFP expression were initially observed in hepatocytes, after 12 weeks gene edited cells were eliminated (Li et al. 2020). Similar immunization models could help mimic the human pre-existing immunity toward the Cas9 of choice and allow addressing important questions of multiple dosing and effect of pre-existing immunity on the long-term editing efficacy.

The detailed immunophenotyping of innate and adaptive responses by flow cytometry will allow to differentiate the immunity toward Cas9 and its delivery vehicle. A large majority of T cells recognize targets of immunity through direct interactions of TCR with antigen-derived peptide presented on the cell surface by HLA molecules. The Cytotoxic $CD8^+$ T cells are important effectors in the adaptive immune response against antigens presented by MHC-I and play an important role in immunosurveillance. The prolonged acceptance of edited cells *in vivo* largely depends on host immunity, and this will affect the duration of the editing intervention in the body. There are evidences that Cas9 peptides could be presented by the MHC-I, triggering a cytotoxic response against the edited cells. This could potentially compromise the efficiency and the overall outcome of the genetic intervention but can also induce a potent immunogenic response with severe safety drawbacks. The *in vivo* TCR sequencing using single-cell RNAseq may help to further understand and mitigate the $CD8^+$ T cell-mediated cytotoxicity. This can further identify clonality and specificity of Cas9-specific T cells. Structural studies of the TCR have revealed that the CDR3 (Complementarity Determining Region) of the Vβ chain is directly involved in the recognition/binding of antigenic peptide-HLA complexes. The CDR3 repertoire diversity reflects clonal composition, the potential antigenic recognition spectrum, and the quantity of antigen-specific T-cell responses, which are critical in assessing the body's immunity status against specific antigens.

19.5 Mitigation Strategies

19.5.1 Targeting Immune-Privileged Organs

Gene therapy with AAV vectors has demonstrated safety and long-term efficacy in a number of trials across target organs, including immune-privileged organ like eye, tolerogenic organ like liver, and the central nervous system. The cellular responses to the AAV capsid are being managed clinically with transient immunosuppression. However, it has remained unclear whether the stable expression of the bacterial Cas9 protein will be tolerated. Minimizing the duration of Cas9 expression should diminish the cellular response and could be achieved with alternative transient delivery vehicles or self-cleaving systems. CRISPR/Cas9 systems from exotic species not typically seen by humans could also provide an easy workaround to pre-existing immunity. Less inflammatory vectors such as AAV, intravascular over intramuscular injections, the lowest efficacious doses for non-liver tissues or tolerogenic doses for liver, and tissue-specific promoters that prevent expression in antigen-presenting cells should be chosen whenever possible (Mays and Wilson 2011).

Targeting younger patients, who are less likely to have developed anti-Cas9 $CD8^+$ cytotoxic T lymphocytes (CTLs), or utilizing novel Cas9 with lower rates of pre-existing CTLs would minimize the number of patients that require exclusion (Mortensen et al. 2015). Tolerance induction could also be utilized in patients with pre-existing anti-Cas9 CTLs. AAV-directed liver expression has been shown to induce tolerance, even in the context of a pre-existing immune response (Crudele and Chamberlain 2018).

19.5.1.1 Evasion from MHC Class I Recognition

There are several strategies being used to avoid detection of Cas9 peptides by MHC class I. One strategy to circumvent an immune response to protein therapeutics is the removal of immunogenic T cell epitopes. Once the epitopes are identified by in silico analysis, substitution of key amino acids may reduce the immunogenicity of the protein because modification of critical anchor residues can abrogate binding to MHC molecules and prevent antigen presentation. This approach may not account for the high level of diversity at the human leukocyte antigen (HLA) loci. The sequential use of immune-orthogonal Cas9 proteins and adeno-associated viruses (AAVs) evades adaptive immune responses and enables effective gene editing using repeated dosing (Moreno et al. 2019).

The transient expression of Cas9 through self-destruction or nonviral delivery of mRNA or protein would shorten the immune recognition. Additionally, short-term (or muscle-restricted) expression would limit the impact of pro-inflammatory DNA damage signals, which could increase the chances of developing CTLs, kill or arrest Cas9-expressing cells, or even theoretically lead to cancer (Haapaniemi et al. 2018).

19.6 Outlook

It is expected in the coming years that therapeutic genome editing will become a reality, at least for some diseases. The complexity of human immune system and its variability among individual's age, gender, HLA variations, pre-exposure to microbes, and genetics makes it extremely difficult to predict the best immunogenicity safety strategy. In order to move closer to the clinics, the CRISPR-based gene editing can find guidance from the classical gene therapy field, which has overcome anti-capsid and anti-transgene CTL responses by carefully considering vector, dose, target tissue, administration route, promoter, and immune suppression. CRISPR/Cas9 platforms that lead to short-term expression of Cas9 and new Cas9 variants from human gut microbiome which humans have mucosal tolerance to, should also continue to be developed. While preclinical studies must address the issue of anti-Cas9 immune responses in immune-competent, large animal models. The immunogenicity toward the Cas9 and its delivery vehicle is at the center of safety and efficacy of future therapeutics.

References

Alloatti, A., Kotsias, F., Magalhaes, J.G., and Amigorena, S. (2016). Dendritic cell maturation and cross-presentation: timing matters! *Immunol. Rev.* 272: 97–108.

Baker, A.H. and Herzog, R.W. (2020). Did dendritic cell activation, induced by adenovirus-antibody complexes, play a role in the death of Jesse Gelsinger? *Mol. Ther.* 28 (3): 704–706.

Chan, Y.K., Wang, S.K., Chu, C.J. et al. (2021). Engineering adeno-associated viral vectors to evade innate immune and inflammatory responses. *Sci. Transl. Med.* 13 (580): eabd3438.

Charlesworth, C.T., Deshpande, P.S., Deverd, P. et al. (2019). Identification of preexisting adaptive immunity to Cas9 proteins in humans. *Nat. Med.* 25: 249–254.

Chew, W.L. (2018). Immunity to CRISPR Cas9 and Cas12a therapeutics: immunity to CRISPR Cas9 and Cas12a therapeutics. *Wiley Interdiscip. Rev. Syst. Biol. Med.* 10: e1408.

Crudele, J.M. and Chamberlain, J.S. (2018). Cas9 immunity creates challenges for CRISPR gene editing therapies. *Nat. Commun.* 9: 3497.

Doudna, J.A. (2020). The promise and challenge of therapeutic genome editing. *Nature* 578: 229–236.

Doudna, J.A. and Charpentier, E. (2014). The new frontier of genome engineering with CRISPR-Cas9. *Science* 346: 1258096-1–1258096-9.

Haapaniemi, E., Botla, S., Persson, J. et al. (2018). CRISPR–Cas9 genome editing induces a p53-mediated DNA damage response. *Nat. Med.* 24: 927–930.

Kim, S., Kim, D., Cho, S.W. et al. (2014). Highly efficient RNA-guided genome editing in human cells via delivery of purified Cas9 ribonucleoproteins. *Genome Res.* 24: 1012–1019.

Kim, S., Koo, T., Jee, H.G. et al. (2018). CRISPR RNAs trigger innate immune responses in human cells. *Genome Res.* 28: 367–373.

Li, A., Tanner, M.R., Lee, C.M. et al. (2020). AAV-CRISPR gene editing is negated by pre-existing immunity to Cas9. *Mol. Ther.*;S1525001620301982 28 (6): P1432–P1441.

Loduca, P.A., Hoffman, B.E., and Herzog, R.W. (2009). Hepatic gene transfer as a means of tolerance induction to transgene products. *Curr. Gene Ther.* 9: 104–114.

Mays, L.E. and Wilson, J.M. (2011). The complex and evolving story of T cell activation to AAV vector-encoded transgene products. *Mol. Ther.* **19**: 16–27.

Mingozzi, F., Liu, Y.-L., Dobrzynski, E. et al. (2003). Induction of immune tolerance to coagulation factor IX antigen by in vivo hepatic gene transfer. *J. Clin. Invest.* 111: 1347–1356.

Mingozzi, F., Meulenberg, J.J., Hui, D.J. et al. (2009). AAV-1-mediated gene transfer to skeletal muscle in humans results in dose-dependent activation of capsid-specific T cells. *Blood* 114: 2077–2086.

Moreno, A.M., Palmer, N., Alemán, F. et al. (2019). Immune-orthogonal orthologues of AAV capsids and of Cas9 circumvent the immune response to the administration of gene therapy. *Nat Biomed Eng.* 3: 806–816.

Mortensen, R., Nissen, T.N., Blauenfeldt, T. et al. (2015). Adaptive immunity against *Streptococcus pyogenes* in adults involves increased IFN-γ and IgG3 responses compared with children. *J. Immunol.* 195: 1657–1664.

Reis e Sousa, C. (2006). Dendritic cells in a mature age. *Nat. Rev. Immunol.* 6: 476–483.

Shirley, J.L., De Jong, Y.P., Terhorst, C., and Herzog, R.W. (2020). Immune responses to viral gene therapy vectors. *Mol. Ther.* 28 (3): 709–722.

Simhadri, V.L., Mcgill, J., Mcmahon, S. et al. (2018). Prevalence of pre-existing antibodies to CRISPR-associated nuclease Cas9 in the USA population. *Mol. Ther. Methods Clin. Dev.* 10: 105–112.

Somanathan, S., Roberto Calcedo, R., James, M., and Wilson, J.M. (2020). Adenovirus-antibody complexes contributed to lethal systemic inflammation in a gene therapy trial. *Mol. Ther.* 28 (3): 784–793.

Verdera, H.C., Kuranda, K., and Mingozzi, F. (2020). AAV vector immunogenicity in humans: a long journey to successful gene transfer. *Mol. Ther.* 28 (3): 723–746.

Wagner, D.L., Amini, L., Wendering, D.J. et al. (2019). High prevalence of *Streptococcus pyogenes* Cas9-reactive T cells within the adult human population. *Nat. Med.* 25: 242–248.

Wagner, D.L., Peter, L., and Schmueck-Henneresse, M. (2021). Cas9-directed immune tolerance in humans—a model to evaluate regulatory T cells in gene therapy? *Gene Ther.*

Wirth, T., Parker, N., and Yla-Herttuala, S. (2013). History of gene therapy. *Gene* 525: 162–169.

20

Specificity of CRISPR-Cas9 Gene Editing

Niklas Selfjord, Alexandra Madsen*, and Pinar Akcakaya*

Genome Engineering, Discovery Sciences, BioPharmaceuticals R&D, AstraZeneca, Gothenburg, Sweden

20.1 Introduction

Discovery of CRISPR as a new tool for genome editing has been met with great enthusiasm and high expectations for possible therapeutic applications in genetic diseases. However, as with every new technology, there are some important safety aspects that need to be taken into account. Major concerns regarding the safety of CRISPR are the unintentional introduction of potentially harmful or pathogenic mutations at nontarget sites in the genome (Fu et al. 2013; Hsu et al. 2013; Pattanayak et al. 2013), as well as induction of bigger chromosomal rearrangements such as translocations, deletions, and inversions that may potentially lead to deleterious outcomes (Cho et al. 2014; Cradick et al. 2013).

Canonical targeted gene editing by CRISPR-Cas nucleases is based on induction of double-strand breaks (DSBs) (Jiang et al. 2013; Jinek et al. 2012). These DSBs are then repaired by the cell predominantly via non-homologous end joining (NHEJ), generating nucleotide insertions and deletions, or via homology-directed repair (HDR), resulting in relatively precise editing of the locus in the presence of a repair template (Takata et al. 1998). For programming Cas9 to target a region of interest for cleavage, two prerequisites must be met: (i) a guide RNA (gRNA) that directs the Cas9 to the DNA target region by complementary base-pairing interactions with a ~20-base pair long protospacer sequence. (ii) A specific sequence motif called a protospacer adjacent motif (PAM) has to be present at the 3′ end of the protospacer for Cas9 recognition, which in the case of the most widely used *Streptococcus pyogenes* Cas9 (SpCas9) is 5′-NGG-3′ (Jiang et al. 2013; Jinek et al. 2012).

Given the size of the human genome and the relatively short length of the gRNA, the probability is high for the presence of a number of sites within the genome displaying high sequence homology to the target sequence. This, together with the reported ability of Cas9 to tolerate up to five mismatches in the protospacer sequence (Fu et al. 2013; Hsu et al. 2013; Jiang et al. 2013) and to recognize non-canonical PAMs (Jiang et al. 2013; Zhang et al. 2014), further increases the likelihood of DNA off-target editing. Not only the number of mismatches between DNA and gRNA, but also their relative position in the protospacer plays a big role in affecting the Cas9 specificity: while mismatches in the protospacer's seed region comprising the 10–12 nucleotides proximal to the PAM greatly reduce cleavage activity, mismatches outside of this region are better tolerated (Cong et al. 2013; Jiang et al. 2013; Jinek et al. 2012; Semenova et al. 2011; Wiedenheft et al. 2011). Moreover, we and others have reported that that the type of mismatch has an impact on Cas9 cleavage as well, with "wobble" mismatches (e.g. G:A mismatches, resembling DNA G-RNA U wobble base pairing) affecting Cas9 activity to a smaller degree than other mismatches (Akcakaya et al. 2018; Doench et al. 2016; Lazzarotto et al. 2020; Tsai et al. 2015). In addition to mismatched bases, Cas9 can also tolerate deletions (gRNA bulge) or insertions (DNA bulge) of single bases in the DNA compared with the gRNA sequence, depending on both the position of the bulge and the GC content of the gRNA (Lin et al. 2014).

Off-target editing has also been shown to be affected by target DNA properties. It was reported that the presence of nucleosomes at target and off-target sites can strongly interfere with Cas9 binding and cleavage activity (Horlbeck et al. 2016; Yarrington et al. 2018). Similarly, the chromatin state was observed to have an impact on cleavage efficiency with open chromatin conformations

*Authors contributed equally.
Correspondence: pinar.akcakaya@astrazeneca.com

Genome Editing in Drug Discovery, First Edition. Edited by Marcello Maresca and Sumit Deswal.

allowing for higher Cas9 activity (Kim and Kim 2018). We have observed higher probability of occurrence of off-target activity in coding compared with noncoding regions (Akcakaya et al. 2018). In addition, both the frequency of guanine and the overall nucleotide diversity have been suggested to influence specificity of Cas9 (Lazzarotto et al. 2020). The same study found a strong correlation of off-target activity with the presence of an accessible chromatin state, as well as of active enhancers and promoters, and highly expressed genes, highlighting the importance of cellular context in predicting off-target effects. In this chapter, we summarize techniques and approaches for both the detection and the reduction of CRISPR off-target activity that have been developed and applied in order to improve safety of CRISPR genome editing.

20.2 Detecting Genome-wide Off-target Effects

Upon realization of CRISPR-induced mutagenesis at unintended sites, a number of approaches have been developed for the detection of nuclease off-target activities. In earlier studies, in silico tools were used as a starting point for identification of potential off-target sites based on the sequence homology to the target site, and potential off-target sites were analyzed for CRISPR-induced indels by targeted sequencing upon editing (Cradick et al. 2013; Fu et al. 2013; Hsu et al. 2013). Similarly, an in vitro screening approach by Pattanayak et al. used synthetic DNA libraries that consist of 10-fold coverage of all sequences in the genome with 8 or fewer mismatches relative to the intended target sequence (Pattanayak et al. 2013). These oligonucleotides were circularized, rolling circle amplified, in vitro cleaved by Cas9 RNP complex, sequencing adapters were added to the free ends of cleaved fragments, followed by selective sequencing to identify Cas9 cleaved sites. Although these early studies have improved our understanding of CRISPR off-target activities, computational tools are unlikely to be able to identify potential genome-wide off-target sites in an unbiased manner. Thus, these approaches carry the risk of missing the bona fide sites that do not fit the computational criteria used (Tsai et al. 2015).

Whole genome sequencing (WGS) has been used to investigate Cas9 off-target mutagenesis in Cas9-engineered single-cell clones (Smith et al. 2014; Veres et al. 2014) and transgenic F1 mice (Iyer et al. 2015). Sequencing the entire genome of single cell clones to deep coverage provides an unbiased means of assaying for genome-wide off-targets. However, with the commonly used WGS read coverage of 50×, the method is only able to detect the most frequently mutated sites and lacks the required sensitivity for assessment of off-targets in heterogenous nuclease-edited bulk populations of cells in a potential therapeutic setting, where even very low-frequency mutagenesis will be of high importance.

ChIP-seq with catalytically inactive Cas9 (dead Cas9, dCas9) was utilized as an off-target detection tool, to pull down and characterize nuclease-bound DNA sequences in order to identify nuclease off-targets (Kuscu et al. 2014). Later, this method was shown to be unable to identify bona fide Cas9 off-target cleavage sites in an efficient and specific manner due to fundamental differences between dCas9 DNA binding sites and Cas9 off-target cleavage sites (Duan et al. 2014; Tsai et al. 2015; Wu et al. 2014).

In order to achieve comprehensive off-target discovery, unbiased genome-wide methodologies with high sensitivity are preferred today. More recently developed genome-wide off-target detection technologies have been mostly evolved around using nucleases in vitro or in cellular and animal models to create DSBs at on- and off-target sites, and detecting these DSBs with high-throughput sequencing using various sequence enrichment strategies.

Overall, genome-wide off-target detection approaches can be divided into four categories: in silico, cell-based, in situ, and in vitro. Importantly, all these methodologies ultimately require an experimental validation step with targeted deep sequencing in the nuclease edited cell or tissue of interest to confirm bona fide off-target mutagenesis at the identified off-target sites.

20.2.1 In Silico Off-target Detection Tools

gRNA design is a critical step for successful gene editing using CRISPR nucleases. Minimization of gRNA off-target binding should already be considered during the design of gene editing approaches. A variety of in silico tools to predict on-target efficiency as well as off-target activity of gRNAs has been made available over the last years (Table 20.1).

Depending on the tool, the user has to provide the designated target region either as a nucleotide sequence, a gene identifier, or genomic coordinates. Possible off-targets are then identified by analysis of homologies between the gRNA and the reference genome by sequence alignment with, e.g. Bowtie (Langmead et al. 2009). The specificity scoring of the available gRNAs is then calculated based on the occurrence of mismatches between the gRNA sequence and potential genomic off-target sites. One of the first described scoring algorithms was the penalty matrix by Hsu et al. which takes not only the number of mismatches into account, but also their distribution and position within the protospacer sequence (Hsu et al. 2013). Even though the corresponding CRISPR Design tool by the Zhang lab

Table 20.1 Summary of in silico off-target detection tools.

Tool	PAM	#MM	Bulge	Specificity scoring	Reference
Cas-OFFinder	>20 PAMs	≤9	DNA/RNA	Mismatch counting	http://www.rgenome.net/cas-offinder/ (Bae et al. 2014)
CCTop	>15 PAMs	≤5	—	Number and position of mismatches, distance to nearest exon (Stemmer et al. 2015)	https://cctop.cos.uni-heidelberg.de:8043/ (Stemmer et al. 2015)
CHOPCHOP v3	Custom PAM	≤3	—	Choose between Hsu-Zhang (2013) and Cong (2013)	http://chopchop.cbu.uib.no/ (Labun et al. 2019)
CRISPOR	>30 PAMs	≤4	—	Hsu-Zhang (2013), CFD (Doench et al. 2016)	http://crispor.tefor.net/ (Concordet and Haeussler 2018)
E-CRISP	Custom PAM	≤20	—	S-score (Heigwer et al. 2014)	http://www.e-crisp.org/E--CRISP/index.html (Heigwer et al. 2014)

Examples of in silico gRNA design and off-target prediction tools including the number of supported PAMs, maximal number of allowed mismatches, bulge inclusion, and method for specificity scoring.

(crispr.mit.edu) has been recently discontinued, the underlying off-target scoring system is still widely used, e.g. in the CRISPOR (Concordet and Haeussler 2018) and CHOPCHOP v3 tools (Labun et al. 2019). Based on the finding that mismatches in the 11 bp proximal to the PAM block all Cas9 cleavage activity, Cong et al. suggested a different scoring method which is implemented, e.g. in CHOPCHOP v3, as an alternative to the Hsu-Zhang scoring (Cong et al. 2013). Another extensively used algorithm was introduced by Doench et al. in form of the cutting frequency determination (CFD) score (Doench et al. 2016). CFD scoring takes mismatch identity into account in addition to number and position of the mismatches, and it is among others used in the CRISPOR tool. Web-based tools like CCTop (Stemmer et al. 2015) and E-CRISP (Heigwer et al. 2014) instead use their self-developed scoring methods for gRNA specificity. While most of the available tools combine the prediction of off-targets with de novo design of gRNAs, some tools like Cas-OFFinder (Bae et al. 2014) and E-CRISP additionally offer the possibility of off-target analysis and scoring of already previously designed gRNAs. In contrast to the other tools, CasOFFinder moreover offers inclusion of DNA and RNA bulges in the prediction of gRNA off-targets, which have been shown to have an impact on Cas9 specificity (Lin et al. 2014). The output from all of these tools helps the user to choose a highly specific gRNA for the designated purpose. In order to allow further integration of previous results into future research, crisprSQL has recently been presented as an interactive CRISPR-Cas9 study database, which has to date collected data on off-target profiles and associated epigenetics marks for more than 150 guides from 17 different studies (Stortz and Minary 2021). The existing data on integration of off-target analyses with epigenetic data is continuously expanded in order to facilitate efficient design of future studies and further improvement of algorithms for computational off-target prediction. Generation of large-scale datasets revealing off-target profiles in which extensive experimental setups are covered such as cell type, delivery methods, modality of nucleases and gRNAs, together with machine learning approaches will potentially enable more accurate in silico predictions in the near future.

20.2.2 Cell-Based Off-target Detection Methods

Cell-based off-target detection methodologies employ delivery of Cas9 nuclease and gRNAs into living cells and utilize the endogenous NHEJ DNA repair pathway to integrate/fuse "known" sequences to nuclease-induced DSBs for genome-wide off-target identification, followed by high-throughput sequencing.

Integrase-defective lentiviral vector (IDLV) capture was the first methodology described to directly identify off-target nuclease cleavage events in living cells (Figure 20.1a). The capture of cellular DSBs was first utilized to assess the genome-wide off-target mutagenesis of zinc finger nucleases (ZFN) (Gabriel et al. 2011), and later adapted to other nucleases such as transcription activator-like effector nuclease (TALEN) and CRISPR-Cas9 (Wang et al. 2015).

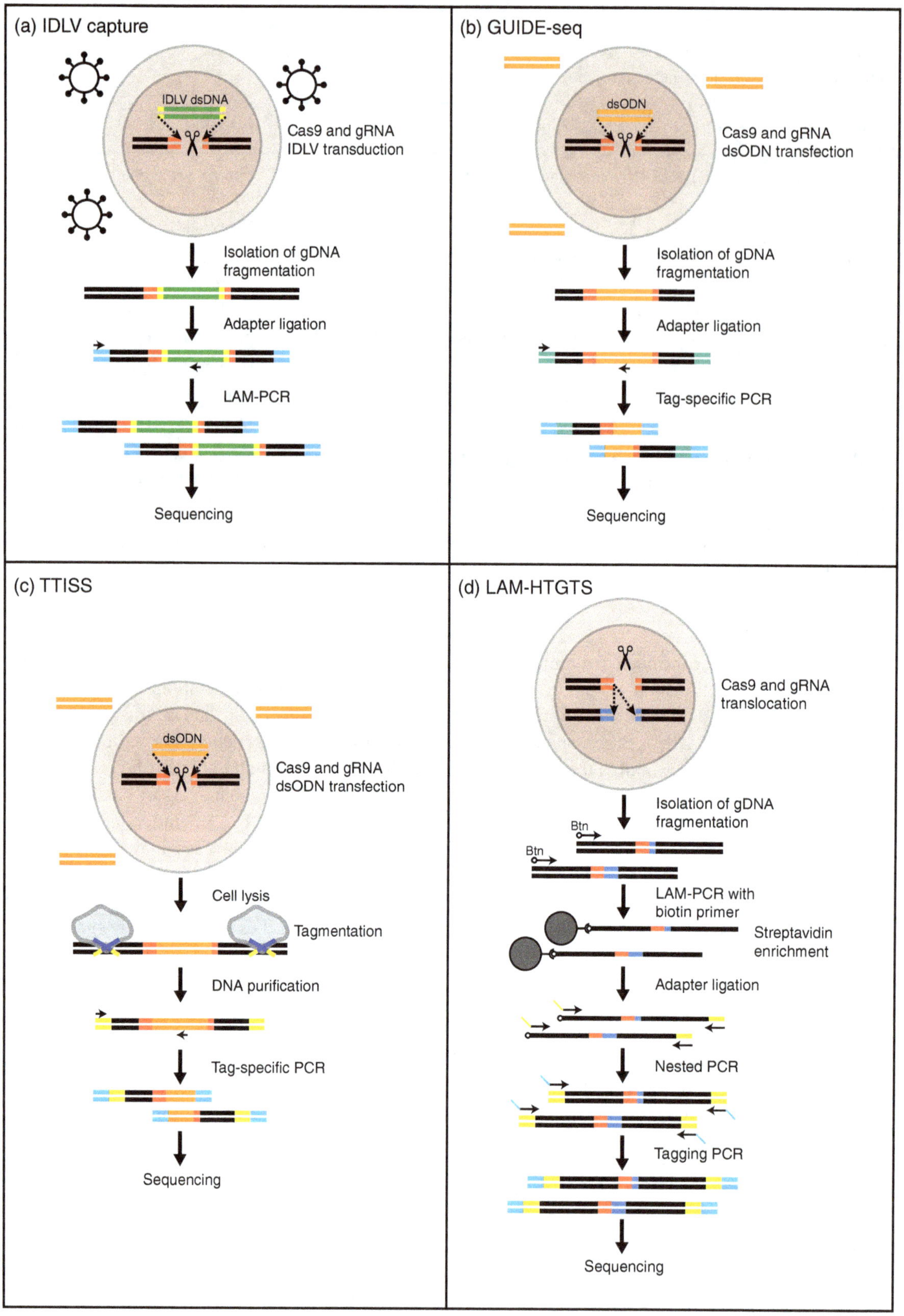

Figure 20.1 Cell-based off-target detection methods. (a) IDLV capture. IDLV genomes are integrated into nuclease-induced DSBs after cellular transduction. After gDNA is purified, fragmented, and ligated to sequencing adapters, IDLV integration sites are retrieved by LAM-PCR followed by sequencing; (b) GUIDE-seq. Short dsODN donors are inserted into Cas9 cleavage sites by NHEJ repair in cells. gDNA is purified and fragmented, and sequencing adapters are ligated to free ends. Adapters and dsODN sequences are used for tag-specific amplification of both sides of integration sites followed by sequencing; (c) TTISS. TTISS builds on the original principle of GUIDE-seq by utilizing NHEJ integration of a dsODN into cellular DSB sites. Cells are then lysed and Tn5 tagmentation is performed directly in the lysates, fragmenting the gDNA and adding adapter sequences to the ends, which are subsequently used for tag-specific

The method is based on delivery of Cas9 nuclease to cells followed by transduction with IDLV, leading to the integration of linear, double-stranded genomes of IDLVs into cellular DSBs by NHEJ-mediated DNA repair. Clustered IDLV integration sites in cellular genomes are then enriched by linear amplification-mediated PCR (LAM-PCR), followed by sequencing of the genomic DNA sequences flanking the integration sites and subsequent mapping to the genome. However, IDLV capture is limited by a relatively low sensitivity (~1%) due to the inherently inefficient integration of IDLVs into DSBs. Moreover, the method suffers from high background integration rates due to the propensity of IDLVs to integrate into naturally occurring DSBs. Mapping the precise nuclease recognition site of the DSB can also be challenging, since IDLV integration often occurs close to but not precisely at the DSB.

Similar to IDLV capture, "Genome-wide unbiased identification of DSBs enabled by sequencing" (GUIDE-seq) is a cellular genome-wide off-target detection assay that relies on integrating an exogeneous tag sequence at nuclease-induced DSBs by NHEJ (Figure 20.1b). GUIDE-seq uses a short, blunt double-stranded oligodeoxynucleotide (dsODN) tag instead of a viral genome for integration, enabling more efficient integration and increased sensitivity (Tsai et al. 2015). Genomic DNA from nuclease-treated cells transfected with dsODNs is purified and randomly sheared to an average length of 500 bp, followed by tag-specific, unbiased amplification of integration sites, deep sequencing, and mapping to the genome. GUIDE-seq is a highly sensitive and experimentally simple method to perform in biologically relevant cell models, but is limited to cells that can be efficiently transfected. As dsODN tags are integrated to DSBs via NHEJ, the method is unlikely to detect DSBs that are repaired through other pathways, which can potentially lead to cell-type specific differences in the off-targets identified depending on the repair pathway choice in the cell type examined. While blunt dsODNs were shown to integrate efficiently into the nuclease-induced DSBs with blunt-ends and 1 bp overhangs (Schmid-Burgk et al. 2020), the integration is not likely to be as efficient at DSBs with staggered-ends induced by other Cas nucleases.

Following a similar approach, "tagmentation-based tag integration site sequencing" (TTISS) builds on the original GUIDE-seq approach of tagging DSBs with a short dsODN, followed by utilizing a Tn5 bulk tagmentation strategy to enrich for integration sites, allowing for greater throughput and scalability (Figure 20.1c) (Schmid-Burgk et al. 2020). Tagmentation is performed directly in the lysates of nuclease-treated cells with integrated dsODNs, generating tag sequence-specific fragments with suitable ends for subsequent adapter ligation and direct sequencing of multiple sites. The rapid sample processing and high enrichment efficiency of TTISS allows multiplexing of >50 gRNAs per transfection while still retaining the majority of sites (71.4%) found in a single-guide experiment. Large-scale multiplex TTISS enables the generation of big datasets that can be used to train machine learning algorithms to predict and evaluate the specificity of a given nuclease-gRNA combination in selected cell types. TTISS has been successfully applied to detect off-targets by Cas enzymes creating 1 bp overhangs, rather than blunt ends. However, one drawback is that TTISS depends on a sequence-similarity algorithm to match detected off-targets to gRNAs, which could potentially lead to complications in multiplex experiments where many guides with similar sequences are used. Similar to GUIDE-seq, TTISS is also reliant on short DNA tag integration, limiting detection to DSBs that are repaired through the NHEJ pathway in cells that can be efficiently transfected.

In contrast to the previously described cell-based DSB detection methods, high-throughput genome-wide translocation sequencing (HTGTS) is an assay designed for the detection of translocation events between cellular DSBs rather than direct detection of DSBs (Chiarle et al. 2011). Similar to endogenous naturally occurring DSBs, multiple simultaneously occurring CRISPR-induced DSBs can be repaired by fusion of two independent free ends, rather than a local repair, leading to chromosomal translocations. Linear amplification-mediated high-throughput genome-wide translocation sequencing (LAM-HTGTS) has been utilized for the detection of nuclease-induced on- and off-targets, by employing LAM-PCR from a known "bait" DSB site (usually on-target) to capture the junctions to other

Figure 20.1 (Continued) amplification and sequencing; (d) LAM-HTGTS. In cells treated with nuclease, the ends of multiple simultaneous DSBs can be joined by translocation. After purification and fragmentation of gDNA, these translocation junctions are captured by LAM-PCR from a known "bait" DSB site, joined to unknown "prey" DSB sites, using a biotinylated primer. Amplified, biotinylated sequences are enriched on streptavidin and ligated to adapters on the distal "prey" end. Fragments with adapters and the original "bait" sequence are processed by a nested PCR, followed by a tagging PCR, to generate fragments with sequencing adapters on both ends, which are subsequently sequenced. Btn, biotin; DSB, double-strand break; dsDNA, double stranded DNA; dsODN, double-stranded oligodeoxynucleotide; gDNA, genomic DNA; gRNA, guide RNA; GUIDE-seq, genome-wide unbiased identification of DSBs enabled by sequencing; IDLV, integrase-defective lentiviral vector; LAM-HTGTS, linear amplification-mediated high-throughput genome-wide translocation sequencing; LAM-PCR, linear-amplification-mediated PCR; NHEJ, non-homologous end joining; TTISS, tagmentation-based tag integration site sequencing.

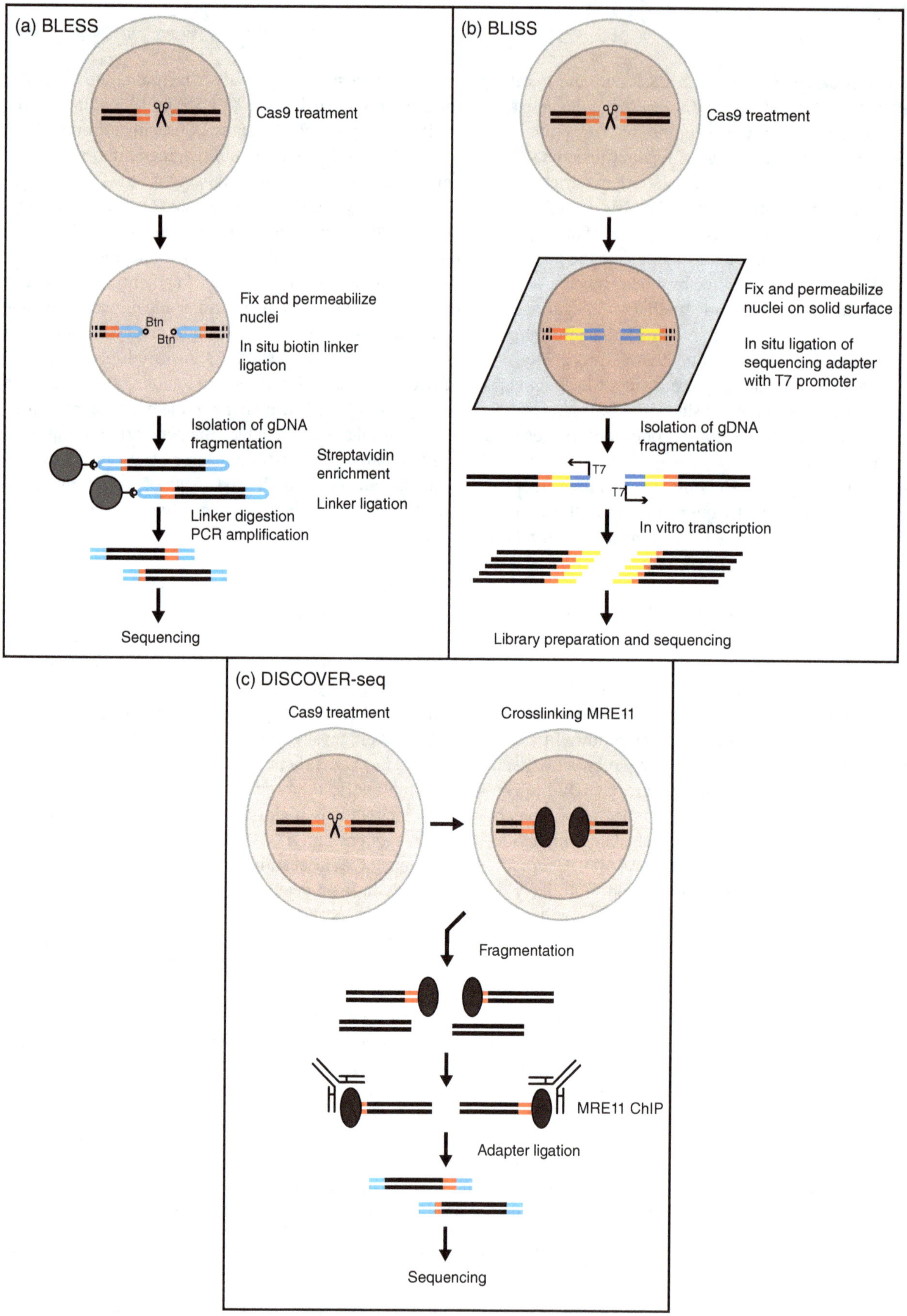

Figure 20.2 In situ off-target detection methods. (a) BLESS. Following treatment with Cas9 and gRNA, cellular nuclei are fixed and permeabilized, and biotinylated hairpin linkers are ligated to ends of unrepaired DSBs in situ. gDNA is then purified and fragmented, and biotinylated fragments are enriched on streptavidin and ligated to another hairpin linker on the distal end. Linkers on both ends are digested, generating linear fragments with adapter sequences used for PCR amplification and sequencing; (b) BLISS. Nuclei of Cas9-treated cells are fixed and permeabilized on a solid surface, such as a slide or multi-well plate, and unrepaired DSB ends are in situ ligated to sequencing adapters containing a T7 promoter and a UMI. gDNA is purified and fragmented, and the T7 promoter on

sequences, called "prey" sequences (off-targets), that have been cleaved and ligated to the bait through translocation (Figure 20.1d) (Frock et al. 2015; Hu et al. 2016). The method is compatible with a wide array of nucleases and cellular systems including in vivo settings, since it does not require the addition of exogenous tags for detection. Furthermore, it provides information on gross-chromosomal changes upon CRISPR editing (e.g. acentric or dicentric fusions) that may lead to deleterious effects, instead of being limited to local indel formations. The major limitation of the method is that it can only detect DSBs that have been translocated, but not DSBs which have been repaired by other means. As translocation events are very rare compared with local end-joining repair events, the method requires large amounts of input DNA for detection and potentially underestimates the frequency of indels generated by end-joining repair at off-target sites. In addition, as translocations occur more frequently between DSBs on the same or nearby chromosomes, the method is biased against off-target sites located more distant to the bait.

20.2.3 In Situ Off-target Detection Methods

In situ off-target detection methodologies are based on direct detection of transient DSBs induced by nucleases in fixed cells/tissues. For this, Cas9 and gRNAs are delivered to living cells/organisms, followed by fixation and permeabilization upon DSB induction, ligation of sequencing adapters into the DSBs in situ, and high-throughput sequencing. As in situ methods do not depend on the delivery and incorporation of an exogenous DNA for detection, they are suitable for off-target detection in vivo. A major drawback is that these methodologies provide only a snap-shot of DSBs at the time the cells/tissues are fixed, instead of detecting the final "end-joined" breaks. Thus, they can only detect the transient DSBs existing at the moment of fixation, but not the DSBs which potentially have been cleaved and already repaired or the ones that have not been generated yet, making the results dependent on the nuclease cleavage kinetics which can vary greatly between different enzymes, gRNAs, and cell types.

In this context, "breaks labeling, enrichment on streptavidin and next-generation sequencing" (BLESS) was initially developed as an unbiased in situ method to directly detect DSBs in cells (Crosetto et al. 2013), but was later also adapted to detect nuclease-induced off-targets (Figure 20.2a) (Ran et al. 2015). Following the fixation of nuclease-edited cells, BLESS marks transient DSBs by ligating a biotinylated proximal linker to the free ends in situ. Genomic DNA is then extracted and fragmented, followed by capture of biotinylated fragments with streptavidin and amplification for next-generation sequencing (NGS). However, the method requires relatively high cell input (five million) due to the limited enrichment and high background signals by DSBs that have been introduced during the technical steps (e.g. fixation and permeabilization).

Several other in situ methodologies have been developed in recent years building on BLESS. In contrast to the blunt-ended ligation of proximal linkers to the DSB free ends and sequential ligation of blunt-ended distal sequencing adapters in BLESS, DSBCapture utilizes enzymatic repair reactions to process the ends before ligating T-tailed cohesive-end sequencing adapters, which increases the ligation efficacy and reduces the number of PCR steps (Lensing et al. 2016). DSBCapture thereby demonstrates increased sensitivity and higher sequence-diversity compared with BLESS, with improved reproducibility between replicates. In another approach, END-seq embeds cells in agarose plugs instead of fixing them, to avoid introduction of background DSBs during the fixation step (Canela et al. 2016). Overall, END-seq was shown to be more sensitive and accurate in identifying DSBs compared with BLESS, likely as a result of avoiding formaldehyde cell fixation and subsequent DNA damage.

Implementing several improvements into the method, "breaks labeling in situ and sequencing" (BLISS) was described as a refinement of the original BLESS protocol (Figure 20.2b) (Yan et al. 2017). For this, cells or tissue sections are fixed and permeabilized on a solid surface, such as a microscopic slide or multiple-well plate, allowing easier processing of samples and higher throughput while lowering the input cell number, minimizing the

Figure 20.2 (Continued) successfully ligated fragments is used for in vitro transcription, generating linearly amplified RNA molecules with sequence information from each unique cellular DSB. RNA is converted to an NGS library and sequenced; (c) DISCOVER-seq. After treatment with Cas9, nuclease-induced DSBs in cells are cross-linked in situ to MRE11, a cellular repair factor involved in the processing of free ends. gDNA is sheared, and fragments with cross-linked protein are captured and enriched by ChIP using an antibody targeting MRE11. Captured fragments are ligated to adapters on both ends, amplified by PCR and sequenced. BLESS, breaks labeling, enrichment on streptavidin and next-generation sequencing; BLISS, breaks labeling in situ and sequencing; Btn, biotin; ChIP, chromatin immunoprecipitation; DISCOVER-seq, discovery of in situ Cas off-targets and verification by sequencing; DSB, double-strand break; gDNA, genomic DNA; gRNA, guide RNA; NGS, next-generation sequencing; UMI, unique molecular identifier.

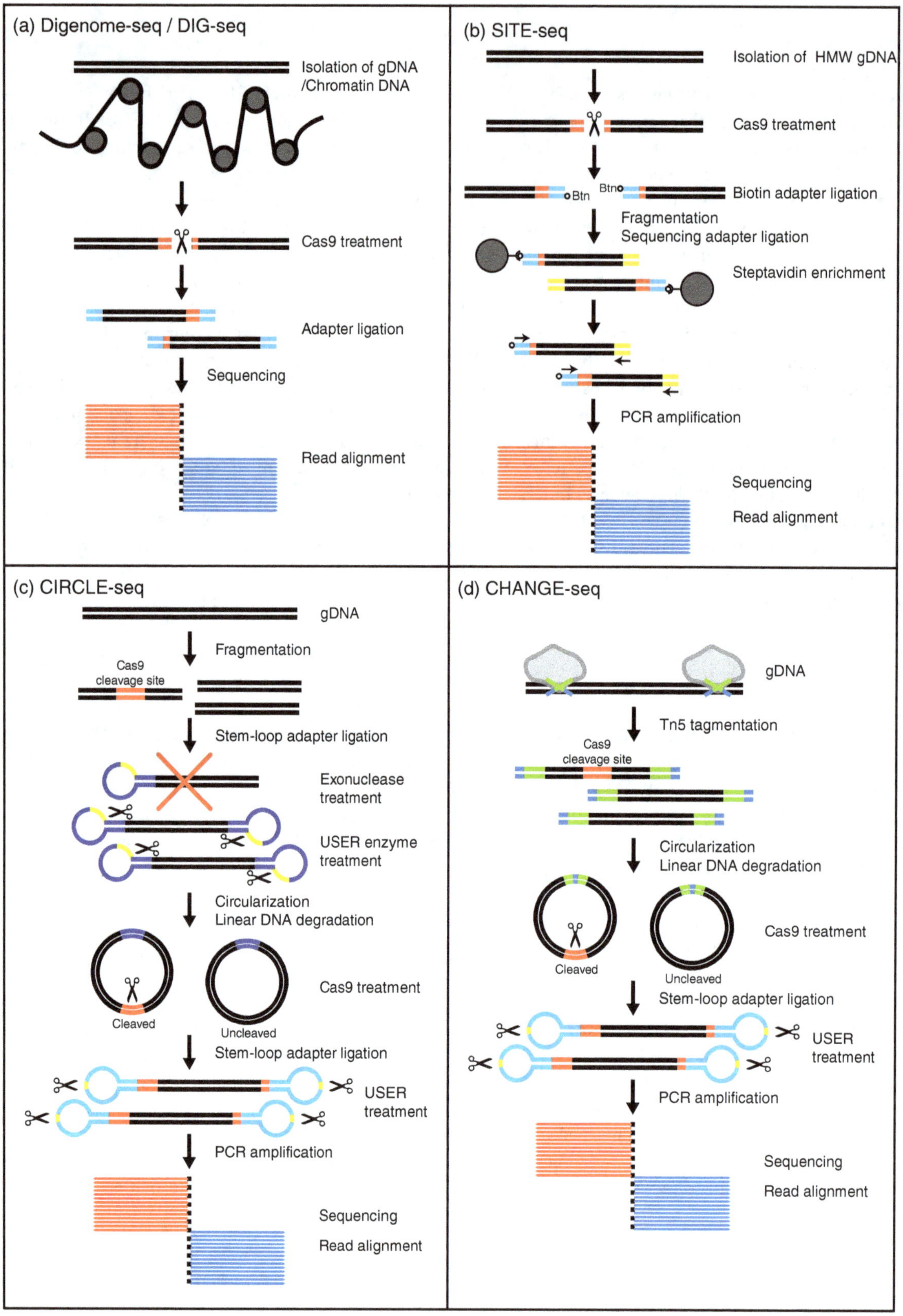

Figure 20.3 In vitro off-target detection methods. (a) Digenome-/DIG-seq. Purified gDNA (Digenome-seq) or chromatin DNA (DIG-seq) is digested by Cas9-gRNA RNP in vitro. RNP digested DNA is then further fragmented and free ends are ligated to sequencing adapters. After sequencing and read alignment, a relative enrichment in reads with the same start and end positions is used to identify nuclease-induced DSBs; (b) SITE-seq. Isolated HMW gDNA is treated with Cas9-RNP in vitro, and the generated free ends of DSBs are ligated to biotinylated adapters. DNA is sheared into uniform fragments and a sequencing adapter is ligated to the opposite end. Biotinylated DNA fragments are enriched on streptavidin, and fragments containing adapters on both ends are amplified by PCR and sequenced; (c) CIRCLE-seq. Purified gDNA is fragmented by sonication, and free ends are used to ligate a stem-loop adapter containing a uracil in the loop. Fragments missing an adapter on either side are degraded by exonuclease

risk of introducing DNA breaks and sample loss during the process. The biotinylated adapter used for enrichment was exchanged with an adapter containing a T7 promoter sequence, enabling linear amplification of captured DSBs through in vitro transcription with the aim of introducing less bias compared with exponential amplification. The adapter moreover includes a unique molecular identifier (UMI) sequence which enables discrimination of PCR duplicates and detection of unique DSB events in individual cells or alleles. Together, these modifications lower the cellular input requirements, while improving the sensitivity and accessibility of the method.

While all aforementioned in situ methods are independent from the cellular repair machinery for detection, "discovery of in situ Cas off-targets and verification by sequencing" (DISCOVER-seq) utilizes localization of endogenous DNA repair factors at the nuclease-induced DSB sites to detect CRISPR-Cas off-targets (Figure 20.2c) (Wienert et al. 2019, 2020). After delivery of the nuclease and gRNAs to the cells, the DNA is cross-linked to localized proteins during DSB repair. A subunit of the MRN complex which is involved in the early processing and repair of free ends in DSBs, MRE11, is recruited to Cas9-induced cleavage sites, enabling capture of the breaks by chromatin immunoprecipitation using an antibody against MRE11, followed by NGS. Enrichment of sequencing reads at specific genomic locations is used to determine the location and molecular characteristics of the Cas9-induced DSB with single-base resolution. DISCOVER-seq-identified sites were all found mutated in the targeted deep sequencing validation step in cells or liver tissue, indicating its high accuracy in finding bona fide off-targets. Nevertheless, the method exhibits a lower degree of sensitivity in discovering off-targets compared with GUIDE-seq or CIRCLE-seq (an in vitro off-target method that is described in the next section). Moreover, DISCOVER-seq requires a high amount of cells (≥5 million) and sequencing depth (≥30 million reads) and may miss off-target sites that lack MRE11 recruitment at the locus.

20.2.4 In vitro Off-target Detection Methods

In vitro methods use purified genomic DNA and Cas9-gRNA ribonucleoprotein (RNP) complexes to assay cleavage in a test tube, rather than delivering the nuclease and gRNAs into living cells. The methods can be applied to DNA purified from all different sources, including live animals and hard-to-transfect cells, as they do not rely on delivery of exogenous components into the cells. Despite the unmatched sensitivity of purifying the DNA and treating it with high nuclease-concentrations in vitro, some biologically relevant contexts such as cellular DNA repair and chromatin structure are lost, often leading to high rates of false-positives. On the other hand, it is possible that these "false-positive" sites are mutated in cells at frequencies below the current detection limit of deep sequencing (0.1%), or cleaved in cells efficiently but repaired seamlessly either via error-free NHEJ or HDR. One drawback of in vitro detection methods is the requirement of purified recombinant Cas9 protein that may not be readily available for novel nucleases.

The first described unbiased in vitro off-target detection assay, "digested genome sequencing" (Digenome-seq) used purified genomic DNA from cells of interest which was treated with a Cas9-gRNA RNP complex in vitro, followed by WGS (Figure 20.3a) (Kim et al. 2015). The reads were aligned to the genome and relatively enriched uniform reads with the same start and end position were considered as Cas9-induced DSBs, while reads that were not uniform or enriched were associated with random breaks. Digenome-seq analysis was further improved, allowing the detection of nuclease-induced 1- or 2-nt overhangs and multiplexing, although with the limitation of requiring guides to have at least an 11 nt-difference to facilitate assignment of off-targets to the correct gRNA (Kim et al. 2016). A web tool for analyzing Digenome-seq data was also made available (Park et al. 2017). Since the in vitro reaction conditions can easily be optimized to maximize off-target cleavage, Digenome-seq is a more sensitive method compared with cell-based methods, detecting

Figure 20.3 (Continued) treatment. The uracils on fragments containing adapters on both ends are subsequently excised by USER enzyme, and the open ends are joined by intramolecular ligation, generating circular fragments. After degradation of residual linear DNA, circular DNA is treated with Cas9-gRNA RNP, and any fragment containing an on- or off-target site will be cleaved open, leaving free ends for ligation of another stem-loop adapter used for PCR amplification, followed by sequencing; (d) CHANGE-seq. Genomic DNA is processed in vitro by Tn5 tagmentation, generating shorter fragments with uniform ends with great efficiency, which are subsequently used for intramolecular ligation. Residual linear fragments are degraded and circular DNA is digested with Cas9. Sequences cleaved by Cas9 will linearize the DNA, leaving free ends for adapter ligation, PCR amplification, and sequencing. CHANGE-seq, circularization for high-throughput analysis of nuclease genome-wide effects by sequencing; CIRCLE-seq, circularization for in vitro reporting of cleavage effects by sequencing; Digenome-/DIG-seq, digested genome sequencing; DSB, double-strand break; gDNA, genomic DNA; gRNA, guide RNA; HMW, high molecular weight; RNP, ribonucleoprotein; SITE-seq, selective enrichment and identification of tagged genomic DNA ends by sequencing; USER, uracil-specific excision reagent.

off-targets with mutation rates as low as 0.1%. As Digenome-seq does not rely on enrichment of DNA ends, the methodology is simple and easy to perform. However, because nuclease-digested sequences are not enriched in any way, the method is sequencing-inefficient and expensive. Moreover, random DSBs may also lead to uniform reads, making it difficult to distinguish low-frequency nuclease-induced cleavage events from the background. As most of the reads are spent on sequences not related to nuclease-induced DSBs, the method requires at least 400 million reads per sample to comprehensively detect break sites, making it unsuitable for comprehensive off-target analysis for large-number gRNA screening experiments.

DIG-seq was developed based on the observation that the DNA chromatin state can have a high influence on Cas9 cleavage activity, especially on off-target sites that show 1–2 bp mismatches to the on-target sequences (Figure 20.3a) (Kim and Kim 2018). To account for the native DNA structures found in eukaryotic cells, DIG-seq uses native chromatin DNA or cell nuclei in the in vitro digestion reaction, as opposed to the histone-free purified gDNA used in the original Digenome-seq and other in vitro methods. Processing and analyzing Cas9-digested chromatin DNA through the established Digenome-seq pipeline, DIG-seq was shown to detect only a small subset of sites detected by Digenome-seq, which have a much higher probability of being mutagenized in cells. As validation of extensive numbers of in vitro detected sites by targeted sequencing is laborious, taking the chromatin structure into account can help to reduce the number of sites to validate. However, it also carries the risk of missing sites that are potentially edited in real cellular contexts.

Two other in vitro off-target detection assays were developed to reduce the background reads observed with Digenome-seq, with the aim of decreasing the sequencing depth required for sensitive detection and overcoming resource limitations: "Selective enrichment and identification of tagged genomic DNA ends by sequencing" (SITE-seq) (Cameron et al. 2017) and "Circularization for in vitro reporting of cleavage effects by sequencing" (CIRCLE-seq) (Lazzarotto et al. 2018; Tsai et al. 2017).

SITE-seq uses Cas9 RNPs to digest high molecular-weight purified genomic DNA to minimize the number of available DNA ends unrelated to nuclease-cleavage (Figure 20.3b). Biotinylated adapters are then ligated to the free ends of Cas9-induced DSBs, DNA is fragmented, and Cas9-cleaved biotinylated sites are enriched by streptavidin pull-down. Fragments containing Cas9 cleavage sites are used for PCR amplification and sequencing. Due to the selective enrichment of nuclease-induced DSBs, SITE-seq provides high sensitivity with minimal read coverage requirements, therefore allowing for higher throughput.

CIRCLE-seq utilizes covalently closed DNA fragments for specific identification and enrichment of genome-wide Cas9 cleavage sites (Figure 20.3c). Purified genomic DNA is first sheared into uniform-sized fragments using sonication and then circularized by intramolecular ligation. Remaining linear DNA molecules are degraded by exonuclease treatment before purified circular DNA molecules are treated with Cas9-gRNA RNP complex. Circular DNA molecules containing Cas9 cleavage sites (on- or off-target) are thereby cleaved and linearized, leaving free ends for adapter ligation, PCR amplification, and paired-end high-throughput sequencing. Similar to SITE-seq, this selective enrichment of Cas9-cleaved sites enables reduction of the number of sequencing reads required. As an advantage over SITE-seq, CIRCLE-seq-generated reads contain both sides of the cleavage site, allowing for identification of off-target sites even without a reference genome sequence. Because of its high sensitivity, CIRCLE-seq provides a comprehensive list of potential off-target sites for a given combination of gRNA and Cas9 nuclease. The identified sites can then be assessed for indels after Cas9 treatment by targeted deep sequencing in a more biologically relevant context such as in cells or in vivo (Akcakaya et al. 2018).

Although CIRCLE-seq and SITE-seq provide a highly sensitive and sequencing-efficient way for identification of a superset of Cas9 off-target sites in vitro, they have the common drawback of a high false discovery rate of in vitro detection assays. The methods are furthermore labor-intensive and require high inputs of genomic DNA, making them unsuitable for high-throughput off-target analyses.

Building on the CIRCLE-seq method, "circularization for high-throughput analysis of nuclease genome-wide effects by sequencing" (CHANGE-seq) implemented modifications to reduce the number of experimental steps and genomic DNA input (Figure 20.3d) (Lazzarotto et al. 2020). CHANGE-seq utilizes a Tn5 tagmentation-based workflow for simultaneous in vitro DNA transposition and fragmentation, enabling efficient, rapid generation of circularized genomic DNA molecules, instead of using mechanical DNA shearing and multiple enzymatic reactions such as in the original CIRCLE-seq protocol. The relatively low genomic DNA input requirement of CHANGE-seq together with the simplified, automatable workflow enable generation of large-scale off-target datasets to further elucidate the mechanisms of nuclease activity and specificity. Additionally, this knowledge can potentially be applied to train machine learning models to improve in silico gRNA design and off-target prediction tools.

20.2.5 Detecting Off-target Effects of CRISPR Tools that Do Not Induce DSBs

Genome editing technologies are constantly being improved in terms of specificity, targeting scope, and deliverability to tissues of therapeutic relevance. In addition to traditional CRISPR-Cas9 gene editing where nuclease-induced DSBs are created to induce mutations in target DNA sites, several other CRISPR-based tools have recently been developed promoting gene editing without induction of DSBs. One example is base editors, which allow the exchange of single nucleotides by deamination. Cytosine base editors (CBE) consisting of a fusion of Cas9 nickase (Cas9n), APOBEC1, or AID deaminase, and a uracil glycosylase inhibitor domain (UGI) specifically deaminate cytosine bases to uracil, promoting a C→T conversion (Komor et al. 2016; Nishida et al. 2016). In addition, adenine base editors (ABE) have been developed by fusing Cas9n and a heterodimer of wild-type and mutated TadA adenine deaminase, specifically deaminating adenine bases to inosine, which are recognized as guanine during DNA replication leading to an A→G conversion (Gaudelli et al. 2017). The development of prime editors (PE) complemented this rather limited range of base editing by facilitating all 12 possible base-to-base conversions, as well as introduction of deletions and insertions in the absence of DSBs (Anzalone et al. 2019). Prime editing relies on a Cas9n-reverse transcriptase fusion protein which is guided by a prime editing guide RNA (pegRNA). In addition to guiding the prime editor to its target and inducing the generation of a single-strand break, the pegRNA also carries the desired mutated sequence which is used as a template for the reverse transcriptase leading to precise mutation of the target locus. Both approaches are independent of the error prone NHEJ pathway; however, several studies have nevertheless reported off-target DNA editing also in the absence of DSBs, in particular for base editing, presumably since nicks are converted to DSBs at certain sites (Jin et al. 2019; Zuo et al. 2019).

Since the established detection methods for CRISPR-Cas-induced off-targets are based on introduction of DSBs, development of novel methods were required for genome-wide off-target profiling of BE and PE. In this context, Kim and colleagues established two in vitro methods to analyze the off-target profiles of CBEs and ABEs (Kim et al. 2017, 2019). For CBEs, genomic DNA is first incubated with the base editor, which catalyzes C-to-U conversions on the nontarget DNA strand and a nick in the target DNA strand. This is followed by incubation with uracil-specific excision reagent (USER) enzyme, which induces a nick in the nontarget strand by recognizing and removing uracil from DNA. The two nicks in the DNA – one in the target DNA strand induced by the base editor and the other in the nontarget DNA strand induced by the USER enzyme – lead to a composite DSB suitable for detection by Digenome-seq (Kim et al. 2017). Similarly, for off-target analysis of ABEs, DSBs can be generated by digesting genomic DNA with ABE and the inosine-specific endonuclease V (Kim et al. 2019; Liang et al. 2019). For genome-wide analysis of prime editing off-targets, a modified Digenome-seq approach, nDigenome-seq, was established in which the sequencing reads were screened for sites which started at the same 5′ nucleotide in at least 10 reads to allow for genome-wide identification of single-strand breaks introduced by the PE (Kim et al. 2020).

Besides the in vitro approaches, genome-wide off-target analysis by two-cell embryo injection (GOTI) has been described as the first in vivo method for base editing off-target analysis (Zuo et al. 2019, 2020). Injection of the base editing components in one of the two blastomeres of an embryo at the two-cell stage allowed for direct comparison of edited and non-edited progeny cells by WGS, facilitating high-sensitivity detection of off-targets. In addition to genomic off-targets, it is, however, important to note that base editors, due to their deaminase's preference for single stranded sequences, can also target and mutagenize RNA molecules leading to introduction of numerous off-targets on the RNA level, which should be considered when interpreting the outcome of base editing safety studies (Grünewald et al. 2019; Rees et al. 2019; Zhou et al. 2019).

In addition to targeted editing of genomic DNA, CRISPR has been utilized to modulate gene expression without changing the genomic sequence. In this context, catalytically inactive dCas9 was fused to a transcriptional activator or repressor domain to target the transcription start site of a desired gene for direct transcriptional activation (CRISPRa) or inhibition (CRISPRi) (Gilbert et al. 2013; Maeder et al. 2013; Perez-Pinera et al. 2013; Qi et al. 2013). Although this approach has overall lower genome-wide off-target effects than gene editing approaches, some changes in the expression of nontarget transcripts have been observed by whole-transcriptome RNA sequencing, which could have an effect on study outcome (Gilbert et al. 2014; Soubeyrand et al. 2019; Stojic et al. 2018; Tycko et al. 2019). Over recent years, there have additionally been endeavors to establish CRISPR-based editing tools targeting RNA transcripts directly, which rely on the activity of the adenosine deaminase acting on mRNA (ADAR) enzymes leading to post-transcriptional A-to-I conversions of the targeted RNA. Using either endogenous or engineered ADAR proteins, several different RNA editing platforms have been established allowing for programmable base-conversions on the targeted RNA (Abudayyeh et al. 2019; Cox et al. 2017; Merkle et al. 2019; Qu et al. 2019; Rauch et al. 2019; Vogel et al. 2018). Even though this approach limits the danger of permanent genomic off-target mutations, extensive transcriptome-wide off-target editing has been observed, especially for platforms using overexpression of exogenous ADAR proteins, which might have implications for potential uses in clinical applications (Montiel-Gonzalez et al. 2019; Vallecillo-Viejo et al. 2018).

20.2.6 Off-target Detection Method Choice

As no currently established single method provides definitive assessment of CRISPR specificity for therapeutic use, we believe that the best strategy is to use multiple approaches to assess genome-wide off-target mutagenesis until such an ideal method is developed. The choice of method depends on the scientific questions to be addressed, and the strengths and weaknesses of these different approaches (Table 20.2). In silico tools are essential when designing the gRNAs. The off-targets detected with experimental methods and in silico tools largely overlap at the sites that show higher sequence homology to the on-target; however, they are not considered an equivalent substitute to experimental detection methods prior to therapeutic usage. Cell-based methods directly and quantitatively measure off-target mutagenesis in cells, but have the limitations of requiring efficient delivery, dependence on the NHEJ repair pathway, and a lack of consistency for detecting off-targets with low mutation frequencies. In situ methods generate a snap-shot of the DSBs in time, rather than detecting the end-repaired products. Recent additions to in situ methods made the input requirements smaller and improved their accuracy by decreasing the background DSBs introduced during the experimental process. These methods are useful for studying off-target kinetics; however, time-sensitivity carries the risk of missing off-target sites that have already been repaired or not been cleaved yet. Although in vitro methods are the most sensitive and comprehensive, they disregard the cellular context, often resulting in high numbers of false positives. Thus, bona fide off-targets in cells must still be verified in subsequent extensive sequencing experiments.

Sensitivity of mutation-detection is currently limited by the error rate of high-throughput sequencing technologies and requires particular attention. Off-target mutagenesis occurring at frequencies of <0.1% are currently not reliably assessed, meaning that the off-target mutagenesis frequencies that occur in less than one cell in a thousand may be missed. In potential therapeutic settings, where heterogenous bulk populations of cells will be edited, even very low-frequency mutagenesis will be of high importance. Further developments in increasing the sensitivity of NGS platforms are therefore warranted.

20.3 Reducing Genome-wide Off-target Effects

Various approaches to reduce off-target effects of Cas9-mediated cleavage have been described (Tsai and Joung 2016). These approaches can be categorized as either increasing the specificity of the Cas9 nuclease target recognition and cleavage, or limiting the activity of the Cas9 nuclease to avoid accumulation of off-target mutations. Examples of strategies used for improving CRISPR-Cas9 specificity are shown in Figure 20.4.

20.3.1 Improving the Specificity of Cas9 Target Recognition and Cleavage

For increasing the specificity of Cas9 target recognition and cleavage, one strategy is to lower the energy barrier between formation and dissociation of the Cas9-gRNA complex to the target DNA, or to increase the energy barrier between the ground state of the RNP-DNA complex and its transition state for cleavage. Modifications on gRNA sequence and chemistry as well as engineering nuclease variants are among the approaches successfully employed to modulate the energy barriers. Another strategy is using engineered dual Cas9-gRNA systems that require binding of two RNP complexes to the target in close proximity for inducing DNA cleavage, therefore decreasing the likelihood of unspecific DSB generation.

20.3.1.1 Modification of gRNA Sequence and Chemistry

Moving beyond the choice of target sequence, a number of strategies have been employed to improve genome-wide specificity of CRISPR by modification of different gRNA characteristics (Figure 20.4a). A common observation is an increased targeting specificity upon weakening the interactions of the Cas9-gRNA complex with the target DNA site. The "excess energy" model rationalizes this observation by postulating that binding energy beyond a certain threshold stabilizes binding to both correct and incorrect targets (Bisaria et al. 2017; Fu et al. 2014; Hu et al. 2016; Kleinstiver et al. 2016; Tsai et al. 2015; Tsai and Joung 2016; Wyvekens et al. 2015). Therefore, reducing the target binding energy to threshold levels can enable reduction of off-target activity while not perturbing on-target cleavage.

In this context, use of truncated gRNA (tru-gRNA) with only 17–18 nucleotides complementary to the target DNA

Table 20.2 Summary of genome-wide methods for detecting CRISPR-Cas9 off-targets.

Method	Off-target discovery using			Detection stage			Sensitivity	Specificity	In vivo applicable	Approach	Reference
	Pure DNA	Cells	Tissue	Without repair	Pre-repair	Post-repair					
Digenome-seq/DIG-seq	+			+			+++/++	+/++	+	In vitro cleavage and WGS of purified gDNA/Chromatin DNA.	Kim et al. (2015) and Kim and Kim (2018)
SITE-seq	+			+			+++	+	+	In vitro cleavage of HMW gDNA and selective enrichment of biotin-tagged Cas9 cleaved sites.	Cameron et al. (2017)
CIRCLE-seq	+			+			+++	+	+	In vitro cleavage of circularized DNA, selective sequencing of Cas9 cleaved sites.	Tsai et al. (2017)
CHANGE-seq	+			+			+++	++	+	In vitro cleavage of circularized DNA generated by tagmentation, selective sequencing of Cas9 cleaved sites.	Lazzarotto et al. (2020)
WGS		+				+	+	++	+	Deep sequencing of the entire genome of interest upon nuclease treatment.	Iyer et al. (2015), Smith et al. (2014), and Veres et al. (2014)
IDLV Capture		+				+	++	++		Viral DNA integration into nuclease-induced cellular DSBs, selective sequencing.	Wang et al. (2015)
GUIDE-seq		+				+	++	++		dsDNA tag integration into nuclease-induced cellular DSBs, selective tag-specific sequencing.	Tsai et al. (2015)
TTISS		+				+	++	++		dsDNA tag integration into cellular DSBs, tagmentation, selective tag-specific sequencing.	Schmid-Burgk et al. (2020)
HTGTS/LAM-HTGTS		+				+	+	++	+	Detection of translocation events in-between the "bait" on-target and off-targets.	Frock et al. (2015) and Hu et al. (2016)
BLESS/DSBCapture/END-seq		+			+		++	++	+	In situ biotin-adapter ligation into DSBs followed by PCR enrichment and selective sequencing.	Canela et al. (2016), Crosetto et al. (2013), and Lensing et al. (2016)
BLISS		+	+		+		++	++	+	In situ adapter ligation into DSBs on a solid surface followed by in vitro transcription enrichment and selective sequencing.	Yan et al. (2017)
DISCOVER-seq		+	+		+		+	+++	+	ChIP-seq detection of MRE11 localized to nuclease-induced DSBs.	Wienert et al. (2019)

BLESS, breaks labeling, enrichment on streptavidin and next-generation sequencing; BLISS, breaks labeling in situ and sequencing; CHANGE-seq, circularization for high-throughput analysis of nuclease genome-wide effects by sequencing; ChIP-seq, chromatin immunoprecipitation-sequencing; CIRCLE-seq, circularization for in vitro reporting of cleavage effects by sequencing; Digenome-/DIG-seq, digested genome sequencing; DISCOVER-seq, discovery of in situ Cas off-targets and verification by sequencing; DSB, double-strand break; dsDNA, double-stranded DNA; gDNA, genomic DNA; GUIDE-seq, genome-wide unbiased identification of DSBs enabled by sequencing; HMW, high molecular weight; IDLV, integrase-defective lentiviral vector; LAM-HTGTS, linear amplification-mediated high-throughput genome-wide translocation sequencing; LAM-PCR, linear-amplification mediated PCR; SITE-seq, selective enrichment and identification of tagged genomic DNA ends by sequencing; TTISS, tagmentation-based tag integration site sequencing.

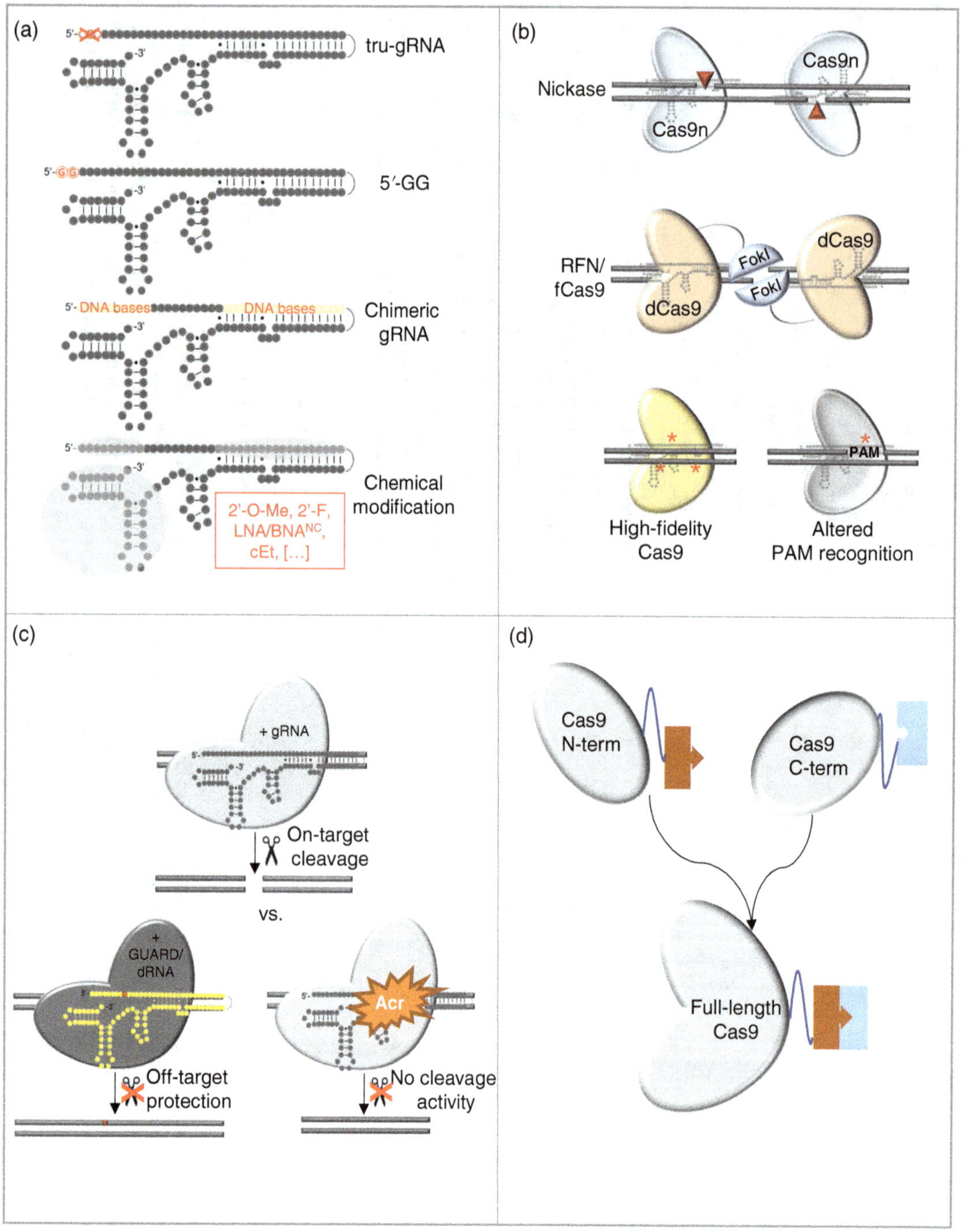

Figure 20.4 Overview of methods for reduction of genome-wide off-targets. (a) Modification of gRNA by truncation of the guide sequence (tru-gRNA), addition of two noncomplementary guanine nucleotides at the 5′ end (5′-GG), replacing residues with DNA bases (chimeric gRNA), or chemical modification of the gRNA backbone can increase binding specificity; (b) Cas9 engineering can mediate lower off-target effects by use of paired nickases, fusion of catalytically inactive dCas9 with dimerization-dependent FokI-nuclease, mutations of, e.g. DNA-binding residues to generate high-fidelity variants, or mutation of the PAM-recognition domain to achieve altered PAM recognition. Asterisks indicate Cas9 mutations; (c) using competitive RNA binding or protein inhibitors can increase Cas9 specificity. GUARD or dRNAs can protect off-targets by complementary binding to the DNA without induction of Cas9 cleavage, whereas Acr proteins act as natural Cas9 inhibitors which disrupt cleavage activity; (d) nuclease exposure can be limited by using split-Cas9 variants, which only form the active full-length Cas9 after induction of dimerization of the two subdomains. Acr, anti-CRISPR protein; Cas9n, Cas9 nickase; dRNA, dead RNA; gRNA, guide RNA; GUARD, Guide RNA Assisted Reduction of Damage; RFN, RNA-guided FokI-dCas9 nuclease; tru-gRNA, truncated gRNA.

has been shown to greatly reduce off-target cleavage activity compared with the commonly used length of 20 nucleotides (Fu et al. 2014). The authors hypothesized 5′-end nucleotides may not be needed for full gRNA activity and that these nucleotides may compensate for mismatches at other positions; thus, shorter gRNAs might be more sensitive to mismatches and more specific. Indeed, Cas9 nucleases guided by tru-gRNA displayed decreased tolerance for mismatches without affecting on-target efficiency. On the contrary, addition of two noncomplementary guanine nucleotides to the 5′ end of the gRNA (5′-GG) was also reported to reduce off-target mutations, although on-target efficiency was also affected to a small degree (Cho et al. 2014). The authors attributed these effects to an impact of the non-pairing nucleotides on RNA stability and/or secondary structure which could affect the concentration of the gRNA.

Additionally, several studies have analyzed the effect of chimeric gRNAs in which several RNA nucleotides were replaced by DNA (Kartje et al. 2018; Rueda et al. 2017; Yin et al. 2018). In this context, Kartje and colleagues found DNA substitutions in the tracrRNA pairing region of the crRNA to be beneficial for cleavage activity and specificity (Kartje et al. 2018), whereas Yin et al. reported reduction of off-target activity with DNA replacements present at the 5′ end of the guide sequence (Yin et al. 2018). Rueda and colleagues observed that a chimeric crRNA containing only seven RNA residues at Cas9-interacting positions greatly reduced off-target activity, while preserving on-target cleavage (Rueda et al. 2017). The authors attributed the improved specificity to altered binding energetics of the chimeric gRNAs and thermodynamic stability of the gRNA/Cas9 complex. In contrast, substitution of nucleotides in the seed region of the guide sequence (Yin et al. 2018) or in the Cas9-interacting positions (Rueda et al. 2017) completely abolished nuclease activity, highlighting the importance of these residues.

In a different approach, several research groups applied chemical modifications to the gRNA in order to increase the specificity. In this context, introduction of 2′-O-methyl-phosphonoacetate modification of the phosphodiester bonds at specific positions within the guide sequence has been shown to increase the cleavage specificity 10-fold compared with non-modified gRNA (Ryan et al. 2018). Incorporation of locked nucleic acids (LNAs) or next-generation bridged nucleic acids (2′,4′-BNA^{NC}[N-Me]) at specific regions of the gRNA also had a positive impact on Cas9 specificity by reduction of off-target cleavage activity (Cromwell et al. 2018). Based on the idea that combinations of different modifications would have a greater effect than single modifications, Rahdar and colleagues developed a shortened synthetic CRISPR RNA (scrRNA) in which phosphorothioate backbone modifications were combined with 2′-O′-methyl and 2′-F modifications at the 5′ and 3′ terminal nucleotides, as well as (S)-constrained ethyl substitutions in the ribose skeleton. Using this scrRNA for gene editing, the authors reported higher gRNA stability due to improved resistance to ribonucleases, an increased affinity of the scrRNA for DNA, as well as increased on-target and decreased off-target cleavage activity (Rahdar et al. 2015).

Taken together, these studies show great potential for improvement of CRISPR genome editing safety by modification of gRNAs. It should, however, be noted that despite their overall beneficial effects at specific locations, almost all of these modifications led to a complete loss of cleavage activity following modification of the seed or Cas9-interacting regions, which should be taken into consideration when designing modified gRNAs.

20.3.1.2 Engineered Cas9 Variants

Not only the gRNA but also the Cas9 enzyme itself has been subjected to vigorous engineering efforts aiming to increase the safety of gene editing (Figure 20.4b). Building on the observation that specific point mutations in either the HNH or RuvC nuclease domains can convert Cas9 into a nickase (Cong et al. 2013; Jinek et al. 2012), a double nicking approach has been developed where the region of interest is targeted by two gRNAs binding to opposite strands at nearby sites (Mali et al. 2013; Ran et al. 2013). The approach reduces the probability of off-target mutations, since binding of both gRNAs in close proximity is needed for induction of DSBs. In contrast to DSBs, single-strand DNA nicks are not repaired via NHEJ but via the less error-prone base excision repair pathway, reducing the probability of unwanted off-target mutations caused by single gRNA targeting (Dianov and Hubscher 2013). Gene editing with Cas9n has been reported to have a 100-fold increased specificity compared with wild-type Cas9, while showing similar on-target efficiency as the nuclease (Ran et al. 2013). Nevertheless, nickase-induced indels and point mutations have been reported (Cho et al. 2014; Fu et al. 2014; Tsai et al. 2014), indicating some nicks may be converted into DSBs, which is why off-target effects of each of the gRNAs used together with nickases still need consideration.

Following a similar rationale, a fusion protein of a catalytically dCas9 with the dimerization-dependent FokI nuclease called dimeric RNA-guided FokI nuclease (RFN; [Tsai et al. 2014] or fCas9 [Guilinger et al. 2014]) has been introduced for gene editing. Since monomeric FokI displays no catalytical activity, binding of two distinct gRNAs at a specific distance to each other is required for dimerization of FokI and subsequent introduction of DSBs at the target site. This also has implications for off-target mutagenesis, as only binding of both gRNAs to the same off-target site would introduce DNA cleavage and mutations. The approach improves upon generation of nicks that can potentially lead to DSBs, further decreasing the risk of off-target mutagenesis. In line with this, RFN/fCas9 has been shown to have an increased specificity with off-target mutations only detectable at the background level, while on-target mutagenesis was similar or slightly lower compared with wild-type Cas9 (Guilinger et al. 2014; Tsai et al. 2014). Moreover, Wyvekens and colleagues showed that combination of RFNs with tru-gRNA (Fu et al. 2014) was able to further reduce frequency of any remaining off-targets (Wyvekens et al. 2015). It should, however, be noted that delivery of dimeric FokI nuclease fusion proteins in a therapeutic setting would be challenging, due to their large size.

In addition to dual-gRNA proximity-based approaches, Cas9 nuclease has been extensively engineered to improve the specificity of target recognition and cleavage. Based on the findings from structural studies, Kleinstiver and colleagues rationally engineered Cas9 by mutagenesis of four DNA-contacting residues, leading to reduced binding energetics of the Cas9-gRNA complex to the target DNA. The new variant, SpCas9-HF1, possesses a decreased energy only allowing efficient binding and cutting of fully complementary targets in line with the "excess energy" model (Kleinstiver et al. 2016). Similarly, Slaymaker and colleagues targeted the positively charged groove of Cas9 in which the nontarget strand is stabilized during DNA binding. Mutation at this region normalized the positive charge and gave rise to eSpCas9 (Slaymaker et al. 2016). In another approach, Chen and colleagues used targeted mutation of residues within REC3 in the target recognition domain giving rise to HypaCas9 which showed better discrimination between target and nontarget DNA (Chen et al. 2017). In contrast to these rationally engineered high-fidelity variants, evoCas9 was generated by randomized screening in yeast, followed by rational optimization of additional residues (Casini et al. 2018; Lee et al. 2018; Vakulskas et al. 2018). Using a directed evolution approach by selection-based bacterial screenings, Sniper-Cas9 (Lee et al. 2018) and HiFiCas9 (Vakulskas et al. 2018) were generated as two additional high-fidelity Cas9 variants. All of these Cas9 variants showed similar on-target efficiency as wild-type Cas9 but were able to reduce occurrence of off-target mutagenesis in their original reports, highlighting the great potential of engineered Cas9 variants in improving the safety of gene editing. However, it became evident that editing efficacies of these variants may vary greatly depending on the target site and protospacer length (Casini et al. 2018; Lee et al. 2018; Vakulskas et al. 2018) and thus, on-target activity needs to be evaluated for the specific site of interest prior to therapeutic applications.

Further Cas9 optimization efforts were made by Kleinstiver and colleagues, who generated several Cas9 variants with altered PAM recognition by combination of a bacterial positive selection-based screening with rational engineering broadening the targeting range within the human genome (Kleinstiver et al. 2015). In a similar approach, Hu and colleagues used a phage-assisted continuous evolution (PAGE) approach generating the xCas9 variant which showed a high targeting specificity despite its broad range of recognized PAM sequences (Hu et al. 2018). Moreover, use of other bacterial CRISPR-Cas9 systems with more stringent PAM requirements as opposed to SpCas9 opens up possibilities to increase gene editing specificity. Gene editing with Cas9 from *Neisseria meningitidis* (NmeCas9; [Lee et al. 2016]), *Streptococcus thermophilus* (StCas9; [Muller et al. 2016]), or *Staphylococcus aureus* (SaCas9; [Ran et al. 2015]) has been shown to generate lower off-target effects as their more complex PAMs have a lower prevalence within the human genome, greatly reducing the overall number of possible off-target sites. In addition, the smaller size of other Cas9 proteins compared with SpCas9 allows for more efficient packaging into AAVs, which has important implications for in vivo delivery of Cas9 components (Lee et al. 2016). Discovery of the Cas12a-based CRISPR systems (Zetsche et al. 2015a) further complemented the toolbox for CRISPR-mediated gene editing. In contrast to Cas9, Cas12a (formerly called Cpf1) is guided by a single crRNA and recognizes a T-rich PAM at the upstream end up the spacer sequence. Moreover, Cas12a possesses only a single nuclease domain which generates staggered cuts with 5′ overhangs at the cleavage site promoting NHEJ-based gene insertion at the target site (Maresca et al. 2013; Zetsche et al. 2015a).

20.3.2 Competitive Binding and Protein Inhibitors

In addition to modification of the components of the Cas9-gRNA complex, there have also been efforts to reduce off-target mutagenesis by targeting the off-target sites themselves instead (Figure 20.4c). The observation that gRNAs with protospacer sequences of ≤16 nucleotides can guide Cas9 to complementary DNA without inducing DNA cleavage (Dahlman et al. 2015; Kiani et al. 2015; Ye et al. 2018), has led to the development of two strategies to specifically protect potential off-target sites from cleavage by Cas9: dead-RNA Off-Target Suppression (dOTS; [Rose et al. 2020]) and CRISPR Guide RNA Assisted Reduction of Damage (CRISPR GUARD; [Coelho et al. 2020]). Together with Cas9 and the full-length gRNA targeting the gene of interest, short gRNAs (14–16 nucleotide-long protospacer) with perfect complementarity to previously identified potential off-target sites are introduced into the cells. Cas9-short gRNA complexes are then directed specifically to the off-target sites where they bind the DNA without introducing DSBs, and thereby outcompete binding of active Cas9-gRNA complexes preventing off-target mutations and increasing the specificity of gene editing.

In another approach to improve safety of CRISPR, researchers have made use of the natural defense mechanism of phages against the bacterial CRISPR-Cas systems: phage-encoded anti-CRISPR proteins (Acrs) which function as CRISPR-Cas inhibitors (Bondy-Denomy et al. 2013; Pawluk et al. 2014). To date, several types of Acrs differing in their mode of action have been described to inhibit Cas9 cleavage activity in vitro by binding directly to the nuclease domain of Cas9 (Dong et al. 2017; Harrington et al. 2017; Zhu et al. 2019), by prevention of gRNA or DNA binding by induction of conformational changes (Zhu et al. 2019), or by interaction with the PAM-recognition site of Cas9 (Dong et al. 2017; Liu et al. 2019). The functionality of Acrs has also been shown in cellular environments where Cas9 cleavage activity could be successfully repressed (Marino et al. 2018; Nakamura et al. 2019; Pawluk et al. 2016; Rauch et al. 2017). Following the observation that Cas9 cleavage at off-target sites occurs more slowly than at on-target sites (Raper et al. 2018; Sternberg et al. 2015), Shin and colleagues showed that addition of Acrs six hours after introduction of Cas9 and gRNA was able to greatly reduce off-target cleavage while retaining on-target editing, highlighting the potential of Acrs to improve Cas9 specificity when timely delivery is feasible (Shin et al. 2017b).

20.3.3 Limiting Nuclease Exposure

In addition to increasing the specificity of single CRISPR components, it is possible to decrease the number of off-targets by limiting the exposure time of active Cas9 within the cells. The choice of delivery method of the Cas9-gRNA components plays a large role in determining the duration of Cas9 activity. While Cas9-encoding plasmids result in continuous expression of Cas9 over the duration of several days, transfection of Cas9 mRNA or preassembled RNPs that are active for shorter durations before being rapidly degraded, has been shown to reduce off-target cleavage (Kim et al. 2014; Liang et al. 2015; Shen et al. 2013; Zuris et al. 2015). Another way of controlling Cas9 activity is the use of inducible Cas9 variants, which are only activated within the cells after delivery for a short duration (Figure 20.4d). In this context, Davis and colleagues engineered an inducible Cas9 variant containing a 4-hydroxytamoxifen (4-HT)-responsive intein which disrupts Cas9 nuclease activity in the absence of 4-HT (Davis et al. 2015). Addition of the small molecule leads to splicing of the intein domain and in turn activation of the Cas9 nuclease, offering a controlled activation of gene editing. In a similar approach, Zetsche and colleagues developed an inducible split-Cas9 by division of the nuclease into a C-terminal fragment fused to the FK506 binding protein 12 (FKBP) domain of mammalian target of rapamycin (mTOR) and an N-terminal fragment fused to FKBP rapamycin binding (FRB) domain of mTOR (Zetsche et al. 2015b). While the two fragments are kept at different cellular locations in an inactive state under baseline conditions, addition of rapamycin leads to dimerization of the two fragments and subsequent translocation to the nucleus where the now-active nuclease can be used for gene editing. Similarly, Nihongaki and colleagues described a photoactivatable Cas9, where the two split-Cas9 fragments were fused to the photoinducible dimerization domains of the Magnets system (Kawano et al. 2015), allowing for controlled Cas9 activation by irradiation with blue light (Nihongaki et al. 2015). Additionally, cell type-specific Cas9 on-switches have been developed based on the discovery of Acrs (see Section 20.3.1). For enabling of cell type-specific Cas9 repression, recognition sites of miRNAs selectively expressed in the target cell type were inserted into the 3′ UTR of Acr genes. Introduction of the modified Acr transgene together with Cas9 and gRNA led to Acr-mediated repression of Cas9 activity in all non-target cell types. In contrast, Acr expression was repressed by the presence of miRNAs in the target cells, resulting in retained Cas9 activity (Hoffmann et al. 2019; Lee et al. 2019).

20.4 Other Unwanted Effects: Translocations and Large Deletions

CRISPR-induced DSBs are not always repaired by local end joining events. Cas9 nuclease-based gene editing can additionally result in undesired gross chromosomal changes, such as translocations, inversions, and large deletions. Comprehensive genomic analysis for evaluation of these events is required prior to the clinical translation of CRISPR.

Introduction of on-target large deletions up to 2.3 kb upon single gRNA-Cas9 editing was first reported in mouse embryos (Adikusuma et al. 2018; Shin et al. 2017a). Following this, Kosicki and colleagues observed evidence for on-target deletions of up to 9.5 kb, as well as complex chromosomal rearrangements, after editing of mouse embryonic stem cells, mouse hematopoietic progenitor cells, as well as a differentiated human cell line (Kosicki et al. 2018). More reports confirmed these observations in several different cellular contexts (Hoijer et al. 2020; Iyer et al. 2018; Song et al. 2020). In-depth analysis of CRISPR-edited human-induced pluripotent stem cells revealed the presence of large on-target deletions in up to 40% of edited clones which had been not detected during standard screening (Weisheit et al. 2020). Large deletions were found frequently located at sites with microhomologies, suggesting involvement of the microhomology-mediated end-joining (MMEJ) pathway in the DSB repair (Owens et al. 2019; Weisheit et al. 2020). Large deletions in embryos may potentially be attributed to the chromosomal instability caused by CRISPR DSBs resulting in error during cell division, or to the inefficacy in cell cycle checkpoint activation by a single DSB to mediate cell cycle arrest at this stage of development (Chia et al. 2017). Additionally, chromosomal truncations have been shown to be a common side effect of CRISPR editing in human cell lines harboring defects in p53, but not in primary fibroblasts with intact p53 signaling (Cullot et al. 2019). This observation suggests an involvement of p53 in prevention of these unwanted events during editing, and might therefore have implications for gene editing in cancer cells with disturbances in the p53 pathway. In line with this, a higher frequency of large deletions was observed when editing chromosomally unstable cancer cell lines compared with more stable lines (Rayner et al. 2019).

Chromosomal translocations may arise upon fusion of two DNA DSBs located on heterologous or homologous chromosomes during DNA repair (Klein et al. 2011; Stephens et al. 2009). CRISPR editing can potentially cause on-target translocations leading to acentric, dicentric, or balanced chromosomes upon end-joining of the DSBs in between homologous chromosomes. Furthermore, DSBs generated at on-target and off-target regions can be repaired via fusion in between free ends, rather than a local repair, leading to chromosomal translocations (Frock et al. 2015). Since translocations may potentially lead to deleterious effects, for example end joining of a highly expressed gene with an oncogene that results in a "first hit" of cancerogenesis, these events need to be carefully evaluated. Targeting multiple sites simultaneously is not advised due to probable generation of translocations or chromosomal loss, although the frequency of translocations was reported to be lowered over time (Bothmer et al. 2020). Using two different nucleases (e.g. one generating a blunt-end DSB, while the other is generating staggered ends) has been employed successfully to reduce the occurrence of translocation events while inducing multiple edits (Bothmer et al. 2020).

Editing outcome evaluations in most studies in the literature have relied on amplification of short regions (<1 kb) around the target site, which are unable to detect large lesions around the DSBs. To accommodate this, a number of methods have been described for the detection of large deletions and translocations. In addition to traditional long-range PCR sequencing (Adikusuma et al. 2018) which are prone to amplification bias toward shorter fragments, WGS (Veres et al. 2014), ddPCR (Owens et al. 2019), quantitative genomic PCR, and SNP genotyping methodologies have been used for identification of large deletions after editing (Weisheit et al. 2020). Moreover, LAM-HTGTS (described in Section 20.2.2) and primer-extension-mediated sequencing (PEM-seq), which combines LAM-HTGTS with targeted sequencing, have been described to successfully identify both large deletions and chromosomal translocations (Ikeda et al. 2020; Yin et al. 2019).

Taken together, although specificity and safety analyses of gene editing often have a strong focus on unintended induction of indels at off-target sites, on-target safety and large-scale chromosomal rearrangements also pose a potential safety hazard for therapeutic application of gene editing approaches. These should especially be taken into consideration when editing less stable, dividing cells (e.g. cancer cells, preimplantation embryos), as the events will be passed on to cell progeny and could potentially lead to deleterious outcomes.

20.5 Clinical Implications and Future Directions

Over recent years, genome-editing technologies have been constantly improved to enable targeted changes in the genome of any living cell or organism, with expanded targeting scope and deliverability to tissues of therapeutic relevance. In parallel, remarkable progress has been made in developing tools to determine genome-wide editing activity of nucleases and improving nuclease specificity. These methods serve the critical purpose of determining locations of unintended mutagenesis in the genome and enabling careful and deliberate selection of therapeutic genome-editing strategies. Increasing the sensitivity of the current methodologies demands attention as even low-frequency events could potentially lead to deleterious outcomes in therapeutic settings that modify millions to billions of cells. Better understanding the effect of the target DNA features such as chromatin state and gene expression levels will advance our capability of designing safer CRISPR medicines. In the future, big datasets of nuclease genome-wide activity profiles with a large number of gRNAs in various experimental conditions will potentially improve our capabilities to predict on-target rearrangements and off-target mutations, which would ultimately lead to development of more efficacious in silico tools for designing effective and specific gRNAs.

Genetic variation in between individuals should be additionally taken into account while designing therapeutic editing strategies. Single nucleotide variants have been reported to affect nuclease activity (Lazzarotto et al. 2020; Tsai et al. 2017; Yang et al. 2014) and should be considered when evaluating the specificity of nucleases. Personalized evaluation of the genome-wide nuclease activity using the individual's genomic DNA may be employed in the future prior to next-generation CRISPR therapies, with the establishment of more practical and scalable off-target detection tools.

Although not directly related to specificity, it is worth mentioning another safety aspect of potential CRISPR medicines. Activation of p53 response upon CRISPR-Cas9-mediated DSBs has been reported in certain cell types, leading to cell cycle arrest and selection against cells with a functional p53 pathway (Haapaniemi et al. 2018; Lareau et al. 2018). This brings the probability of more efficient editing in cells with less p53 activity. The most commonly used therapeutic gene editing approaches do not depend on p53 inhibition or genetic selection of modified cells, therefore they would not affect the frequency of p53 vulnerable cells in cell populations. However, if the functional editing outcome gives a selective growth advantage to edited cells over unedited cells, this can potentially lead to enrichment of p53 vulnerable cells in a bulk cell/tissue population upon editing. Therefore, this aspect should be carefully evaluated in preclinical research, prior to employing gene editing therapeutics in the clinic.

We have gained a greater understanding of the specificity of Cas9 nucleases in recent years, with the latest technological developments in the field. An important future challenge is to interpret the functional biological consequences of genome-wide editing activities and translate this knowledge to clinical settings. Advancements in our capability to determine and interpret genome-wide editing activity will build confidence in the safety of future gene-editing therapies.

References

Abudayyeh, O.O., Gootenberg, J.S., Franklin, B. et al. (2019). A cytosine deaminase for programmable single-base RNA editing. *Science* 365: 382–386.

Adikusuma, F., Piltz, S., Corbett, M.A. et al. (2018). Large deletions induced by Cas9 cleavage. *Nature* 560: E8–E9.

Akcakaya, P., Bobbin, M.L., Guo, J.A. et al. (2018). In vivo CRISPR editing with no detectable genome-wide off-target mutations. *Nature* 561: 416–419.

Anzalone, A.V., Randolph, P.B., Davis, J.R. et al. (2019). Search-and-replace genome editing without double-strand breaks or donor DNA. *Nature* 576: 149–157.

Bae, S., Park, J., and Kim, J.S. (2014). Cas-Offinder: a fast and versatile algorithm that searches for potential off-target sites of Cas9 RNA-guided endonucleases. *Bioinformatics* 30: 1473–1475.

Bisaria, N., Jarmoskaite, I., and Herschlag, D. (2017). Lessons from enzyme kinetics reveal specificity principles for RNA-guided nucleases in Rna interference and CRISPR-based genome editing. *Cell Systems* 4: 21–29.

Bondy-Denomy, J., Pawluk, A., Maxwell, K.L., and Davidson, A.R. (2013). Bacteriophage genes that inactivate the CRISPR/Cas bacterial immune system. *Nature* 493: 429–432.

Bothmer, A., Gareau, K.W., Abdulkerim, H.S. et al. (2020). Detection and modulation of DNA translocations during multi-gene genome editing in T cells. *CRISPR Journal* 3: 177–187.

Cameron, P., Fuller, C.K., Donohoue, P.D. et al. (2017). Mapping the genomic landscape of CRISPR–Cas9 cleavage. *Nature Methods* 14: 600–606.

Canela, A., Sridharan, S., Sciascia, N. et al. (2016). Dna breaks and end resection measured genome-wide by end sequencing. *Molecular Cell* 63: 898–911.

Casini, A., Olivieri, M., Petris, G. et al. (2018). A highly specific SpCas9 variant is identified by in vivo screening in yeast. *Nature Biotechnology* 36: 265–271.

Chen, J.S., Dagdas, Y.S., Kleinstiver, B.P. et al. (2017). Enhanced proofreading governs CRISPR–Cas9 targeting accuracy. *Nature* 550: 407–410.

Chia, G., Agudo, J., Treff, N. et al. (2017). Genomic instability during reprogramming by nuclear transfer is DNA replication dependent. *Nature Cell Biology* 19: 282–291.

Chiarle, R., Zhang, Y., Frock, R.L. et al. (2011). Genome-wide translocation sequencing reveals mechanisms of chromosome breaks and rearrangements in B cells. *Cell* 147: 107–119.

Cho, S.W., Kim, S., Kim, Y. et al. (2014). Analysis of off-target effects of CRISPR/Cas-derived RNA-guided endonucleases and nickases. *Genome Research* 24: 132–141.

Coelho, M.A., De Braekeleer, E., Firth, M. et al. (2020). CRISPR GUARD protects off-target sites from Cas9 nuclease activity using short guide RNAs. *Nature Communications* 11: 4132.

Concordet, J.P. and Haeussler, M. (2018). CRISPOR: intuitive guide selection for CRISPR/Cas9 genome editing experiments and screens. *Nucleic Acids Research* 46: W242–W245.

Cong, L., Ran, F.A., Cox, D. et al. (2013). Multiplex genome engineering using CRISPR/Cas systems. *Science* 339: 819–823.

Cox, D.B.T., Gootenberg, J.S., Abudayyeh, O.O. et al. (2017). RNA editing with CRISPR-Cas13. *Science* 358: 1019–1027.

Cradick, T.J., Fine, E.J., Antico, C.J., and Bao, G. (2013). CRISPR/Cas9 systems targeting beta-globin and CCR5 genes have substantial off-target activity. *Nucleic Acids Research* 41: 9584–9592.

Cromwell, C.R., Sung, K., Park, J. et al. (2018). Incorporation of bridged nucleic acids into CRISPR RNAs improves Cas9 endonuclease specificity. *Nature Communications* 9: 1448.

Crosetto, N., Mitra, A., Silva, M.J. et al. (2013). Nucleotide-resolution DNA double-strand break mapping by next-generation sequencing. *Nature Methods* 10: 361–365.

Cullot, G., Boutin, J., Toutain, J. et al. (2019). CRISPR-Cas9 genome editing induces megabase-scale chromosomal truncations. *Nature Communications* 10: 1136.

Dahlman, J.E., Abudayyeh, O.O., Joung, J. et al. (2015). Orthogonal gene knockout and activation with a catalytically active Cas9 nuclease. *Nature Biotechnology* 33: 1159–1161.

Davis, K.M., Pattanayak, V., Thompson, D.B. et al. (2015). Small molecule–triggered Cas9 protein with improved genome-editing specificity. *Nature Chemical Biology* 11: 316–318.

Dianov, G.L. and Hubscher, U. (2013). Mammalian base excision repair: the forgotten archangel. *Nucleic Acids Research* 41: 3483–3490.

Doench, J.G., Fusi, N., Sullender, M. et al. (2016). Optimized sgRNA design to maximize activity and minimize off-target effects of CRISPR-Cas9. *Nature Biotechnology* 34: 184–191.

Dong, G.,.M., Wang, S., Zhu, Y. et al. (2017). Structural basis of CRISPR-SpyCas9 inhibition by an anti-CRISPR protein. *Nature* 546: 436–439.

Duan, J., Lu, G., Xie, Z. et al. (2014). Genome-wide identification of CRISPR/Cas9 off-targets in human genome. *Cell Research* 24: 1009–1012.

Frock, R.L., Hu, J., Meyers, R.M. et al. (2015). Genome-wide detection of DNA double-stranded breaks induced by engineered nucleases. *Nature Biotechnology* 33: 179–186.

Fu, Y., Foden, J.A., Khayter, C. et al. (2013). High-frequency off-target mutagenesis induced by CRISPR-Cas nucleases in human cells. *Nature Biotechnology* 31: 822–826.

Fu, Y., Sander, J.D., Reyon, D. et al. (2014). Improving CRISPR-Cas nuclease specificity using truncated guide RNAs. *Nature Biotechnology* 32: 279–284.

Gabriel, R., Lombardo, A., Arens, A. et al. (2011). An unbiased genome-wide analysis of zinc-finger nuclease specificity. *Nature Biotechnology* 29: 816–823.

Gaudelli, N.M., Komor, A.C., Rees, H.A. et al. (2017). Programmable base editing of A*T to G*C in genomic DNA without DNA cleavage. *Nature* 551: 464–471.

Gilbert, L.A., Larson, M.H., Morsut, L. et al. (2013). CRISPR-mediated modular RNA-guided regulation of transcription in eukaryotes. *Cell* 154: 442–451.

Gilbert, L.A., Horlbeck, M.A., Adamson, B. et al. (2014). Genome-scale CRISPR-mediated control of gene repression and activation. *Cell* 159: 647–661.

Grünewald, J., Zhou, R., Garcia, S.P. et al. (2019). Transcriptome-wide off-target RNA editing induced by CRISPR-guided DNA base editors. *Nature* 569: 433–437.

Guilinger, J.P., Thompson, D.B., and Liu, D.R. (2014). Fusion of catalytically inactive Cas9 to FokI nuclease improves the specificity of genome modification. *Nature Biotechnology* 32: 577–582.

Haapaniemi, E., Botla, S., Persson, J. et al. (2018). CRISPR-Cas9 genome editing induces a p53-mediated DNA damage response. *Nature Medicine* 24: 927–930.

Harrington, L.B., Doxzen, K.W., MA, E. et al. (2017). A broad-spectrum inhibitor of CRISPR-Cas9. *Cell* **170**: 1224–1233.e15.

Heigwer, F., Kerr, G., and Boutros, M. (2014). E-CRISP: fast CRISPR target site identification. *Nature Methods* 11: 122–123.

Hoffmann, M.D., Aschenbrenner, S., Grosse, S. et al. (2019). Cell-specific CRISPR-Cas9 activation by microRNA-dependent expression of anti-CRISPR proteins. *Nucleic Acids Research* 47: e75.

Hoijer, I., Johansson, J., Gudmundsson, S. et al. (2020). Amplification-free long-read sequencing reveals unforeseen CRISPR-Cas9 off-target activity. *Genome Biology* 21: 290.

Horlbeck, M.A., Witkowsky, L.B., Guglielmi, B. et al. (2016). Nucleosomes impede Cas9 access to DNA in vivo and in vitro. *eLife* 5: e12677.

Hsu, P.D., Scott, D.A., Weinstein, J.A. et al. (2013). DNA targeting specificity of RNA-guided Cas9 nucleases. *Nature Biotechnology* 31: 827–832.

Hu, J., Meyers, R.M., Dong, J. et al. (2016). Detecting DNA double-stranded breaks in mammalian genomes by linear amplification–mediated high-throughput genome-wide translocation sequencing. *Nature Protocols* 11: 853–871.

Hu, J.H., Miller, S.M., Geurts, M.H. et al. (2018). Evolved Cas9 variants with broad PAM compatibility and high DNA specificity. *Nature* 556: 57–63.

Ikeda, M., Taniguchi-Ikeda, M., Kato, T. et al. (2020). Unexpected mutations by CRISPR-Cas9 CTG repeat excision in myotonic dystrophy and use of CRISPR interference as an alternative approach. *Molecular Therapy – Methods & Clinical Development* 18: 131–144.

Iyer, V., Shen, B., Zhang, W. et al. (2015). Off-target mutations are rare in Cas9-modified mice. *Nature Methods* 12: 479–479.

Iyer, V., Boroviak, K., Thomas, M. et al. (2018). No unexpected CRISPR-Cas9 off-target activity revealed by trio sequencing of gene-edited mice. *PLoS Genetics* 14: e1007503.

Jiang, W., Bikard, D., Cox, D. et al. (2013). RNA-guided editing of bacterial genomes using CRISPR-Cas systems. *Nature Biotechnology* 31: 233–239.

Jin, S., Zong, Y., Gao, Q. et al. (2019). Cytosine, but not adenine, base editors induce genome-wide off-target mutations in rice. *Science* 364: 292–295.

Jinek, M., Chylinski, K., Fonfara, I. et al. (2012). A programmable dual-RNA-guided DNA endonuclease in adaptive bacterial immunity. *Science* 337: 816–821.

Kartje, Z.J., Barkau, C.L., Rohilla, K.J. et al. (2018). Chimeric guides probe and enhance Cas9 biochemical activity. *Biochemistry* 57: 3027–3031.

Kawano, F., Suzuki, H., Furuya, A., and Sato, M. (2015). Engineered pairs of distinct photoswitches for optogenetic control of cellular proteins. *Nature Communications* 6: 6256.

Kiani, S., Chavez, A., Tuttle, M. et al. (2015). Cas9 gRNA engineering for genome editing, activation and repression. *Nature Methods* 12: 1051–1054.

Kim, D. and Kim, J.-S. (2018). Dig-seq: a genome-wide CRISPR off-target profiling method using chromatin DNA. *Genome Research* 28: 1894–1900.

Kim, S., Kim, D., Cho, S.W. et al. (2014). Highly efficient RNA-guided genome editing in human cells via delivery of purified Cas9 ribonucleoproteins. *Genome Research* 24: 1012–1019.

Kim, D., Bae, S., Park, J. et al. (2015). Digenome-seq: genome-wide profiling of CRISPR-Cas9 off-target effects in human cells. *Nature Methods* 12: 237–243.

Kim, D., Kim, S., Kim, S. et al. (2016). Genome-wide target specificities of CRISPR-Cas9 nucleases revealed by multiplex Digenome-seq. *Genome Research* 26: 406–415.

Kim, Y.B., Komor, A.C., Levy, J.M. et al. (2017). Increasing the genome-targeting scope and precision of base editing with engineered Cas9-cytidine deaminase fusions. *Nature Biotechnology* 35: 371–376.

Kim, D., Kim, D.E., Lee, G. et al. (2019). Genome-wide target specificity of CRISPR RNA-guided adenine base editors. *Nature Biotechnology* 37: 430–435.

Kim, D.Y., Moon, S.B., Ko, J.-H. et al. (2020). Unbiased investigation of specificities of prime editing systems in human cells. *Nucleic Acids Research* 48: 10576–10589.

Klein, I.A., Resch, W., Jankovic, M. et al. (2011). Translocation-capture sequencing reveals the extent and nature of chromosomal rearrangements in B lymphocytes. *Cell* 147: 95–106.

Kleinstiver, B.P., Prew, M.S., Tsai, S.Q. et al. (2015). Engineered CRISPR-Cas9 nucleases with altered PAM specificities. *Nature* 523: 481–485.

Kleinstiver, B.P., Pattanayak, V., Prew, M.S. et al. (2016). High-fidelity CRISPR–Cas9 nucleases with no detectable genome-wide off-target effects. *Nature* 529: 490–495.

Komor, A.C., Kim, Y.B., Packer, M.S. et al. (2016). Programmable editing of a target base in genomic DNA without double-stranded DNA cleavage. *Nature* 533: 420–424.

Kosicki, M., Tomberg, K., and Bradley, A. (2018). Repair of double-strand breaks induced by CRISPR-Cas9 leads to large deletions and complex rearrangements. *Nature Biotechnology* 36: 765–771.

Kuscu, C., Arslan, S., Singh, R. et al. (2014). Genome-wide analysis reveals characteristics of off-target sites bound by the Cas9 endonuclease. *Nature Biotechnology* 32: 677–683.

Labun, K., Montague, T.G., Krause, M. et al. (2019). CHOPCHOP v3: expanding the CRISPR web toolbox beyond genome editing. *Nucleic Acids Research* 47: W171–W174.

Langmead, B., Trapnell, C., Pop, M., and Salzberg, S.L. (2009). Ultrafast and memory-efficient alignment of short DNA sequences to the human genome. *Genome Biology* 10: R25.

Lareau, C.A., Clement, K., Hsu, J.Y. et al. (2018). Response to "unexpected mutations after CRISPR–Cas9 editing in vivo". *Nature Methods* 15: 238–239.

Lazzarotto, C.R., Nguyen, N.T., Tang, X. et al. (2018). Defining CRISPR–Cas9 genome-wide nuclease activities with CIRCLE-seq. *Nature Protocols* 13: 2615–2642.

Lazzarotto, C.R., Malinin, N.L., Li, Y. et al. (2020). CHANGE-seq reveals genetic and epigenetic effects on CRISPR–Cas9 genome-wide activity. *Nature Biotechnology* 38: 1317–1327.

Lee, C.M., Cradick, T.J., and Bao, G. (2016). The *Neisseria meningitidis* CRISPR-Cas9 system enables specific genome editing in mammalian cells. *Molecular Therapy* 24: 645–654.

Lee, J.K., Jeong, E., Lee, J. et al. (2018). Directed evolution of CRISPR-Cas9 to increase its specificity. *Nature Communications* 9: 3048.

Lee, J., Mou, H., Ibraheim, R. et al. (2019). Tissue-restricted genome editing in vivo specified by microRNA-repressible anti-CRISPR proteins. *RNA* 25: 1421–1431.

Lensing, S.V., Marsico, G., Hänsel-Hertsch, R. et al. (2016). DSBCapture: in situ capture and sequencing of DNA breaks. *Nature Methods* 13: 855–857.

Liang, X., Potter, J., Kumar, S. et al. (2015). Rapid and highly efficient mammalian cell engineering via Cas9 protein transfection. *Journal of Biotechnology* 208: 44–53.

Liang, P., Xie, X., Zhi, S. et al. (2019). Genome-wide profiling of adenine base editor specificity by EndoV-seq. *Nature Communications* 10: 67.

Lin, Y., Cradick, T.J., Brown, M.T. et al. (2014). CRISPR/Cas9 systems have off-target activity with insertions or deletions between target DNA and guide RNA sequences. *Nucleic Acids Research* 42: 7473–7485.

Liu, L., Yin, M., Wang, M., and Wang, Y. (2019). Phage AcrIIA2 DNA mimicry: structural basis of the CRISPR and Anti-CRISPR arms race. *Molecular Cell* 73: 611–620.e3.

Maeder, M.L., Linder, S.J., Cascio, V.M. et al. (2013). CRISPR RNA-guided activation of endogenous human genes. *Nature Methods* 10: 977–979.

Mali, P., Aach, J., Stranges, P.B. et al. (2013). CAS9 transcriptional activators for target specificity screening and paired nickases for cooperative genome engineering. *Nature Biotechnology* 31: 833–838.

Maresca, M., Lin, V.G., Guo, N., and Yang, Y. (2013). Obligate ligation-gated recombination (ObLiGaRe): custom-designed nuclease-mediated targeted integration through nonhomologous end joining. *Genome Research* 23: 539–546.

Marino, N.D., Zhang, J.Y., Borges, A.L. et al. (2018). Discovery of widespread type I and type V CRISPR-Cas inhibitors. *Science* 362: 240–242.

Merkle, T., Merz, S., Reautschnig, P. et al. (2019). Precise RNA editing by recruiting endogenous ADARs with antisense oligonucleotides. *Nature Biotechnology* 37: 133–138.

Montiel-Gonzalez, M.F., Diaz Quiroz, J.F., and Rosenthal, J.J.C. (2019). Current strategies for site-directed RNA editing using ADARs. *Methods* 156: 16–24.

Muller, M., Lee, C.M., Gasiunas, G. et al. (2016). *Streptococcus thermophilus* CRISPR-Cas9 systems enable specific editing of the human genome. *Molecular Therapy* 24: 636–644.

Nakamura, M., Srinivasan, P., Chavez, M. et al. (2019). Anti-CRISPR-mediated control of gene editing and synthetic circuits in eukaryotic cells. *Nature Communications* 10: 194.

Nihongaki, Y., Kawano, F., Nakajima, T., and Sato, M. (2015). Photoactivatable CRISPR-Cas9 for optogenetic genome editing. *Nature Biotechnology* 33: 755–760.

Nishida, K., Arazoe, T., Yachie, N. et al. (2016). Targeted nucleotide editing using hybrid prokaryotic and vertebrate adaptive immune systems. *Science* 353: aaf8729-1–aaf8729-8.

Owens, D.D.G., Caulder, A., Frontera, V. et al. (2019). Microhomologies are prevalent at Cas9-induced larger deletions. *Nucleic Acids Research* 47: 7402–7417.

Park, J., Childs, L., Kim, D. et al. (2017). Digenome-seq web tool for profiling CRISPR specificity. *Nature Methods* 14: 548–549.

Pattanayak, V., Lin, S., Guilinger, J.P. et al. (2013). High-throughput profiling of off-target DNA cleavage reveals RNA-programmed Cas9 nuclease specificity. *Nature Biotechnology* 31: 839–843.

Pawluk, A., Bondy-Denomy, J., Cheung, V.H. et al. (2014). A new group of phage anti-CRISPR genes inhibits the type I-E CRISPR-Cas system of *Pseudomonas aeruginosa*. *MBio* 5: e00896.

Pawluk, A., Amrani, N., Zhang, Y. et al. (2016). Naturally occurring off-switches for CRISPR-Cas9. *Cell* 167: 1829, e9–1838.

Perez-Pinera, P., Kocak, D.D., Vockley, C.M. et al. (2013). RNA-guided gene activation by CRISPR-Cas9-based transcription factors. *Nature Methods* 10: 973–976.

Qi, L.S., Larson, M.H., Gilbert, L.A. et al. (2013). Repurposing CRISPR as an RNA-guided platform for sequence-specific control of gene expression. *Cell* 152: 1173–1183.

Qu, L., Yi, Z., Zhu, S. et al. (2019). Programmable RNA editing by recruiting endogenous ADAR using engineered RNAs. *Nature Biotechnology* 37: 1059–1069.

Rahdar, M., Mcmahon, M.A., Prakash, T.P. et al. (2015). Synthetic CRISPR RNA-Cas9-guided genome editing in human cells. *Proceedings of the National Academy of Sciences of the United States of America* 112: E7110–E7117.

Ran, F.A., Hsu, P.D., Lin, C.-Y. et al. (2013). Double nicking by RNA-guided CRISPR Cas9 for enhanced genome editing specificity. *Cell* 154: 1380–1389.

Ran, F.A., Cong, L., Yan, W.X. et al. (2015). In vivo genome editing using *Staphylococcus aureus* Cas9. *Nature* 520: 186–191.

Raper, A.T., Stephenson, A.A., and Suo, Z. (2018). Functional insights revealed by the kinetic mechanism of CRISPR/Cas9. *Journal of the American Chemical Society* 140: 2971–2984.

Rauch, B.J., Silvis, M.R., Hultquist, J.F. et al. (2017). Inhibition of CRISPR-Cas9 with bacteriophage proteins. *Cell* 168: 150–158.e10.

Rauch, S., HE, E., Srienc, M. et al. (2019). Programmable RNA-guided RNA effector proteins built from human parts. *Cell* 178: 122–134.e12.

Rayner, E., Durin, M.A., Thomas, R. et al. (2019). CRISPR-Cas9 causes chromosomal instability and rearrangements in cancer cell lines, detectable by cytogenetic methods. *CRISPR Journal* 2: 406–416.

Rees, H.A., Wilson, C., Doman, J.L., and Liu, D.R. (2019). Analysis and minimization of cellular RNA editing by DNA adenine base editors. *Science Advances* 5: eaax5717.

Rose, J.C., Popp, N.A., Richardson, C.D. et al. (2020). Suppression of unwanted CRISPR-Cas9 editing by co-administration of catalytically inactivating truncated guide RNAs. *Nature Communications* 11: 2697.

Rueda, F.O., Bista, M., Newton, M.D. et al. (2017). Mapping the sugar dependency for rational generation of a DNA-RNA hybrid-guided Cas9 endonuclease. *Nature Communications* 8: 1610.

Ryan, D.E., Taussig, D., Steinfeld, I. et al. (2018). Improving CRISPR-Cas specificity with chemical modifications in single-guide RNAs. *Nucleic Acids Research* 46: 792–803.

Schmid-Burgk, J.L., Gao, L., LI, D. et al. (2020). Highly parallel profiling of Cas9 variant specificity. *Molecular Cell* 78: 794–800.e8.

Semenova, E., Jore, M.M., Datsenko, K.A. et al. (2011). Interference by clustered regularly interspaced short palindromic repeat (CRISPR) RNA is governed by a seed sequence. *Proceedings of the National Academy of Sciences of the United States of America* 108: 10098–10103.

Shen, B., Zhang, J., Wu, H. et al. (2013). Generation of gene-modified mice via Cas9/RNA-mediated gene targeting. *Cell Research* 23: 720–723.

Shin, H.Y., Wang, C., Lee, H.K. et al. (2017a). CRISPR/Cas9 targeting events cause complex deletions and insertions at 17 sites in the mouse genome. *Nature Communications* 8: 15464.

Shin, J., Jiang, F., Liu, J.-J. et al. (2017b). Disabling Cas9 by an anti-CRISPR DNA mimic. *Science Advances* 3: e1701620.

Slaymaker, I.M., Gao, L., Zetsche, B. et al. (2016). Rationally engineered Cas9 nucleases with improved specificity. *Science* 351: 84.

Smith, C., Gore, A., Yan, W. et al. (2014). Whole-genome sequencing analysis reveals high specificity of CRISPR/Cas9 and talen-based genome editing in human iPSCs. *Cell Stem Cell* 15: 12–13.

Song, Y., Liu, Z., Zhang, Y. et al. (2020). Large-fragment deletions induced by Cas9 cleavage while not in the BEs system. *Molecular Therapy--Nucleic Acids* 21: 523–526.

Soubeyrand, S., Lau, P., Peters, V., and Mcpherson, R. (2019). Off-target effects of CRISPRA on interleukin-6 expression. *PLoS One* 14: e0224113.

Stemmer, M., Thumberger, T., Del Sol Keyer, M. et al. (2015). CCTop: an intuitive, flexible and reliable CRISPR/Cas9 target prediction tool. *PLoS One* 10: e0124633.

Stephens, P.J., Mcbride, D.J., Lin, M.-L. et al. (2009). Complex landscapes of somatic rearrangement in human breast cancer genomes. *Nature* 462: 1005–1010.

Sternberg, S.H., Lafrance, B., Kaplan, M., and Doudna, J.A. (2015). Conformational control of DNA target cleavage by CRISPR-Cas9. *Nature* 527: 110–113.

Stojic, L., Lun, A.T.L., Mangei, J. et al. (2018). Specificity of RNAi, LNA and CRISPRi as loss-of-function methods in transcriptional analysis. *Nucleic Acids Research* 46: 5950–5966.

Stortz, F. and Minary, P. (2021). crisprSQL: a novel database platform for CRISPR/Cas off-target cleavage assays. *Nucleic Acids Research* 49: D855–D861.

Takata, M., Sasaki, M.S., Sonoda, E. et al. (1998). Homologous recombination and non-homologous end-joining pathways of DNA double-strand break repair have overlapping roles in the maintenance of chromosomal integrity in vertebrate cells. *The EMBO Journal* 17: 5497–5508.

Tsai, S.Q. and Joung, J.K. (2016). Defining and improving the genome-wide specificities of CRISPR–Cas9 nucleases. *Nature Reviews Genetics* 17: 300–312.

Tsai, S.Q., Wyvekens, N., Khayter, C. et al. (2014). Dimeric CRISPR RNA-guided FokI nucleases for highly specific genome editing. *Nature Biotechnology* 32: 569–576.

Tsai, S.Q., Zheng, Z., Nguyen, N.T. et al. (2015). GUIDE-seq enables genome-wide profiling of off-target cleavage by CRISPR-Cas nucleases. *Nature Biotechnology* 33: 187–197.

Tsai, S.Q., Nguyen, N.T., Malagon-Lopez, J. et al. (2017). CIRCLE-seq: a highly sensitive in vitro screen for genome-wide CRISPR–Cas9 nuclease off-targets. *Nature Methods* 14: 607–614.

Tycko, J., Wainberg, M., Marinov, G.K. et al. (2019). Mitigation of off-target toxicity in CRISPR-Cas9 screens for essential non-coding elements. *Nature Communications* 10: 4063.

Vakulskas, C.A., Dever, D.P., Rettig, G.R. et al. (2018). A high-fidelity Cas9 mutant delivered as a ribonucleoprotein complex enables efficient gene editing in human hematopoietic stem and progenitor cells. *Nature Medicine* 24: 1216–1224.

Vallecillo-Viejo, I.C., Liscovitch-Brauer, N., Montiel-Gonzalez, M.F. et al. (2018). Abundant off-target edits from site-directed RNA editing can be reduced by nuclear localization of the editing enzyme. *RNA Biology* 15: 104–114.

Veres, A., Gosis, B.S., Ding, Q. et al. (2014). Low incidence of off-target mutations in individual CRISPR-Cas9 and TALEN targeted human stem cell clones detected by whole-genome sequencing. *Cell Stem Cell* 15: 27–30.

Vogel, P., Moschref, M., Li, Q. et al. (2018). Efficient and precise editing of endogenous transcripts with SNAP-tagged ADARs. *Nature Methods* 15: 535–538.

Wang, X., Wang, Y., Wu, X. et al. (2015). Unbiased detection of off-target cleavage by CRISPR-Cas9 and TALENs using integrase-defective lentiviral vectors. *Nature Biotechnology* 33: 175–178.

Weisheit, I., Kroeger, J.A., Malik, R. et al. (2020). Detection of deleterious on-target effects after HDR-mediated CRISPR editing. *Cell Reports* 31: 107689.

Wiedenheft, B., Van Duijn, E., Bultema, J.B. et al. (2011). RNA-guided complex from a bacterial immune system enhances target recognition through seed sequence interactions. *Proceedings of the National Academy of Sciences of the United States of America* 108: 10092–10097.

Wienert, B., Wyman, S.K., Richardson, C.D. et al. (2019). Unbiased detection of CRISPR off-targets in vivo using DISCOVER-Seq. *Science* 364: 286.

Wienert, B., Wyman, S.K., Yeh, C.D. et al. (2020). CRISPR off-target detection with DISCOVER-seq. *Nature Protocols* 15: 1775–1799.

Wu, X., Scott, D.A., Kriz, A.J. et al. (2014). Genome-wide binding of the CRISPR endonuclease Cas9 in mammalian cells. *Nature Biotechnology* 32: 670–676.

Wyvekens, N., Topkar, V.V., Khayter, C. et al. (2015). Dimeric CRISPR RNA-Guided FokI-dCas9 nucleases directed by truncated gRNAs for highly specific genome editing. *Human Gene Therapy* 26: 425–431.

Yan, W.X., Mirzazadeh, R., Garnerone, S. et al. (2017). Bliss is a versatile and quantitative method for genome-wide profiling of DNA double-strand breaks. *Nature Communications* 8: 15058.

Yang, L., Grishin, D., Wang, G. et al. (2014). Targeted and genome-wide sequencing reveal single nucleotide variations impacting specificity of Cas9 in human stem cells. *Nature Communications* 5: 5507.

Yarrington, R.M., Verma, S., Schwartz, S. et al. (2018). Nucleosomes inhibit target cleavage by CRISPR-Cas9 in vivo. *Proceedings of the National Academy of Sciences of the United States of America* 115: 9351–9358.

Ye, L., Wang, C., Hong, L. et al. (2018). Programmable DNA repair with CRISPRa/i enhanced homology-directed repair efficiency with a single Cas9. *Cell Discov* 4: 46.

Yin, H., Song, C.Q., Suresh, S. et al. (2018). Partial DNA-guided Cas9 enables genome editing with reduced off-target activity. *Nature Chemical Biology* 14: 311–316.

Yin, J., Liu, M., Liu, Y. et al. (2019). Optimizing genome editing strategy by primer-extension-mediated sequencing. *Cell Discov* 5: 18.

Zetsche, B., Gootenberg, J.S., Abudayyeh, O.O. et al. (2015a). Cpf1 is a single RNA-guided endonuclease of a class 2 CRISPR-Cas system. *Cell* 163: 759–771.

Zetsche, B., Volz, S.E., and Zhang, F. (2015b). A split-Cas9 architecture for inducible genome editing and transcription modulation. *Nature Biotechnology* 33: 139–142.

Zhang, Y., GE, X., Yang, F. et al. (2014). Comparison of non-canonical PAMs for CRISPR/Cas9-mediated DNA cleavage in human cells. *Scientific Reports* 4: 5405.

Zhou, C., Sun, Y., Yan, R. et al. (2019). Off-target RNA mutation induced by DNA base editing and its elimination by mutagenesis. *Nature* 571: 275–278.

Zhu, Y., Gao, A., Zhan, Q. et al. (2019). Diverse mechanisms of CRISPR-Cas9 inhibition by type IIC anti-CRISPR proteins. *Molecular Cell* 74: 296, e7–309.

Zuo, E., Sun, Y., Wei, W. et al. (2019). Cytosine base editor generates substantial off-target single-nucleotide variants in mouse embryos. *Science* 364: 289–292.

Zuo, E., Sun, Y., Wei, W. et al. (2020). GOTI, a method to identify genome-wide off-target effects of genome editing in mouse embryos. *Nature Protocols* 15: 3009–3029.

Zuris, J.A., Thompson, D.B., Shu, Y. et al. (2015). Cationic lipid-mediated delivery of proteins enables efficient protein-based genome editing in vitro and in vivo. *Nature Biotechnology* 33: 73–80.

Part 5

Intellectual Property Aspects and Future Prospects

The emergence of CRISPR Cas technology has transformed the biotechnology business sector. Biotechnology companies' share in venture capital funds has increased dramatically and new companies are being launched that use genome editing for therapeutically targeting genes in patients or use genome editing in diagnosis or other applications. This part includes a chapter on the ethical and societal aspects of intellectual property landscape for genome editing technologies. In the final chapter, emerging technologies for genome editing are discussed.

21

Key Socio-Economic and (Bio)Ethical Challenges in the CRISPR-Cas9 Patent Landscape

Franc Mali

Faculty of Social Sciences, University of Ljubljana, Ljubljana, Slovenia

21.1 Introduction

The recent speed at which human genome editing technologies have been developing has sparked considerable optimism about their socioeconomic benefits and expectations of short-term profits among certain stakeholders. This helps explain the greater social interest in issues connected with the intellectual property rights (IPR) of CRISPR-Cas9. The CRISPR-Cas9 technique is one "niche" in genome editing technology that has seen the most exciting progress in the last few years. Genome editing is a powerful new technology that enables very accurate alterations to the genetic material of all living organisms. It is a technique that allows the genetic background of living organisms to be modified very efficiently, precisely, and simply. For example, CRISPR components are now routinely included in the kits of BioBricks (Martin et al. 2020). BioBricks are DNA sequences which serve a defined biological function and can be readily assembled with any other BioBrick parts to create new BioBricks with novel properties. BioBricks are typically used and distributed by the well-known International Genetically Engineered Machines (iGEM) competition.

Put simply, CRISPR-Cas9 works as a type of molecular scissors that can selectively trim away unwanted parts of the (human) genome and replace them with new stretches of deoxyribonucleic acid (DNA). In recent times, this revolutionary genetic engineering technology has become hugely popular. Genetic engineering research labs around the world are quickly adopting this new approach. Of course, CRISPR-Cas9 is not the only genome editing technology today. Researchers are still relying on other technologies such as zinc finger nucleases (ZFNs) and transcription activator-like effector nucleases (TALENs). Still, CRISPR-Cas9 is a special genome-editing approach because it is relatively simple to use compared with other approaches and also much cheaper and more efficient (Martin et al. 2020; Samy 2018; Van Erp et al. 2015; Zhang et al. 2020).

With the rise of CRISPR-Cas9, especially the IPR issues connected with human genome editing technologies are emerging in a quite different context than 40 years ago when the industrial sector was making its first efforts to commercialize genetic engineering research outcomes. Today, the range of potential uses of CRISPR-Cas9 and number of parties interested in commercializing the genetic engineering inventions is becoming much bigger.

Many challenging questions accompany the IPR issues of CRISPR-Cas9: Who should own this technology, and what should "ownership" mean? Do the existing patent systems as the most frequent form of IPR in human genome editing adequately address the complex bioethical issues of CRISPR technology? In which conditions should various stakeholders employ which forms of protection for inventions in the field of CRISPR technology? How should patent systems in the case of CRISPR technology be structured?

In my contribution, two dimensions of IPR connected with recent progress with CRISPR Cas9, i.e. socioeconomic and socio-ethical, are addressed in particular. In practice, these two dimensions do not necessarily operate separately.

I first focus on the question of what might be (what are) the socioeconomic consequences of the strict application of IPR for the free flow of information in human genome editing. Namely, CRISPR technology is transforming the "ivory tower" of genetic engineering into a multibillion-dollar technological enterprise built on entrepreneurship, venture capital, start-ups, and wide-ranging changes to academic research culture. As noted by many authors, a critical point is being reached at the start of the twenty-first century with respect to who will own this revolutionary

Genome Editing in Drug Discovery, First Edition. Edited by Marcello Maresca and Sumit Deswal.

technology and who might benefit from its applications (Halilem et al. 2017; Jasanoff et al. 2015; Martin et al. 2020; Mitschel 2019; Sherkow 2016). The financial and symbolic stakes in CRISPR-Cas9 are enormous. The best indicator of these trends is the prolonged and costly patent legal proceedings involving research groups from the leading American universities. Described simply, patent litigation is the legal process that unfolds when someone who owns the patent for a particular invention enforces their right by suing another for manufacturing or selling the invention without permission.

Last but not least, there is the question of whether any extension of the aggressive patent wars could slow the progress of basic research in human genome editing because the academic science sector is often more worried about possible patent infringement than focusing on the benefits of improved human genome editing technology for the whole field of genetic engineering.

Second, I address the bioethical dimension of the CRISPR-Cas9 patent landscape. While the newest patent battles over CRISPR technology may be seen as part of the controversial (bioethical) discussions over bio-patents that have been going on for over 40 years, the revolutionary peculiarities of the CRISPR technology mean that some traditional approaches are insufficient for ensuring suitable protection. Namely, it is sometimes necessary to interpret the IPR issues connected with CRISPR technology in a completely new way. Although this unprecedented advance in germline engineering holds great promise for next-generation therapeutics, it has sparked an ethical debate.

This article has the following structure. In the next section, I explain why CRISPR-Cas9 is today attracting such great interest from both the business-enterprise and academic sectors. In Section 21.3, I then turn to the issues of the business-enterprise sector in terms of the crucial question: why is the business-enterprise sector increasingly interested in relying on the new and less restrictive forms of IPR when it comes to CRISPR-Cas9? In Section 21.4, I provide insight into the extremely competitive CRISPR-Cas9 patent landscape in academic institutions. In Section 21.5, I describe in more detail certain socioeconomic and (bio) ethical concerns associated with the mentioned landscape. Finally, some concluding words are provided.

21.2 CRISPR-Cas9 is Attracting Great Interest from Both the Business-enterprise and Academic Sectors

The academic research sector is today extremely oriented to making a profit from the commercialization of human genome editing technologies. Many universities are very interested in monopolizing their CRISPR Cas9 inventions with the help of patenting as the strictest form of IPR. This interest is in line with the general trend of expanding ownership models in the public research domain, which has today produced "a spiralling effect" (Winickoff 2015, p. 15). A consequence of this trend is that research groups from universities and other academic institutions with the possibility to become patent owners of CRISPR are now emerging in the very lucrative position of owner of this ground-breaking technology.

In the area of human genome editing research at academic institutions, the number of patents applied to "research tools" is growing. This means that, unlike traditional patents focused on commercial end-products which are rarely infringed upon by academic scientists, patents of such research tools are leading to frequent patent infringements. Patent infringement occurs when academic scientists breach the patent law by using a patent without permission (Churi and Taylor 2020; Feeney et al. 2018).

Academic institutions feature strongly in the CRISPR-Cas9 patent landscape. For example, of the top 15 patent owners in the CRISPR-Cas9 patent space in the world, companies account for approximately 40% while 5 universities and 4 research institutes are ranked among the top 15 patent owners (CRISPR-Cas9 Tools 2020).

According to the latest IPStudies, the second-most important source of data on CRISPR patents, patent landscape trends in this technology have been remarkably stable in past years, with continued patenting competition seen chiefly among academic players in China and the United States (IPStudies 2020). One may especially notice the extreme growth in patent applications over the last few years. When in 2014, contributors of IPStudies started to analyze the genome editing patent landscape, they recorded 96 patent families in relation to CRISPR technology. In the most recent volume of IPStudies in 2020, one finds 7424 patent families. (Patent family means the same invention disclosed by a common inventor(s) and patented in more than one country.) Since 2014, 1850 academic institutions and business companies around the world have filed at least one patent application related to CRISPR technology. For example, on the global level, the leading academic institutions with the most patent applications are: Massachusetts Institute of Technology (286 patent applications), Chinese Academy of Sciences (265), Chinese Academy of Agricultural Sciences (260), Broad Institute from the USA (251), University of California (229), and Harvard University (217).

The data showing the academic sector's dominant position as concerns CRISPR patent applications seem a little surprising. Do they show that the business-enterprise sector is not particularly interested in such revolutionary and ground-breaking technology in genetic engineering? Not at all! The business-enterprise sector is also interested in commercializing CRISPR-Cas9 and aware of the extremely wide range of possible uses of the technology. The speed of progress of the applied potential of this revolutionary technology is also creating huge expectations in the economic sector. Through various applications, the global CRISPR market generated a revenue of USD 361 million in 2016 and is anticipated to contribute USD 5966 million by 2025 (Kwon 2019).

Although very young, the genome editing industry is today becoming a very fast-developing and extensive industrial branch. Efforts to commercialize genome editing technologies are especially prominent in biotechnology and pharmacy. In biotech and pharma companies, CRISPR is becoming a go-to tool for drug discovery. Human genome editing technology has already been integrated into mainstream pharmaceutical and biotech companies around the world whose main interest is to offer the market new treatments for genetic diseases created with the help of CRISPR technology. Large companies like AstraZeneca, Millipore Sigma (earlier Sigma Aldrich), Thermo Fisher Scientific, Toolgen, Genecopeia, Horizon Discovery, and Agilent Technologies and pharmaceuticals like Novartis, Vertex, Bayer AG, Evolva, Juno Therapeutics, and GE Healthcare have skin in the CRISPR game by exploring novel techniques to evolve their drug discovery and development processes.

Still, many start-up companies can also be found operating in the area of human genome editing technologies. Many of these were founded on the initial CRISPR IPR granted to various individuals and research institutions. Their primary characteristics are that they are very young, extremely well-financed, and mostly oriented to developing products in human healthcare. Several deal with gene-based therapies for various genetic diseases and selling research tools and services, such as enzymes. The start-up companies working on human therapeutics concentrate on two main areas: genetic diseases, such as hemophilia, and CART-cell therapy (using engineered immune cells) for cancer. One important reason for the active collaboration seen between the small start-up companies and the much larger integrated pharmaceutical companies is that few start-up firms have the resources and expertise to take their products all the way to market without investment and support from a bigger partner (Ferreira et al. 2018; Martin et al. 2020; Mitschel 2019).

21.3 The Business-enterprise Sector is Ever More Interested in Using the New and Less Restrictive Forms of IPR for CRISPR-Cas9

Although we expect big pharmaceutical companies to have an exclusive interest in protecting inventions in human genome editing technologies with patents because they incur great financial costs to develop a novel drug or product which consists of hundreds and thousands of lead compounds, this is not always the case. Many of such companies are starting to look for new, less restrictive forms of IPR. Namely, they are increasingly aware that it is insufficient to ensure their rapid and widespread (economic) growth by only relying on the availability of exclusive patent rights.

If we look back at the history of genetic engineering, similar less restrictive IPR forms were already being encouraged when the recombinant DNA technology was emerging in the late 1970s. Already then came the first awareness of the importance of following open innovation models which use external ideas as well as internal ideas and paths to market. For example, at that time, Herbert W. Boyer and Stanley N. Cohen had discovered a method to produce recombinant DNA in bacteria. Upon filing the Cohen-Boyer patent, Stanford University created a nonexclusive licensing programme that benefits both the business-enterprise sector and academic institutions. The academic world has embraced this nonexclusive licensing programme of Stanford University as a best-practice model for the commercialization of biotechnology (Egelie et al. 2016; Feldman et al. 2007).

The main difference from past times is that industrial companies are these days encouraging the proliferation of various forms of nonexclusive and cross-licensing policies, material transfer agreements, etc. Put very simply, a nonexclusive licensing policy means the owner of IPR grants a licensee the right to use the intellectual property, while the licensor remains free to exploit the same intellectual property and to allow any number of other licensees to exploit the same intellectual property. Cross-licenses generally occur between companies that hold patents over different aspects of the same product. By entering into an agreement, each company involved can avoid litigation over infringement disputes. Transfer agreements govern the transfer of materials (e.g. cell lines, plasmids, nucleotides, proteins, other materials with scientific or commercial value) from the owner or their authorized licensee (usually known as "the provider") to a third party ("the recipient"), who may wish to use the material for research purposes.

Today, industrial companies are much more aware than before that the less restrictive forms of regulation of pre-commercial research can provide a reasonable likelihood of their market competition and economic development. In this sense, they tend to fashion themselves as part of the "vanguards" (Hilgartner 2017, p. 3) by creating a condition for a change in IPR but while competing with each other at the same time.

It seems that this is the only right way to proceed. The concept of property rights has been considerably broadened in recent decades and industrial stakeholders that have found success in the market are trying to learn from each other and simultaneously implement good ideas. Many of such interlinks have created the grounds for the biomedical and pharmacy industry to develop products and services of enormous benefit to society (Egelie et al. 2016; Feeney et al. 2018; Hilgartner 2017).

CRISPR technology holds great potential to introduce a change in the discourse on IPR. The changing conditions in which this new revolutionary technology is prospering mean that there are better chances of reopening the traditional rigid patent systems and stimulating the introduction of new forms of licensing policy. In recent times, licensing is no longer based on an external knowledge-sourcing strategy. Whereas traditional forms of licensing give the licensee an exclusive right to use the knowledge in exchange for money but without mutual interactions and resource-sharing between the licensee and the licensor, the new models based on a broader partnership or alliance are including the mutual sharing of intellectual resources and joint R&D efforts (Klueter et al. 2017).

Last but not least, such open forms of license policy assure a more socially impartial and balanced proliferation of CRISPR-Cas9 technology across the globe and marginalize the negative impacts of patent litigation which has mostly ended in a zero-sum game (Graff and Johansen 2016; Summerfield 2015).

The licensing policy described above is part of the "open innovation" models. They generally refer to major global changes in the behavior of the business-enterprise sector. They are known as "the use of purposive inflows and outflows of knowledge to accelerate internal innovation, and expand the markets for external use of innovation, respectively" (Chesbrough 2006, p. 9). In such models, progress in innovativeness occurs on the basis of internal and external sources of knowledge and hence in collaboration with several R&D actors (Bogers et al. 2018; West et al. 2014). The models lead to stronger collaboration between companies with the aim of intensifying innovation and bringing in new resources not available internally. In the context of open innovation, intellectual property is playing a new role which no longer reflects the usual defensive proprietary mechanism adopted by companies (Enkel et al. 2009).

21.4 The Academic Research Sector has Created an Extremely Competitive CRISPR-Cas9 Patent Landscape

While many companies in the business-enterprise sector have quite a flexible approach to IPR policy, this is not always the case in some academic institutions. With regard to human genome editing technology, a very conservative approach to IPR is often followed, despite the fact that even at the state level it has been suggested that academic institutions be oriented to more open licensing strategies. For example, the National Institutes of Health in the United States recommend to academic institutions that patents on "research tools" developed using federal funding be licensed nonexclusively so as to promote their greatest utilization, commercialization, and public availability (Cockbain and Sterckx 2020).

A clear indicator of this conservative approach to IPR is the legal proceedings over patents involving academic institutions. In simple terms, patent litigation describes the legal processes that unfold when someone who owns the patent for a particular invention enforces their right by suing another person for manufacturing or selling the invention without permission. Namely, disputes over ownership of CRISPR technology are currently underway at various patent offices across the world. These patent battles are typically based on the extreme mutual exclusivity of the involved parties (Cockbain and Sterckx, 2020; Egelie et al. 2016; Kapoora and Klueterb 2020; Sherkow 2017b, 2017c).

The new situation characterized by aggressive patent fights is nowhere more evident than in the patent litigation involving Jennifer Doudna's research group (The University of California) and Feng Zhang's research group (The MIT/Broad Institute). There was a brief period in which it seemed these two research groups would be able to partner and share intellectual property. Yet, it seems that disagreements over academic credit, company locations, loyalty, ego, and financial gain splintered the two groups of researchers who are considered to be CRISPR pioneers (Mitschel 2019). The case of patent litigation in which scientists from two academic institutions, i.e. the University of California and the MIT/Broad Institute, are involved has also attracted enormous public attention. One group led by Jennifer Doudna, together with Emmanuelle Charpentier (both scientists were awarded the 2020 Nobel Prize in Chemistry), published the first results of CRISPR gene editing in prokaryotes. The second group, led by Feng Zhang from the MIT/Broad Institute, claimed that it was the first to successfully implement the CRISPR technique in eukaryotes. The most important distinction between prokaryotic and eukaryotic cells is that eukaryotic cells have a membrane-bound nucleus, which stores the cell's genetic information. For example, the genetic information of eukaryotic cells, DNA, is organized in chromosomes, of which humans have 46. The University of California group filed a patent in early 2012 at the USPTO to cover the basic contours of CRISPR-Cas9. Its patent claim referred to the use of a genome-editing tool in any type of cell. Six months later, the MIT/Broad Institute group filed a claim for a patent at the USPTO where it was demonstrated that CRISPR-Cas9 can be used on eukaryotic cells. Feng Zhang from the MIT/Broad Institute argued that the patent claim of his group is sufficiently different from that of the University of California and therefore both parties involved in the interference process should be allowed to independently pursue their patent claims. This interpretation by the MIT/Broad Institute was not supported by the University of California. Zhang's patent claim to use genome editing technology in any noncellular or cellular setting (including in human cells) was opposed by the University of California, which retaliated by filing an application with the Patent Trial and Appeal Board at the USPTO to investigate interference. After quite a long process of interference proceedings, in 2017 the Patent Trial and Appeal Board at the USPTO declared that the patents granted to the MIT/Broad Institute do not interfere with the patent claims of UC Berkeley (Ku 2017; Sherkow 2017a). In September 2018, the US Federal Appeals Court also ruled in favor of the Broad Institute, confirming an earlier US patent board decision that the patents from the lab of the investigator Feng Zhang did not "interfere" with those sought by the University of California. This should have meant that Zhang and his team had succeeded in obtaining the patent rights in the United States (United States Court of Appeals for the Federal Circuit 2018).

Concomitantly, the patent fight among the two parties, i.e. Jennifer Doudna's research group and Feng Zhang' research group, was also underway at the EPO (European Patent Office). However, in Europe, the patent dispute was following a different trajectory. While in February 2017 the Patent Trial and Appeal Board at the USPTO declared that the CRISPR editing of eukaryotic genomes by the MIT/Broad Institute did not interfere with the University of California's patent claims, only a few months later the EPO ruled the very opposite. Contrary to the USPTO declaration, the EPO revoked the first of several patents concerning CRISPR-Cas9 technology obtained by the MIT/Broad Institute, citing a clear lack of novelty (Akst 2017; Cockbain and Sterckx 2020). Most of Broad's other patent applications in Europe suffer from the same problem, meaning a serious degrading of Broad's patent position in Europe.

This situation will probably not see any winner-take-all solution and it is likely that patent interference processes among academic research groups will continue and become even more complicated. The different positions held by the US and European patent offices reveal several disparities in the outcomes of international patenting. These differences show the global human genome editing patent landscape is very confusing. Moreover, it seems that we are today withdrawing from any harmonization of the various patent practices connected with human genome editing technologies, even though the TRIPS (Trade Related Aspects of Intellectual Property Rights) agreement states that all World Trade Organization (WTO) members should adopt a set of minimum standards on IPR, including patents. On the global level, the TRIPS agreement does not oblige WTO member countries to legally provide for the patentability of genes and DNA sequences (Churi and Taylor 2020; OECD 2014; Van den Belt 2013).

For example, although Article 27.2 of the TRIPS agreement introduces a "necessity test" to assess whether the protection of inventions due to their conflict with a social interest (e.g. the need to protect human, animal or plant life, or to avoid serious prejudice to the environment) is justified, thus far in the case of human genome editing technologies neither the USPTO nor the EPO have used this legal instrument in a decision to grant (or not grant) a patent (Duncan 2020).

The lack of harmony among patent regimes around the world is leading to the situation that different lobbying groups are obtaining monopoly positions in various parts of the world. In the recent period, the case of China has often been mentioned. As a serious contender in the race to further develop CRISPR technology, the vast majority of CRISPR patents has been awarded to Chinese applicants (Churi and Taylor 2020). If this situation is to worsen in the near future and the procedures for processing patent applications vary considerably depending on the regulatory framework of a particular state/region, then parties that submit patent applications or encourage interferences will be involved in prolonged and costly legal entanglements. In addition, due to the arrival on the scene of many new players interested in the newly emerging technology (venture capitalists, IP fund managers, patent auction houses, lawyers, etc.), national and international patent offices will experience backlogs of unexamined patent applications, which will add to the legal uncertainty.

Does the fact that many academic research groups are in the middle of fierce patent wars mean that the progress of such a promising technology as human genome editing technology could be slowed down? This threat is realistic, although a number of authors warn that the leading academic institutions in CRISPR technology are also slowly changing their licensing strategy and started to license third parties.

For example, the University of California and the MIT/Broad Institute have concomitantly created spin-offs through which they have formed more complex licensing mechanisms. The University of California has delegated its licensing rights for CRISPR-Cas9 to Caribou Biosciences, a profit-based "surrogate" company which, in turn, has granted an exclusive license to Intellia Therapeutics to develop human therapies. Meanwhile, the MIT/Broad

Institute is using the company Editas Medicine as its surrogate for human therapeutics (Egelie et al. 2016; Martin et al. 2020; Van Erp et al. 2015). Some differences exist between the licensing policies of Editas Medicine and Intellia Therapeutics. Editas Medicine's strategy is to pool patents with other companies directed at developing CRISPR-Cas9. In this regard, in 2014, Editas Medicine developed the inclusive innovation model. In this model, Editas Medicine has the right to exclusively use the technology on targets of its choosing for the development of genomic medicines. After an initial period, other companies may apply to license certain CRISPR for use against genes of interest not being pursued by Editas Medicine (Döring and Lim 2017; Mathias et al. 2018). Still, the inclusive innovation model is associated with several deficiencies. According to some critics, the model is based on significant moral hazard (Feeney et al. 2018). Namely, while a third party can apply to license a use which is neither being pursued by Editas Medicine nor is an attractive prospect for Editas Medicine, it will have to wait an undefined period for a decision by Editas Medicine.

Unlike Editas Medicine, Intellia Therapeutics, the "surrogate" company established by the University of California, has announced the Global Agreement on the Foundational Intellectual Property for CRISPR-Cas9 Gene Editing Technology. Under this agreement, Intellia Therapeutics is committed to maintaining and coordinating the prosecution, defense, and enforcement of the CRISPR-Cas9 foundational patent portfolio worldwide, and each of the co-owners of the intellectual property grants cross-consents to all existing and future licenses and sub-licenses based on the rights of another co-owner. The main goal of this patent pool is to protect the share of IPR among companies by approaching a global agreement (Mathias et al. 2018; Samy 2018).

The biggest problem of the CRISPR spin-offs created at the leading universities is that the universities, i.e. the patent owners, have transferred the licensing rights to their spin-off companies, which function as if they are a private "commercial arm." The outcome is that "such profit 'surrogate' companies by universities for managing university licensing are still primarily oriented to short-term profit rather than, say, the free flow of information and public access to knowledge" (Sherkow 2017a, p. 565).

Although the licensing strategies of Editas Medicine and Intellia Therapeutic herald new approaches in the complex academic IPR landscape, which could encourage the free flow of information in CRISPR technology, they are currently not functioning as "vanguards" driving a wave of changes in the existing patent discourse. Just the opposite, the patent litigation among the leading CRISPR research groups mean that many technological "niches" in human genome editing are *unnecessarily* being forced to wait and see the final result of these unproductive patent wars. To conclude this part of the discussion, if we compare the situation with the business-enterprise sector, the lack of good cases among academic institutions is apparent. It seems they are less inclined to reconceptualize their interest to follow any less strict IPR policy.

21.5 Certain Socioeconomic and (Bio)ethical Concerns Connected with the CRISPR-Cas9 Patent Landscape

A critical challenge facing academic patent systems today is finding the right balance between providing sufficient openness for the further progress of basic scientific research and, simultaneously, sufficient control to provide incentives for innovativeness and technological development. With human genome editing technologies it is not easy to establish such a balance because academic science groups are particularly interested in coming into ownership of broad patents and patent thickets. Broad patents are patents that grant broad rights to the patent holder and may be seen as covering applications subsequently invented by someone else.

A huge challenge with broad patents is that their claims may exceed what the inventor actually discovered. They contain fewer details than narrow claims and therefore give the patent owner protection over a wider range of activities. Such owners seek property rights that extend beyond uses of their inventions that they originally anticipated or predicted, but also over any new uses that are developed (Nuffield Council for Bioethics 2002). Another issue is patent thickets. A patent thicket is an overlapping set of patent rights requiring that those seeking to commercialize new technology obtain licenses from multiple patentees. Here, the progress of new technology is being inhibited by multiple overlapping patents (Singh 2015).

Patents connected with CRISPR-Cas9 are typically drafted very broadly because human genome editing technologies fall in the category of "enabling technology," i.e. their use does not directly provide a product but enables a product to be made using other knowledge and probably technology (Grens 2016; Nuffield Council for Bioethics 2016; Sherkov and Greely 2013). For instance, Doudna and Charpentier's original patent application contained over 150 claims and was notably nonspecific with respect to cell type (Sherkow 2017b). Broad patents and patent thickets are in fact already posing a considerable challenge to the whole field of genetic engineering (König et al. 2013; Sampat and Shadlen 2017; Van Zimmeren et al., 2011). If the number of broad patents and patent thickets grows in the near future, they could become a "bottleneck" hindering any faster progress with CRISPR technology and bring many other indirectly negative consequences. For example, in the public health system, the omnipotence of broad patents and patent thickets holds the potential to inflate access prices to biogenetic drugs because the high cost to license these types of patents will be passed on to consumers. This will probably lead to less affordable treatments for patients, exacerbating ill-health or prolonging disability (Feeney et al. 2018; Perkins 2019).

In recent times, one may find at the top of expert and wider social reflexivity about the CRISPR patent landscape not only its socioeconomic aspects, where attention is paid not simply to the issue of how aggressive patenting forms can represent an obstacle for any faster progress in CRISPR technology. The same attention must be paid to the (bio) ethical dilemmas which are arising upon the breakthrough of this revolutionary technology. The bioethical dilemmas surrounding CRISPR patents also represent a continuation of controversial discussions about bio-patents that have been underway for 40 years. Although ethical dilemmas concerning bio-patents are not new and appeared already at the birth of modern genetic engineering, patent-based monopolies over CRISPR are producing totally new ethical dilemmas. Here, the central question is the risk connected with editing the human germline. This specific application of CRISPR raises several questions: are strong enough arguments given in favor of encouraging such inventions (even if they can eradicate certain hereditary diseases from society) or even in favor of permitting patents on altering the germline identity of human beings? Should such inventions be consistent with fundamental human rights, particularly the right of future generations to human dignity?

If we look at the past, the modification of living organisms with genetic engineering and the commercial interest to hold ownership of such genetically modified organisms opened the "black box" of controversial discussions about bio-patents already at the end of 1970s and in the early 1980s. The patent revolution in genetic engineering first occurred in the United States, and its consequences went further in Europe (Parthasarathy 2015). The Diamond v. Chakrabarty case represents the beginning of the discussion of whether genetically modified organisms should be treated as a product of nature and thus not patentable or like any other composition of (unnatural) matter. After a lengthy series of lawsuits, the US Supreme Court awarded Chakrabarty a patent on *Pseudomonas putida*, a strain of bacterium he had transformed with several plasmids. It was the first patent application for a recombinant bacterium. The US Supreme Court held that anything new under the sun that is made by man, whether living or nonliving, can in principle be patented. This decision established a precedent for the patentability of living micro-organisms modified through human intervention (Bhutkar 2005).

During the 1980s, the patentability of living organisms was further extended from bacteria to multicellular organisms. Along with the advance of genetic sequencing, from 1980 onward the pressure to submit everything to patent rights grew strongly. Patent rights also included genetically engineered life forms and genes that had been isolated and cloned (Hilgartner 2017; Jasanoff 2005). In 1984, Harvard University filed for a patent of the first animal, i.e. a laboratory genetically altered mouse. The mouse had been genetically modified to be particularly susceptible to cancer. The laboratory case is known as the »Harvard oncomouse« (Jasanoff 2005, p. 210). The USPTO awarded the patent for the oncomouse in 1988, being the first time a patent had been granted to a transgenic nonhuman mammal whose germ cells and somatic cells contain a recombinant activated oncogene sequence. The argument for granting the patent for the »Harvard Oncomouse« was again that in this case the isolation and purification of a particular DNA sequence from the body turns it into something radically different from its natural state.

In this disreputable case, this means that the awarding of the patent relied on argumentation drawn from the »patentable subject matter doctrine« (Sherkow and Greely 2015, p. 164). With this doctrine, an "invention" is understood as something that someone creates or develops which did not exist previously; on the other hand, a "discovery" is understood as the acquisition of knowledge of a new but already existing fact about the world. In the case of a genome, it was accepted that genetic substances which are subjected to laboratory manipulations are "artificial" matters, i.e. they do represent an invention, although ". . . on the usual interpretation of the words, it seems apparent that the identification of a gene is a discovery, since genes exist in the world, in our bodies" (Nuffield Council for Bioethics 2002, p. 23).

In the United States, patent courts were institutionally mandated to interpret the patentable subject matter doctrine as they found it (Kleinman and Kinchy 2003). "Major legal disputes were disposed of as narrower questions of statutory interpretation, in accordance with technical criteria for granting patents, interpreted case-by-case by the courts" (Jasanoff 2005, p. 209). In that sense, decisions made at the patent court have mostly avoided »the philosophical and ethical discussions« (Calvert 2012, p. 172), even in the most controversial cases where the meaning of patent law was most in doubt.

The legal discourse in the United States called »patent eligibility« (Sherkov and Greely 2013, p. 1569), which ignores the ethical issues surrounding bio-patents, was applied in practice, at least until the well-known »Myriad case« in 2013. The Myriad case led to an important change in American patent doctrine. In this case, the US Supreme Court did not confirm a patent claim that supposedly covered isolated genomic DNA, i.e. DNA fragments of various sizes which had simply been removed from the surrounding genome. It was declared that the separation of a gene from its surrounding genetic material cannot be defined as an invention (Calvert 2012; Perkins 2019; Singh 2015; Winickoff 2015). The outcome is that, after several years of controversy, the US Supreme Court no earlier than in 2013 established a rule that human genes cannot be patented because DNA is a "product of nature" (Sherkow and Greely 2015; Zhang et al. 2020). Before the »Myriad case«, the USPTO had granted at least 60 000 patents relating to genes and genetics (Perkins 2019).

Whereas in the United States, the strong patent courts have led to a weakness in the broader ethical reconsiderations of the function of bio-patents, the situation in the European Union (EU) has been slightly different. Although in the EU considerable discord has been observed over the issue of gene patents between the European Parliament, the EPO and specialist law reform advisory bodies (Rimmer 2008), under EU patent law ethical objections have more often been recognized as a reason for not granting patent rights (Cook-Deegan and Heaney 2010; Jasanoff 2005; Parthasarathy 2015). The crucial reason is that in the mid-1990s, the European Parliament accepted the European Biotechnological Directive on the legal protection of biotechnological inventions (European Biotechnological Directive 1998).

This was later implemented in the regulations of the European Patent Convention (EPC), an intergovernmental treaty which established a common legal framework for patent regimes in EU member states. The EU's patent regime should in principle exclude from patentability any biotechnological invention whose commercial exploitation would be contrary to "public order of morality" (Hübel et al. 2012; OECD 2014; Van den Belt 2013).

Still, while the European Biotechnological Directive has been an important element of European patent law that binds national governments, the patent regimes in certain EU member states have sometimes followed a different course. They have quite freely interpreted how to use the guidelines of the EPC in their patent regimes (Gold and Gallochat 2001; Mali 2004; Schneider 2009). Although the EPC also follows other guidelines (e.g. the Oviedo Convention), one cannot always expect harmony in interpretations of biopatents. This is perhaps the most highly accepted only in the case of the prohibition on heritable genetic modification. Article 13 of the Oviedo Convention, which is an international legal instrument, is expressly prohibited to introduce any modification in the genome of any descendants (Oviedo Convention 1997). Even those countries that have not signed or ratified the Oviedo Convention, including the United Kingdom and Germany, have taken regard of the prohibition on heritable genetic modification and framed their domestic provisions in many areas of biomedicine such as patient rights, consent, and privacy, the protection of biomedical research participants or living donors, and in relation to applications of biomedicine like genetics.

It seems that despite the fact that Europe has considered bio-patents as part of the existing moral and social order with the responsibility for protecting public morality, inequity, national security, and human rights, this requirement may often be hard to satisfy (Jasanoff 2005; Parthasarathy 2017).

This is also shown in the case of CRISPR patents, which represent the most recent phase in the long history of bio-patent conflicts. Namely, the recent fights over CRISPR patents may not only be seen as a continuation of the bio-patent discussions underway for 40 years, but as quite an unprecedented scenario in the patent landscape that involves a plethora of (mostly academic) players, each with their own partial interest in technology.

The patenting of CRISPR-Cas9 technology as such does not pose any specific bioethical issue. CRISPR is an illustration of a technology that interacts with human composition, but is itself not human (Perkins 2019). Cas9 is a naturally occurring protein and part of a naturally occurring bacterial process but, unlike the BRCA genes in the Myriad case, CRISPR-Cas9 technology is subjected to patenting because scientists are able to alter, control, and modify this technology to function in animal and human cells, namely a cellular system in which CRISPR-Cas9 does not function naturally (Beale 2015; Ku 2017). One can also detect opposite views. Some experts might say that CRISPR was discovered in a bacterial system and is thus possibly not an invention. Accordingly, it is not entirely clear why the discovery of CRISPR and the subsequent steps required to develop this functioning genome editing tool remain the subject of patent claims (Mulvihill et al. 2017).

From the bioethical point of view, much more critical is CRISPR-Cas9's ability to power gene drives which alter normal patterns of inheritance such that engineered genes are always passed on to future generations. This makes it extremely important that all stakeholders involved in the patent landscape pay attention to both the socioeconomic and ethical aspects of CRISPR technology. The new technology is leading us to a totally new situation. CRISPR-Cas9 can be used for germline interventions which might be aimed at altering a genome in a way that would not only affect the resulting child but potentially some of the child's descendants as well (Esvelt 2016; Sherkow 2017b). The combination of germline intervention and patent protections could very quickly lead to forms of ownership that span an entire species (National Academies of Science, Engineering, and Medicine, 2016). In this situation, it is essential that CRISPR innovations intended to relate to processes for modifying the germline genetic identity of human beings are not rewarded with patents. The question arises of whether reflexivity exists about how, and by whom, the ownership of inventions using CRISPR-Cas9 technology belongs to, taking the germline human editing risk into account. Unfortunately, this reflexivity is still not developed strongly enough.

Although countries playing the lead role in CRISPR technology progress have already developed some regulatory laws or guidelines for human germline gene editing and while certain international activities have been initiated to rethink a prohibition on possible germline gene editing, taken together, one can identify the lack of more comprehensive reflections and more efficient actions on the global level with respect to the risk of germline genome editing.

In the complex CRISPR technology patent landscape, we encounter a deficiency in more standardized rules on the global level. In the context of the progress of human genome editing, there has been little standardization of WTO members in the sense of excluding the patenting of inventions on the grounds of "order public" throughout the whole period (Duncan 2020; Martin et al. 2020). In some senses, this is understandable because the "germline" category in various patent systems has not been very well defined. The scientific audience has initiated discussions on this precarious bioethical issue relatively late.

Globally, scientists assembled for the first time to discuss the ethical and regulatory issues of germline gene editing in 2015. This single platform was organized in Washington, DC by the American National Academy of Sciences and the National Academy of Medicine's Human Gene Editing Initiative, in collaboration with the Chinese Academy of Sciences and the UK's Royal Society. In this Human Gene Editing Initiative, all three institutions agreed that if no babies are to result from the experiments, then gene editing research can go ahead. In accordance with this, the global Human Gene Editing Initiative National Institute of Health reaffirmed in 2015 that it would not be funding any use of gene-editing technologies in human embryos. In the same year, the German National Academy of Sciences (*Leopoldina*) endorsed a call for an international moratorium on all kinds of genome editing interventions in the human germline that affect the offsprings' genomes. The International Bioethics Committee (IBC) of UNESCO quite similarly declared that genetic modification should only be allowed for prevention, diagnostic, and therapeutic purposes without the transmission of any artificial genetic alteration to the next generation (Heidari et al. 2017).

The World Health Organization (WHO) has lately also been playing an increasingly active role in coordinating global policy and (bio)ethical responses to CRISPR genome editing worldwide. On 29 August 2019, the WHO expert advisory committee announced that it had approved the first phase of a new global registry to track research on human genome editing, using the International Clinical Trials Registry Platform (ICTRP), a WHO entity (World Health Organization 2019).

However, in the opposite direction of efforts to arrive at a global agreement, in November 2018, the Chinese scientist He Jiankui announced that he had edited human embryos to produce two genetically altered children, i.e. twin girls with the goal to confer resistance to HIV on them. Jiankui's research group has created the world's first gene-edited babies (Tyagia et al. 2020). While the birth of these "CRISPR babies" led to an international outcry and calls for a moratorium on such applications, it seems that it will take some time until the best socio-ethical regulation is found.

It is expected that the ever-rising importance of human genome editing technologies will generate ongoing questions on how intellectual property law should treat such progressive technologies. Even after many decades of controversies around bio-patents, important contested issues remain very challenging. This means that a deliberative and global discussion is still needed to reconsider and reframe the existing ethical guidelines and regulatory policies to develop more scientific and technical worldwide criteria for genome editing with the CRISPR-Cas system. Expectations will thus grow that nations first strictly commit themselves to international genome editing laws before proceeding with any genome editing application with the CRISPR-Cas system in order to ensure the safety of humankind.

21.6 Conclusion

The main aim of my discussion was to address two particular dimensions of IPR connected with the recent progress of CRISPR Cas9, i.e. the socioeconomic and socio-ethical challenges. Concerning the socioeconomic dimension of IPR, the big challenge is how to establish a suitable CRISPR patent landscape that ensures a balance between providing sufficient openness for further progress with this breakthrough technology, while also giving sufficient incentives for innovators. Today, it is hard to arrive at such a balance because academic science groups are greatly interested in protecting their inventions through patents. What is perhaps surprising is that some companies in business-enterprise sector are following a less conservative approach to IPR. Pharmaceutical companies are a good case. Although one might expect big pharmaceutical companies to have an exclusive interest in protecting inventions in human genome editing technologies with patents due to the high costs they incur to develop a novel drug or product which consists of hundreds and thousands of lead compounds, this is not always the case. Many of them have started to look in the recent period for new and less restrictive forms of IPR. The article presented some cases. Unlike such positive cases in the business-enterprise sector, academic institutions are mostly committed to the strict patenting regimes. CRISPR technology is turning the ivory tower of genetic engineering into a multibillion-dollar technological enterprise and certain academic institutions are becoming extremely oriented to protecting the ownership of their inventions. As I have described in detail, the prolonged and costly patent legal proceedings involving research groups from the leading American universities are the best indicator of these trends.

The bioethical challenge is interlinked with the socioeconomic challenge. Namely, the socioeconomic discourse on the CRISPR patent landscape will impact the future progress of human genome editing technologies at large and may, in turn, generate proper responses to the increasing bioethical concerns. The CRISPR-Cas9 patent landscape also raises socio-ethical dilemmas which on one hand represent a continuation of controversial discussions about bio-patents that have been underway for 40 years. But, on the other hand, today's patent-based monopolies over CRISPR are producing totally new ethical dilemmas. The article considered the question of the human germline as a new ethical dilemma connected also with CRISPR patent landscape was shortly elaborated. This specific application of CRISPR opens up the issue of whether we should encourage such inventions (even if they can eradicate certain hereditary diseases from society) or even allow patents on altering the germline identity of human beings. My position is clear: human germline editing should never be part of the CRISPR patent landscape!

To conclude, the main goal of my contribution was not to add to the stockpile of views on the commercial benefits of CRISPR-Cas9 patents. I have instead tried to address the broader question of how to find at the intersection of recent social, economic, and ethical dilemmas connected with the CRISPR patent landscape a new way which will ensure the further progress of this revolutionary technology.

References

Akst, J. (2017). UC Berkeley receives CRISPR patent in Europe. *The Scientist Magazine* (24 March). https://www.thescientist.com/?articles.view/articleNo/48987/title/UC-Berkeley-Receives-CRISPR-Patent-in-Europe/ (accessed 21 July2020).

Beale K (2015). The CRISPR patent battle: who will be "cut" out of patent rights to one of the greatest scientific discoveries of our generation? *Boston College Intellectual Property & Technology Forum* (9 February). http://bciptf.org/wp-content/uploads/2016/02/KBeale-CRISPR.pdf (accessed 30 August 2019).

Bhutkar, A. (2005). Synthetic biology. Navigating the challenges ahead. *The Journal of Biolaw & Business* 8 (2): 1–11.

Bogers, M., Fossb, N., and Lyngsiec, J. (2018). The "human side" of open innovation: the role of employee diversity in firm-level openness. *Research Policy* 47 (1): 218–231.

Calvert, J. (2012). Ownership and sharing in synthetic biology: a 'diverse ecology' of the open and the proprietary? *BioSocieties* 7 (2): 169–187.

Chesbrough, H. (2006). Open innovation: a new paradigm for understanding industrial innovations. In: *Open Innovation: Researching a New Paradigm* (eds. H. Chesbrough, W. Vanhaverbecke and J. West), 1–15. Oxford: Oxford University Press.

Churi, A. and Taylor, S. (2020). Continuing CRISPR patent disputes may be usurped by its potential role in fighting global pandemics. *Biotechnology Law Report* 184 (3): 184–189.

Cockbain, J. and Sterckx, S. (2020). Patenting foundational technologies: recent developments in the CRISPR patent struggle. *The American Journal of Bioethics* 20 (4): 11–12. https://doi.org/10.1080/15265161.2020.1735865.

Cook-Deegan, R. and Heaney, C. (2010). Patents in genomics and human genetics. *Annual Review of Genomics and Human Genetics* 11 (September): 383–425.

CRISPR-Cas9 Tools (2020). CRISPR-Cas9 Tools – Global Market and Patent Landscape Report till 2025. https://www.aheadintel.com/product/crispr-cas9-tools-global-market-and-patent-landscape-report-till-2025/ (accessed 1 September2020).

Döring M and Lim D (2017). Questions about CRISPR. *Intellectual Property Magazine* (29 April). http://www.allenovery.com/SiteCollectionDocuments/

Questions%20about%20CRISPR.pdf (accessed 19 December 2018).

Duncan, M. (2020). Access to CRISPR genome editing technologies: patents, human rights and the public interest. Legal Studies Research Paper No. 332/2020. Queen Mary University of London, School of Law.

Egelie, K., Graff, G., and Strand, S. (2016). The emerging patent landscape of CRISPRCas gene editing technology. *Nature Biotechnology* 34 (10): 1025–1031.

Enkel, E., Gassmann, O., and Chesbrough, H. (2009). Open R&D and open innovation: exploring the phenomenon. *R&D Management* 39 (4): 311–314.

Esvelt K (2016). Strategies for responsible gene editing. *Project Syndicate* (25 January). https://www.project-syndicate.org/commentary/crispr-gene-drive-editing-rules-by-kevin-m–esvelt-2016-01 [https://perma.cc/ZP3A-89CX] (accessed 31 January 2019).

European Biotechnological Directive (1998). Directive 98/44/EC of the European Parliament and of the Council of 6 July 1998 on the Legal Protection of Biotechnological Inventions. http://europa.eu.int/eurlex/pri//en/oj/dat/1998/1_213199807en00130021.pdf (accessed 8 January 2018).

Feeney O., Cockbain J., Morrison M., Diependaele L., Van Assche K., Sterckx S. (2018). Patenting foundational technologies: lessons from CRISPR and other core biotechnologies. *American Journal of Bioethics* 8 (12): 36–48.

Feldman, M., Colaianni, A., and Kang, C. (2007). Lessons from the commercialization of the Cohen-Boyer patents: the Stanford University Licensing Program. In: *Intellectual Property Management in Health and Agricultural Innovation: A Handbook of Best Practices*, vol. 1 and 2 (eds. A. Krattiger, R.T. Mahoney, L. Nelsen, et al.), 1797–1806. Oxford:MIHR.

Ferreira, R., David, F., and Nielsen, J. (2018). Advancing biotechnology with CRISPR/Cas9: recent applications and patent landscape. *Journal of Industrial Microbiology & Biotechnology* 45 (7): 467–480.

Gold, R. and Gallochat, A. (2001). The European Biotech Directive: past as prologue. *European Law Journal* 7 (3): 331–366.

Graff, G. and Johansen, B. (2016). The emerging patent landscape of CRISPR–Cas gene editing technology. *Nature Biotechnology* 34 (10): 1025–1031.

Grens, K. (2016). That other CRISPR patent dispute. The Broad Institute and Rockefeller University disagree over which scientists should be named as inventors on certain patents involving the gene-editing technology. *The Scientists* (31 August). https://www.thescientist.com/?articles.view/articleNo/46921/title/That-Other-CRISPR-Patent-Dispute/ (accessed 8 January 2018).

Halilem, N., Amaraa, N., Olmos-Peñuelab, J., and Mohiuddin, M. (2017). To own, or not to own: a multilevel analysis of intellectual property right policies' on academic entrepreneurship. *Research Policy* 46 (8): 1479–1489.

Heidari, R., Shaw, D.M., and Elger, B.S. (2017). CRISPR and the rebirth of synthetic biology. *Science and Engineering Ethics* 23 (2): 351–363.

Hilgartner, S. (2017). *Reordering Life: Knowledge and Control in the Genomics Revolution*. Cambridge, London: MIT Press.

Hübel, A., Schmelcher, T., and Storz, U. (2012). *Biopatent Law: Patent Strategies and Patent Management*. Heidelberg Dordrecht London New York: Springer.

IPStudies (2020). CRISPR patent analytics. https://www.ipstudies.ch/crispr-patent-analytics/ (accessed 19 October 2020).

Jasanoff, S. (2005). *Designs on Nature. Science and Democracy in Europe and in United States*. Princeton: Princeton University.

Jasanoff, S., Hurlbut, B.J., and Saha, K. (2015). CRISPR democracy: gene editing and the need for inclusive deliberation. *Issues in Science and Technology* 32 (1): 25–32.

Kapoora, R. and Klueterb, T. (2020). Progress and setbacks: the two faces of technology emergence. *Research Policy* 49 (1): 103874.

Kleinman, D. and Kinchy, A. (2003). Why ban bovine growth hormone? Science, social welfare, and the divergent biotech policy landscapes in Europe and the United States. *Science as Culture* 12 (3): 375–414.

Klueter, T., Monteirob, F., and Dunlapc, D. (2017). Standard vs. partnership-embedded licensing: attention and the relationship between licensing and product innovations. *Research Policy* 46 (9): 1629–1643.

König, H., Frank, D., Heil, R., and Coenen, C. (2013). Synthetic genomics and synthetic biology applications between hopes and concerns. *Current Genomics* 14 (1): 11–24.

Ku, D. (2017). The patentability of the Crispr-Cas9 genome editing tool. *Chicago-Kent Journal of Intellectual Property* 16 (2): 408–439.

Kwon, D. (2019). A brief guide to the current CRISPR landscape. *The Scientist* (15 July). https://www.the-scientist.com/news-opinion/a-brief-guide-to-the-current-crispr-landscape--66128 (accessed 8 October 2020).

Mali, F. (2004). Recent dilemmas in the social and legal regulation of biotechnology in the European Union. *Vest – Journal of Science Technology Studies* 17 (3–4): 39–60.

Martin, P., Morrison, M., Turkmendag, I. et al. (2020). Genome editing: the dynamics of continuity, convergence, and change in the engineering of life. *New Genetics and Society* 39 (2): 219–242.

Mathias T, Ropka S and Clegg R (2018) The CRISPR tug of war. *IPWatchdog* (17 August). https://www.ipwatchdog.com/2018/08/17/crispr-tug-of-war/id=100378/ (accessed 1 February 2019).

Mitschel, F. (2019). With the recent patent news, who owns CRISPR now? *Synbiobeta* (March 19). https://synbiobeta.com/with-the-recent-patent-news-who-owns-crispr-now/ (accessed 25 August 2020).

Mulvihill, J., Capps, B., Joly, Y. et al. (2017). Ethical issues of CRISPR technology and gene editing through the lens of solidarity. *British Medical Bulletin* 122 (1): 17–29.

National Academies of Sciences, Engineering, and Medicine (2016). *Gene Drives on the Horizon: Advancing Science, Navigating Uncertainty, and Aligning Research with Public Values*. Washington, DC: The National Academies Press.

Nuffield Council for Bioethics (2002). *The Ethics of Patenting DNA: A Discussion Paper*. London: Nuffield Council of Bioethics.

Nuffield Council for Bioethics (2016). *Genome Editing and Ethical Review*. London: Nuffield Council on Bioethics.

OECD (2014). *Emerging Policy Issues in Synthetic Biology*. OECD Publishing http://www.oecd-ilibrary.org/science-and-technology/emerging-policy-issues-in-synthetic-biology_9789264208421-en (accessed 5 August 2020).

Oviedo Convention (1997). Convention for the Protection of Human Rights and Dignity of the Human Being with regard to the Application of Biology and Medicine: Convention on Human Rights and Biomedicine. Council of Europe: European Treaty Series – No. 164. https://www.coe.int/en/web/conventions/full-list/-/conventions/rms/090000168007cf98 (accessed 4 October 2020).

Parthasarathy, S. (2015). Co-producing knowledge and political legitimacy: comparing life form patent controversies in Europe and the United States. In: *Science and Democracy*, 1e (eds. S. Hilgartner, C. Miller and R. Hagendijk), 20. Routledge. ISBN: 9780203564370 https://www.taylorfrancis.com/chapters/edit/10.4324/9780203564370-9/co-producing-knowledge-political-legitimacy-comparing-life-form-patent-controversies-europe-united-states-shobita-parthasarathy (accessed 29 August 2020).

Parthasarathy, S. (2017). *Patent Politics: Life Forms, Markets, and the Public Interest in the United States and Europe*. Chicago: University of Chicago.

Perkins, A. (2019). Gene[ie] in a Bottle: Regulating the Future of Man. *Tulane Journal of Technology & Intellectual Property* 21 (Spring 2019): 91–107.

Rimmer, M. (2008). *Intellectual Property and Biotechnology. Biological Inventions*. Cheltenham: Edward Elgar Publishing Limited.

Sampat, B. and Shadlen, K. (2017). Secondary pharmaceutical patenting: a global perspective. *Research Policy* 46 (3): 693–707.

Samy, M. (2018). CRISPR genome-editing market opportunity and key players. *Research White Paper* New York: ARK Invest. https://research.ark-invest.com/hubfs/1_Download_Files_ARK-Invest/White_Papers/ARK%20Invest_081318_White%20Paper_CRISPR%20Opportunity.pdf?hsCtaTracking=3e9ae410-326d-4658-9a4a-fd6c4cb7b263%7Cfa54a728-0144-4138-a518-b0051eae3b7f (accessed 6 July 2020).

Schneider, I. (2009). Governing the patent system in Europe: the EPO's supranational autonomy and its need for a regulatory perspective. *Science and Public Policy* 36 (8): 619–629.

Sherkov, J. and Greely, H. (2013). The future gene patents and the implications for the medicine. *AMA Internal Medicine* 173 (17): 1569–1570.

Sherkow, J. (2016). Who owns gene editing? Patents in the time of CRISPR. *Biochemist* 38 (April 5): 26–30.

Sherkow, J. (2017a). How much is a CRISPR patent license worth? *Pharma & Healthcare* (21 February). https://www.forbes.com/sites/jacobsherkow/2017/02/21/how-much-is-a-crispr-patent-license-worth/#6c6541ab6b77 (accessed 16 November 2017).

Sherkow, J. (2017b). Patent protection for CRISPR: an ELSI review. *Journal of Law and the Biosciences* 4 (3): 565–576.

Sherkow, J. (2017c). Inventive steps: the CRISPR patent dispute and scientific progress. *EMBO Reports* 18 (7): 1047–1050.

Sherkow, J. and Greely, H. (2015). The history of patenting genetic material. *Annual Review of Genetics* 49 (1): 161–182.

Singh, T. (2015). Open source business models and synthetic biology. *Chicago – Kent Journal of Intellectual Property* 14 (2): 455–484.

Summerfield, M. (2015). CRISPR – will this be the last great US patent interference? *Patentology* (11 July). https://blog.patentology.com.au/2015/07/crispr-will-this-be-last-great-us.html. (accessed 11 July 2020).

Tyagia, S., Kumarb, R., Dasc, A. et al. (2020). CRISP R-Cas9 system: a genome-editing tool with endless possibilities. *Journal of Biotechnology* 319 (August): 36–53.

United States Court of Appeals for The Federal Circuit (2018). Appeal from the United States Patent and Trademark Office, Patent Trial and Appeal Board in No. 106,048. 10 September. http://www.cafc.uscourts.gov/sites/default/files/opinions-orders/17-1907.Opinion.9-10-2018.pdf (accessed 12 August 2020).

Van den Belt, H. (2013). Synthetic biology, patenting, health and global justice. *Systems and Synthetic Biology* 7 (3): 87–98.

Van Erp, P., Bloomer, G., Wilkinson, R., and Wiedenheft, B. (2015). The history and market impact of CRISPR RNA-guided nucleases. *Current Opinion in Virology* 12 (June): 85–90.

Van Zimmeren, E., Vanneste, S., and Matthijs, G. (2011). Patent pools and clearinghouses in the life sciences. *Trends in Biotechnology* 29 (11): 569–576.

West, J., Salter, A., and Vanhaverbeke, W. (2014). Open innovation: the next decade. *Research Policy* 43 (5): 805–811.

Winickoff, D. (2015). Biology denatured: the public-private lives of lively things. In: *Science and Democracy: Making Knowledge and Making Power in the Biosciences and Beyond* (eds. S. Hilgartner, C.A. Miller and R. Hagnedijk), 33–56. Routladge: New York and Abingdon.

World Health Organization (2019). WHO Launches Global Registry on Human Genome Editing (29 August 2019). www.who.int/news-room/detail/29-08-2019-who-launches-global-registry-on-human-genome-editing (accessed 30 August 2020).

Zhang, D., Hussain, A., Manghwar, H. et al. (2020). Genome editing with the CRISPR-Cas system: an art, ethics and global regulatory perspective. *Plant Biotechnology Journal* 18 (8): 1651–1669.

22

Emerging Technologies for Genome Editing

Martin Peterka

Genome Engineering, Discovery Sciences, BioPharmaceuticals R&D, AstraZeneca, Gothenburg, Sweden

22.1 Introduction

In less than three decades since the inception of nuclease-based targeted genome editing, we have now reached an era where the first patients have been successfully treated using targeted modification of their genomes. During this time, we have witnessed several generational leaps in genome editing technologies. These include, but are not limited to meganucleases, zinc-finger nucleases, and transcription activator-like effector nucleases (TALENs) (Urnov 2018). While all of these technologies are still utilized today and have had a profound impact on our understanding of genome editing mechanisms, the current landscape has been virtually monopolized by CRISPR (clustered regularly interspaced short palindromic repeats)-based approaches, mainly due to their accessibility, ease of use, and cost-effectiveness.

The goal of this chapter is to provide a brief overview of some of the current trends regarding newly emerging genome editing technologies. Due to the sheer size of the field of genome editing and the speed of its evolution, this chapter does not attempt to provide a complete picture of the current trends (many of which are comprehensively described elsewhere in this book) but instead focuses on three distinct topics.

The first part provides a short summary of the work that has been performed on Cas9 to make it a better genome editing tool by improving its targeting scope, specificity, reducing its size and immunogenicity. Any improvements of these general properties of Cas9 automatically apply to and extend the versatility and clinical potential of many Cas9-based techniques, such as base editing or Cas9-based transcriptome modulation (to learn about these, the reader is referred to other parts of this book). To achieve these goals, two principal strategies will be mentioned – protein engineering of Cas9 from *Streptococcus pyogenes* (SpCas9) and exploration of different Cas9 orthologs with potentially favorable properties compared with the widely used SpCas9.

The second part is centered around prime editing, a recently emerged Cas9-based technology, which holds great promise to reduce difficulties in performing precise genome modifications such as point mutations or defined insertions – one of the areas where the traditional technologies are lacking in efficiency and/or specificity.

The final part of the chapter will consider less conventional techniques, such as targeted transposition, which, despite being at the early exploratory stages of development, might become invaluable genome editing tools in the years ahead.

22.2 Improving and Expanding the Cas9 Toolbox

The current workhorse of genome editing is the Cas9 protein from *S. pyogenes,* a Cas9 variant described in the landmark studies that laid the foundations for the field of CRISPR-Cas-based genome engineering (Cong et al. 2013; Jinek et al. 2012). SpCas9 is currently widely used for numerous applications and has recently reached a remarkable milestone by being used for the first time in patients to successfully treat sickle cell disease and β-thalassemia (Frangoul et al. 2021). Nevertheless, the properties of SpCas9 are far from ideal and there is much room for improvements that would enhance its utility as both therapeutic and research tool.

Genome Editing in Drug Discovery, First Edition. Edited by Marcello Maresca and Sumit Deswal.

The primary shortcomings of SpCas9 that can also be generalized to some other Cas nucleases are:

1) Propensity of Cas9 nucleases to induce off-target mutations in the genome (also described in detail elsewhere in this book).
2) Targeting scope limited by the requirement for a protospacer adjacent motif (PAM) sequence that is essential for target binding by Cas proteins and thus is required at any targeted site.
3) Relatively large size of SpCas9, which is approaching the limits of AAV packaging capacity, thus limiting applications of SpCas9 that requires *in vivo* delivery.

To realize the full potential of Cas9-based genome editing in therapeutic and research applications, two principal approaches to engineer better Cas9 variants have been applied: rational, structure-guided design and unbiased *in vitro* evolution strategies. In addition, the push to discover novel natural Cas9 orthologs and explore their genome editing capabilities had been equally fruitful.

This chapter aims to provide a brief global overview of the state of the field. For deeper insights into specific aspects of Cas9 editing regarding safety, specificity, and delivery, the reader is referred to separate chapters elsewhere in this book that elaborate on these topics.

22.2.1 Engineering the Specificity of Cas9

Rational, structure-guided mutagenesis of SpCas9 has been enabled by the plethora of structural and mechanistic studies on SpCas9 that have accumulated over the past several years. In particular, early crystallographic studies of SpCas9-gRNA-DNA structures (Anders et al. 2014; Nishimasu et al. 2014) unveiled the functions and organization of different SpCas9 domains and precisely mapped amino acid residues that make contact with the gRNA and the target DNA. Together with elegant biochemical assays, these studies provided invaluable insights into mechanisms of PAM recognition by SpCas9, gRNA pairing with the target DNA, and the extensive allosteric interplay between the two nuclease domains of Cas9 that finally leads to coordinated cleavage of both strands of the target DNA (Jiang and Doudna 2017).

Studies guided by these observations yielded engineered variants of SpCas9 with greatly reduced off-target activities. This was first achieved by mutating residues that form a positively charged groove which interacts with the non-target strand (DNA-strand that is not pairing with the gRNA). Neutralization of these charged residues lead to a more stringent requirement for gRNA:DNA base-pairing of the complex. This reduced the likelihood of SpCas9 binding to imperfect target sites. These "enhanced specificity" SpCas9 variants (eSpCas9) indeed show greatly reduced off-target activity, while retaining on-target activity comparable to that of wild-type SpCas9 (Slaymaker et al. 2016).

In a different study with a similar objective, it was reasoned that the affinity of the Cas9-sgRNA complex toward its target DNA is larger than needed for optimal cutting. Thus, decreasing this affinity would in turn decrease the chances of Cas9 binding and cutting at mismatched sites, while preserving binding to perfectly matched sites. Although is not entirely clear whether this exact mechanism is true, mutating residues in SpCas9 that make non-specific contacts with target DNA resulted in a SpCas9 variant (SpCas9-HF1) with remarkable on-target specificity (Kleinstiver et al. 2016).

Less biased discovery of more specific SpCas9 variants was carried out using a screening of randomly mutated SpCas9 variants in a yeast reporter system. Here, the mutagenesis of REC3 domain of SpCas9, which acts as a sensor of gRNA:DNA pairing, gave rise to evoCas9, the specificity of which is superior to both wild-type SpCas9 and the rationally engineered variants described previously (Casini et al. 2018).

22.2.2 Increasing the Targeting Scope of Cas9

An inherent limitation of Cas9 proteins in genome editing is their requirement for a specific sequence that allows binding of Cas9 to any given target DNA target site. This sequence, the so-called "protospacer adjacent motive" (PAM) thus limits the targeting scope of Cas9. This can hinder precise genome editing applications, especially where the targeted window can be very narrow. Therefore, to widen the targeting range of Cas9, there is a great need for relaxing or even removing the stringent requirements for PAM sequence presence.

Structural studies uncovered the mechanisms of recognition and binding of SpCas9 to its NGG PAM sequence (N is any nucleobase) (Jiang and Doudna 2017). Understanding these mechanisms opened the way to engineer variants of SpCas9 with altered PAM recognition properties and thus to expand the number of DNA sites that can be targeted by this nuclease.

One such example is SpCas9-NG, which is a product of structure-guided engineering based on the crystal structures of the SpCas9-gRNA-DNA complex. This structure revealed two arginine residues (Arg1335 and Arg1333) within the so-called PAM-interacting (PI) domain, which is responsible for recognition of the NGG PAM sequence by SpCas9. Mutating Arg1335, which recognizes the third G in the PAM, abolished PAM binding of SpCas9.

Simultaneous mutagenesis of other residues surrounding the PAM also revealed substitutions that could compensate for loss of Arg1335 and thus restore the mutant SpCas9 binding to DNA, while simultaneously changing PAM sequence preference. This mutagenesis yielded SpCas9-NG that can target genomic sites containing NGN PAM sequences. Although the activity of SpCas9-NG at these sites is generally lower than that of wtCas9 at a NGG PAM, this mutant greatly expanded the size of Cas9-targetable genome (Nishimasu et al. 2018).

The rational, structure-guided approach was recently pushed to its limit by extensive mutagenesis and optimization of the PI domain of SpCas9 including all possible amino acid substitutions at Arg1333. This yielded the most PAM-relaxed SpCas9 variant to date, SpRY, which is able to target almost any NNN PAM sequence, although with a large target-to-target variability (Walton et al. 2020).

A different strategy toward relaxing the PAM requirements, as well as improving off-target properties of Cas9 is an unbiased approach where new variants are generated using molecular evolution. The primary method of molecular evolution of Cas9 proteins has been phage-assisted continuous evolution (PACE). This ingenious system is based on continuous propagation of bacteriophages in *Escherichia coli*, wherein only bacteriophages carrying a Cas9 variant with desired properties are able to propagate. PACE has proven to be a fruitful approach that has led to the discovery of a number of Cas9 variants with expanded PAM compatibility, including non-G PAMs (Hu et al. 2018; Miller et al. 2020). A recent review provides an excellent and exhaustive overview of engineering and discovery of PAM relaxed Cas variants (Collias and Beisel 2021).

22.2.3 The Search for Small Cas Proteins

The size of SpCas9 is reaching the maximum payload capacity of adeno-associated viruses (AAVs), thus complicating AAV-mediated delivery of SpCas9 for *in vivo* applications (Xu et al. 2019). Due to their specificity and low immunogenicity, AAVs are one of the most promising delivery tools and the ability to package Cas9 including all the necessary regulatory elements and a gRNA expression cassette in one vector would be highly desirable.

This problem is driving the efforts to discover more compact Cas proteins through characterization of SpCas9 orthologs as well as mining metagenomics datasets in search of entirely novel CRISPR-Cas proteins. Several promising smaller orthologs of SpCas9 have been described that show genome editing activity in mammalian systems. Notable examples are SaCas9 from *Staphylococcus aureus* (Ran et al. 2015) and CjCas9 from *Campylobacter jejuni* (Kim et al. 2017). Both are significantly smaller than SpCas9 and, when delivered using AAVs, show specific editing activity at levels similar to those of SpCas9. However, both SaCas9 and CjCas9 have more complex PAM requirements than Cas9. While this limits the targeting flexibility of these orthologs, it may also reduce the number of potential off-target sites. Notably, AAV-mediated delivery of SaCas9 was successfully used to treat mouse models of Duchenne muscular dystrophy (Nelson et al. 2016; Tabebordbar et al. 2016).

Given the massive number and diversity of different CRISPR-Cas systems identified in nature (Makarova et al. 2020), it is expected that CRISPR-Cas enzymes superior to SpCas9 in various aspects are waiting to be characterized. To date, only a tiny fraction of these candidates have been studied regarding their genome editing potential. A metagenomic analysis of DNA sourced from groundwater and sediment samples identified several new Cas enzymes in uncultivated bacteria and Archaea (Burstein et al. 2017). One of these proteins – CasX, has been characterized in detail and holds promise as a genome editing tool mainly due to its small size – at 986 amino acids, CasX is approximately 30% smaller than SpCas9 (Liu et al. 2019). Another fascinating example of emerging genome editors are CasPhi proteins found in huge bacteriophages. At 700 amino acids, CasPhi represent the smallest Cas enzymes known to have genome editing activity in human cells. In addition, these proteins also have relaxed PAM requirements (Pausch et al. 2020).

22.2.4 Mitigating the Immunogenicity of Cas9

One of the lesser understood issues is a potential immunogenicity of Cas9 proteins. *Streptococcus pyogenes* is a common human pathogen and the possibility has been raised of pre-existing adaptive immunity against SpCas9 in human population. Indeed, recent studies revealed reactive T-cell populations as well as antibodies against SpCas9 in humans at relatively high frequencies (Charlesworth et al. 2019; Wagner et al. 2019). This could pose a serious roadblock toward the use of SpCas9-based therapeutics *in vivo*; however, it remains to be seen if SpCas9 elicits significant immune responses in humans. Nevertheless, the first steps to tackle this issue were taken by engineering "deimmunized" Cas9 variants with mutated immunoreactive epitopes (Ferdosi et al. 2019).

The synergy between structure-based protein engineering, molecular evolution, and metagenomic-driven discovery of novel proteins has led to a greatly expanded repertoire of both engineered and natural variants of CRISPR-Cas proteins. While these efforts are ongoing and the expansion of new as well as engineered Cas proteins will continue at ever-increasing speed, it is already now possible to target almost any genomic locus due to expanded PAM compatibility of Cas proteins. Furthermore, a remarkable on-target specificity has been achieved and Cas9 variants that show no detectable off-target editing are now available. All of these combined with the improvements of the immunogenic profile of Cas9 and discovery of small Cas proteins to facilitate *in vivo* delivery are moving us toward an era where CRISPR-Cas-based therapeutics will be applicable to treat a wide spectrum of genetic diseases.

22.3 Prime Editing

One of the most exciting recent expansions of the genome editing toolbox is prime editing (Anzalone et al. 2019). This elegant technique employs a reverse transcriptase (RT) fused to the Cas9 nickase, and a modified gRNA (prime editing gRNA – pegRNA) with an extended 3′ end, which serves as an RT template and encodes the desired edit to be incorporated into the genome. Briefly, upon Cas9-mediated nicking of the target site, the extended tail of the pegRNA hybridizes with the free 3′ end of the nontarget DNA strand, which is used as a primer by RT. Primed RT then catalyzes the extension of the 3′ end of the DNA strand according to the pegRNA template (Figure 22.1). This extension ("flap") is then incorporated by hybridization with the target strand of the DNA locus. However, even after being successfully extended by RT, the flap which contains the desired edit can be recognized as an extra structure and removed by the cellular DNA repair factors before being fully incorporated into the genome. The mechanisms that decide whether the flap will be removed or incorporated are yet to be fully delineated and their understanding and possible modulation may provide ways to remove this roadblock and maximize editing efficiencies in the upcoming generations of this technique.

The current generation of prime editors (designated PE2) is based on SpCas9 nickase (H840A) and an engineered variant of the Moloney Murine Leukemia Virus Reverse Transcriptase (M-MLV RT). The PE2 system can be combined with a second gRNA designed to cut the target strand in close proximity to the edit itself. This system (designated PE3) can result in up to several-fold higher editing efficiencies compared with PE2, presumably by preventing the cell from removing the newly incorporated edit, which can be recognized as a mismatch and targeted for removal by the cellular mismatch repair machinery. However, since PE3 introduces two adjacent DNA nicks, it also induced increased the rate of unwanted indels (Aida et al. 2020; Anzalone et al. 2019).

Prime editing is remarkably flexible and the pegRNA design can be easily adapted to introduce small genomic deletions, insertions, or substitutions. Since the intended modification is fully encoded in the pegRNA, prime editing eliminates the need for a donor DNA and thus circumvents a major hurdle of delivering an extra component to perform precise genome editing. This will be of great importance especially for *in vivo* applications.

It remains to be seen how well prime editing performs in different systems and especially *in vivo*, but the initial results are very encouraging for both germline and somatic genome editing in mice (Aida et al. 2020; Gao et al. 2020b; Liu et al. 2020).

Looking ahead, the improvements of prime editing will likely go into several directions. First, prime editors utilizing different DNA polymerases and/or DNA nucleases could lead to prime editors with expanded targeting scope and improved efficiency. Second, delineating pegRNA design rules for different types of edits will be important to maximize the efficiency of prime editing. The first steps have been undertaken in this direction and a high-throughput evaluation of pegRNAs allowed to define basic design rules of efficient pegRNAs (Kim et al. 2021). Last but not least, the understanding of DNA repair mechanisms that take place during prime editing will be crucial to modulate these pathways in favor of successful prime editing outcomes.

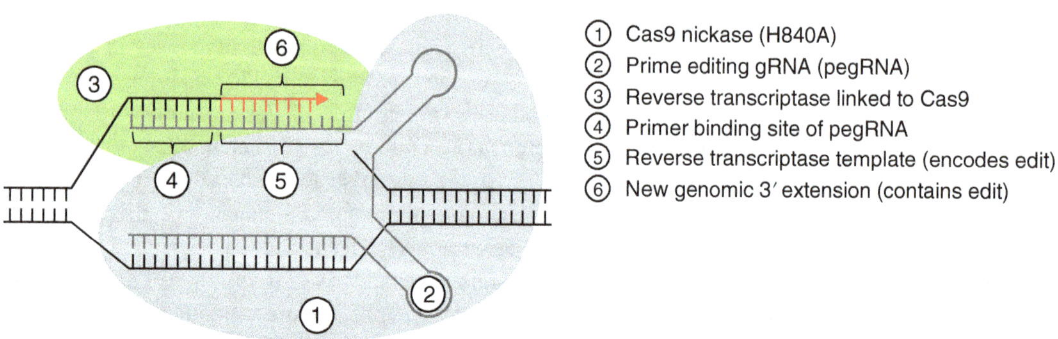

Figure 22.1 Schematic representation of prime editing.

22.4 Targeted Transposition and Beyond

An inherent shortcoming of nuclease-based genome editing techniques is that they fully rely upon the cellular DNA repair machinery which ultimately decides whether any desired genome editing outcome will be successful or not. For example, the traditional way of introducing a specific mutation or tag into the genome is based on homologous recombination (HR) and uses a nuclease and DNA donor template, which contains the intended modification flanked by arms that are homologous to the targeted locus. In this scenario, the nuclease only initiates the editing process by introducing a DNA double-strand break (DSB) at the targeted locus and the final outcome is fully dependent on the capacity of the cell to preform HR using the provided DNA template. Thus, this particular approach is only suitable in cell types that are capable of performing HR and even then, the time window of the cell cycle where HR can proceed is limited. Significant progress has been made by integrating our knowledge of DNA repair pathways with mechanisms of nuclease-based editing, and there are now ways of harnessing different modes of DNA repair to maximize the chances of successful genome editing in various contexts (Yeh et al. 2019).

One of the attractive emerging approaches that could circumvent the need for cellular DNA repair factors to perform precise genome editing is targeted transposition. Transposons are mobile DNA elements that can move through the genome mostly by a cut-and-paste mechanism. To this end, they utilize a transposase – an enzyme often encoded by the transposon itself that can excise the transposon from its original location and reintegrate it into a different genomic locus. Traditional transposon systems, such as piggyBac or Sleeping Beauty, have been in development for over two decades as gene delivery alternatives to viral vectors and early clinical trials of these systems are underway to assess their therapeutic potential in genome editing applications such as T cell engineering (Kebriaei et al. 2017). Despite having several favorable properties over more traditional viral gene delivery, such as low immunogenicity and manufacturing costs, the remaining issue of transposons is their genome-wide, close to random integration profile, vastly limiting the genome editing potential of current transposon-based technologies. Since the emergence of programmable nucleases, the concept of combining transposases with a DNA-binding component of either ZFNs, TALENs, or CRISPR-Cas systems emerged. Such hybrid systems, if functional, would allow for targeted integrations of large DNA fragments (>10 kb) and thus would open up new possibilities for therapeutic gene corrections, such as targeted introduction of entire genes or exons, a feat difficult to achieve with currently available tools. Importantly, perhaps the most appealing property of these systems is their independence on HR or DSBs. This is highly desirable, since introduction of DSBs using current nuclease-based gene editing methods was shown to cause unwanted chromosomal aberrations such as large deletions, translocations (Kosicki et al. 2018; Stadtmauer et al. 2020), or even chromothripsis – a catastrophic event, in which an entire chromosome falls apart and is reassembled in a disordered manner (Leibowitz et al. 2020). Additionally, Cas9 editing was shown to activate p53-mediated cell cycle arrest, which could lead to enrichment of p53 loss-of-function mutants in edited cell population, raising potential safety concerns regarding Cas9-based therapies (Haapaniemi et al. 2018).

Several aspects of targeted transposition are nevertheless unfavorable. Transposition requires the cargo to be flanked by transposon repeats, which are short repetitive sequences recognized by the transposase and co-inserted alongside the cargo into the genome. Thus, transposition currently does not allow for seamless genome editing. Additionally, transposases have certain sequence requirements for transposon insertion, which in conjunction with the nuclease binding sequence requirements limits the targeting scope of these systems. Moreover, unwanted whole donor plasmid integrations were observed in different transposon systems (Rice et al. 2020; Saha et al. 2015; Strecker et al. 2020).

Several attempts at performing targeted transposition by combining well-characterized transposases Sleeping Beauty (Kovač et al. 2020) and piggyBac (Hew et al. 2019) with catalytically dead Cas9 show early promise, although the low efficiency and specificity is currently prohibitive for their widespread use as precise genome editing tools.

Excitingly, naturally occurring Tn7-like transposons containing CRISPR-Cas systems have been bioinformatically predicted (Peters et al. 2017) and two of these systems were recently experimentally validated (Klompe et al. 2019, Strecker et al. 2019). In these biologically fascinating entanglements, transposons appear to have recruited different types of CRISPR-Cas systems, which allow for RNA-guided integration of the transposon into specific locations such as plasmids, which can for example facilitate their inter-bacterial propagation (Faure et al. 2019). The first structural insights have been recently gained regarding RNA-guided DNA transposition (Halpin-Healy et al. 2020) and one of the systems have been optimized for efficient engineering of various bacteria strains (Vo et al. 2020). Already transformative for biotechnology, these efforts will certainly help to pave the way toward the adaptation of CRISPR-associated transposons for genome editing of mammalian cells.

Last but not least, it is apparent that the repertoire of CRISPR-unrelated systems with potential applications in genome editing is by no means exhausted. This is exemplified in a recent study, which has predicted and validated a variety of enzymes involved in antiviral defense of a wide spectrum of bacteria and Archaea. These systems, some previously unknown, contain a number of enzymes including reverse transcriptases, deaminases, and ATPases among others (Gao et al. 2020a). One particularly promising group of enzymes with genome editing potential are Argonaute proteins. Prokaryotic Argonaute proteins are typically much smaller than Cas proteins and some of them have been demonstrated to carry DNA-guided nuclease activity (Hegge et al. 2018).

It is clear that CRISPR-Cas-based tools will be the dominant genome editing technology in both laboratory and clinics in years to come. Ultimately, this success was only allowed by basic research which led to discovery and understanding of these systems. It will be the continuation of these curiosity-driven efforts that will ensure CRISPR-Cas was not the last revolution we have witnessed in the turbulent and exciting field of genome editing.

References

Aida, T., Wilde, J.J., Yang, L. et al. (2020). Prime editing primarily induces undesired outcomes in mice. *bioRxiv*, 2020.08.06.239723.

Anders, C., Niewoehner, O., Duerst, A., and Jinek, M. (2014). Structural basis of PAM-dependent target DNA recognition by the Cas9 endonuclease. *Nature* 513: 569–573.

Anzalone, A.V., Randolph, P.B., Davis, J.R. et al. (2019). Search-and-replace genome editing without double-strand breaks or donor DNA. *Nature* 576: 149–157.

Burstein, D., Harrington, L.B., Strutt, S.C. et al. (2017). New CRISPR–Cas systems from uncultivated microbes. *Nature* 542: 237–241.

Casini, A., Olivieri, M., Petris, G. et al. (2018). A highly specific SpCas9 variant is identified by in vivo screening in yeast. *Nature Biotechnology* 36: 265–271.

Charlesworth, C.T., Deshpande, P.S., Dever, D.P. et al. (2019). Identification of preexisting adaptive immunity to Cas9 proteins in humans. *Nature Medicine* 25: 249–254.

Collias, D. and Beisel, C.L. (2021). CRISPR technologies and the search for the PAM-free nuclease. *Nature Communications* 12: 555.

Cong, L., Ran, F.A., Cox, D. et al. (2013). Multiplex genome engineering using CRISPR/Cas systems. *Science* 339: 819.

Faure, G., Shmakov, S.A., Yan, W.X. et al. (2019). CRISPR–Cas in mobile genetic elements: counter-defence and beyond. *Nature Reviews Microbiology* 17: 513–525.

Ferdosi, S.R., Ewaisha, R., Moghadam, F. et al. (2019). Multifunctional CRISPR-Cas9 with engineered immunosilenced human T cell epitopes. *Nature Communications* 10: 1842.

Frangoul, H., Altshuler, D., Cappellini, M.D. et al. (2021). CRISPR-Cas9 gene editing for sickle cell disease and β-thalassemia. *New England Journal of Medicine* 384: 252–260.

Gao, L., Altae-Tran, H., Böhning, F. et al. (2020a). Diverse enzymatic activities mediate antiviral immunity in prokaryotes. *Science* 369: 1077.

Gao, P., Lyu, Q., Ghanam, A.R. et al. (2020b). Prime editing in mice reveals the essentiality of a single base in driving tissue-specific gene expression. *bioRxiv*, 2020.11.07.372748.

Haapaniemi, E., Botla, S., Persson, J. et al. (2018). CRISPR-Cas9 genome editing induces a p53-mediated DNA damage response. *Nat Med* 24: 927–930.

Halpin-Healy, T.S., Klompe, S.E., Sternberg, S.H., and Fernández, I.S. (2020). Structural basis of DNA targeting by a transposon-encoded CRISPR–Cas system. *Nature* 577: 271–274.

Hegge, J.W., Swarts, D.C., and Van Der Oost, J. (2018). Prokaryotic Argonaute proteins: novel genome-editing tools? *Nature Reviews Microbiology* 16: 5–11.

Hew, B.E., Sato, R., Mauro, D. et al. (2019). RNA-guided piggyBac transposition in human cells. *Synth Biol (Oxf)* 4: ysz018.

Hu, J.H., Miller, S.M., Geurts, M.H. et al. (2018). Evolved Cas9 variants with broad PAM compatibility and high DNA specificity. *Nature* 556: 57–63.

Jiang, F. and Doudna, J.A. (2017). CRISPR–Cas9 structures and mechanisms. *Annual Review of Biophysics* 46: 505–529.

Jinek, M., Chylinski, K., Fonfara, I. et al. (2012). A programmable dual-RNA–guided DNA endonuclease in adaptive bacterial immunity. *Science* 337: 816.

Kebriaei, P., Izsvák, Z., Narayanavari, S.A. et al. (2017). Gene therapy with the sleeping beauty transposon system. *Trends in Genetics* 33: 852–870.

Kim, E., Koo, T., Park, S.W. et al. (2017). In vivo genome editing with a small Cas9 orthologue derived from *Campylobacter jejuni*. *Nature Communications* 8: 14500.

Kim, H.K., Yu, G., Park, J. et al. (2021). Predicting the efficiency of prime editing guide RNAs in human cells. *Nature Biotechnology* 39: 198–206.

Kleinstiver, B.P., Pattanayak, V., Prew, M.S. et al. (2016). High-fidelity CRISPR–Cas9 nucleases with no detectable genome-wide off-target effects. *Nature* 529: 490–495.

Klompe, S.E., Vo, P.L.H., Halpin-Healy, T.S., and Sternberg, S.H. (2019). Transposon-encoded CRISPR–Cas systems direct RNA-guided DNA integration. *Nature* 571: 219–225.

Kosicki, M., Tomberg, K., and Bradley, A. (2018). Repair of double-strand breaks induced by CRISPR–Cas9 leads to large deletions and complex rearrangements. *Nature Biotechnology* 36: 765–771.

Kovač, A., Miskey, C., Menzel, M. et al. (2020). RNA-guided retargeting of sleeping Beauty transposition in human cells. *eLife* 9: e53868.

Leibowitz, M.L., Papathanasiou, S., Doerfler, P.A. et al. (2020). Chromothripsis as an on-target consequence of CRISPR-Cas9 genome editing. *bioRxiv*, 2020.07.13.200998.

Liu, J.-J., Orlova, N., Oakes, B.L. et al. (2019). CasX enzymes comprise a distinct family of RNA-guided genome editors. *Nature* 566: 218–223.

Liu, P., Liang, S.-Q., Zheng, C. et al. (2020). Improved prime editors enable pathogenic allele correction and cancer modelling in adult mice. *bioRxiv*, 2020.12.15.422970.

Makarova, K.S., Wolf, Y.I., Iranzo, J. et al. (2020). Evolutionary classification of CRISPR–Cas systems: a burst of class 2 and derived variants. *Nature Reviews Microbiology* 18: 67–83.

Miller, S.M., Wang, T., Randolph, P.B. et al. (2020). Continuous evolution of SpCas9 variants compatible with non-G PAMs. *Nature Biotechnology* 38: 471–481.

Nelson, C.E., Hakim, C.H., Ousterout, D.G. et al. (2016). In vivo genome editing improves muscle function in a mouse model of Duchenne muscular dystrophy. *Science* 351: 403.

Nishimasu, H., Ran, F.A., Hsu, P.D. et al. (2014). Crystal structure of Cas9 in complex with guide RNA and target DNA. *Cell* 156: 935–949.

Nishimasu, H., Shi, X., Ishiguro, S. et al. (2018). Engineered CRISPR-Cas9 nuclease with expanded targeting space. *Science* 361: 1259.

Pausch, P., Al-Shayeb, B., Bisom-Rapp, E. et al. (2020). CRISPR-CasΦ from huge phages is a hypercompact genome editor. *Science* 369: 333–337.

Peters, J.E., Makarova, K.S., Shmakov, S., and Koonin, E.V. (2017). Recruitment of CRISPR-Cas systems by Tn7-like transposons. *Proceedings of the National Academy of Sciences* 114: E7358.

Ran, F.A., Cong, L., Yan, W.X. et al. (2015). In vivo genome editing using *Staphylococcus aureus* Cas9. *Nature* 520: 186–191.

Rice, P.A., Craig, N.L., and Dyda, F. (2020). Comment on "RNA-guided DNA insertion with CRISPR-associated transposases". *Science* 368: eabb2022.

Saha, S., Woodard, L.E., Charron, E.M. et al. (2015). Evaluating the potential for undesired genomic effects of the piggyBac transposon system in human cells. *Nucleic Acids Research* 43: 1770–1782.

Slaymaker, I.M., Gao, L., Zetsche, B. et al. (2016). Rationally engineered Cas9 nucleases with improved specificity. *Science* 351: 84.

Stadtmauer, E.A., Fraietta, J.A., Davis, M.M. et al. (2020). CRISPR-engineered T cells in patients with refractory cancer. *Science* 367: eaba7365.

Strecker, J., Ladha, A., Gardner, Z. et al. (2019). RNA-guided DNA insertion with CRISPR-associated transposases. *Science* 365: 48.

Strecker, J., Ladha, A., Makarova, K.S. et al. (2020). Response to comment on "RNA-guided DNA insertion with CRISPR-associated transposases". *Science* 368: eabb2920.

Tabebordbar, M., Zhu, K., Cheng, J.K.W. et al. (2016). In vivo gene editing in dystrophic mouse muscle and muscle stem cells. *Science* 351: 407.

Urnov, F.D. (2018). Genome editing B.C. (Before CRISPR): lasting lessons from the "old testament". *The CRISPR Journal* 1: 34–46.

Vo, P.L.H., Ronda, C., Klompe, S.E. et al. (2020). CRISPR RNA-guided integrases for high-efficiency and multiplexed bacterial genome engineering. *bioRxiv*, 2020.07.17.209452.

Wagner, D.L., Amini, L., Wendering, D.J. et al. (2019). High prevalence of *Streptococcus pyogenes* Cas9-reactive T cells within the adult human population. *Nature Medicine* 25: 242–248.

Walton, R.T., Christie, K.A., Whittaker, M.N., and Kleinstiver, B.P. (2020). Unconstrained genome targeting with near-PAMless engineered CRISPR-Cas9 variants. *Science* 368: 290.

Xu, C.L., Ruan, M.Z.C., Mahajan, V.B., and Tsang, S.H. (2019). Viral delivery systems for CRISPR. *Viruses* 11: 28.

Yeh, C.D., Richardson, C.D., and Corn, J.E. (2019). Advances in genome editing through control of DNA repair pathways. *Nature Cell Biology* 21: 1468–1478.

Index

Genome Editing in Drug Discovery, First Edition. Edited by Marcello Maresca and Sumit Deswal.

www.ingramcontent.com/pod-product-compliance
Lightning Source LLC
LaVergne TN
LVHW060056220926
848484LV00008BA/1954